World Sustainability Series

Due to its scope and nature, sustainable development is a matter which is very interdisciplinary, and draws from knowledge and inputs from the social sciences and environmental sciences on the one hand, but also from physical sciences and arts on the other. As such, there is a perceived need to foster integrative approaches, whereby the combination of inputs from various fields may contribute to a better understanding of what sustainability is, and means to people. But despite the need for and the relevance of integrative approaches towards sustainable development, there is a paucity of literature which address matters related to sustainability in an integrated way.

Notes on the quality assurance and peer review of this publication

Prior to publication, the quality of the works published in this series is double blind reviewed by external referees appointed by the editor. The referees are not aware of the author's name when performing the review; the referees' names are not disclosed.

More information about this series at http://www.springer.com/series/13384

Walter Leal Filho · Ulisses Miranda Azeiteiro ·
Andréia Faraoni Freitas Setti
Editors

Sustainability in Natural Resources Management and Land Planning

Springer

Editors
Walter Leal Filho
FTZ-ALS
HAW Hamburg
Hamburg, Germany

Ulisses Miranda Azeiteiro
Department of Biology
University of Aveiro
Aveiro, Portugal

Andréia Faraoni Freitas Setti
Department of Biology
University of Aveiro
Aveiro, Portugal

ISSN 2199-7373 ISSN 2199-7381 (electronic)
World Sustainability Series
ISBN 978-3-030-76623-8 ISBN 978-3-030-76624-5 (eBook)
https://doi.org/10.1007/978-3-030-76624-5

This Springer imprint is published by the registered company Springer Nature Switzerland AG
The registered company address is: Gewerbestrasse 11, 6330 Cham, Switzerland

Preface

To achieve the policy objective of sustainable production and conservation of natural resources, governments should pursue strategies which actively promote forms of land use which are both attractive to the people and sustainable in terms of their impacts on land resources.

In addition, the noticeable growth in both the frequency and intensity of wildfires, whose occurrence is also associated with changing climate conditions, is also a cause of global concern. Apart from the human suffering caused by loss of property and income and damages to wildlife, wildfires are leading to permanent changes in landscapes in some regions.

But despite the relevance of sustainable land planning and use, there is a paucity of publications in this field.

It is against this background that this book has been produced. It is a truly interdisciplinary publication, useful to teaching staff and scholars on the one hand, but also to members of governmental agencies on the other, as well as to all those undertaking research and/or executing projects focusing on land planning and sustainable land use from across the world. The book is structured around two parts:

Introduction: this part entails papers describing approaches to conserving and restoring ecosystems, the contribution of tools such as geographical information systems and barriers to sustainable land management, among others.

Handbook of Wildlife Fires: Monitoring, Control and Management Under a Changing Climate: this part contains papers related to fires, with case studies, interdisciplinary initiatives, and papers which describe the use of approaches and tools to foster the cause of fire control and management.

This book presents various examples of intersectoral practices linked to the participation in and the implementation of multi-strategic actions to promote changes aiming at the sustainability and the resilience of territories. The papers are innovative, cross-cutting and many have practice-based experiences, some of which may be replicable elsewhere.

We thank the authors and reviewers for their contribution. We hope that the contributions on this volume will provide a timely support towards the implementation of initiatives on sustainable land use and management and will foster the global efforts towards seeking solutions for the problems related to fires, and especially the prevention of forest fires.

Enjoy the reading.

Hamburg, Germany — Walter Leal Filho
Aveiro, Portugal — Ulisses Miranda Azeiteiro
Aveiro, Portugal — Andréia Faraoni Freitas Setti
Winter 2021/2022

Contents

Introduction

Conserving and Restoring Water-Related Ecosystems World-Widely: Have We Met the 2020s the Benchmarks of the Sustainable Development Goal Six?

Bila-Isia Inogwabini

1 Introduction

This paper is about the Sustainable Development Goal 6 on water and water management and deals specifically the objective's benchmark 6 whose focus is to protect and to restore water-related ecosystems. The metric to evaluate this goal was set so that by 2020 the world should be able to protect and restore all sorts of water-related ecosystems. The paper discusses what it means to protect and restore water ecosystems and what it would take to get to a point where claims of satisfaction can be genuinely made. The paper starts with a brief sketch of theoretical knowledge of how water ecosystems are laid across the globe and how they operate. That brief presentation is deemed necessary to ensure that talks over protecting and restoring waters come forth, everyone is clear about what is being said. Indeed, conserving water equates to keeping it in its functionally optimal conditions while restoring water-related ecosystems equals bringing deteriorated waters back to the normalcy of their ecological functions.

Because of the need for theoretical knowledge on water ecosystems across the world, the paper starts with generalities of water distributions, water availability and then moves on to dissecting broad water ecosystems. This includes a view on global water cycles. Only after these general concepts that water conservation issues are raised and possible solutions are identified. Then, comes questions related to the restoration of water bodies that had been significantly impacted over the last decades. By water-related ecosystems it is principally meant to include aquifers, lakes, rivers, oceans and wetlands. However, it should be remembered that waters are parts of the systems where they occur. So, ecologically speaking, aquifers, lakes, rivers, oceans and wetlands cannot be separated of forests, mountains and even soils

B.-I. Inogwabini (✉)
Department of Aquatic Sciences and Assessment, Swedish University of Agricultural Sciences, Uppsala, Sweden
e-mail: inogwabini.bila-isia@slu.se; bila_inogwabini@biari.brown.edu

W. Leal Filho et al. (eds.), *Sustainability in Natural Resources Management and Land Planning*, World Sustainability Series,
https://doi.org/10.1007/978-3-030-76624-5_1

of the regions where they are respectively located. That is why water ecology goes beyond the physics and chemistry of waters and embraces other natural sciences (such as forestry and environmental sciences), social and economical sciences (such as anthropology and natural resources management). For that very reason, some indications will be given of forests, mountains and other habitats even when they are seemingly not directly related to water ecosystems defined above.

2 Water Abundance, Distribution and Water Circulations

From geo-physical perspectives, the global water cycle has been maintained constant for millennia and water remains constant on earth. To state the obvious, most of the earth (71%) is composed of liquid water, which in principles, should mean that water is plenty all over the world. Nevertheless, the repartition of surface water is that *ca.* 97.5% is salted water of the oceans while the remaining (*ca.* 2.5%) is made of freshwater. Of this small portion consisting of freshwater, close to 79% is made of solid state water of the glaciers and 20% consists of ground water. This means that surface water (lakes, rivers, and the permanent water on the floors of wetlands) is made only of about 1% of 39×10^6-km^3 of non-oceanic waters (inclusive of 0.035% of atmospheric water).

However, waters move from one area to another, they can go from the oceans to lakes and rivers and vice versa. The mechanisms that are used to go from one area to another include evaporation, infiltration, transpiration, percolation, precipitation and water physical runoffs. There is no need to go through definitions and how each of these mechanisms works here. But suffice it to be said that all of these mechanisms constitute what is known as water cycle. The global water cycle goes from surface water that is heated by the sun rays and goes up to the atmosphere by evaporation; the water stored in plants is transported back to the atmosphere via plants' transpiration. On the other hand, the water in the atmosphere is brought back down to earth via precipitations and then into soil to feed the rivers by infiltration, percolation or physical runoffs. With the knowledge of the global water cycle in mind, there are two important elements that are sufficient to ensure that waters are conserved both in quantities and quality. The first of these, quantity-wise, is that the health of water ecosystems is intricately tied to the conditions of ecosystems and physical environment that keep these pathways in good working conditions. Secondly, quality-wise, good water health is an intrinsic condition; it is a function of water chemistry. Rudimentarily stated, the less chemicals are added to the waters, the better (or closer to natural) waters are or inversely, the more chemically waters deviate from their 'natural's conditions, the more they are impaired. This is called pollution. Aquatic and marine biodiversity play an important role in this respect. In fact, the aquatic and marine biomasses are known to smooth some of the chemical changes happening in waters (Relyea and Hoverman 2006) allowing, therefore, waters to be qualitatively resilient by bringing them back to their 'normal' chemical conditions (e.g. Liberati et al. 2018). It should also, immediately, be added that quality of waters can also be, and it is indeed, significantly affected by mechanical objects that are not part of the water-scapes.

3 The Health of and Threats to Water Ecosystems

The first paragraph of the section above section of this chapter clearly would conclude that since freshwater is the part of waters that is readily available and usable, despite waters being ubuquous on earth, utilizable water availability differs significantly. The availability of utilizable water (aka freshwater) depends on ecological and physical environmental conditions of each locality on earth. Because the whole objective of this sustainable development goal (SDG 6) is to ensure 'access to safe and affordable drinking water' and freshwater is the only part of the world's waters that is readily available and usable for that purpose, most of that follows will concentrate heavily on freshwater. Of course, this does not mean dropping all the salted oceanic waters aside; when necessary, references will be made to the latter type of waters.

Quantity-wise, the distribution of water can become disturbed when water ecosystems' health are injured. This means when evaporation and transpiration, infiltration, percolation, physical runoffs, lakes and rivers are dented. Evaporation and transpiration can be damaged, principally, in two ways: it can be increased or reduced. Evaporation is a function of the exposed surface of water and quantities of solar radiations, movement of air above the water surface and relative humidity (Magin and Kandall 1960). Because of this, any work that would increase the sun-exposed water surface will automatically increase water evaporation, changing therefore the speed at which water in a given environment goes back to the atmosphere. This can have an impact on available quantities of water and, at least, on the duration a certain water quantity spends in a given location. On the other hand, activity that diminishes the water exposure to sunrays will, effectively, reduce evaporation. An example of this case would be when significant quantities of physical bodies float on water surface and prevent water from evaporating up. As it has been long known (e.g. Adam et al. 1939; Allan and Alexander 1954; Archer and LaMer 1954), examples of such phenomena can include fine yet large layers of chemicals spilled on water bodies (Magin and Kandall 1960). Oils spilled over water bodies and plastics bags deposed over water offer such a possibility. Human-induced phenomena can impact precipitations is many diverse ways. One of the most important cases of these human-induced activities is deforestation (Aragão et al. 2008). Studies across the globe show that deforestation has direct effects precipitations because deforestation acts on tree transpiration, directions of wind and microclimates. Along with evaporation, transpiration is another pathway through water is returned back to the atmosphere and winds transport humidity from lower temperature zones to hotter areas where it transformed into precipitations. A good example substantiating this latter effect of deforestation is the documented effects it has on African monsoons, which directly impede on movement of moisture in the continental area of Sahel where it is needed the most for rainfall. Indeed, Impacts on precipitations are in the form of either droughts or heavy rains.

Other water pathways are infiltration, percolation, physical runoffs and rivers. Physical engineering such as compaction of soil for building infrastructure can affect infiltration, percolation, physical runoffs and rivers. Logging, which contributes to deforestation, beyond negatively impacting transpiration, does also compact soils and, therefore affect water infiltration down to soil, water percolation and physical runoffs. It does, in some cases, physically prejudice even rivers course. Indeed, the logging activities come along with infrastructure building, including roads, camps, etc. All these activities have collateral effects on the ecosystems not only through fragmenting forests into smaller different blocs (Wlikie et al. 2000) but also in compacting soils, redressing water basins and opening up large geographical spaces to sun.

Qualitatively, water's health can be wounded through different mechanisms. Though instances where freshwaters can be impaired by natural causes are possible, most of causes of numerous problems with waters that are known to humans stem from anthropogenic disturbance of the terrestrial water cycle (e.g. Vörösmarthy and Sahagian 2000; Seitz et al. 2018). To repeat the same examples as above, logging and agriculture are part of such anthropogenic disturbances on terrestrial ecosystems, which lead to massive consequences on waters. Mining activities are the third type of anthropogenic disturbances, which have similar consequences, and sometimes can be even more devastating for forest and water ecosystems than agriculture and logging. Indeed, agriculture, logging and mining denude large stretches of lands. Denuded soils contaminate surface waters through different mechanisms. The first and most obvious of this is through contaminated water charges that are carried over by water runoffs that, ultimately, change water chemistry and physics. Particularly parameters such as acidity, colour, conductivity, hardness, etc. are affected. Physical water runoffs in deforested areas, for example, to carry the soil humus and other chemical and mechanical debris into the water streams that are responsible for changes in water chemistry and for bringing loads of solid state debris that cannot dissolve into water (Vörösmarthy and Sahagian 2000). These outcomes are to be expected whether logging is clear-cut or not because one way or the other even minor logging activities end up denuding forest soils. Denuded soil loses the abilities to filter water through infiltration as it does increase the speed of water toward streams. By the same token, denuded soil also carries excessively load waters of nutrients and toxins which, then, change and eutrophy or even poison many waters so much that they cannot support their natural biotic communities (Strayer and Dudgeon 2010). The description given here about deforestation does also apply for urbanized areas where cemented infrastructures prevent urban soils from playing their role to sponge the water through back to the aquifers. Equally cultivated zones expose soil to direct rainfalls, which lead to the same consequences as deforestation. Indeed, agriculture, at least in its first phases, consists of denuding land to sow selected seeds. More than its effects on water runoff's speed and increasing the water ability to carry loads of soil to water bodies, agriculture is known to directly impact water chemistry because phyto-chemicals it uses are, in the end, carried out to basins, streams, lakes and rivers where they threaten the quality of water and its biodiversity (e.g. Liberati et al. 2018; Reno et al. 2018).

4 Water Ecosystems Across the Continents: Current Status

The current situation of water word-widely is that no single water body or basin can claim to have remained in pristine conditions. In a world where the human populations have increased dramatically over the last half century, it is rather snobbish to look for parts of the planet that would have been kept untouched. The United Nations Environmental Program (UNEP 2016) produced its sixth Global Environment Outlook (GEO-6), which paints a comprehensive picture of the environmental factors contributing to human health and well-being across six continents. While UNEP (2016) reports advances toward sustainable development goals made in the five continents, it does also shows a globally bleak perspective in renewable resources conservation. According to UNEP (2016), expanding African economies and the continent's rapidly growing human populations demand greater quantities of freshwater, which leads to decreases in its quantity and quality. UNEP (2016) projects that, even though African groundwater is still largely under-exploited water resource, over-exploitation of African freshwaters, due largely bourgeoning increases in human demographics and its subsequent pollution, combined with climate change will conduce to average internal renewable water resources continued dwindling. For Asia and Pacific, UNEP (2016) found that rapid economic growth caused by rapid and intensified industrialization increased proportions of the regional population (ca. 60% of the world's total) entering into economical middle classes. However, that fast-paced economic development has come through relatively high ecological price. The ecological footprints for these advances were sharp, unsustainable and inefficient increase in the usages of natural resources. Expectedly, results of such huge, unsustainable and inefficient usages of natural resources include pollution, depletion of biological diversity and decreases in both renewable and non-renewable natural resources. As specifically as water is concerned, UNEP (2016) reports of widespread contamination of ground water by human and industrial waste. Human wastes that contaminate water are mostly composed of human faeces and personal care products while industrial wastes are made of pharmaceutical, runoff of agrochemicals, nano-materials, and organochlorins. To illustrate how serious the situation of water is in Asia and Pacific, UNEP (2016) indicated that *ca.* 30% of *ca.* 60% of the world's total residing in that part of the world drinks human faeces- contaminated water. While access to drinkable water was nearly complete its populations, UNEP (2016) found, as it would have been expected a variable picture in water prospects for Western, Central Europe and Eastern Europe. UNEP (2016)'s projections were to expect more than 50% decrease in river discharge in Southern Europe (inclusive of Israel and Turkey) while expecting a rather lower (30%) decrease for South Eastern Europe, Eastern Europe and Caucasus. Great intensified water scarcity was to be expected in the Mediterranean region. Profound changes in coastal fringes such as losses of dune systems, erosions and overall coastal squeeze (UNEP 2016) were also to be expected. Reasons for this disparity of prospects for waters in the near future Western, Central Europe and Eastern Europe are also varied. Excessive water consumption in agriculture, with average water withdrawals for irrigation higher (*ca.* 159%) than the world global average cubic metres per hectare per year for Southern Europe, South Eastern Europe, Eastern Europe and Caucasus. Climate change drives the intensification of water scarcity in the Mediterranean. Climate change, with its induced changes in weather conditions and more frequent and intense storm surges, combined with sea level rise were suggested to cause the overall coastal squeeze. As agriculture drives the above-mentioned excessive water consumption, its intensive forms also spearhead water pollution as it injects loads of agrochemicals, nano-materials, and organochlorins in water bodies. Despite the fact that more than 90% of Western-Europeans and Israelis being connected to a sewer system that treats wastewaters before it is reinjected to the natural systems since 2010, populations in urban agglomerations in the rest of Europe generate colossal quantities of wastewaters, which pollute waters and water ecosystems. Indeed, despite some noticeable recent improvements in West Europe, chemically harmful substances continue to degrade coastal areas and open oceans, nutrient loads remain high and the impacts of new pollutants, including plastic wastes forming marine litter, steadily grow. Ecologically, about 50% of rivers and lakes in the European Union perform rather poorly. According to UNEP (2016), with only 7% of marine species is good conservation status, the remaining aquatic (freshwater rivers and lakes) and marine (oceans and seas) and estuarine biological diversity is in poor ecological conditions Europe-widely and, overall, overfishing shot high records over global long-time. Europe-wide, oceans and seas have generally improved but major physical, geochemical and biological disturbances also accompanied that improvement. Multiple consequences of these major physical, geochemical and biological turbulences caused by warmed seawaters are that oxygen is depleting in the European seas and marine food webs are disrupted in scales that are still to be documented. Warmed seawaters, secondly, provide thriving environments for invasive species that further encroach on seas and marine food webs. Warmed waters deplete the ice cover in the Arctic Ocean; this single fact releases new and unexpected species of planktons into liquid seas and oceans without knowledge of how these new species will impact the ecological processes of waters.

UNEP (2016) indicated that drinking water quality is extremely good in North America while expressing increasing concerns about decreasing water quantities. UNEP (2016) lauded North America about well-kept natural landscapes that provide clean freshwater, healthy habitats for wildlife and fish, quality outdoor recreation opportunities. Yet, it was also revealed that large-scale disruptive land-use and land cover changes fragment the same natural landscapes, principally because of natural causes, such as wildfires and pest outbreaks, and decisions made about land management activities, ownership transfers to heirs, and development decisions. However, UNEP (2016) missed to point out an emerging situation caused by hydraulic fracking which extracts millions of liters of local ground waters to generate extensive fracture networks within these low-permeability reservoirs, allowing extraction of the trapped hydrocarbons (Burton et al. 2014). Generally, Burton et al. (2014) concluded that hydraulic fracking operations increased erosion and sedimentation, increased risk to aquatic ecosystems from chemical spills or runoff, habitat fragmentation, loss of stream riparian zones, altered biogeochemical cycling, and reduction of available surface and hyporheic water volumes because of withdrawal-induced lowering of local groundwater levels. The US Environmental Protection Agency (EPA 2016) suggested that decreasing water quantities, which UNEP (2016) identified as an issue of increasing concern, may be caused by hydraulic fracturing in times or areas of low water availability. Findings (EPA 2016), despite being overly cautious, led support to what Food & Water Europe found in 2012. For Food & Water (2012) fracking contaminated water supplies across the United States; more than 1000 cases of water contamination near drilling sites documented by courts, states and local governments around the country prior to 2009. As indicated above, close to 79% of freshwater is solid state water of the glaciers; a significant part of it is in the Arctic of which large parts are under the sovereign states of North America. The Arctic, UNEP (2016) argues, currently experiences profound climate-induced and human-driven transformations whose impacts will be felt not only on North America but world-widely. According to UNEP (2016), warming in the Arctic has increased at twice the global average since 1980 leading to glacier and ice sheet melt, altered salinity concentrations and ocean circulation patterns, sea level rise, and ocean acidification.

Quantity-wise, UNEP (2000) described Latin American as a paradoxical region in terms of waters. While the region is extremely rich in water resources, two-thirds of the Latin America is arid or semi-arid. Regional major rivers (Amazon, Orinoco, São Francisco, Paraná, Paraguay and Magdalena) carry >30% of the world's continental surface water but large parts of central and northern Mexico, north-eastern Brazil, Argentina, Chile, Bolivia and Peru are under huge water stress. Unsustainable patterns of water withdrawals, such as pumping from aquifers at rates far greater than they are recharged, which UNEP (2000) identified nearly two decades ago are to continue because of growing populations, rapidly increasing industrial activity and expanding irrigated agriculture (the largest use) and increasing pasture for export-oriented food industry will continue increase the water stress (Mekonnen et al. 2014). Many people still lack an adequate water supply and a sewage system. Of course, this situation varies; it is far better in Central American whereas it is rather bad in Latin America as where only 2% of sewage is properly treated. Qualitatively, industrial wastes, mining industry and agricultural chemicals contaminate rivers by chemicals and heavy metals (UNEP 2000, 2010). Despite noticed significantly reductions in mercury emissions (compared to levels of 1980s) UNEP (2000) continues to believe that probably quantities of removed mercury are equal to those still emitted as gold is produced. Phenomena such as aquifer depletion, improper use and disposal of heavy metals, salt water intrusion, synthetic chemicals and hazardous wastes also pollute groundwater. The situation is worsened by human populations' concentration in large metropolitan centers and industrial areas that erode soils and, subsequently produce sediments, and the bad habit of directly discharging untreated domestic and industrial wastes to surface water bodies. These contaminate not only the surface water bodies but adjacent groundwater aquifers. Additionally, structural economic changes, with emphasis on manufacture, led to large paved areas in the cities and artificially regulated stream flows. This subsequently modified natural water run-off systems and seriously compromised water cycles in the region. The production of hydroelectricity that began in the 1970s in most of the river basins also changed the water dynamics in the region and caused some ecologically serious problems over the last years (UNEP 2010). Housing developments continue to be sited in sensitive areas such as steep slopes, upper parts of water catchment areas, and too close to sensitive groundwater aquifers. Freshwater resources are thus being damaged at the same time as demand for water is increasing.

Globally, the situation in the Oceania is not that different from other continents described above. For example, reports are that New Zealand's freshwaters are stressed (Gluckman 2017). But, once more, the stress varies in terms of quantity and quality. According to Gluckman (2017), some water bodies are in a good state but others are currently significantly compromised. Same reasons as for other parts of the world do apply here too. As Gluckman (2017) reports, these reasons include agricultural intensification, urban expansion and industrial pollution, hydroelectric development, or the effects of drought. Gluckman (2017) goes on to conclude that New Zealand's wetlands have been greatly reduced and many river catchments are significantly affected by dam systems; over recent decades, flows of foothill catchment or spring-fed rivers and streams have declined substantially, particularly in lowland areas on the eastern sides of both islands. Overall, the water situation in the Oceania is rendered more complicated by the decline of marine water quality associated with land-based run-off from the adjacent catchments (Waterhouse et al. 2017). Coastal ecosystems (for Australia and New Zealand) are highly modified and exposed to a range of pressures from catchment development. Historically, marine ecosystems in the Oceania were thought to be resilient and supposed to have the ability to at least partially recover from previous losses during periods of low disturbance and reduced catchment pollutant loads. This resilience capacity was felt to augment by possible existence of reefs' refugia, which would have to allow reefs to recover from somewhat severe environmental stresses (Riegl and Piller 2003). Unfortunately, the hopes for this capacity to bounce back from disturbances on the marine water of that region have been watered down because recent events severely impacted systems (Waterhouse et al. 2017). Indeed, events such as un-normally prolonged periods of extreme sea surface temperatures and tropical cyclones nearly broke the resilience of the marine ecosystems to the point that coral cover had regionally significantly declined of (Bruno and Selig 2007; De'ath et al. 2012). Furthermore, Waterhouse et al. (2017) predicted climate change to increase the frequency of large-scale bleaching events and the intensity of extreme weather events.

Mountains are globally an important kind of ecosystems in relationship to water management. Mountains capture moisture from air masses and precipitate that moisture down in the form of snow, which can melts into liquid water when heat increases. People who reside at the mountain foot directly use the melted freshwater to drink and other livelihood activities such as agriculture but the role of mountains in global water cycles goes beyond mountain ecosystems. Without going into details for each continent, suffice it here to say that, according to the Albertine Rift Conservation Society (ARCOS 2014), all of Africa's major rivers originate from mountains and African mountains offer water sources to more than half of the continental human population. In Europe, the Alps cover 11% of the Rhine's basin and provide 31% of the total Rhine's annual flow (Preusser 2008). It also worth noting that mountains, as high as 4000-m above sea level, traverse Central and South America plays a key role in temporal and spatial distribution of water resources (Global Water Partnership 2016). Nearly all Central and South American States share substantial these mountains with neighboring countries; as these are sources of freshwater, water resources are also shared continentally (Global Water Partnership 2016). In Asia, mountains such as Tienshan Mountain are known to be water towers and are main water sources playing key ecological roles (Chen et al. 2016). Unfortunately, the current scientific knowledge (e.g. Barnett et al. 2005; Immerzeel et al. 2010; Kääb et al. 2012; Sorg et al. 2012; Lutz et al. 2014; Chen et al. 2016) clearly and consistently indicates that rapid warming is affecting and will continue to affect fractions of precipitated moisture in forms of snow, snow melting water processes and increase mountain water runoff. Obviously, this will bite on the water storage, which are expected to show deficits in the long run. Consequently large parts of glaciers are retreating and would continue with such trends. In Central Asia, for example, about 98% of glaciers witness this situation whereas the globally, changes in cryosphere are likely to increase river runoff trends in glacier-dependent catchments.

5 Conserving Water Ecosystems

The conclusions that can drawn from the sketchy presentation of the current status of water across globe presented above is that waters of the world are undergoing significant shifts and changes that will affect their geographical distribution, their intrinsic quality and their ecological functions. That is why action is needed to conserve what can be conserved and to restore the lost quality wherever that will be possible. First, conserving aquatic and marine ecosystems in good functioning conditions, ecologically means to ensure that water cycle can be maintained as unimpaired as possible. Secondly, it also means to ensure that water chemistry and physics remain as safe as possible of additional chemicals as possible. This equates water ecosystems conservation's work to addressing causes of physical impairments and external chemical 'impurities'. To sum up what follows from the section on water ecosystems health, major concerns about water health stem from (1) destruction of natural habitats (deforestation) around water bodies and their basins,

(2) chemically soiled dejections into either soil or directly to water bodies (chemical pollution), (3) rejecting torrential quantities of mechanical fragments into water bodies (mechanical pollution), particularly into rivers and oceans and (4) losing aquatic and marine biodiversity. Agriculture is part of all the most important threats to waters.

That word-widely there is no single water body or basin that remains in pristine conditions is a fact the world needs to live with is no surprising. Some areas are ecologically worn out and can be said to have gone far beyond their resilience thresholds (Waterhouse et al. 2017). However, there are still aquatic and marine areas in good ecological conditions. Taken from the conservation perspectives, these areas would need to have special status and should be subtracted from extractive or any other type of activities that would negatively impact habitats around water bodies and their basins, chemically tarnish soil around water bodies or directly infect water bodies, reject heavy quantities of mechanical fragments into waters and speed the losses of aquatic and marine biodiversity. To avoid these negative impacts, the world should not need to re-invent the wheel; the way that has been paved by the International Union for Conservation of Nature (IUCN) seems to prove the most realistic and the most comprehensible. Indeed, to prevent the above four major causes to continue threatening the integrity of water bodies, IUCN has been suggesting the creation of aquatic and marine protected areas of dissimilar IUCN categories across the world (Kelleher and Kenchington 1992). Each of these comes with specific requirements to ensure that not only their ecological integrity is preserve but also that all other social and economic values (e.g. Inogwabini 2020) are asserted without jeopardizing water ecosystems themselves. For Kelleher and Kenchington (1992) the need to devise methods to manage and protect marine environments and resources became apparent during the course of the 1950s and early 1960s and serious considerations to the demand protect coastal and marine areas were given at the First World Conference on National Parks (1962). But it took too long before first Marine Protected were officially created and accepted world-widely. The Ross Sea Marine Protected Area (1.55 million km^2) is the World's largest marine protected area (Ballard et al. 2012); other marine protected areas include Coral Sea Natural Park, Coral Sea Commonwealth Marine Reserve, Marine National Monument, South Sandwich Islands Marine Protected Area, etc. The current situation is that about 3.4% of the world's oceans are protected one way or the other. This remains below the 2014 target of 10% of the world's marine and coastal areas to be protected (Juffe-Bignoli et al. 2014). Protection status varies great and 50% of the world's protected oceans are legally fully protected, which means they are supposed to be exempted of any type of off-take. Data on protected areas of inland waters are difficult to disentangle from terrestrial ones but freshwater biodiversity and freshwater features such as rivers and lakes are often underrepresented in protected area networks. To give a sense of where things stand with freshwater protected areas globally, suffice it to repeat with Juffe-Bignoli et al. (2014) that although about 21% of the world's lakes and wetlands are part of protected areas, efforts to conserve wetlands through the Ramsar convention have been unsuccessful as between 64 and 71% of wetlands of the world have been lost since 1900.

Losses of wetlands and the disappearance of aquatic and marine biodiversity indicate that even though the way paved by IUCN proves to be the most realistic and the most comprehensible, there is something more that is needed to achieve the goals assigned to the sustainable development goal 6 in its 6th goal, which is to protect and restore water-related ecosystems, including mountains, forests, wetlands, rivers, aquifers and lakes by 2020. Clearly, the 2020 is fast approaching and many actions needed to preserve water-related ecosystems are, at best, in their infancy. The clear message one gets from the assessment of the current situation is that beyond guidelines and many international overarching agreements, the world needs more political will and a stronger commitment to change the current negative trends on ecosystems globally and water-related ecosystems particularly. The message here is that to protect and restore water-related ecosystems is to be achieved only through an integrated approach whereby of other milestones of other sustainable development goals, particularly SDGs (14) and (15) (respectively Life below Water and Life on Land) are reached.

To conclude with the issue of conserving water ecosystems, it is important to recall that despite efforts to preserve these biomes, the picture continues to become gloomier and gloomier. This sends a message that protected areas are not sufficient to preserve water habitats and protecting water habitats needs to be extended beyond the boundaries of the waterscapes we wish to preserve. Indeed, to protect rivers and lakes, for example, one needs to also include the mountain habitats where water sources are located. Equally, as it has been argued by Waterhouse et al. (2017), the troubles in the oceans and costal habitats come from the mess people create on continents. It is, therefore, a sort of short sighting to think of preservation of water quality in rivers, lakes and oceans as isolated from what is done on land. Indeed, the whole issue of water pollution, whether through agriculture, hydraulic fracking or inappropriate sewage management are to be treated as part of the holistic package to preserve functions and the quality of waters.

6 Restoring Injured Water Ecosystems

Following the logics of the above section, most of the world waters now need some sort of restoration for them to continue being of sufficient good quality. Ecologically, restoring an ecosystem means doing whatever is necessary to bring the injured ecosystem back to its initial conditions (McCarty and Zedler 2002). Without background information on what the initial conditions were for each of soiled water bodies, it would be nearly impossible to know up to which point restoration should be taken. The same lack of knowledge on initial conditions runs the temptation of homogenising water chemistry and water physics across the world. The good thing to know, however, is the ecosystems, whenever left on their own will naturally strive to bounce back to some good biological conditions. Undeniably, life sustains life provided humans do not meddle with the recuperating process. But recovering naturally will take time. Hence, restoration whereby humans assist the recovery and management of ecological integrity (McCarty and Zedler 2002) remains the only option for some of the most important ecosystems of the world. The whole thing here is to identify areas where waters are in important quantities that have been spoiled because of human activities and stop being intrusive into them. Protecting these areas from being further impaired would lead, in the long run, to improved water quality, which is the objective of restoration, per se. Of course, there are those areas where water quality has been so sullied that leaving the water to recuperate its own quality on its own may reveal fatal to the ecosystems themselves. In this case, careful water treatment program, inclusive of the re-insertion of lost biodiversity, recreation of catchment's habitats, etc. can be tried. As McCarty and Zedler (2002) reported, restoration has been increasingly attempted across the world in freshwater and wetland but the results of these efforts have been mixed. According to McCarty and Zedler (2002) mixed results are caused by the fact that restoration projects involve high inputs of time and money for each hectare restored and ongoing maintenance is frequently required. In many cases, the species that make up a community can be established but healthy ecosystem functioning is more difficult to restore. Finally, the length of time necessitated by ecological restoration is likely to increase if things have to be done properly, which means ecological restoration should be done only after experiments are scientifically designed, tested and results significantly able to show differences between the destroyed status and the incremented improvements.

7 Sustainability of Good Water Quality Across the Globe

Water ecosystems, water quantities and its quality are part of the world's current affairs. This means ecologically, economically, politically and socially. Stating so says that water issues dealt with above should not be limited to ecology; problems of water, as it is also the case for all other ecosystems, are caused by humans and should be addressed by inclusion of human sciences into the frameworks of their solutions. Basically, this boils down to saying that for water conservation and restoration to work, not only ecological and technical solutions are needed. As it has been argued in this article, water management is to been taken holistically. This, politically, means bringing water to the forefront of international cooperation. This is an issue of water governance and citizens' participation. Technically, the holistic approach to water management to build capacities to manage water where there is lack of capacities. As it has been shown above, these capacities are not needed only areas such as hydrology, water ecology, water distribution, etc. that are directly related to water but also in areas such as treatments of solid wastes, ecological agriculture and global sanitation; these are intricately linked to water quality and should be part of the whole package. Technically, another huge need is to make all water related technologies accessible to all parts of the world. These technologies include not only technical apparatuses to ensure efficiency in harvesting water and the treatment of wastewater treatment, and to desalinate ocean water and to recycling and reuse technologies but should also include forecasting techniques to ensure that people are prepared to cope with future water shortages.

References

Adam NK, Askew FA, Pankhurst KGA (1939) Interaction between adsorbed substances of simple constitution and insoluble monolayers. R Soc Lond Proc A: Math Phys Eng Sci 170:485–500

Albertine Rift Conservation Society (ARCOS) (2014) African mountains status report. Albertine Rift Conservation Society, African Mountains Programme (Kampala, Uganda)

Allan AJG, Alexander AE (1954) Monolayers at low surface pressures. Faraday Soc Trans 50:863–873

Aragão LEO, Malhi Y, Barbier N, Lima A, Shimabukuru Y, Anderson L, Saatchi S (2008) Interactions between rainfall, deforestation and fires during recent years in the Brazilian Amazon. Philos Trans R Soc B 363:1779–1785

Archer RJ, LaMer VK (1954) The effect of monolayers on the rate of evaporation of water. Ann N Y Acad Sci 58:807–829

Ballard G, Jongsomjit D, Veloz SD, Ainley DG (2012) Coexistence of mesopredators in an intact polar ocean ecosystem: The basis for defining a Ross Sea marine protected area. Biol Cons 156:72–82

Barnett TP, Adam JC, Lettenmaier DP (2005) Potential impacts of a warming climate on water availability in snow-dominated regions. Nature 438:303–309

Bruno JF, Selig ER (2007) Regional decline of coral cover in the Indo-Pacific: timing, extent, and subregional comparisons. PLoS ONE 2:

Burton GA, Basu N, Ellis BR, Kapo KE, Entrekin S, Nadelhoffer K (2014) Hydraulic "fracking": are surface water impacts an ecological concern? Environ Toxicol Chem 33(8):1679–1689

Chen Y, Li W, Deng H, Fang G, Li Z (2016) Changes in Central Asia's water tower: past, present and future. Nat Sci Rep 6:35458. https://doi.org/10.1038/srep3545

De'ath G, Fabricius KE, Sweatman H, Puotinen M (2012) The 27-year decline of coral cover on the Great Barrier Reef and its causes. Proc Natl Acad Sci USA 109(44):17995–17999

Environmental Protection Agency (EPA 2016) Hydraulic fracturing for oil and gas: impacts from the hydraulic fracturing water cycle on drinking water resources in the United States. EPA-600-R-16-236ES

Food & Water (2012) Fracking: the new global water crisis. Food & Water Europe, Brussels (Belgium)

Global Water Partnership (2016) Integrated water resources management in Central America: the over-riding challenge of managing transboundary waters. Technical Focus Paper, Global Water Partnership, Sweden

Gluckman P (2017) New Zealand's fresh waters: values, state, trends and human impacts. Office of the Prime Minister of New Zealand

Immerzeel WW, Van Beek LPH, Bierkens MFP (2010) Climate change will affect the Asian Water Towers. Science 328:1382–1385

Inogwabini BI (2020) The changing water cycle: freshwater in the Congo. WIRES Water. https://doi.org/10.1002/wat2.1410

Juffe-Bignoli D, Burgess ND, Bingham H, Belle EMS, De Lima MG, Deguignet M, Bertzky B, Milam AN, Martinez-Lopez J, Lewis E, Eassom A, Wicander S, Geldmann J, Van Soesbergen A, Arnell AP, O'Connor B, Park S, Shi YN, Danks FS, MacSharry B, Kingston N (2014) Protected planet report 2014. UNEP-WCMC, Cambridge, UK

Kääb A, Berthier E, Nuth C, Gardelle J, Arnaud Y (2012) Contrasting patterns of early twenty-first-century glacier mass change in the Himalayas. Nature 488:495–498

Kelleher G, Kenchington R (1992) Guidelines for establishing marine protected areas. A Marine Conservation and Development Report. The International Union for Conservation of Nature (IUCN), Gland, Switzerland

Liberati D, Sconocchia P, Ricci A, Gigliotti G, Tacconi C, Grenni P, Tariciotti S, Caracciolo AB, Massacci A, Lagomarsino A, De Angelis P (2018) Phyto-dehydration of confined sludge: a sustainable approach for the management of polluted ponds. J Soils Sediments 18:2292–2304

Lutz AF, Immerzeel WW, Shrestha AB, Bierkens MFP (2014) Consistent increase in High Asia's runoff due to increasing glacier melt and precipitation. Nat Clim Change 4:587–592

Magin GB, Kandall LE (1960) Review of literature on evaporation suppression. Studies of evaporation. Geological survey professional paper 272-C. US Department of the Interior. US Government printing office, Washington DC

McCarty JP, Zedler JB (2002) Restoration, ecosystem. In: Mooney HA, Canadell JG (eds) Encyclopedia of global environmental change volume 2: the earth system: biological and ecological dimensions of global environmental change. Wiley, Chichester, pp 532–539

Mekonnen MM, Pahlow M, Aldaya MM, Zarate E, Hoekstra AY (2014) Water footprint assessment for Latin America and the Caribbean: an analysis of the sustainability, efficiency and equitability of water consumption and pollution, Value of Water Research Report Series No. 66, UNESCO-IHE, Delft, the Netherlands

Preusser F (2008) Characterisation and evolution of the River Rhine system. Neth J Geosci (Geologie en Mijnbouw) 87(1):7–19

Relyea R, Hoverman J (2006) Assessing the ecology in ecotoxicology: a review and synthesis in freshwater systems. Ecol Lett 9:1157–1171

Reno U, Doyle SR, Momo FR, Regaldo L, Gagneten AM (2018) Effects of glyphosate formulations on the population dynamics of two freshwater cladoceran species. Ecotoxicology 27:784–793

Riegl B, Piller WE (2003) Possible refugia for reefs in time of environmental stress. Int J Earth Sci 92:520–531

Seitz RD, Knick KE, Davenport TM, Saluta GG (2018) Human influence at the coast: upland and shoreline stressors affect coastal macro-fauna and are mediated by salinity. Estuaries Coasts 41 (Supplement 1):114–130

Sorg A, Bolch T, Stoffel M, Solomina O, Beniston M (2012) Climate change impacts on glaciers and runoff in Tien Shan (Central Asia). Nat Clim Change 2(10):725–731

Strayer DL, Dudgeon D (2010) Freshwater biodiversity conservation: recent progress and future challenges. J North Am Benthol Soc 29(1):344–358

United Nations Environmental Programme (UNEP 2000) GEO-2000 global environmental outlook. Chapter Two: The State of the Environment - Latin America and the Caribbean. United Nations Environment Programme (Nairobi, Kenya)

United Nations Environmental Programme (UNEP 2010) State of the biodiversity in Latin America and Caribbean. United Nations Environment Programme Regional Office for Latin America and the Caribbean (ROLAC) (Panama City, Panama)

United Nations Environmental Programme (UNEP 2016) GEO-2016 Global Environmental Outlook. United Nations Environment Programme (Nairobi, Kenya). Details given in this text come principally from continental factsheets produced by United Nations Environmental Assembly, which accompany each regional assessment (Geo-6 Africa Factsheet, Geo-6 Asia and Pacific Factsheet, GEO-6 Assessment for the pan-European region, GEO-6 North America Factsheet)

Vörösmarthy C, Sahagian D (2000) Anthropogenic disturbance of the terrestrial water cycle. Bioscience 50:753–765

Waterhouse J, Schaffelke B, Bartley R, Eberhard R, Brodie J, Star M, Thorburn P, Rolfe J, Ronan M, Taylor B, Kroon F (Lead Authors) (2017) 2017 Scientific consensus statement: land use impacts on Great Barrier Reef water quality and ecosystem condition. The State of Queensland

Wlikie D, Shaw H, Rotberg F, Morelli G, Auzel P (2000) Roads, development and conservation in the Congo Basin. Conserv Biol 14(6):1614–1622

The Contribution of Geographical Information Systems—GIS in Water and Sewage Companies for Water Sustainability

Thomas R. Aquino Ficarelli and Helena Ribeiro

1 Introduction

Water is essential to human life and activities. To guarantee its sustainability, this indispensable natural resource requires good management and land planning, based on well-crafted policies and actions. In 2015, among the 17 Sustainable Development Goals (SDGs), the United Nations (UN) dedicated SDG 6 (*To guarantee accessible and sustainable water and sanitation systems for all*) specifically to water. Additionally, all other SDGs depend on and concern actions related to good water management. Globally, close to 2.2 billion people lack safely managed drinking water and 4.4 billion people lack safely managed sanitation (UN 2020). Despite progress, 2 in 5 health care facilities worldwide lack soap and water, and 3 billion people worldwide lack basic handwashing facilities at home, with serious health implications, especially given that this is the most effective method for COVID-19 prevention (UN 2020). In addition, 700 million people might be displaced by water scarcity by 2030. Thus, the Sustainable Development Goal 6 is an urgent task and requires the development and widespread use of new technologies to support land planning and effective resource management that might allow for providing efficient infrastructure and services for more inhabitants at lower costs, amid huge funding gaps.

The world's urban population is growing rapidly, increasing from 36.56% in 1970 to 55.71% in 2020 according to the World Bank. In rural areas, on a global scale, one concern is the high percentage (48%) of communities that lack of access to improved sanitation facilities (UN 2014). Human waste can lead to contamina-

T. R. Aquino Ficarelli (✉)
Global Health and Sustainability Program, University of São Paulo, São Paulo, Brazil
e-mail: teraviz@teraviz.com.br

H. Ribeiro
Department of Environmental Health, University of São Paulo, São Paulo, Brazil
e-mail: lena@usp.br

W. Leal Filho et al. (eds.), *Sustainability in Natural Resources Management and Land Planning*, World Sustainability Series,
https://doi.org/10.1007/978-3-030-76624-5_2

tion of surface water resources and water wells and the proliferation of communicable diseases, which especially affect the gastrointestinal system. These include, but are not limited to, diarrhea. In such communities, there is a vicious cycle of poverty and waterborne diseases that only providing good quality water and adequate sanitation can break.

Sustainability in the water cycle consists of ensuring its multiple uses, such as public supply, irrigation, industrial production, power generation and transportation, but without removing the responsibility of citizens, technicians and governments to preserve the resource and optimize its use. In this case, actions related to upgrading technologies that avoid water waste or reduce consumption, with equal or better efficiency, and which encourage policies that ensure supply, are appropriate and essential. Additionally, stakeholders' counterparts are needed to improve quality, such as pollution control in watersheds and springs in large cities, where there is a greater risk of eutrophication (Oliver et al. 2019).

The main factors that lead water use to unsustainability are lack of funds for investment in the sector, embezzlement and corruption, environmental pollution, disorderly land use and occupation and water losses in hydraulic systems. Water scarcity requires better care and efficiency from the services that use it, especially water and sewage companies responsible for urban areas. Challenging scenarios have led governments and companies to seek solutions aimed at conserving and recovering natural resources and improving the resilience of water systems to extreme events (floods or droughts), reducing costs under certain operations (water treatment and distribution; sewage collection and treatment) and including new populations into public water and sewage systems.

In this chapter, we highlight the Geographic Information Systems (GIS) that store large volumes of data and allow for the visualization of graphic and alphanumeric information simultaneously through the global positioning of objects, phenomena and territories at different scales. In addition to these platforms, it is necessary to qualify human capital to create information, diagnose problems, formulate solutions, prioritize actions and plan expenses. Obtaining reliable data benefits these actions and facilitates dialogue between stakeholders.

In the urban environment, GIS applications can and should also include informal settlements. According to the UN, about 1 billion people live under these conditions (United Nations 2020), often unattended by water and/or sewage services, which can aggravate health conditions. Governments, communities and sanitation companies have sought to face these difficulties and, for this, use registration and geospatial information, often with the help of communities. GIS enables the application of semi-automatic methods to identify buildings (Mayunga et al. 2007) and conduct partial information collection in the field for statistical analysis (Paul 2015). In Fiji, Saunders et al. (2016) invited the population of an informal settlement in the capital Suva to use a mapping platform called *Households Family Mapping* to identify places of interest, their own residences and basic sanitation equipment. Later, this database was delivered to the National Development Public Consultation to facilitate planning of basic sanitation policies and actions in subnormal housing sites.

Other studies in Palestine, Austria and India highlighted the usefulness of GIS in performing hydraulic modeling and controlling distribution losses, including automatization, leading to greater efficiency of operational resources and lower rates of energy consumption and water leaks or losses (Eljamassi and Abeiad 2013; Sitzenfrei et al. 2013; Varade et al. 2017).

It is desirable for water and sanitation companies to invest in training their employees to work with GIS and in developing a corporate GIS, given the potential and real operational improvement, the greater speed in customer service and to prevent incidents (Shamsi 2005). The possibility of crossing a GIS focused on water resources with other corporate systems like Enterprise Resource Planning (ERP) or with SCADA (Supervisory Control and Data Acquisition), which brings operational data of equipment and operating units, allows for an even greater range of uses (Baumann 2007).

Trust, training, motivation, integration and budget planning are essential for such projects to progress and, consequently, achieve results such as greater protection of water sources, greater durability of hydraulic structures, prioritization of projects and sampling points for water quality, optimization of vehicle paths and teams for maintenance and reductions in leakage rates in public networks. The resources saved with this improved planning can encourage projects aimed at expanding services.

The guaranteed long-term return on investment has convinced many companies to create their own customized GIS platforms to facilitate employee routines and communication. In addition to routine operations, complex analyses can be conducted by the staff and/or with partners, through further training and qualification in GIS (Mäkelä et al. 2010; Bearman et al. 2016), often developed in partnership with universities, research institutes, public agencies and consultancies.

The objective of the study was to verify if and how geoprocessing technologies have been used by water and sanitation companies in cities located in different countries and continents and to investigate how projects and tasks fit in one or more of the three pillars of sustainability (Society, Economy and Environment) as well as in the SDGs of the UN (besides the SDG 6). The results are reported in this chapter.

2 Methodology

We studied sanitation companies in several cities around the world to understand how corporate GIS sanitation services have been developed. Among 14 sanitation companies contacted, five welcomed this research.

We conducted on-site visits to the cities and their companies between 2014 and 2018, preceded by drawing up a research form in Portuguese, English, Italian and Hungarian. The locations studied had different income levels, and information was gathered through interviews with their employees. The interviews took place in person and in the same languages as the forms. Respondents were employees with experience with the subject, from the companies' different departments. The main issues addressed, in addition to the history of the system implementation, were:

- Diagnosis mapping for natural resources management and planning, land use policies, locational alternatives, and protection of water springs and sources;
- Geospatial models for efficiency in water and energy uses for pumping and supplying;
- Water and sanitation in informal settlements;
- Partnerships for geospatial information sharing.

The companies selected allowed us to understand the use of GIS of a global sample of countries ranging from Very High Human Development Index—HDI (Italy and Hungary), High Human Development (Brazil) and Medium Human Development (Kenya), according to UN classification (Table 1).

The five studied companies work directly with water and sewage services at the metropolitan level. Considering that they operate in several local territories (municipalities, autonomous governments, etc.), they can ultimately provide only one of these services. The exception was Budapest Waterworks that, despite working with sewage collection and treatment in several territories of the metropolitan cluster, works in Budapest exclusively with water and partly with sewage treatment, through Budapest Sewage Works, responsible for sewage collection and part of the treatment.

Table 2 presents the 68 employees interviewed in the 5 companies and other relevant operational information.

Table 1 Cities investigated

City	São Paulo	Fortaleza	Nairobi	Rome	Budapest
Country	Brazil	Brazil	Kenya	Italy	Hungary
National HDI (2018)[a]	0.761 (79th)	0.761 (79th)	0.579 (147th)	0.883 (29th)	0.845 (43th)
City's founding year	1554	1597	1899	753 (B.C.)	89
Population (millions)	21.571	4.074	4.397	4.354	2.457
% population in slums	11	21.7	40	None	None
Year of 1st public modern water supply system	1877	1920	1906	1870	1868
Farthest water spring for supply and its name	83 km (São Lourenço)	283 km (Orós)	60 km (Sasumua)	130 km (Peschiera-Capore)	20 km (Csepel)

[a]The Human Development Index is a set of indexes related to life expectancy, education and per capita income applied by the United Nations (UN). Every year, 189 countries are ranked and rated *Source* United Nations (2018); Brazilian Institute of Geography and Statistics (IBGE 2010; 2018); Kenyan National Bureau of Statistics (2009). Italian National Institute of Statistics (ISTAT) (2013); Hungarian Central Statistical Office (KSH) (2018). *Source* Compiled by the authors

Table 2 Utilities companies investigated

Utilities' company	SABESP (São Paulo)	CAGECE (Fortaleza)	NCWSC (Nairobi)	ACEA ACT 2 (Rome)	Budapest waterworks
Type of company[a]	Public (51%) Private (49%)	Public	Public	Public (30%) Private (70%)	Public
Employees[b]	17,000	3400	3400	1500	1730
Population served (millions)	20.541	4.054	4.397	4.354	1.876
Operating territories	38	18	1	112	12
Water service (%)[c]	98	75.4	76	n/a	95.4
Sewage service (%)[d]	90	45.7	47.7	n/a	89.9
Employees interviewed for this research	6	17	16	15	14

[a]When defined as "public," there was no distinction as to whether ownership belongs to local and/or regional government. [b]Employees of the two Brazilian companies also deal with services in the countryside of the respective states of the federation; [c]Isolated systems such as small flow water collections, septic tank and autonomous supply and depletion systems; [d]All other companies' sewage is fully treated. Regarding SABESP, 85% of the sewage collected in the metropolitan region of São Paulo is treated

Source Compiled by the authors. Respective utilities mentioned: Brazilian National Sanitation Information System (SNIS) (2018); Kenyan National Bureau of Statistics (2009); Italian National Institute of Statistics (ISTAT) (2013); Hungarian Central Statistical Office (KSH) (2018)

3 Results

Both free and high cost GIS technologies have been applied in different socioeconomic contexts. Despite the diversity, the companies were able to develop georeferenced databases, according to their teams' maturity and familiarity, and also invested in technologies and training. These allowed for activities that facilitate diagnostics and decision-making in many aspects, sometimes together with other public agencies but mainly on their own. We briefly report what has been done in each company based on the four aspects highlighted in the methodology.

3.1 Mapping for Diagnoses and Land Use Policies, for Locational Alternatives, and Protection of Water Springs

Maps are helpful in decision making to prevent negative impacts′ buffering and by providing useful information for decision making on projects and policies about natural resources and water management. This aspect is mainly on the Environment pillar of sustainability and it can be also related to the SDG 14 (Life Below Water), in coastal areas to avoid pollution of the sea, as the case of Fortaleza, and the SDG 15 (Life on Land) by supporting efforts to protect areas near water springs and dams. All studied cities applied GIS Technologies successfully for these purposes.

Specific environmental laws on land use and occupation regulate watersheds in São Paulo. The maps created by environmental agencies, by consultancies, and by SABESP have contributed to direct sanitation projects in areas not yet served according to water quality goals and parameters. Staff commented that GIS has contributed to minimizing the impacts of large projects, with the aid of geographic information and spatial analysis. In addition, it helped to comply with environmental legislation, and to increase project efficiency, making them more acceptable by affected populations and environmental authorities, speeding up environmental assessment and licensing.

At CAGECE, in Fortaleza, research prepared by employees has also contributed to the development of GIS applications for environmental purposes. Among them, a motivated biologist, with GIS tools, completed a PhD research project on dispersion and purification of pollutants from an underwater effluent outlet, from which 2.5 m^3/s of preconditioned effluents were released (Pereira 2012). This same employee has been collaborating with the company on studies of alternative sites for large enterprises, such as the Desalination Plant, which should provide 1 m^3/s of treated water.

The Geospatial Database (GDB) of NCWSC, in Nairobi, makes many resources available, such as shapefiles of 2 m resolution topographic contour lines, hydrography and the limits of watersheds at a 1:50,000 scale and satellite images with a 25 cm/pixel resolution to support environmental analysis. Regarding geographic information for water springs, an employee carried out, as part of her doctorate, a modeling of organic pollution dispersion in the Ruiru basin, the water source closest to Nairobi, where population has grown in recent years. The modelling was performed through a partnership between the University of Nairobi and the German Agency of International Cooperation (GIZ).

To preserve the main water supply spring in the Rome metropolitan area, located in Rieti, as well as its pipeline, ACEA signed a partnership in 2010 with the University of Cassino to perform territorial surveillance, by satellite images. The system, called SatGuardian, updated satellite images within 3–6 months and automatically identified changes in land use and occupation. The partnership, which lasted for years, identified changes in vegetation and possible invasions, unauthorized uses and dumping of waste in areas owned by the company. At the national scale in Italy, a specific law to define protection of groundwater areas has existed since 1923. At that time, areas in Rieti were already of concern but maps have been updated constantly in order to support watershed protection actions in the province and throughout the region (Lazio).

In Budapest, the Ministry of the Environment undertook a GIS modeling of Csepel Island, the city's main water source, on the hydrogeological dispersion of agricultural contaminants in the groundwater. This study led to approval of a law restricting land use and occupation and the application of pesticides for agricultural crops, based on their proximity to wells for public water supply. According to the company's staff, generally, the government demands high-level diagnoses and cost evaluations based on GIS for environmental and water resource planning.

3.2 Geospatial Models for Water and Energy Use Efficiency for Pumping and Supplying

Models are a valuable tool to predict scenarios and to optimize engineering and operational costs of water structure implementation and maintenance and, mainly, to control water leaks. Technology is now available (both proprietary and open source) to design georeferenced networks for whole cities for both water and sewerage systems. The benefits of such a tool concern mainly the Economical pillar of sustainability and fall within the SDG 9 (Industry, Innovation and Infrastructure) by supporting resiliency and sustainability for long term use of pipe networks, treatment stations, reservoirs and other structures. Because modelling requires highly accurate geospatial data, not all companies could model their systems even if willing, according to the interviews. Nevertheless, they have been using other means to map water leakage, such as locating demands through customer care and operational platforms.

SABESP has developed research on maintenance and loss control in its regional units, to prioritize interventions in water networks, and to support the management and the selection of companies for outsourced services. SIGNOS, its GIS system, has allowed for the implementation of projects and methods that are more efficient, achieving greater sustainability and durability for the structures in place, permitting intelligent control of pressure for public supply and consequently better diagnoses of distribution losses (Silva Júnior and Cabral 2015). The company has gradually integrated SIGNOS with other ERP platform systems and has been experimenting with localized hydraulic modeling to optimize supply.

Fortaleza suffers from water scarcity given the metropolis' high demand, and the fact that raw water is fetched from 283 km away. CAGECE has been using automatic heat maps by locating demands and complaints about water leaks in the city to accelerate maintenance and reduce water losses. Employees are trying to improve diagnoses of losses through hydraulic modeling. However, such modeling requires precision and topological fitting between the georeferenced vectors that represent the networks. This step is still under development at the corporate level. Even though, an employee conducted a study on sewage interceptor obstructions in the eastern region of Fortaleza, using hydraulic modeling (Storm Water Management Modeling—SWMM) of geographic data from the corporate GIS "IGEO" (Fernandes 2017).

In Nairobi, all water and sewage networks were mapped and are updated monthly, but still lack maximum metric accuracy and the extensions to buildings, which makes hydraulic modeling impossible for now. In spite of several advantages of modeling, such a project was not considered a GIS priority in the short term due to other required urgent tasks such as mapping new clients, pipes, city areas in expansion, and capacity building for GIS in the company.

In ACEA, high levels of interest in consulting geographic information led to the creation of an ERP system integrated to the corporate GIS, by the Workforce Management Project (WFM). WFM started in 2016 and was a pioneering initiative, not yet required by the service regulator. However, it sped up all incident responses and communication between employees and service providers. Its employees use field devices to update, in real time, the status of requests, including their geolocation. Between 2017 and 2018, WFM completed 14,000 maintenance tasks in the water network and reduced water losses from 45 to 38% in the city. Nevertheless, modeling for pipe networks was not performed on an urban scale to support decision making on controlling water leaks and substitution of equipment and pipes.

In Budapest, in partnership with students from the Budapest University of Technology (BME), the necessary adjustments were made to implement the water and sewage networks topology, thus allowing for the hydraulic distribution modeling. Today the company has a sector dedicated to modeling, which assists the operational units on the chlorine content in the water, the design of structures according to diameter, material, pressure and flow required, and on larger interventions and the populations they affect. Currently, the levels of leakage in the water supply are 15%. This was possible due to the MIR platform. It helped the city demarcate 100 district-metered areas (DMA) for pressure control in each district. They were categorized based on land occupation, population and especially on terrain and altitude. Districts are under constant review, as demographic and water demand changes, or varies over the months and years.

3.3 Water and Sanitation in Informal Settlements

There are several difficulties of obtaining a precise geospatial and demographic diagnosis for informal settlements in urban areas. Therefore, designing a map with water and sanitation systems requires expertise in making the local population as a stakeholder in this process. This aspect relates mostly to the Social pillar of sustainability and the SDG 3 (Good Health and Well Being) by formally providing freshwater and/or sewerage to reduce risk of disease and child mortality for those underserved. The SDG 11 (Sustainable Cities) is also connected, since this infrastructure is more than necessary to upgrade urban life and environmental quality in slums. The companies investigated address this reality and showed that projects are implemented successfully once there is popular acceptance and engagement of government at different levels.

In Sao Paulo, the implementation of services in irregular occupations has been monitored by the City government, which eventually obtains registrations (georeferenced or not) of these areas and participate in projects in collaboration with the population, and in studies of alternative sites that facilitate implementation of sanitary network. SABESP staff, together with local NGOs and outsourced consultants, can enter these areas, and formalize services for expanding pipe networks with better agreement from the community, with the support of maps. A Program called *Corrego Limpo* (Clear Streams) was created in 2007 to use GIS to identify hundreds of water streams, communities and informal settlements that would be the focus of a partnership between SABESP and the municipality. Many projects were implemented with success.

CAGECE aims to provide services for informal settlements but there are difficulties in working with other public institutions. Once staff is inside the slums, residents are often helpful and participate in conceiving alternatives and giving information. The residents have to collect all information from the area in a participatory GIS in areas where the municipality does not have a base map. There isn't yet an agreement on state or municipal level to map these areas in detail and consequently there are more difficulties to develop and to implement projects and pipe installation.

NCWSC has a specific department and staff for informal settlements, which supports water and sanitation projects in obtaining information about the slums and their populations. To provide sanitation for these populations, the company installed public toilets with toilet bowls, showers and taps with treated water. The company itself produced registration and mapping information for the areas to design the projects, in partnerships with residents from the local communities. The company consults the population to get more creative and efficient ideas for implementing projects and discuss location alternatives, in a shared mapping initiative.

In the cities of Budapest and Rome, there were no such informal settlements or areas of extreme poverty according to the employees and official data available. Thus, this aspect is absent from the research in both cities.

3.4 Partnerships for Geospatial Information Sharing

On a governmental and strategic level, all information concerning a territory is very important for decision-making. The same applies for public services since they involve most citizens and urban areas. Spatial Data Infrastructures (SDI) have been implemented in many regions and countries across the world to provide fast and accurate data and maps for all interested stakeholders, including common citizens and the press, in some cases.

Engineering projects for water and sanitation often require geospatial data about underground networks (telecommunication, electricity, gas, tunnels, subway, etc.) to verify any interference during the implementation. Public environmental protection and urban development agencies also need geospatial information to diagnose and predict projects and activities. Health issues concerning waterborne diseases also ask for information to map vulnerable areas and people. Thus, a common and current sharing of information is important for all parties.

Partnerships for sharing geospatial information overlaps with the three pillars of sustainability (Economy, Environment and Society) because spatial analysis is important for all purposes, depending on the scope of each institution, agency or company (public or private). This aspect is related to the SDG 16 (Peace, Justice and Strong Institutions) by providing more transparency, participation and public access to information.

SABESP has a steady corporate GIS (SIGNOS) to support all employees, by providing 140 geospatial layers created by its staff, and another 70 created by other public institutions (like municipalities and cartography agencies) or required for outsourced private firms. On the state level, there is a Decree requiring all agencies (including SABESP) to share geospatial data as soon as it is generated, as a way to save costs. This is helpful for all agencies and for common projects. SABESP also provides data to outsourced engineering and maintenance companies. Citizens can ask for data through some means, like formal letters, and set agreements for small underground interferences to avoid damaging the pipes.

The corporate geospatial platform of CAGECE (IGEO) provides available data from many state and federal agencies, on the internet and also through formal demand. There is an agreement with the State Financial Secretariat to register new clients and map their location. On the local level, information is shared with the gas distribution company and the municipality. CAGECE provides geospatial data to any public agency interested and to citizens, although they have to pay for it. Staff predict that some tools and data will be available in the IGEO for common citizens to submit information about water and sewerage in the metropolitan area of Fortaleza.

In Nairobi, the company has a strong relationship with private entities (from the areas of energy and telecommunications), NGOs and other official cartography bodies in sharing geospatial information. Despite this openness, the exchange often takes place bureaucratically, internally or with other agencies, requiring several steps for approval, and is shared via flash drives or CDs after one or many days. A spatial data infrastructure (IDE) is expected at the national level to facilitate exchanges in an immediate and automated manner. Citizens cannot directly access the geospatial information because there is not a specific platform provided by NCWSC even for internal use, but they can ask for it by justifying its purpose.

In Rome, ACEA and the metropolitan area of Rome have a common agreement to share all underground structures in which any update must be reported among them and vice versa. The company has a specific office in which this work is developed, but employees can access this geospatial data only on paper and for small scale projects after one or two days waiting. Some geospatial data are available for common citizens through their websites, like water sample chemistry analysis in different pipes across the city. They can also request geospatial data for specific purposes. If the request is for private interests they have to pay for it. For research purposes, an agreement can be signed between ACEA and the institution.

Hungary already has a type of Geospatial Data Infrastructure, through which different public services make their information available. In this case, Budapest Waterworks provides its water and sewage network vectors for consultation, just as its employees consult electricity, gas, roads and railway structures across the country. The development of the platform still faces difficulties related to intellectual property over geospatial data, the non-obligation to share and the metric inaccuracy about certain data. However, as a benefit, it already includes open access for society about public service data and reduced bureaucracy in exchanging and demanding this information.

4 Final Considerations

The five companies presented initiatives for using geographic information for greater protection of water sources and for complying with environmental legislation. These spatial analyses facilitate surveillance studies, clearly define better alternative sites for structures and provide greater knowledge about pollutant dispersion and purification, based on real natural environment conditions.

The companies have been using GIS to plan repairs to their networks and thus optimize team time in the field. In Budapest, where it is already possible to perform hydraulic modeling across the city, the greater precision offers service diagnostics and, consequently, results in reduced water consumption for supply, reduced leakages and lower power consumption for the control of pumping and pressure in networks.

In the cities of Fortaleza and Nairobi and, to a lesser extent in São Paulo, we observed the companies' difficulty in mapping and registering populations in informal settlements. This effort must come from the entire public sector and not only from these companies. According to our research, the structured collaboration between public agencies falls short of what is necessary, which hinders providing services in these areas by formal means. The partnership with the community, made in Nairobi, for a participatory cartography represents an interesting and low-cost social technology.

Larger spatial analysis studies received support from external entities, such as universities, government agencies and consultancies. Such initiatives guarantee better water quality in the springs and, consequently, lower use of coagulants in the treatment process for water distribution, making the production chain more sustainable in the long term and with reduced costs.

More than simply providing technologies and information, companies should invest and encourage the training of teams about GIS, to achieve more accurate data editing and creative analysis by employees. A motivated employee can benefit the entire company with a spatial analysis, avoiding expenses and/or negative impacts on the natural environment and water resources.

GIS allows for the communication between different public authorities, bringing greater confidence to their joint actions and, consequently, inviting entrepreneurial and intersectional partnerships, covering the three pillars of sustainability (Environmental, Economic and Social) for regional sustainable development. GIS for water management and water and sewerage services can also positively influence more than the SDG 6, as shown in this chapter. GIS can contribute to achieving SDGs 3, 9, 11, 14, 15 and 16.

Lastly, GIS of Water and Sanitation Companies can be used in different urban, social and economic contexts worldwide. Its viability on a global scale in public agencies across different economic levels was demonstrated. Therefore, it can contribute to the expansion of sanitary services, public health and the sustainable development.

GIS support facilitates protecting the springs that supply cities, by launching adequately treated effluents into water bodies to allow for the biodiversity of aquatic fauna and flora. GIS play an important role in diagnosing and defining protection measures, based on multicriteria analyses that enable identifying land use and occupation, sanitation indexes, and the degree of vulnerability of water bodies. From these, the restrictive and protective management of surface or underground water bodies can be defined, contributing to the sustainability of natural resources.

References

Baumann J (2007) Cape town's emphasis on systems integration exemplifies "SmartCity" goals. ArcNews Mag (summer)

Bearman N, Jones N, André I, Herculano AC, Demers M (2016) The future role of GIS education in creating critical spatial thinkers. J Geogr High Edu 40(3):394–408

Eljamassi A, Abeiad A (2013) A GIS-based DSS for management of water distribution networks (Rafah City as Case Study). J Geogr Inf Syst 5:281–291

Fernandes LB (2017) Modelagem hidráulica do interceptor leste de Fortaleza utilizando o SWMM. Final Graduation Work for Bcs. In: Environmental Engineering. Federal University of Fortaleza—UFC

Mäkelä J, Vaniala R, Ahonen-Rainio P (2010) Competence management within organizations as an approach to enhancing GIS maturity. Int J Spat Data Infrastruct Res 5:267–285

Mayunga SD, Coleman DJ, Zhang Y (2007) A semi-automated approach for extracting buildings from Quickbird Imagery applied to informal settlements mapping. Int J Remote Sens 28:2343–2357

Oliver SL, Corburn J, Ribeiro H (2019) Challenges regarding water quality of eutrophic reservoirs in urban landscapes: a mapping literature review. Int J Environ Res Public Health 16:40

Paul S (2015) Analyzing and mapping urban poverty of English Bazar slum: an approach of micro level planning perspective from a developing country (India). Bangladesh. E-J Sociol 12:2

Pereira SP (2012) Modeling of bacteriological quality of coastal water in Fortaleza (Northeast of Brazil). PhD Thesis for Civil and Sanitary Engineering. Federal University of Ceará—UFC

Saunders SG, Barrington BJ, Sridharan S, Meo S, Hadwen W, Shields KF, Souter R, Bartram JK (2016) Addressing WaSH challenges in Pacific Island Countries: a participatory marketing systems mapping approach to empower informal settlement community action. Habitat Int 55:159–166

Shamsi UM (2005) GIS applications for water, wastewater and stormwater systems. CRC Press Book

Silva Junior EG, Cabral RC (2015) Indicador de vulnerabilidade da infraestrutura – uma proposta para o diagnóstico e tomada de decisões no combate às perdas reais. Revista SANEAS ano XIII. 55

Sitzenfrei R, Mödelr M, Rauch W (2013) Automatic generation of water distribution systems based on GIS data. Environ Model Softw 47:138–147

United Nations—UN (2014) World water development report

United Nations—UN (2020) The sustainable development goals report 2020. https://unstats.un.gov/sdgs/report/2020/The-Sustainable-Development-Goals-Report-2020.pdf

Varade AM, Khare YD, Dongre KP, Muley S, Wasnik G (2017) Integrated geographical information system (GIS)-based decision support system (DSS) approach to identify the site-specific water conservation structures in a watershed of Nagpur district, Central India. Sustain Water Resour Manage 3:141–155

Towards Adaptive Water Governance in South America: Lessons from Water Crises in Argentina, Brazil, and Uruguay

Micaela Trimble, Pedro Henrique Campello Torres, Pedro Roberto Jacobi, Natália Dias Tadeu, Franco Salvadores, Lara Mac Donnell, Tomás Olivier, Gabriel Giordano, Lidiane Alonso Paixão dos Anjos, Igor Matheus Santana-Chaves, Miguel Pascual, Néstor Mazzeo, and Esteban Jobbágy

1 Introduction

Although there are numerous definitions of governance, the concept usually includes forms of government and regulation involving the market and state-centric decision-making processes with the participation of civil society, although it goes beyond these. The convergence of these three sectors would be based on a set of practices that combine hierarchical structures, participatory dynamics, and associative actions from multilevel and multiscale perspectives in the formulation and implementation of public policy (Driessen et al. 2012; Jacobi et al. 2009; Zurbriggen 2014).

M. Trimble (✉) · P. R. Jacobi · N. Dias Tadeu · L. Mac Donnell · G. Giordano · N. Mazzeo · E. Jobbágy
South American Institute for Resilience and Sustainability Studies, SARAS, Bella Vista - Maldonado, Uruguay
e-mail: mica.trimble@gmail.com; mica.trimble@saras-institute.org

P. H. C. Torres
Institute of Energy and Environment, University of São Paulo, São Paulo, Brazil

P. H. C. Torres · P. R. Jacobi
IEE/USP, São Paulo, Brazil

N. Dias Tadeu
Faculty of Social Sciences, Universidad de la República, UDELAR, Maldonado, Uruguay

F. Salvadores
Patagonian Institute for the Study of Continental Ecosystems, IPEEC-CONICET-CENPAT, Puerto Madryn, Argentina

L. Mac Donnell
National Technological University-UTN, Buenos Aires, Argentina

W. Leal Filho et al. (eds.), *Sustainability in Natural Resources Management and Land Planning*, World Sustainability Series,
https://doi.org/10.1007/978-3-030-76624-5_3

The issue of "water governance" is not new. Several studies from the most diverse perspectives and fields of knowledge have sought to contribute to one of the most significant challenges of the new century—the sustainable management of water resources (Castro 2007). Water governance can be defined as "the social function that regulates [the] development and management of water resources and provisions of water services at different levels of society and guiding the resource towards a desirable state and away from an undesirable state" (Pahl-Wostl 2015, p.26). Latin America and the Caribbean follow diverse approaches (as discussed in the next section). Water governance is a complex contemporary global problem, with significant impacts on the poorest and most vulnerable countries, which face the greatest difficulty in adapting to climate change.

This chapter deals with water governance in South America, focusing on the experiences of Argentina, Brazil, and Uruguay. The objective is to analyse the consequences of the recent water crises in three watersheds (Chubut river in Argentina, Piracicaba-Capivari-Jundiaí rivers in Brazil, and Laguna del Sauce lake in Uruguay), and the potential lessons they offer. The chapter starts with an analytical overview of water governance in Latin America, after which the three case studies are presented. Finally, we discuss the findings in light of the need for adaptive and anticipatory water governance in the context of climate change and uncertainty.

2 Water Governance in Latin America

Contemporary State reform processes in most Latin American countries are characterized by privatization, decentralization, and delegation of public service provision to private sector and civil society actors, resulting in significant changes in the policy-making process (Zurbriggen 2014). According to the Organisation for Economic Co-operation and Development (OECD), the prevailing governance models (2012) are characterized by (1) multiple actors at the central level and few implementers at the subnational level (Chile, Costa Rica, and El Salvador); (2) multiple actors at both central and subnational levels (Brazil, Colombia,

T. Olivier
School of Public Administration, Florida Atlantic University, Boca Raton, USA

L. A. P. dos Anjos · I. M. Santana-Chaves
Federal University of ABC, UFABC, São Paulo, Brazil

M. Pascual
IPEEC-CONICET-CENPAT, Puerto Madryn, Argentina

N. Mazzeo
Eastern Regional University Center, CURE-UDELAR, Maldonado, Uruguay

E. Jobbágy
San Luis Institute of Applied Mathematics, IMASL-CONICET, National University of San Luis, San Luis, Argentina

Mexico, Peru, and Uruguay); and (3) few central government actors and multiple subnational authorities (Argentina, Bolivia, Mexico, and Panama). The main trend has been a gradual transition from a conventional centralized management model towards decentralization with subnational governments, mostly based on the Integrated Water Resources Management (IWRM) principles of decentralization, integration, and participation. Some countries have focused on environmental sustainability, accountability, transparency, creation of water use levies, and tariffs for cost recovery (De Stefano 2014).

Many of the institutional reforms included privatization of water services, mostly pushed by international organizations, considering the critical financial conditions of public operators (Budds and McGranahan 2013), but many private operators did not flourish. The reforms were introduced mainly through the modification of the legal system, often with the formulation of new water policies. In Latin America, different institutional models of water management co-exist, implying heterogeneity in the institutionalization of the human right to water and water systems governance. Social conflicts have emerged in the most vulnerable social groups due to the impact of privatization (Poupeau et al. 2018). In several Latin American countries (including Argentina, Brazil, and Uruguay), multistakeholder forums composed of government actors, water users, and civil society organizations have been formed. Trimble et al. (2021) compare the main institutional changes in water governance in these three countries during the past few decades, emphasizing polycentric arrangements and forums for participation as attributes providing resilience to watersheds as social-ecological systems.

Challenges related to water governance in Latin America include the weak capacity of institutions in terms of management, lack of transparency, and accountability. The OECD (2008, 2012) mentions six gaps: accountability, funding, capacity, information, administrative, and objectives (financial, economic, social, and environmental areas for collective enforcement of water policy). Lack of coordination among government agencies, fragmented competencies, concentration of power in the central government, and delay in responding to the universal drinking water and sanitation services have all undermined the credibility of water sector institutions and led to the deterioration and pollution of many water resources (UNESCO 2015).

According to the United Nations Educational, Scientific and Cultural Organization (UNESCO), almost 77% of the Latin American population lack safe sanitation and only 28% of the region's wastewater is discharged into sewers (2019). This has a strong impact on the most vulnerable sectors of the population and is linked to the human rights dimension of access to clean water. This is directly related to the enforcement of more advanced legal water systems, as the lack of adequate application of environmental laws to protect water quality is one of the main environmental challenges. In this regard, Zurbriggen (2014) addresses the most recent processes of re-statization of services in some municipalities, and the multiplication of managed independent co-ops, informal committees, or locally elected councils.

Groundwater also demands attention for its governance in the region, involving the issues of accumulating use rights and land property (Wester and Hoogesteger 2011). Latin America often faces overexploitation, quality degradation, and uneven use of groundwater (Yacoub et al. 2015). With at least 29 transboundary aquifers, the transboundary governance of surface and underground water resources is still a major challenge (Villar et al. 2018). De Stefano and Lopez-Gunn (2012) point to a lack of effective national policies to control overuse.

A relevant issue that also needs to be stressed is the increasing recurrence of water crises, which reveals that societies are badly and in many cases wrongly informed about the existing problems of water management, the lack of preventive measures, and the risks involved. This implies that public authorities provide the necessary response to water scarcity in the changing and uncertain scenarios of climate change (Torres et al. 2020).

The Latin American reality—as in most of the Global South—indicates the need to reduce the gaps in access to water and basic sanitation as a human right, and improve the quality of services. These initiatives to improve water security need to be considered. Stimulating the collaborative dynamics of collective action to promote sustainable water use and actions to reduce its degradation is a public policy challenge (UNESCO 2015). The most important changes in water governance in Latin America illustrate the classical transition from hierarchical and fragmented (command-control) schemes to integrated systems (Pahl-Wostl 2015). This chapter is based on the concept of adaptive governance (Folke et al. 2005), which includes the need to promote the capacity for dialogue and social learning, collaboration among multiple actors, multilevel coordination, polycentricity, management of uncertainty, and flexibility to deal with shocks and surprises.

3 Lessons from Water Crises in Argentina, Brazil, and Uruguay

This section focuses on the crises analysed in the three cases—Río Chubut (Argentina), Piracicaba-Capivari-Jundiaí (PCJ) (Brazil), and Laguna del Sauce (Uruguay) (Fig. 1).

Official data, legislation, and other secondary sources were used to characterize these watersheds as well as to analyse the water crises. Secondary data were complemented by semi-structured interviews with key stakeholders, conducted from October 2019 to May 2020 (the exact field period varied from case to case). A general description of the three cases is presented in Table 1.

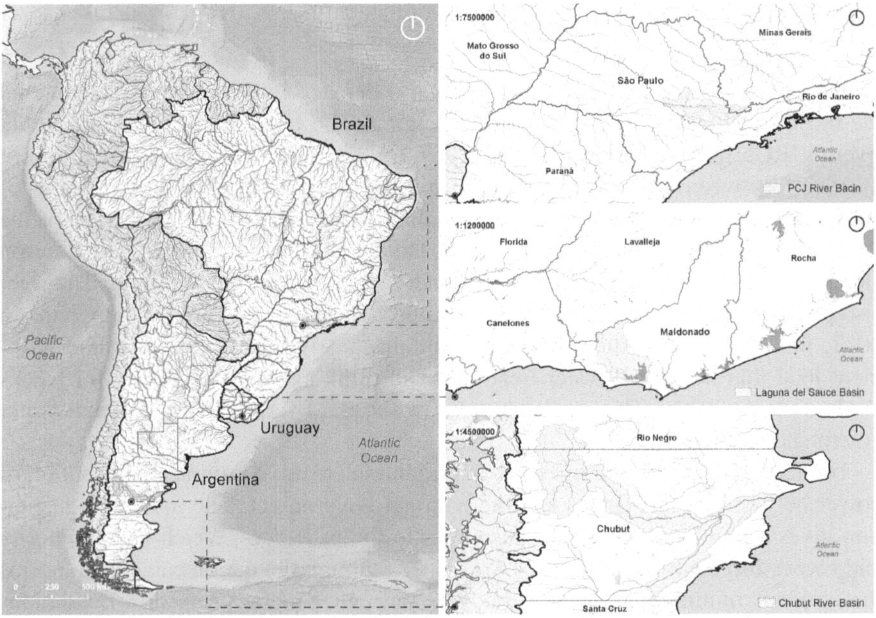

Fig. 1 Location of the study areas: Piracicaba-Capivari-Jundiaí Rivers (Brazil), Laguna del Sauce Lake (Uruguay) and Chubut River (Argentina). Elaboration: LaPlan (UFABC), 2020. Projection: Universal Transverse Mercator—Spindle 23/Datum SIRGAS 2000. *Source* Bases Latin America and Brazil (ANA 2019), Argentina (IGN 2019), Uruguay (MVOTMA 2011)

Table 1 Description of the three studied basins in South America

Characteristics	Chubut river basin (Argentina) (a)	PCJ rivers basin (Brazil) (b)	Laguna del Sauce Basin (Uruguay) (c)
Size (km^2)	5,967	15,000	722
Population	165,400	5.8 million	11,500
Users	280,000	Not available	300,000
Water uses (activities)	Agriculture, human consumption, industry, livestock, mining, tourism	Human consumption, industry, agriculture, tourism	Human consumption, forestry, agriculture, livestock, tourism
Studied crises	Provision of drinking water due to turbidity that affected the purification process after heavy rainfall (in 2017)	Water supply crisis due to reduced water availability (in 2014–2015)	Provision of drinking water due to a bloom of cyanobacteria that affected the purification process (in 2015)

Source Authors' elaboration based on Caporale et al. (2018), Consórcio Profill-Rhama PCJ (2020), Giordano et al. (2020), Mac Donnell et al. (2019), Pascual et al. (2020), Steffen and Inda (2010)

3.1 Chubut River (Argentina)

The Chubut river basin is the second largest watershed in Argentine Patagonia and the most variable in terms of flow in the region. The Lower Valley of the Chubut river (VIRCh from its Spanish name; 5,967 km^2) extends along the last 150 km of the river, between the Ameghino Dam and the Atlantic Ocean. A population of approximately 280,000 inhabitants is supplied with water from this section of the river, including the towns of 28 de Julio, Dolavon, Gaiman, Trelew, and Rawson along the valley, and the city of Puerto Madryn located outside the basin. The river also provides water for irrigation to the second largest irrigated area in Patagonia. In 2017, following extraordinarily heavy rainfall, the river carried a very high sediment load that hampered water treatment and drinking water provision for weeks (Kaless et al. 2019). This crisis revealed a series of flaws in the regional water governance system.

The federal legislative framework in Argentina creates a complex distribution of power between the provinces and the national government. The national constitution gives dominion over natural resources to the provinces, which have authority and ownership control over water (Art. 124), although the nation can dictate norms that contain minimum protection budgets through general protection regulations (General Environmental Law No. 25,675, Law No. 25,688 Environmental Water Management Regime 25,688, Law No. 26,348) but without altering local jurisdictions (Art. 41). At the local level, municipalities exercise an important degree of autonomy with respect to water management, including the provision of public services (OECD 2020).

In Chubut, dominion over public waters is asserted through Article 101 of the Provincial Constitution and regulated by a vast legal framework (Law XVII-53 Water Code of the Province of Chubut, Law XVII-88 of Provincial Water Policy, Law XVII-74 of Management Units in watersheds and Law XI-35 Environmental Code of the Province of Chubut). The water administration structure is centred around two institutions: the Provincial Water Institute (*Instituto Provincial del Agua*, IPA) and the Ministry of the Environment and Control of Sustainable Development (*Ministerio de Ambiente y Control del Desarrollo Sustentable*). In this context, local governments enforce regulations and provide drinking water and sewage services directly or through concessions.

A series of nongovernmental, private, and research organizations also play a role in this water management structure. Hydroelectric production and management of the dam in the valley is carried out by the company *Hidroeléctrica Ameghino*, managing the Ameghino Dam. The irrigation system that uses Chubut river water is managed by the Irrigation Company of the VIRCh (*Compañía de Riego del VIRCh*), and drinking water is provided by local cooperatives, organized under a federation, the *Federación Chubutense de Cooperativas* (FECHCOOP). In addition, three scientific-technical organizations present in the region provide technical advice regarding water problems.

All these organizations participate in the Chubut River Basin Committee, a forum created in 2013 to coordinate water management actions in the basin. The operation of this committee, however, has been irregular, with the forum meeting for the last time in 2018. The absence of formal spaces for articulation has been partly filled by alternative fora, such as semi-formal consultation among water service cooperatives and research institutions. The lack of a more general, permanent decision-making apparatus to resolve water problems with the participation of different actors was apparent during the sediment crisis of 2017.

In April 2017, the water treatment systems of three major cities in the lower valley collapsed due to an extraordinary storm that swelled the Chico river, a non-permanent tributary of the Ameghino Dam, which contributed an enormous amount of sediment causing a turbidity peak in the Chubut river. Cuts and restrictions in the supply of drinking water lasted three months, causing discontent and turmoil among the community, which demanded the reestablishment of the service (El Chubut 2018). The cooperatives faced enormous social pressure and struggled to immediately reinstate water supply. Local governments took provisional measures, such as conditioning old boreholes, distributing water in trucks, or using the reserves of public buildings.

Most of the actors in the governance scheme were mobilized during the crisis. At the provincial level, the IPA called meetings between local governments and cooperatives to assess the situation and define actions. The municipalities established supply protocols and, together with the cooperatives, sought technical solutions such as the construction of pre-sedimentation ponds, incorporation of flocculants, or use of irrigation canals as pre-sedimentation devices. The IPA created a dynamic vertical communication channel to ensure continuous contact among actors (mainly with local governments), but this did not lead to a permanent structure after the crisis. At the horizontal level, communication among local governments was limited, with the cooperatives taking the lead in sharing information with other cooperatives and local decision-makers in the region. For instance, FECHCOOP promoted unified communication with the media. Although these organizations participated in the river basin committee, the committee itself neither participated nor was convened during the crisis.

Despite the salience of the crisis, no permanent structures, procedures, or legislation emerged at the provincial or municipal level. Existing emergency and communication mechanisms have been somewhat strengthened, but they have not been formalized in any way. No social movements emerged, nor were new formal spaces for participation created, so no changes in power relations resulted from the crisis.

Interviews with key stakeholders show that the main lessons from the crisis concern a search for alternative water treatment technologies or water sources, the importance of working in tandem, and the need to improve vertical coordination for the construction of a common vision of the problems and the future of the basin. However, these have shown no formal expression.

3.2 *Piracicaba-Capivari-Jundiaí Rivers (Brazil)*

Water policy history in Brazil crosses the pathways of the PCJ river basin participatory processes. Popular mobilization in the region began in the 1960s, resulting in the constitution of the Intermunicipal Consortium of PCJ in 1989. The process directly led to the emergence of a new water management model from the discussions held on the adoption of a decentralized, inclusive, and sustainable management system (Hernández 2019). At the same time, water supply in the São Paulo Metropolitan Region (SPMR) was entrusted to the Basic Sanitation Company of the State of São Paulo (SABESP). The year 1991 witnessed the establishment of the Water Resources Policy of the São Paulo State (Law No. 7,663), as well as the Integrated Water Resources Management System. The policy defines a watershed as a "physical-territorial unit of planning and management" (Art. 3, II).

The PCJ Basin Committee was first implemented in 1993, and it was the first committee in the State of São Paulo. Given that the watershed comprises more than one state (São Paulo and Minas Gerais) with federal and state river jurisprudence, it is divided into three committees: São Paulo, Minas Gerais (created in 2008), and federal (created in 2002). It also holds a general plenary session with the members of the three committees, including representatives of the three government levels, civil society, and water users. The PCJ watershed area covers 76 municipalities and more than 15,000 km^2, with a population of approximately 5.8 million (Consórcio Profill-Rhama PCJ 2020).

With the population growth in the SPMR from the mid-1960s through the 1970s and the permanent industrial and agribusiness stimulus packages introduced during the period, the state government decided to strengthen water capture in the region. The State planned several new dams (Jaguari, Jacareí, Cachoeira, and Atibainha) across the Piracicaba river, thus laying the foundation for the Cantareira System, one of the largest in the world—covering more than 2000 km^2 across 12 municipalities. However, although this model of urban-centric governance, in which the SPMR was the articulator of the economic and water resource system, provided for water supply to the SPMR, it aggravated and triggered scarcity crises, especially in the surrounding municipalities, and contributed to the contamination of river beds (Hernández 2019).

The Water Resources Policy (WRP) (Law No. 9,433) formulated in 1997 established institutional frameworks (such as for granting and collection of resources) and created river basin committees at the national level. According to Empinotti et al. (2019), the changes brought about by the WRP provided for increased allocation of water, such as the use of the Cantareira System by SABESP, and the need to follow the recommendations under a watershed plan designed by each committee.

Several crises have occurred in these watersheds. The crisis from 2013 to 2015—considered the worst in recent years—was due to rainfall decrease and governance limitations, among other factors. It ended up exposing the management deficiencies of the PCJ and Cantareira System, both run by the São Paulo State

Government. Until then, the measures to mitigate water crises favoured industry and agribusiness stakeholders, while penalizing civil society with restrictions ranging from water rationing to consumption taxes. These measures reinforce the social dimensions of policies that allow privileges to specific regions or neighbourhoods with higher socioeconomic status (Empinotti et al. 2019). Examples of these locational inequalities were highlighted by the local media during the crisis. Several residents supplied by the PCJ rivers were seriously affected. Local media reported a total absence of water in some neighbourhoods (G1 2014), especially in the most vulnerable regions (Empinotti et al. 2019), leading to water rationing in one year, with water supplied for 3–15 h per day. However, official data from water companies have never indicated these facts. Access to transparent and real-time data has always been a challenge for civil society organizations, emphasizing the need for more transparency in water governance.

Hernández (2019) discusses the water crisis in São Paulo as just one example of the change in the scale of the problems of scarcity, poor management, and contamination of water resources, as well as their new dimensions. Empinotti et al. (2019) add that under the so-called good governance, the PCJ watershed crises—often described as a natural consequence of climate change—masked a hegemonic model of management with a centralized and technocratic (top-down) agenda which prioritizes economic sectors.

However, despite the negative consequences of centralization and the lack of autonomy in decision-making during the crisis, the PCJ committees led several initiatives. One of these was the "Operation Drought PCJ", which encouraged water users and municipalities to adopt mitigation measures, resulting in 12 municipal laws and decrees in 2014 and 2015. In 2015, the PCJ Water Supply Policy was approved, and the PCJ Basin Agency adhered to the Paris Agreement. The PCJ Water Supply Policy aims at the recovery, conservation, and protection of springs through programmes such as payment for environmental services, preservation of the Atlantic Forest and "Cerrado" biomes, management of areas with use restriction, and ecological restoration of areas of interest (Agência PCJ 2019).

With widespread social mobilization, a change in the water use classification for the Jundiaí river was approved in 2017 to allow for public water supply in a specific stretch of the river. In addition, the Cantareira System's concessional facility for SABESP was renewed with greater autonomy for the PCJ and Alto Tietê committees following discussions between the parties. This allows for the possibility of setting a maximum limit for the amount of water transferred to the SPMR. As early as 2018, the PCJ Basin Agency became the first in the world to sign the United Nations Global Compact and adhere to the Sustainable Development Goals (Agência PCJ 2019).

Although the legal and institutional process of water governance in Brazil has had considerable civil society participation, and provides an example of a decentralized system in South America, the recent crises (2013–2015) have demonstrated the river basin committees' failure to prevent centralized decision-making, which benefits only a small group of stakeholders. The 2020 revision of the PCJ Watershed Plan projects an important role for its climate component with participatory workshops and use of climate scenarios. Whether the recent crises have caused the inclusion of the climate components in the PCJ Watershed Plan review is difficult to conclude, but they may have had an effect, given the increasing attention devoted to climate exigencies in planning and policy-making.

3.3 Laguna del Sauce Lake (Uruguay)

Over the past 15 years, water governance in Uruguay has been gradually moving from a centralized, hierarchical model to a decentralized, participatory, and integrated model (Mazzeo et al. 2019; Trimble et al. 2021). The constitutional reform of Article 47 in 2004 and the National Water Policy in 2009 (Law No. 18.610) were the main drivers of this transition. Among the principles promoted in the legislation are user and social participation in management and the adoption of the watershed as the "unit for action". Since then, multistakeholder and multilevel advisory bodies have been formed, with the participation of government actors, users, and civil society. The basin commissions and regional water resource councils (13 and 3 in the country, respectively) are overseen by the Ministry of Housing, Land Planning and Environment (MVOTMA). The MVOTMA (MA—Ministry of Environment since 2020) is the highest water authority in Uruguay, in terms of water use, management, and control, through the National Water Directorate (DINAGUA) and the National Environment Directorate (DINAMA) (Trimble et al. 2021).

The first basin commission was established in 2010 for the Laguna del Sauce lake in the Department of Maldonado (one of Uruguay's 19 administrative units), where an interinstitutional coordination mechanism for the watershed was already under way (Giordano et al. 2020). This basin commission (CCLS—Comisión de Cuenca de Laguna del Sauce) is the main setting for participation and coordination among multiple stakeholders, including government actors at the three levels (national, departmental, and municipal), the State-owned water and sewerage company (OSE—Obras Sanitarias del Estado), social organizations (of neighbours, farmers, etc.), and academia. The water governance in this watershed is based on a combination of hierarchical and network models (as defined by Pahl-Wostl 2017) since the decision-making and coordination processes either follow a top-down approach (e.g. from MVOTMA to the basin) or are implemented through the basin commission, with knowledge exchange and collective agreements among the participating actors.

In March 2015, a water crisis occurred in Laguna del Sauce. It was characterized by an important bloom of cyanobacteria with potential risks to human health and by the presence of metabolites which produced a taste and odour in the drinking water. This water was supplied from the OSE to more than 140,000 people during a period of around one month. Steffen and Inda (2010) describe the main characteristics of Laguna del Sauce and its eutrophication problem.

The crisis resulted in, among others, complaints by the population about the tap water quality. OSE and its Deconcentrated Management Unit of Maldonado

(OSE-UGD) adopted technical and operational measures in the water purification plant and in the water supply network to improve the purification process. This was also accompanied by intense monitoring. Operational costs thus increased.

At the government level, the crisis led to numerous decision-making situations involving coordination among OSE-UGD, OSE, MVOTMA (e.g. DINAMA and DINAGUA), the Maldonado government, and the Eastern Regional University Centre (CURE-UDELAR), in addition to the Regulatory Unit for Energy and Water Services (URSEA). Governmental organizations released official communications and held press conferences during the crisis. In addition, basin commission members and other actors who wanted to understand the situation better engaged in informal coordination. Furthermore, OSE-UGD coordinated with voluntary groups of the society to supply water to some urban areas of Maldonado through water tankers.

Soon after the crisis, in June 2015, MVOTMA approved the "Action Plan for Laguna del Sauce", containing 12 measures to control, stop, and reverse the water quality deterioration in Laguna del Sauce (associated with land uses such as agriculture, livestock, and urbanization—Table 1) and ensure its sustainable use as a source of drinking water in terms of both quality and quantity. Most of these measures were part of a plan proposed by the basin commission to MVOTMA in 2011. It appears that the plan was not considered relevant at the time, but the crisis led to its approval. The Laguna del Sauce plan became a key legal instrument for national water management.

Over the months following the crisis, the basin commission also discussed the "Precautionary Measures for Laguna del Sauce watershed", a departmental regulation for land-use management. Furthermore, the commission performed a self-critical analysis of how it had acted during the crisis. As a result, it recognized the need for a communications plan on a priority basis. Accordingly, a plan was prepared with the participation of multiple actors and accepted by the basin commission. However, the plan was not approved by the MVOTMA, and the need for a communications plan remains a concern as of today.

The water crisis and the associated water governance issues were included in the political agenda for the municipal elections in Maldonado of the same year (which is unusual). An environmental conflict unfolded around the controversy over water potability and water safety, also involving the management of information by authorities. Different actors were called to explain aspects of the crisis at the national and subnational (departmental) legislatures. An important consequence of the crisis was the mistrust created in the population regarding the quality of water supplied by OSE, leading to new water consumption patterns (household water filters and bottled water are now commonplace, for example).

The crisis led to numerous initiatives, but it also holds lessons for different organizational levels. Technical personnel from OSE and OSE-UGD believe that they are now better prepared to face similar crises with improved equipment in the plant and better monitoring processes in the lake. The crisis allowed basin commission members to recognize the value of this space and the importance of the measures adopted since its occurrence (e.g. formation of numerous relationships and contributions to planning and monitoring). The shared knowledge accumulated in the basin commission was considered valuable for navigating the crisis. The lessons learned in Laguna del Sauce transcended the borders of this watershed and reached numerous parts of the country and multiple organizational levels. The ecosystemic conditions of the lake may lead to further algal blooms, so similar crises could recur unless the lessons learned from previous events are put into practice and anticipatory capacities strengthened.

4 Challenges Under Uncertainty

In the current global climate emergency context, extreme events tend to be more frequent and massive, probably resulting in unusually long periods of drought as well as more intense storms, both affecting water supply systems and the population (IPCC 2018), as shown in this chapter. Faced with this challenge, policy-makers need to break out of existing water resource planning and management paradigms (Quay 2010).

In this sense, adaptive and anticipatory water governance are two important approaches for dealing with unpredictability and uncertainty. Anticipatory governance "involves changing short-time decision making to a longer-term policy vision", thus allowing "managing events instead of waiting until a climate-related or regulatory socio-economic event results in crisis" (Boyd et al. 2015, p.153). Anticipatory governance provides the necessary elements to anticipate climate change using the best knowledge available and not just reacting to undesirable effects of climate change (Quay 2010). Anticipatory governance offers essential insights into changes in the form of planning to deal with climate change and tend to be more successful in a flexible institutional architecture necessary to deal with complex problems.

Adaptive governance (Folke et al. 2005) has emerged as an environmental management approach to be developed and pursued worldwide, including in South American countries, where the challenges are enormous and institutional arrangements often act as strong barriers. Adaptive governance must seek an inclusive and decentralized model without a top-down approach and with a large number of stakeholders directly and indirectly involved (Young 2002). The interdependencies and cross scales of ecosystem services must be considered in complex social-ecological systems (Cumming et al. 2006). Another key aspect towards adaptive governance is that this flexible model would fit multilevel and polycentric arrangements, essential in situations such as the cases studied, which involve different levels of government and various actors. The basin committees could play an important role in this regard.

In this chapter, we addressed the experiences of the recent water crises in three South American countries' river basins. The objective was not a comparative study of the crises, but an analysis of the respective legal frameworks, institutional advances, participation, and lessons learned. An important consequence of the crisis in the Uruguayan case (algal blooms) was the mistrust that emerged in the population regarding drinking water quality, leading to new patterns of water consumption. In addition, the value of the multistakeholder basin commission's contributions became evident after the crisis. In the Argentinian case (sediment crisis), vertical communication between the provincial and local governments was enhanced, and collaborations between the cooperatives and research actors emerged for sediment monitoring. In the Brazilian case (drought), climate components have been included in the basin management plan for the PCJ rivers, probably as a consequence of the crisis and other factors.

The monitoring and evaluation of the measures, strategies, plans, and policies adopted constitute one of the key aspects that are absent or weakly present in all the cases analysed, preventing a wider understanding of the causality of successes and failures. This hampers the emergence of social learning and an adequate management of uncertainty, which are essential components of adaptive governance.

Building water systems capable of adapting to climate-driven events is still a challenge (Lemos et al. 2020). Governance based on anticipatory models, adaptive capacities, climatic scenarios, and spaces for social participation with active citizenship—which not only establishes the citizen as a bearer of rights and duties but essentially creates rights to open new spaces for political participation (de Benevides 1994)—remains a goal to be pursued. Within the context of growing uncertainty, the increase in spaces or forums of participation at multiple scales is becoming an essential means to strengthen resilience and adaptation in water governance (Trimble et al. 2021).

5 Conclusions

Water crises are complex problems; they occur due to a combination of factors (including ecological, climate, economic, and governance variables), and involve multiple actors and, thus, multiple interests and perspectives. Water quality and quantity issues are usually intertwined, so they should be addressed in combination. Moreover, water crises are not just about "water"; they are closely related to land-use practices and management, and involve multiple sectors of society (generally affected unequally by the water problem). Overcoming the fragmentation in management functions and actions (e.g. between institutions in charge of environmental protection vs. those in charge of agricultural productivity) is still a challenge in Argentina, Brazil, and Uruguay. The analysis of water crises in the studied cases has shown that they tend to trigger actions and changes. However, reactive responses prevail, suggesting that anticipatory capacities should be strengthened to be better prepared to deal with uncertainty.

Acknowledgements This work was carried out with the aid of a grant from the Inter-American Institute for Global Change Research (IAI): SGP-HW 056 (GovernAgua Project). The support from the São Paulo Research Foundation (FAPESP; 2018/06685-9, 2019/05644-0, 2015/03804-9) is also appreciated. The authors thank the stakeholders interviewed in each watershed for sharing their knowledge and perspectives.

References

Agência PCJ—Agência das Bacias Piracicaba, Capivari e Jundiaí (2019) Relatório institucional 2018 da Agência das Bacias PCJ. Available at: http://www.agenciapcj.org.br/docs/relatorios/relat-inst-agencia-2018.pdf. Accessed 20 July 2020

ANA—Agência Nacional de Águas e Saneamento Básico, Brasil (2019) Metadados. ANA. Available at: https://metadados.ana.gov.br/geonetwork/srv/pt/main.home. Accessed 15 Aug 2020

Boyd E, Nykvist B, Borgström S, Stacewicz IA (2015) Anticipatory governance for social-ecological resilience. Ambio 44(Suppl. 1):149–161

Budds J, McGranahan G (2013) Are the debates on water privatization missing the point? Experiences from Africa, Asia and Latin America. Environ Urbanization 15(2):87–114

Castro JE (2007) Water governance in the twentieth-first century. Ambiente & Sociedade 10 (2):97–118

Consórcio Profil-Rhama PCJ (2020) Relatório Síntese: Plano de Recursos Hídricos das Bacias Hidrográficas dos rios Piracicaba, Capivari e Jundiaí, 2020 a 2035. Consórcio Profil-Rhama PCJ, Porto Alegre

Caporale M, Vallvé E, Bazet MA, Hareau I (2018) Plan Local de Ordenamiento Territorial y Desarrollo Sostenible de la Cuenca de Laguna del Sauce. Available at http://cultura.maldonado.uy/documentos/pdf/2020/otds/6_Caporale%20M_%20Patrimonio%20Cultural_02_2019.pdf. Accessed 24 Nov 2020

Cumming GS, Cumming DHM, Redman CL (2006) Scale mismatches in social-ecological systems: Causes, consequences, and solutions. Ecol Soc 11(1):14

de Benevides MVM (1994) Cidadania e democracia. Lua Nova: Revista de Cultura e Política 33:5–16

De Stefano L, Lopez-Gunn E (2012) Unauthorized groundwater use: Institutional, social and ethical considerations. Water Policy 14(1):147–160

De Stefano L (2014) Reforming water and governance structures. In: Willaarts B, Garrido A, Llamas MR (eds) Water for food security and well being in Latin American America and the Caribbean—social and environmental implications for a globalized economy. Earthscan, London

Driessen PPJ, Dieperink C, van Laerhoven F, Runhaar HAC, Vermeulen WJV (2012) Towards a conceptual framework for the study of shifts in modes of environmental governance—experiences from the Netherlands. Environ Policy Governance 22:143–160

Empinotti VL, Budds J, Aversa M (2019) Governance and water security: The role of the water institutional framework in the 2013–15 water crisis in São Paulo, Brazil. Geoforum 98:46–54

El Chubut (2018) Se agudiza la crisis y recién a medianoche volvería el agua. Available at https://www.elchubut.com.ar/nota/2017-4-18-se-agudiza-la-crisis-y-recien-a-medianoche-volveria-el-agua/amp. Accessed 20 July 2020

Folke C, Hahn T, Olsson P, Norberg J (2005) Adaptive governance of social-ecological systems. Annu Rev Environ Resour 30:441–473

G1 (2014) Ano de Estiagem teve cidades sem águas e agonia do rio piracicaba. Available at: http://g1.globo.com/sp/piracicaba-regiao/noticia/2014/12/ano-de-estiagem-teve-cidades-sem-agua-e-agonia-do-rio-piracicaba.html. Accessed 11 Sept 2020

Giordano G, Dias Tadeu N, Trimble M (2020) Análisis de la gobernanza y aprendizajes de las crisis en las Cuencas de Laguna del Sauce (Maldonado) y Laguna del Cisne (Canelones), Uruguay. Informe Técnico en el marco del Proyecto GovernAgua (SGP-HW 056), Instituto SARAS

Hernandez M (2019) Movimiento de redención ecológica de la cuenca del Río Piracicaba: una experiencia de acción colectiva. Cadernos Metrópole 21(46):903–928

IGN—Instituto Geográfico Nacional, Argentina (2019) *GEOSERVICIOS*. IGN. Available at: https://www.ign.gob.ar/NuestrasActividades/InformacionGeoespacial/ServiciosOGC. Accessed 15 Aug 2020

IPCC—Intergovernmental Panel on Climate Change (2018). Global Warming of 1.5°C. World Meteorological Organization, Geneva

Jacobi PR, de Almeida Sinisgalli P, Medeiros Y, Ribeiro Romeiro A (2009) Governança da água no Brasil: dinâmica da política nacional e desafios para o futuro. In: Jacobi PR, Sinisgalli PA (eds) Dimensões político institucionais da governança da água na América Latina e Europa. São Paulo, Annablume

Kaless G, Pascual M, Flaherty S, Liberoff AL, Garcia Asorey MI, Brandizi L, Pessacg N (2019) Echoes of the Comodoro Rivadavia Storm in the Lower Valley of the Chubut River. Sediment contribution to the Chubut River from the Chico River Basin. In: Comodoro Rivadavia and the Catastrophe of 2017: Multiple Visions for a City at Risk. Editorial University of Patagonia

Law No. 25,675. General Environmental Law, 2002. Available at: http://www.infoleg.gob.ar. Accessed 20 July 2020

Law No. 25,688. Environmental Water Management Regime 25,688, 2002. Available at: http://www.infoleg.gob.ar. Accessed 20 July 2020

Law No. 26,348. Available at: http://www.infoleg.gob.ar. Accessed 20 July 2020

Law XVII-74 of Management units in watersheds and Law XI-35

Environmental Code of the Province of Chubut. Available at: www.legischubut.gov.ar. Accessed 20 July 2020

Law No. 7,663, 1991.Water Resources Policy and the Integrated Water Resources Management System. Available at: https://www.al.sp.gov.br/repositorio/legislacao/lei/1991/lei-7663-30.12.1991.html. Accessed 20 July 2020

Law No. 9,433 of 8 January 1997. Institutes the National Water Resources Policy. 21 of the Federal Constitution, as amended by Art. 1 of Law No. 8,001, of March 13, 1990. Available at: http://www.planalto.gov.br/ccivil_03/LEIS/L9433.htm. Accessed 20 July 2020

Law No. 7,990 of December 28, 1989. Available at: http://www.planalto.gov.br/ccivil_03/LEIS/L9433.htm. Accessed 20 July 2020

Lemos MC, Puga BP, Formiga-Johnsson RM, Seigerman CK (2020) Building on adaptive capacity to extreme events in Brazil: water reform, participation, and climate information across four river basins. Reg Environ Change 20:53–71

Mac Donnell L, Olivier T, Salvadores F, Pascual M (2019) Análisis de la gobernanza y aprendizajes de la crisis en la cuenca del Valle Inferior del Río Chubut – VIRCH. *Informe Técnico en el marco del Proyecto GovernAgua* (SGP-HW 056), Instituto Patagónico para el Estudio de los Ecosistemas Continentales (IPEEC-CONICET-CENPAT)

Mazzeo N, Zurbriggen C, Steffen M, Barquín J, Gadino I, Díaz I, Ciganda AL, Goyenola G, Trimble M (2019) Combinando estrategias. Descentralización y centralización en la gestión del agua en Uruguay. In: Cardarello A, Ferla P (eds) Descentralización en Uruguay. Propuestas para avanzar en la agenda. Montevideo: Konrad-Adenauer-Stiftung

MVOTMA - Ministerio de Vivienda Ordenamiento Territorial y Medio Ambiente, Uruguay (2011) *DINAMA Data Portal*. MVOTMA. Available at: https://www.dinama.gub.uy/geonetwork/srv/eng/catalog.search#/home. Accessed 15 Aug 2020

OECD—Organisation for Economic Co-operation and Development (2008) En la búsqueda de esquemas apropiados de participación del sector privado en el suministro de agua potable y saneamiento. Experiencias recientes de América Latina. México: Instituto Mexicano de Tecnología del Agua (IMTA)

OECD—Organisation for Economic Co-operation and Development (2012) Water governance in Latin America and the Caribbean—A multi-level approach. OECD Publishing, Paris
OECD—Organisation for Economic Co-operation and Development (2020) Water governance in Argentina. OECD Publishing, Paris
Pahl-Wostl C (2015) Water governance in the face of global change—From understanding to transformation. Springer International Publishing
Pahl-Wostl C (2017) An evolutionary perspective on water governance: from understanding to transformation. Water Resour Manage 31(10):2917–2932
Pascual MA, Olivier T, Brandizi L, Rimoldi P, Malnero HA, Kaless G (2020) Cuenca del Río Chubut. Análisis de Factibilidad para Fondo de Agua. Mayo 2020. Alianza Latinoamericana de Fondos de Agua. 197p
Poupeau F, Razafimahefa L, Robert J, Mercier D, Massardier G, Jacobi PR (2018) Water conflicts hydrocracy in the Americas: coalitions, networks, policies. IEE-USP, São Paulo
Quay R (2010) Anticipatory governance—a tool for climate change adaptation. J Am Plan Assoc 67(4):496–511
Steffen M, Inda H (2010) Bases técnicas pàra el manejo integrado de Laguna del Sauce y cuenca asociada. Universidad de la República—South American Institute for Resilience and Sustainability Studies
Torres PHC, Cortes PL, Jacobi PR (2020) Governing complexity and environmental justice: lessons from the water crisis in metropolitan São Paulo (2013-2015). Revista Desenvolvimento e Meio Ambiente 53:61–77
Trimble M, Jacobi PR, Olivier T, Pascual M, Zurbriggen C, Garrido L, Mazzeo N (2021) Reconfiguring water governance for resilient social-ecological systems in South America. In: Baird J, Plummer R (eds) Water resilience: management and governance in times of change. Springer Nature, Switzerland AG
UNESCO—United Nations Educational, Scientific and Cultural Organization (2015) Desafíos del agua urbana en América Latina. UNESCO, Paris
UNESCO—United Nations Educational, Scientific and Cultural Organization (2019) World water development report—leaving no one behind. UNESCO, Paris
Villar PC, Ribeiro WC, Sant'Anna FM (2018) Transboundary governance in the La Plata river basin: status and prospects. Water Int 43(7):978–995
Water Code of the Province of Chubut, Law XVII-88 of Provincial Water Policy. Available at: www.legischubut.gov.ar. Accessed 20 July 2020
Wester P, Hoogesteger J (2011) Uso intensivo y despojo del agua subterránea: hacia una conceptualización de los conflictos y la concentración del acceso al agua subterránea. In: Boelens R (ed) Justicia hídrica: acumulación, conflicto y acción social. IEP Fondo Editorial PUCP & Alianza Justicia Hídrica, Lima
Yacoub C, Duarte B, Boelens R (eds) (2015) Agua y Ecología Política: El extractivismo en la agroexportación, la minería y las hidroeléctricas en Latinoamérica. Abya-Yala, Quito
Young OR (2002) The institutional dimensions of environmental change: fit, interplay, and scale. MIT Press, Cambridge
Zurbriggen C (2014) Governance a Latin America perspective. Policy Soc 33(4):345–360

Polycentric Water Governance in the Urban Global South

Klaus Frey, Ruth Ferreira Ramos, Lidiane Alonso Paixão dos Anjos, Beatriz Milz, and Pedro Roberto Jacobi

1 Introduction

Our paper is focused on environmental governance in the urban context, particularly water resources, in a conjuncture of increasing demand for water use, and at the same time climate-change-related water scarcity and deterioration of water quality, a common condition of most urban metropolises in the Global South. Based on an empirical case study on urban water governance in the São Paulo Macrometropolis (SPMM), this chapter analyses emerging polycentric water governance practices from the conceptual perspective of Elinor and Vincent Ostrom, *inter alia*, emphasizing the need for interdisciplinarity, intersectorality, politicization and the integration of water resources and land use planning as fundamental elements of sustainable governance.

The SPMM comprises a widely urbanized region with intensively used agricultural landscapes, as well as environmentally protected areas and other ecologically valuable places, providing the several metropolitan regions of the SPMM with ecosystem services of fundamental importance for the people and their socio-economic activities, being water the major life-sustaining natural resource in the cities. Such urban agglomerations, economically integrated and ecologically interdependent, call for innovative approaches of integrative land use planning and governing across multiple scales (Sayre 2009) associated with high sensitivity regarding the political interests, conflicts and asymmetries involved (Eraydin and

K. Frey (✉)
Federal University of ABC, Alameda da Universidade, São Bernardo do Campo/SP 09606-070, Brazil
e-mail: klaus.frey@ufabc.edu.br

R. Ferreira Ramos · L. Alonso Paixão dos Anjos
Federal University of ABC, UFABC, São Bernardo do Campo, Brazil

B. Milz · P. R. Jacobi
Institute of Energy and Environment, University of São Paulo, IEE/USP, São Paulo, Brazil

W. Leal Filho et al. (eds.), *Sustainability in Natural Resources Management and Land Planning*, World Sustainability Series,
https://doi.org/10.1007/978-3-030-76624-5_4

Frey 2019). This chapter aims to contribute with the identification of such obstacles to sustainable and resilient water governance, land use planning and public policy-making adjusted to the interdependencies and interconnectedness that characterize such territories.

Based on the conceptual underpinnings of rational choice theory, and sustained by their extensive empirical research, mainly related to "small-scale, self-regulatory systems" (Toonen 2010:193), the researchers of the Indiana School came to the common understanding that cooperation on the community level used to turn out a valuable alternative development strategy to hierarchic state intervention and pro-competitive market mechanisms. The question, therefore, arises to what extent these findings apply to the context of metropolises in the Global South, subject of our empirical study? In view of the evidently major complexity and diversity in biophysical, institutional, and cultural terms and the patent fact of longstanding exclusionary processes and extreme social and economic inequalities, what kinds of implications we have to expect concerning the applicability of Ostrom's perspective on "governing the commons" to the management of natural resources in the metropolises of the Global South?

Vincent Ostrom, with his studies on metropolitan regions, on American federalism, and on Natural Resources Management, adds an important perspective to this questions with the concept of polycentricity "as a normative ideal and practical form of governance" (McGinnis and Ostrom 2012:16), which has increasingly gained importance also in Elinor Ostrom's work in the fields of common-pool resource (CPR) management (Ostrom 1990), co-production of public services and infrastructure (Ostrom 1996), as well as in metropolitan governance (Toonen 2010), being "both scholars [...] complementary and mutually reinforcing" (ibid.:194).

In this chapter, we put emphasis on water as the most fundamental natural resource for sustaining life in densely populated urban agglomerations. Therefore, in a context of increasing pressure for water privatization (Bakker 2010), running against the community-based natural resources management approach as idealized by the Indiana School, we wonder whether polycentric governance (PCG) based on cooperation and commonly decided rules could be a promising approach to adequately cope with water resources in an apparently adverse socio-political context. In the following section, we present the concept of PCG as the theoretical underpinning of our empirical research on water governance in the SPMM.

2 Theoretical Concept of Polycentric Governance: Potentialities and Limits

A first major contribution to the concept of PCG from the part of political science and public administration has been given by Vincent Ostrom and his collaborators in their article on "The Organization of Government in Metropolitan Areas", in which they stated that "polycentric' connotes many centers of decision-making which are formally independent of each other" (Ostrom et al. 1961:831). Applied to metropolitan areas in the US, they identified, on the one hand, a lack of a clear central leadership function and an informal interplay of "many local public authorities, each pursuing its own aims in a seemingly uncoordinated manner" (Stephan et al. 2019:21).

On the other hand, the authors argued that the cooperation between organizations across different scales could in fact be the appropriate way to provide public services in metropolitan areas, forming a properly functional and effective "polycentric political system". At that time, this was a quite innovative and challenging diagnosis, in view of the predominant conceptions in political science as well as urban studies, which used to construe the complex political topography of metropolises as a synonym for pathological "organized chaos" (Ostrom et al. 1961:831; Ostrom 2005:13).

PCG, therefore, goes beyond the idea of a community-based governance practice as far as more complex governance arrangements are envisioned, involving numerous "autonomous authorities with overlapping jurisdictions and, thus, involving multiple, diverse and interdependent actors, at different scales and levels, interacting continuously and performing sometimes more competitive, sometimes more cooperative relations (Ostrom 2005, 2010)" (Frey et al. 2019:4). The relationships among the numerous governance actors involved in collaborative practices at different scales as well as the problems at stake determine how interactions occur and which results are finally obtained (Andersson and Ostrom 2008).

Thus, what is central for characterizing PCG is the absence of a central authority able to impose its will unilaterally upon other collaborating authorities and political actors, as well as cooperation across governmental levels and geographic scales. This means that polycentricity implies a continuous interplay across scales and sectors, between, on the one hand, authorities and actors entrusted with specific tasks, responsibilities, and assignments, and, on the other, general-purpose authorities with overarching coordinative functions. Correspondingly, water management as a specific public duty performed by specific administrative units, public agencies and, increasingly, private companies should necessarily be integrated in and accountable to the overarching land use planning.

According to Elinor Ostrom (2005, 2010; McGinnis and Ostrom 2012), effective PCG requires a regular exchange of ideas and information between these different actors from the public, private and voluntary sectors, an ongoing adjustment of positions and assessments, so that the various actors and institutions are in a real condition to effectively exercise their autonomy, which can always only be relative due to the necessity of these ongoing adjustments. And in the case that one level fails, the other one is expected to come into play, getting the things done. Therefore, the expectation that polycentricity enhances institutional resilience, adaptability, and robustness (Thiel et al. 2019). Thus, Heikkila et al. (2018:207) point out that "polycentricity offers a flexible enough conceptual framework to accommodate current environmental governance solutions, as well as inspire new ones".

For the Indiana School, a monocentric model of governance is thus not compatible with the complex challenges we are confronted with in dealing with the commons. The PCG concept confronts the usual argument, highly appreciated by traditional urban planners, that action undertaken by centralized governments tends to be more effective. PCG involves multiple actors, opinions, scales and levels, and complexity is considered a fundamental, positive and necessary governance characteristic (Ostrom 2005, 2010). Related to metropolitan areas, Ostrom et al. (1961) already discussed the criteria of control, efficiency, political representation and self-determination as important for defining the appropriate boundaries within which public goods can be provided most effectively, thus calling for polycentric political arrangements.

Moreover, Vincent Ostrom, in his discussion on American federalism, already stated the necessity to look beyond governmental levels, considering the "various amounts of overlap" (1973:204) and the "rich structure of overlapping jurisdictions with substantial autonomy among jurisdictions". Substantial institutional mechanisms to ensure democratic control he considers fundamental for "a 'highly federalized' political system" (ibid.:205) able to adapt to existing biophysical conditions. Herewith Ostrom can be considered a forerunner of the currently prominent network governance approach (McGinnis and Ostrom 2012).

In fact, what current research, mostly conducted in the Global North, indicates for most governance contexts, is "a continuum of horizontal dispersion of authority from monocentric to polycentric solutions, with hybrid solutions lying somewhere in the middle" (Paavola 2016:145). Despite the quite common conflictive clashes between growth coalitions and environmentalists in the Global North, and the life-threatening resistance against "the process of development through destruction" (Gupta 2014:159) by the "environmentalism of the poor" (Martinez-Alier 2003) predominant in the Global South, the idea that consensus-oriented democratic participation and cooperation will bring about sustainable development has dominated global conferences and reports on environmental issues in the last decades, despite "uneven or fragile" accomplishments since the publication of the Brundtland Report on "Our Common Future" (Meadowcroft et al. 2019). On the contrary, whereas even the most developed countries turned out unable to address successfully basic issues of the sustainability agenda, new issues like "biodiversity loss, plastics, and other waste, pandemics, religious fundamentalism, and cyber security" emerged menacing a sustainable future for humanity (ibid.).

What seems to have been ignored is that environmental decision-making used to be highly conflictive, involving apparently unbridgeable gaps due to contradictory and incompatible values and interests, which definitely "cannot be satisfied simultaneously—a choice has to be made regarding which interests to affirm and which to block, and to what degree is their balancing possible" (Paavola 2016:144).

What therefore seems rather questionable, at least from a Global South perspective, is Ostrom's insistence upon the mere rule-enforcing nature of these interactive governance practices. What if the existing rules are simply the result of asymmetric power relations and exclusionary practices, incidentally, a very common feature, not only in countries of the south marked by extreme social and economic inequalities?

Water provision has definitely become a wicked policy problem, a "complex, and contested, social problem(s)" (Head 2019:180), particularly evident in the context of huge agglomerations. On the one hand, environmental and water governance have become more and more polycentric with multiple levels and scales involved, basically due to the increasing recognition of governments regarding their incapacity to handle such complex problems unilaterally. Then, there are mandatory interdisciplinary and intersectoral approaches in order to successfully face the multiple economic, social, territorial, and environmental interdependencies. Moreover, the plurality of prevailing values and interests, as well as the demand for long-term approaches to sustainability, have to be taken into consideration (Rydin 2008). On the other hand, it has also to be taken into account that all these emergent institutional structures change only gradually, due to institutional path dependence, and may therefore imply in situations of non-decisions, of strategic mutually blocking behavior, obstructing sustainable solutions (Benz 2007).

Moreover, frequently political decisions on environmental issues are highly conflictive, above all when "the abuse of natural environments and the loss of livelihoods" (Martinez-Alier 2003:ix) are at stake, and finally, problems of democratic legitimacy tend to emerge insofar as decisions are increasingly taken in governance arenas not democratically legitimized, making institutional design particularly challenging and fundamental to reach social acceptance and, thus, legitimacy of "sustainable and innovative policies in a multilevel context" (Schmitter 2002:51).

3 São Paulo Macrometropolis Water Governance in the Context of Climate Change and Water Crisis

Despite having at its disposal 12% of the planet's freshwater resources, Brazil faces immense challenges concerning resilient and sustainable water management, in order to reach universal access to clean water and adequate sanitation services for its population of over 209 million people. In several densely populated regions of the country infrastructure for water supply is still precarious. In the SPMM, the largest urban agglomeration in Brazil with 174 municipalities and a population of approximately 30 million people, the existing infrastructure for water supply is not sufficient to meet the increased demand for industrial and domestic use, and for irrigation of agricultural activities, according to the SPMM Water Resources Master Plan (Jacobi et al. 2020).

Territories like the SPMM concentrate large contingents of people, which depend on the provision of ecosystem services for their well-being. In the SPMM, as in many metropolises worldwide, pressure on natural assets has continuously increased in so far as urbanization and industrial development proliferated (Torres et al. 2019), often catalyzed by poor governance and increased climatic variability.

The current Brazilian model of water management was mainly inspired by the French model, where the participation of civil society in water management was structured in the 1960s. The Integrated Water Resources Management System was established in 1997 by the Federal Water Policy. The recognition of the hydrographic basin as the most appropriate scale for water management resulted in the delimitation of Water Resources Management Units, whose advisory and deliberative management bodies are called Hydrographic Basin Committees, responsible for the regional coordination and planning of water-related policies and measures (Jacobi et al. 2009). Hence, the water basin committees became the central institutional innovation to put into practice the three guiding principles—decentralization, integration, and participation.

Previously, the water resources policies favored the use of water for electric power generation and industrialization (Murtha et al. 2015). Use conflicts were solved centrally by governments without substantial policies and goals regarding environmental and sanitation access (Gomes and Barbieri 2004). For this reason, the model of integrated and decentralized water resources management has been considered by the literature as a privileged space for participation, articulation, and conflict resolution (Martins 2015).

In the period from 2014 to 2015, the southeastern region of the country was affected by a severe drought and the largest water supply system, the Cantareira, was badly affected by an intense decrease in water volume. The existing supply infrastructure, extremely dependent on rainwater and responsible at that time for the supply of almost half the population for the São Paulo Metropolis, has reached its limits (Côrtes et al. 2015). It is important to highlight that this water supply system is not controlled by a hydrographic basin committee. Instead, the Basic Sanitation Company of the State of Sao Paulo (SABESP), a mixed capital company created in 1973, is responsible for water and sanitation services in the majority of the municipalities in the SPMM. Although there is a formal structure of participatory and decentralized water management in basin committees in SPMM, the main conflicts in the decision-making process are not necessarily resolved at the meetings of the basin committees, even though SABESP has a seat guaranteed in the basins where it is operating water and sanitation services.

The severe water crisis not only revealed the fragility of the water supply system but also the shortcomings of urban and regional water governance in the Metropolitan Region, as well as a lack of compliance regarding the general land use planning, itself precarious at the macro-metropolitan scale. The water policy during the period of the water crisis was marked by great conflicts, with the São Paulo State Government becoming the final decision-making authority. With the mechanisms of social control suspended, it was the poor periphery of the cities which most suffered from the interruptions of water supply during the crisis. Powerful private economic interests were privileged to the detriment of the poorest and most vulnerable social groups (Fracalanza and Freire 2015; Torres et al. 2020).

According to the Ostroms, the existence of multiple autonomous decision-making centers, operating at multiple governmental levels and involving various administrative sectors in a relatively coordinated way, is, first and foremost, a positive aspect for the political dispute process and for the governance of common-pool resources (Ostrom 2010). However, in the case of the SPMM, it is observed that irrespective of the fact that the urban water governance arrangements possess elements of polycentricity, the political arena, permeated by asymmetric power relations, is marked by intense conflicts due to the multiple and distinct interests at play, especially between the river basin committee, civil society, the São Paulo State Government, and the basic sanitation company.

The organizational chart of Fig. 1 presents the polycentric and multilevel governance structure that has been in force at the time of the 2014–2015 water crisis. It demonstrates how integrated water management in the SPMM is embedded in an institutional and regulatory framework at the level of the provincial state, which itself follows the regulatory determinations established at the national level. Apart from that, municipalities are responsible for providing services of basic sanitation to their citizens, for wastewater treatment, water management and drainage, and together with the provincial and the federal state, for the monitoring and supervision of the management of water resources in its territory (Whately and Neves 2017). But in order to fulfil adequately their task, municipalities are participating, together with representatives from the local and regional civil society, in regional water governance at the scale of the watersheds, mainly in the water basin committees, backed by the regulatory and operational agencies. Therefore, an effective cooperation between the different governmental and societal levels represented in Fig. 1 is crucial for successful polycentric water governance in the SPMM.

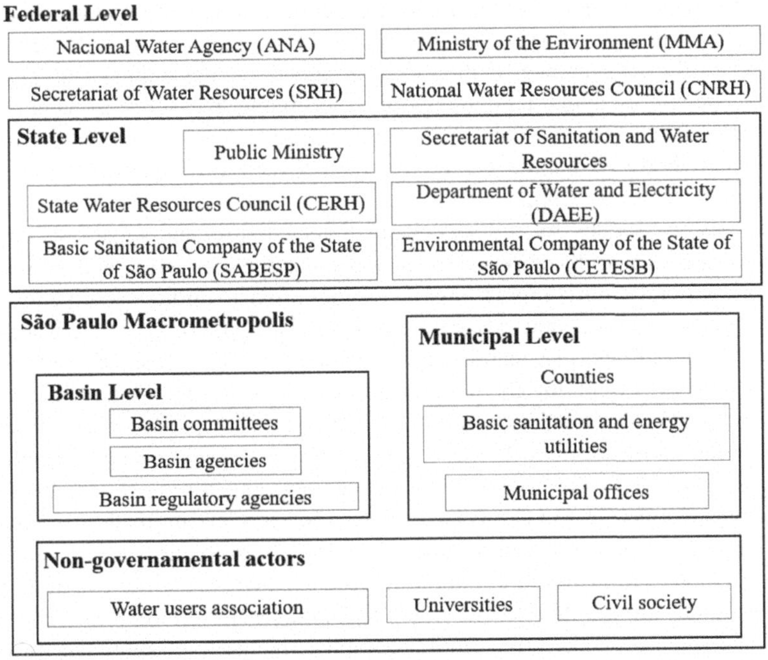

Fig. 1 Political and administrative actors involved in water governance at the time of the water crisis 2014–2015 (own elaboration; modified from Puga et al. 2020)

The chart reveals the high institutional complexity and significant interdependencies of the formal system of water governance in the SPMM. The multiple stakeholders and wide range of institutions involved call for a system of collaborative governance involving the municipalities, the provincial state, the private sector, and local social organizations as the central premise for effective decision-making processes in face of the multiple agendas, interests and contradictions that inevitably arise in the political arena. Nevertheless, it also demonstrates a potential lack of interaction with neighboring policy sectors and planning processes, particularly related to land use. From a PCG perspective, the main challenge is to supplant the predominant sectoral logic by an integrated cross-sectoral approach, taking into account the multiple scales involved.

It is appropriate to recall that part of the water used in the SPMM has its source in the neighboring state of Minas Gerais, thus increasing the need for articulations and intermediation of interests between the different stakeholders of the different water basins that compose the system in the SPMM. Besides, part of the water used comes from municipalities located in the state of Rio de Janeiro, which implies in articulation with an interstate committee that aggregates São Paulo, Rio de Janeiro and Minas Gerais thus taking the analysis to the federal scale, implying an even more complex scenario of competition for water.

During the water crisis, both institutional arrangements, the basin committees and the State Water Resources Council (CERH), the major deliberative forum of the state water policy, were—at least temporarily—disempowered, implying a lack of transparency and democratic legitimacy of the decision-making process (Jacobi et al. 2018). One of the main reasons for this shift, as they were mainly taken by a technical governmental crisis committee under the direct authority of the governor, is related to the interference of the electoral agenda at the State level. The government wanted to avoid taking unpopular rationing measures in view of the imminent elections (Jacobi et al. 2015). Thus, the state governor chose to nominate an executive body to handle the crisis on an exceptional basis, bypassing the competent State Water Council.

The constant denials by the São Paulo State Government regarding the seriousness of the crisis reduced possibilities and space for society to get involved and, moreover, several shortcomings have been identified regarding emergency planning, its implementation, and public relations (Jacobi et al. 2015). As a consequence of these shortcomings, peripheral neighborhoods were the most affected, as they were to a great extent hit by permanent interruptions of supply. The lack of transparency and accountability provoked reactions from part of the civil society organizations and the Public Prosecutors Office demanding comprehensive information. Nevertheless, the government blamed, first and foremost, the absence of rain for the calamity, trying at the same time to foster a less dramatic perception of the situation from part of the population. The drought problem was seen mainly as linked to a lack of infrastructure. Therefore, measures were taken to guarantee water security based on the traditional hydraulic paradigm, ignoring critical voices demanding the adoption of a more complex integrative approach to water governance.

The water crisis exposed both poor crisis management and insufficient long-term planning by the government of the richest state in the Brazilian federation. Two important aspects to understand the shortcomings in the governance process have to be stressed. First, the lack of transparency and information not only related to technical data about the conditions of the water supply system, but also referred to the political decisions taken by the state government (Empinotti et al. 2016). More detailed information about the water crisis was only released after the intervention of the Public Prosecutor's Office. The lack of transparency obstructs an informed and qualified participation and democratic control from part of civil society, a fundamental dimension of effective polycentric water governance (Ostrom 1973). This refers to the second flaw of the management of the water crisis. As stressed by Fracalanza (2017), the decisions were taken unilaterally and without the participation of civil society. Although integrated and decentralized management by river basins is recognized as a fundamental part of water governance, what happened in practice during the water crisis was a process of centralization of decision-making power and authority in favor of the state government and SABESP, contradicting the principles of both decentralized river basin management and PCG. The consequence was the subordination of water governance to technocratic and economistic reasons, in detriment to democratic participation, as well as social, environmental and territorial justice.

4 The Challenge of Water Governance in the SPMM—Between Crisis Management, Integrated Long-Term Planning, and Democratic Deliberation

From the conceptual perspective of Elinor and Vincent Ostrom, the existence of multiple autonomous decision centers, located at different scales and levels, and thus the complexity of the political decision-making process can be, or used to be, beneficial for sustainable and resilient governance of natural resources. And in fact, in our study we were able to verify that the structure and practice of water governance in the SPMM follow basically the principles of PCG as defined by the Ostroms. At least, the governance by water basins allows for civil society to participate in decision-making on water issues, contributes to a more integrative view of water and environmental policies, and brings the decision-making process closer to those who use the water and eventually suffer from poor water quality or from water scarcity, being therefore in line with the principles of the federal water policy.

However, the study also revealed that in times of crisis the real underlying asymmetric power-relations come to light, that is to say, the central authority of the executive power ends up in unilaterally imposing authoritarian rule. Justifying recentralization with the argument of technical and scientific requirements, the crisis management mode ultimately benefited economic and technocratic reason in detriment to alternative holistic approaches committed to social, environmental and territorial justice.

Therefore, the institutional arrangements, based on the principles of integrated and participatory water governance, have proven less resilient as could have been expected by the theoretical framework of PCG. On the contrary, it showed that water governance in complex political, economic, and societal settings is first and foremost a question of power distribution and political dispute, embedded in a complex "web" of political actors, interests, and values. The shortcomings in terms of integrated long-term planning and democratic deliberation become particularly evident in times of crisis. This perspective reinforces why it is important to analyze the scope and limits of PCG in SPMM, being attentive to the ongoing political processes and conflicts.

Pahl-Wostl et al. (2012) point out that polycentric water governance experiences have shown that sharing responsibilities and coordination structures may promote good outcomes. Water governance implies the need to reduce unequal access to water. The analysis of the crisis in the SPMM revealed the importance of effective and consolidated participatory governance structures and decision-making processes, in order to increase resilience and therefore establish a sufficient capacity to deal with water scarcity in a region highly vulnerable to persistent periods of drought. The fact is that the authoritarian centralized decision-making of the crisis amplified the size of existing problems, which were not adequately addressed, as a consequence of the suspension of the fundamental principles of integrated water resources management, above all the principles of decentralization, participation and policy integration.

Moreover, the study demonstrated that possible solutions or mitigation of the problems that came up during the water crisis are linked to the need to strengthen the role of intersectoral cooperation in water governance, articulating municipalities with water management agencies and other relevant state agencies. We understand that the main problem in the SPMM is the lack of a comprehensive integrated governance system that considers the wider macro-metropolitan context as well as goes beyond the water management itself, in order to tackle the water resources in strongly interconnected territories. The different responsibilities are still too fragmented across the different state organizations, and often lack effective coordination (Araújo et al. 2020; Gonçalves et al. 2020).

Some progress has indeed been made. Among the instruments provided by the national and the state legislation, the river basin committees have to elaborate a long-term plan for each watershed, with the projection of scenarios and an agenda of goals to be achieved. Notwithstanding, considering the scale of the SPMM supply system, governance at the basin level is not enough, as the territory encompasses water transpositions between different basins. In 2012, the recently extinguished São Paulo Metropolitan Planning Company (EMPLASA) launched the São Paulo Macrometropolis Action Plan (PAM), as an attempt to better integrate different sectoral planning activities in an overall strategic plan to guide the development of the SPPM. Nevertheless, our studies revealed a by and large top-down practice of planning, as well as the predominance of economic concerns in this plan in detriment to issues related to the environment and social justice (Jacobi et al. 2020). Yet, although PAM exists on paper, there is, at the time being, no evidence of its concrete application and effectiveness.

5 Conclusions

The concern to what extent PCG could be a more appropriate political and strategic approach to achieve transformational sustainable change in metropolises of the Global South—as opposed to the traditionally prevalent top-down and authoritarian management practices—was, from the outset, guiding our study on water governance in the São Paulo Macrometropolis.

Considering issues arisen from our research some general conclusions can be drawn. First of all, from *a technical, planning and scientific perspective*, there is a need to adopt a more complex "integrated approach to water management and land use planning, which has consideration for the sustainable management of water resources" (Carter et al. 2005:115). Whereas decisions about water, as said before, the most important resource to sustain urban life, have impact on land use, by the same token, land use changes in the urbanized areas tend to affect the natural environment, for instance through increased concentration of pollutants in water courses, as well as the social environment, in as much as open spaces or water courses are made available or not for leisure activities in favor of the local population (Campbell and Corley 2015). Without the adoption of such a holistic integrative view in planning and management, the unavoidable conflicts between land use planning and the protection of natural resources could hardly be resolved in the spirit of sustainability.

Consequently, as a second point, we have to bear in mind *the political dimension* of urban water governance. Land use conflicts, particularly if they involve water scarcity, or other types of natural resources shortages, are strongly political, that is to say, unequal access to power implies unequal access to water. Climate change and, hence, the multiplication of extreme events (lack and excess of rain) will still aggravate this situation of socioenvironmental injustice in the future. Failures in achieving sustainable integrative governance, and the continued insistence upon an one-sided policy of massive expenditures on storage reservoirs and inter-basin water transfers (Braga 2001), tend to favor overuse and degradation of water sources and the maintenance of already existing deficits in basic sanitation, threatening water security and public health (Carter et al. 2005), as tragically evidenced in the recent pandemic crisis in São Paulo and elsewhere, where the already disadvantaged poor living in the peripheries of the metropolises are those most affected due to lack of water availability and poor hygienic conditions (Acuto 2020). Resorting to the ideas of Jacques Rancière, we can say that only insofar as these societally disadvantaged and excluded, "the 'part which has no part', is enabled to assume an active political role, challenging the established asymmetrical political order in the name of equality" (Eraydin and Frey 2019:45), there is hope that socioenvironmental justice becomes a central element of transformational change. Besides, more far-reaching changes depend on the politicization of planning and governance (Randolph and Frey 2019), the confrontation of the asymmetrical power relations and, thus, on the creation of a proper political arena at the scale of the Macrometropolis, where these political disputes can be effectively fought out and eventually resolved.

Thirdly, from an *administrative and managerial point of view*, multiple measures are urgently needed in order to enhance the adaptive capacity of water governance in the SPMM. Thus, it is vital to improve coordination between the state government, SABESP and the municipalities, to harmonize land use planning with water policy, to control land occupation, and to amplify the provision of sanitary services adapted to the expected climate change-related water shortage in the future. In addition, the shared use of ecosystem resources demands capacity building for the public agents involved in intersectoral activities in order to strengthen shared visions for sustainable water management, to enhance the response capacity of local institutions, and to create stronger commitment between local city managers and the public, inclusively favoring communitarian bottom-up initiatives.

In a wider *institutional governance perspective*, a general reevaluation of existing governance mechanisms and decision-making processes is indispensable. The water crisis showed that in order to reduce risks of water scarcity, there is a need for increased institutional resilience, particularly the strengthening of democratic instances of social participation. This requires the enforcement of environmental norms, a strengthening of the role of the State Council on Water Resources and the River Basin Committees in planning and political decision-making and, thus, better public control of Sabesp for the public benefit, in order to close the investment gap related to the poor peripheral areas (McLeod 2016; Jacobi et al. 2018).

The challenge of institutional design has been put by Vincent Ostrom and his collaborators quite accurately, already anticipating the deliberative turn in planning, public policies and democratic theory, when they pointed out:

> the most essential institutional arrangements are those that enable human beings to maintain an open public realm where people can freely communicate with one another, explore alternatives, engage in critical assessment, and consider contestable arguments in reaching an understanding about the shortcomings of existing institutions, and what might be done to alter the structure of human relationships and improve the conditions of life in the society (Ostrom et al. 1988:456–457).

Yet, in comparison with the perspective of the "Ostrom school" on CPR management, our approach on the metropolitan or macro-metropolitan territory emphasizes the complex institutional and social, economic and territorial inequalities that demand for a consistent and coordinated interplay between water security, water governance and land use planning as the basis for strengthening the emerging adaptive logic in the SPMM, as well as in other metropolises of the Global South.

Nevertheless, the existing mainstream logic of governance is still mainly developmentalist and growth-oriented in the developing world, and as we saw in our case study, this becomes particularly threatening in crisis situations when the lack of institutional resilience becomes explicit and the powerful economic interests, supported by governments, win through. Our research indicates that only in the case that citizens and local communities, civil society and social movements assume, step by step, a more relevant and proactive role as agents of change within broader multilevel and PCG arrangements, overcoming hence the institutional and cultural constraints identified in our case study on water governance in the SPMM, we can expect that natural resources management will evolve in a sustainable and fair manner. Much more research, however, is necessary in order to better understand if and how these kinds of interactions within polycentric governance arrangements could indeed converge towards effective transformational change.

Acknowledgements The authors are grateful to The São Paulo Research Foundation (FAPESP) for the support given to carry out this research (processes no. 2018/23771-6, 2015/03804-9 & 2019/06536-6) and to the Inter-American Institute for Global Change Research (IAI): SGP-HW 056 (GovernAgua Project), for a grant given to one of the authors.

References

Acuto M (2020) COVID-19: lessons for an Urban(izing) World. One Earth 2(4):317–319. https://doi.org/10.1016/j.oneear.2020.04.004

Andersson KP, Ostrom E (2008) Analyzing decentralized resource regimes from a polycentric perspective. Policy Sci 41(1):71–93

Araújo GP, Rodrigues LS, Dunder BD, Zanirato SH (2020) Planejamento e sustentabilidade urbana: uma análise do Plano de Ação da Macrometrópole Paulista. Revista Brasileira de Meio Ambiente 8(1)

Bakker K (2010) Privatizing water: governance failure and the World's urban water crisis. Cornell University Press, Ithaca, NY, EUA, London, UK

Benz A (2007) Multilevel governance. In: Benz A, Lütz S, Schimank U, Simonis G (eds) Handbuch Governance: Theoretische Grundlagen und empirische Anwendungsfelder. Verlag für Sozialwissenschaften, Wiesbaden, pp 297–310

Braga BPF (2001) Integrated urban water resources management: a challenge into the 21st century. Water Resour Dev 17(4):581–599

Campbell HE, Corley EA (2015) Urban environmental policy analysis. Abingdon, Oxon; Routledge, New York

Carter N, Kreutzwiser RD, de Loë RC (2005) Closing the circle: linking land use planning and water management at the local level. Land Use Policy 22(2):115–127

Côrtes PL, Torrente M, Pinto Alves Pinto A, Ruiz MS, Dias AJG, Rodrigues R (2015) Crise de abastecimento de água em São Paulo e falta de planejamento estratégico. Estudos Avançados 29(84):7–26

Empinotti VL, Jacobi PR, Fracalanza AP (2016) Transparência e a governança das águas. Estudos Avançados 30(88):63–75

Eraydin A, Frey K (2019) The political in governance and planning. In: Eraydin A, Frey K (eds) Politics and conflict in governance and planning. Routledge, New York, London, pp 1–17

Fracalanza AP (2017) Crise de governança da água: a recentralização na gestão da água no Estado de São Paulo (Brasil). In: Jacobi A, Fracalanza P, Empinotti VL (eds) Governança da água no contexto da escassez hídrica, IEE-USP, UFABC e GovAmb, São Paulo, pp 163–187

Fracalanza AP, Freire TM (2015) Crise da água na Região Metropolitana de São Paulo: a injustiça ambiental e a privatização de um bem comum. Geousp Espaço e Tempo (online) 19(3):464–478

Frey K, Jacobi PR, Torres PHC, Ramos RF (2019) Ostrom visits the urban global south: a discussion on environment governance and planning of São Paulo Macrometrolis. Paper presented at the Ostrom Workshop (WOW6) conference, "Governance: Past, Present, and Future", Indiana University Bloomington. https://dlc.dlib.indiana.edu/dlc/bitstream/handle/10535/10478/Paper_Ostrom_Workshop%20Final.pdf?s

Gomes JDL, Barbieri JC (2004) Gerenciamento de recursos hídricos no Brasil e no Estado de São Paulo: um novo modelo de política pública. Cadernos EBAPE. BR 2(3):01–21

Gonçalves LR, Fidelman P, Turra A, Young O (2020) The dynamics of multiscale institutional complexes: the case of the São Paulo Macrometropolitan Region. Environ Manage 67(1):109–118

Gupta R (2014) Environmentalism: a global history. Penguin Books, Haryana, India

Head BW (2019) Forty years of wicked problems literature: forging closer links to policy studies. Policy Soc 3(2):180–197

Heikkila T, Villamayor-Tomas S, Garrick D (2018) Bringing polycentric systems into focus for environmental governance. Environ Policy Gov 28:207–211

Jacobi PR, Sinisgalli PA, Medeiros Y, Romeiro A (2009) Governança da Água no Brasil: dinâmica da política nacional e desafios para o futuro. In: Jacobi PR, Sinisgalli PA (2009) Governança da Água e Políticas Públicas na América Latina e Europa. Annablume, São Paulo, pp 49–82

Jacobi PR, Cibim JC, Souza AN (2015) Crise da água na região metropolitana de São Paulo–2013/2015. GEOUSP Espaço E Tempo (online) 19(3):422–444

Jacobi PR, Cibim J, Leão R (2015) Crise hídrica na Macrometrópole Paulista e respostas da sociedade civil. Estudos avançados 29(84):27–42

Jacobi PR, Bujak NL, Souza A (2018) Pénurie hydrique et crise de gouvernance dans la Région métropolitaine de São Paulo. Brésil (s). Sciences humaines et sociales (13)

Jacobi PR, Frey K, Ramos RF, Côrtes P (2020) ODS 6: Água potável e saneamento. In: Frey K, Torres PHC, Jacobi PR, Ramos RF (eds) Objetivo do desenvolvimento sustentável: desafios para o planejamento e a governança ambiental na Macrometrópole Paulista. Edufabc, São Paulo

Martinez-Alier J (2003) The Environmentalism of the poor: a study of ecological conflicts and valuation. Edward Elgar Publishing

Martins RC (2015) Fronteiras entre desigualdade e diferença na governança das águas. Ambiente & Sociedade 18(1):211–228

McGinnis MD, Ostrom E (2012) Reflections on Vincent Ostrom, public administration, and polycentricity. Public Adm Rev 72(1):15–25

McLeod CA (2016) Managing water (In)security in Brazil: lessons from a megacity. LSE. Working paper series 2016 No.16-179

Meadowcroft J, Banister D, Holden E, Langhelle O, Linnerud K, Gilpin G (2019) Introduction. In: Meadowcroft J, Banister D, Holden E, Langhelle O, Linnerud K, Gilpin G (eds) What next for sustainable development? Our common future at thirty. Edward Elgar, Cheltenham, UK; Northampton, MA, USA, pp 1–8

Murtha NA, Castro JE, Heller L (2015) Uma perspectiva histórica das primeiras políticas públicas de saneamento e de recursos hídricos no Brasil. Ambiente & Sociedade 18(3):193–210

Ostrom V (1973) Can federalism make a difference? Publius J Federalism 3(2):197–237. https://doi.org/10.1093/oxfordjournals.pubjof.a038279

Ostrom E (1990) Governing the commons: the evolution of institutions for collective action. Cambridge University Press, Cambridge, UK

Ostrom E (1996) Crossing the great divide: coproduction, synergy, and development. World Dev 24(6):1073–1087

Ostrom E (2005) Understanding institutional diversity. Princeton University Press, Princeton, NJ

Ostrom E (2010) A long polycentric journey. Annu Rev Polit Sci 13:1–23

Ostrom V, Tiebout CM, Warren R (1961) The organization of government in metropolitan areas: a theoretical inquiry. Am Political Sci Rev 55(4):831–842. https://doi.org/10.2307/1952530

Ostrom V, Feeny D, Picht H (1988) Institutional analysis and development: rethinking the terms of choice. In: Ostrom V, Feeny D, Picht H (eds) Rethinking institutional analysis and development: issues, alternatives and choices. International Center for Economic Growth, San Francisco, pp 439–466

Paavola J (2016) Multi-level environmental governance: Exploring the economic explanations. Environ Policy Gov 26(3):143–154

Pahl-Wostl C, Lebel L, Knieper C, Nikitina E (2012) From applying panaceas to mastering complexity: toward adaptive water governance in river basins. Environ Sci Policy 23:24–34

Puga BP, Garcia JR, Maia AG (2020) Governança dos recursos hídricos na bacia do rio Jundiaí (São Paulo). Revibec: revista iberoamericana de economía ecológica 32:93–101

Randolph R, Frey K (2019) Planning and governance: towards radical political approaches. In: Eraydin A, Frey K (eds) Politics and conflict in governance and planning: theory and practice. Routledge, New York & London, pp 38–55

Rydin Y (2008) Sustainable development and governance. In: Cox KR, Low M, Robinson J (eds) The SAGE handbook of political geography. SAGE, London, pp 579–593

Sayre NF (2009) Scale. In: Castree N, Demeritt D, Liverman D, Rhoads B (eds) A companion to environmental geography. Wiley-Blackwell, Chichester, England, pp. 95–108

Schmitter PC (2002) Participation in governance arrangements: Is there any reason to expect it will achieve "sustainable and innovative policies in a multilevel context"? In: Grote JR, Gbikpi B (eds) Participatory governance. Political and societal implications. Leske + Budrich, Opladen, pp. 51–69

Stephan M, Marshall G, McGinnis M (2019) An introduction to polycentricity and governance. In: Thiel A, Garrick DE, Blomquist WA (eds) Governing complexity: analyzing and applying polycentricity. Cambridge University Press, Cambridge, UK, pp 21–44

Thiel A, Blomquist WA, Garrick DE (2019) Introduction. In: Thiel A, Garrick DE, Blomquist WA (eds) Governing complexity: analyzing and applying polycentricity. Cambridge University Press, Cambridge, UK, pp 1–17

Toonen T (2010) Resilience in public administration: the work of Elinor and Vincent Ostrom from a public administration perspective. Public Adm Rev 70(2):193–202. https://doi.org/10.1111/j.1540-6210.2010.02147.x

Torres PHC, Ramos RF, Gonçalves LR (2019) Environmental conflicts at São Paulo Macrometropolis: Paranapiacaba and São Sebastião. Ambiente & Sociedade 22

Torres PHC, Côrtes PL, Jacobi PR (2020) Governing complexity and environmental justice: lessons from the water crisis in Metropolitan São Paulo (2013–2015). Desenvolvimento e Meio Ambiente 53

Whately M, Neves EMSC (2017) O município e a governança da água: subsídios para a agenda municipal de cuidado com a água. Retrieved from São Paulo: https://www.aliancapelaagua.com.br/wp-content/uploads/2017/04/relatorio-municipios.pdf

Land Use Change in Tropical Watersheds: Will It Support Natural Resources Sustainability?

Pamungkas Buana Putra, Cahyono Agus, Rahardyan Nugroho Adi, Pranatasari Dyah Susanti, and Yonky Indrajaya

1 Introduction

Indonesia is located in a strategic geographical position on the equator with abundant sunlight throughout the year and relatively high rainfall. Indonesia is also considered a country rich in natural resources with a large forest cover area and acknowledged as one of the megadiversity countries. Indonesia has successfully increased the GDP per capita from oil and non-oil exports with its rich natural resources after 1973 (Toto Same 2009). However, with its rich natural resources, the overexploitation or lack of sustainable management caused the degradation of natural resources and, in some areas, triggered conflicts between parties, mostly categorized as separatism and inter-communal conflicts (Tadjoeddin 2007).

Owing to its mountainous terrain, Indonesia has rugged land and high slopes. With high rainfall and steep slopes, the potential for land degradation is considerably high, especially in open areas not complying with soil and water conservation principles. Land use that is not based on its designation has resulted in the degradation of natural resources, threatening land sustainability, increasing greenhouse gas emissions, and impacting global climate change (Braimoh and Osaki 2009; Junaidi and Indrajaya 2018). The rate of soil erosion in the agricultural land on a 3–30% slope is high, ranging from 60 to 625 tones/ha/year. As agricultural

P. B. Putra · C. Agus (✉)
Faculty of Forestry Universitas Gadjah Mada, Yogyakarta 55281, Indonesia
e-mail: cahyonoagus@gadjahmada.edu; acahyono@ugm.ac.id

P. B. Putra · R. N. Adi · P. D. Susanti · Y. Indrajaya
Watershed Management Technology Centre, Surakarta, Indonesia

C. Agus
The Association of Tamansiswa Society (PKBTS), Yogyakarta 55151, Indonesia

W. Leal Filho et al. (eds.), *Sustainability in Natural Resources Management and Land Planning*, World Sustainability Series,
https://doi.org/10.1007/978-3-030-76624-5_5

land with a more than 15% slope is common, it has a very high potential for soil erosion (Sutrisno and Heryani 2013). Indonesia's total national GHG emissions were 1800 Mt CO_2e in 2005, land use, land-use change, and forestry (LULUCF) contributing to 63% of the total GHG emissions (Government of Indonesia 2015).

One management unit in natural resource management is a watershed or river basin. A watershed is an integral part of rivers and tributaries wherein water originating from rainfall is accommodated, stored, and made to flow into lakes or seas. The boundaries on land are topographical, and those at seas and other water areas are still affected by activities on Land (Ministry of Forestry 2009). A watershed is a hydrological management area where hydrological processes such as infiltration, runoff, and evapotranspiration occur. These processes are interrelated and can be precisely measured within their boundaries. Watershed degradation by erosion and sedimentation due to land-use change can be accurately assessed at the watershed scale (Lal 1999).

The watershed processes are hydrological; however, several other processes and mechanisms, such as the energy cycle, nutrient cycle, and carbon cycle, also exist. These processes relate to vegetation community factors and other factors such as the climate, soil, animals, and humans (Chiti et al. 2018; Silva et al. 2019). With this in mind, watershed management can contribute to the management of tropical forest ecosystems by determining the land suitable for conversion to non-forest land and the rehabilitation of degraded land. This chapter describes the changes in land use in watersheds in Indonesia and the context for their sustainability.

2 Land-Use Change and Natural Resource Degradation

Most of the land cover in the world comprised forests before deforested by humans for food, homes, and other requirements. The leading causes of deforestation in developing countries are population growth and agricultural expansion, which are exacerbated by the extraction of wood for fuel and export requirements (Allen and Barnes 1985). However, (Lambin et al. 2001) found that population growth is not the primary factor causing deforestation. Angelsen and Kaimowitz (1999) found that the roads, agricultural prices, wage rates, and off-farm shortage cause deforestation through economic modeling. Moreover, they criticize that the factors affecting deforestation, namely, population growth, poverty reduction, national income, economic growth, and foreign debt, are ambiguous. (Seymour and Harris 2019) found that Indonesia's main driver of deforestation is the complex interaction between selective logging and conversion to palm oil and pulpwood plantations.

With a total area of 191 Mha, Indonesia is considered one of the most extensive tropical rain forests (Tsujino et al. 2016) characterized by rich mangrove forests, lowland tropical rainforests, mountains, and subalpine forests. The history of deforestation in Indonesia has been reported that in 1950, Indonesia's forest cover was about 159 Mha or 87% of the total land area (182 Mha) (Tsujino et al. 2016). From 1950 to 1997, 59 ha of the forested land was deforested. According to

Ratnasari et al. (2020), in general, Indonesia's rate of deforestation has shown a decreasing trend from 1990 to 2019, reaching 3.51–0.46 million ha in the 2018–2019 period. The area of forest cover in Indonesia has decreased from 105 Mha in 1990 to 97 Mha in 2013 (Djaenudin et al. 2018).

Deforestation has resulted in increasing areas of degraded land and critical watersheds in Indonesia. The critical land area increased from 10.75 Mha in 1974 to 23.73 Mha in 1998 (Nugroho 2011). The number of critical watersheds increased due to deforestation, from 22 critical watersheds in 1984 to 39 in 1992. Subsequently, the number of critical watersheds increased to 62 in 2003 and 108 in 2009 (Junaidi and Indrajaya 2018). Meanwhile, with the growing watershed rehabilitation program, the critical land area decreased from 30 Mha in 2006 to 14 Mha in 2018 (National Statistics Bureau 2019). The decrease in critical land was also caused by a relatively low rate of deforestation between 2006 and 2018. This critical land area, however, is still quite large, i.e., 14 Mha.

Bradshaw et al. (2007) found that deforestation is positively correlated with flood frequency. Similarly, Rafiei (2016) stated that land-use changes due to deforestation and climate change would increase peak flow through modeling analysis. Tropical deforestation also affects the local climate (Lawrence and Vandecar 2015). This decade in Indonesia, the existence of extreme climate events due to global climate change coupled with relatively high rates of deforestation can further increase hydro-meteorological disasters such as floods, landslides, flash floods, drought, and forest and land fires (Agus et al. 2020). Based on the National Agency for Disaster Management (BNPB) data from 2000 to 2015, more than 78% of disasters were hydro-meteorological, and only about 22% were geological (Amri et al. 2016). Deforestation and land-use change are two factors that describe the land-cover/land-use parameter and land suitable parameters of a watershed condition and explain the critical role of vegetation and forests in supporting the existence of land a processor that will provide ecosystem services (Wahyuningrum and Putra 2019).

Due to national development during the last four decades, changes in land use have significantly impacted watersheds' hydrological function (Pawitan 2004). Moreover, land-use changes, especially in Java during the last 50 years, have decreased the rainfall and increased the frequency of extreme events, including floods and droughts (Pawitan 2004). In contrast, the number of very critical watersheds in Java Island is 16, with vegetated land use of <20% or <30% as required by the Spatial Planning Law Number 26 of 2007 and the Forestry Law Number 41 of 1999 (Mawardi 2010). The forest cover area on Java Island is approximately 16% (Ministry of Environment and Forestry 2018). With such less forest area, the values of river flow in Java vary 10- to 100-fold, indicating the critical condition of Java's watersheds.

3 Land-Use Change and Watershed Hydrology

Two essential factors that influence hydrological conditions in a watershed are land-use and climate (Li et al. 2009). Moreover, climate change and land use significantly impact watershed hydrological/water system conditions such as water yield, surface runoff, evapotranspiration (ET), and water balance in the watershed (Kundu et al. 2017). Of these two factors, land cover is a factor that can be controlled, while climate factors are a given factor.

Various natural disasters, such as flooding, often raise questions about forests' function as water regulators. The flood event is thought to be due to decreased environmental carrying capacity, such as damage to forest areas and reduced forest land cover. Land-use change is expected to have impacts on watershed hydrology (Setyorini et al. 2017), for instance: flood frequency (Brath et al. 2006), severity (Roo et al. 2001), surface runoff, and groundwater or base flow (Ridwansyah et al. 2020).

Changes in land cover will impact hydrological characteristics such as flow coefficients, discharge, and flood hydrograph characteristics (Rukmi et al. 2013). Forest damage indicators can be seen from the hydrograph characteristics. Some experts argue that the increase in peak discharge, apart from being caused by climate change, is also caused by changes in a land cover that have increased the runoff coefficient (Kodoatie and Sjarief 2008; Rukmi et al. 2013).

Changnon and Demissie (1996) report that climate may affect the flow routing time, peak flow, and volume. In line with this, (Asdak 2002) also states that floods and flash floods are caused by high rain intensity in many cases. During the rainy season, when there is high-intensity rain, some areas will be submerged by water due to flooding.

The sustainability of water resources depends on the forest's condition in the area. When a forest is cut down, the water yield will increase due to reduced evapotranspiration; however, the water yield will decrease over time because the amount of water stored in the soil also decreases. The reduction in water storage in the soil is caused by rainwater that falls on open forest areas, most of which directly become surface runoff (Adi 2012). Besides, forests can also act as water regulators and protect against soil degradation by rain because forests can encourage water infiltration into the soil (Siswamartana et al. 2002).

The effect of land cover on the hydrological response of a watershed has been widely demonstrated (Zhang and Wei 2012). Changes in land cover and management factors applied to a forest area will change the watershed hydrological response (Guzha et al. 2018). For example, research on changes in land cover to hydrological responses, among others, is Anwar et al. (2011) stated that a decrease in land cover in the Barito Hulu watershed by 9.51% from its original condition would increase water yield, evapotranspiration, and erosion, respectively, by 8.52%, 5.94% and 1.73 tons/ha/year respectively, and water savings decreased by 14,46%. Prasena and Shrestha (2013) report that the reduction of mixed garden land to residential areas by 4.28% from the original would increase surface runoff by 3.42–4.67%.

Flood and drought can be likened to two sides of a coin that are always sided by the side. However, all of them are the opposing sides of a watershed because both are disastrous events. On the one hand, floods occur during the rainy season, and on the other hand, there is drought during the dry season. The largest river on Java Island in Central Java and Eastern Java of Bengawan Solo with ±12% of the entire area of Java Island, the drought and seawater intrusion problems occur in the dry season and vice versa in various parts during the rainy season, floods often cause significant property losses and human losses (Ministry of Public Works and Housing 2010).

Apart from influencing floods and drought, changes in land cover also affect erosion and sedimentation. In the tropics, erosion by water is a very dominant form of land degradation. Deforestation and land-use change are the main causes of land degradation in production forests and community forests. Besides that, farming practices that do not pay attention to conservation principles can also lead to a decline in land resources, resulting in the expansion of critical Land (Atmojo 2008). Control efforts need to be made to preserve and sustain water resources to meet current and future needs.

4 Land Use Change and Water Quality

Land-use changes will affect the composition of land use and impact the quality of water produced due to water pollution (Susanti and Wahyuningrum 2020). Water pollution is an event that pollutants enter a water body, which causes a decrease in water quality so that it cannot be used for specific purposes (Owa 2013) and affects an ecosystem (Inyinbor 2018). Water pollutants can be chemical, physical, or biological components (Schweitzer and Noblet 2018). Based on the source, water pollution is divided into two categories, namely point source pollution and non-point source pollution.

Water pollution from point source pollution includes industrial waste, pollutants from oil spills, pollutants from refineries factories, and the garbage being dumped into a river. In contrast, non-point source pollution includes urban or domestic, agriculture, and the atmosphere (Singh et al. 2020). Geng and Sharpley (2019) explained that the most influential non-point pollution source is agriculture because it is the primary source of eutrophication both in lakes and surface rivers. Besides, Viman et al. (2010) show that pollutants will affect the photosynthesis process in aquatic ecosystems, aquatic plants and terrestrial will absorb water pollutants. They can even cause death in various plants due to herbicide pollutants and other chemicals that are dangerous.

Land-use change from the forest into agricultural and the settlements areas in the Madiun River Sub Watershed in East Java, during five years (2010–2015), increases the water pollution index (Susanti and Miardini 2017). Meanwhile,

Effendi et al. (2018) found that urban areas' activities would increase Biological Oxygen Demand (BOD), Chemical Oxygen Demand (COD), and Total Suspended Solid (TSS). Anggana and Susanti (2020) found that based on the evaluation of water quality in the swamp river border, the water quality index's value has increased during restoration and has decreased when land fires occur in the upstream part the North Hulu Sungai Regency, South Kalimantan, Indonesia. Susanti and Wahyuningrum (2020) added to their research in Bengawan Solo Watershed, and Brantas Watershed, Central Java and East Java, Indonesia, that land use for mixed gardens and paddy fields activities will cause phenol, Total Dissolved Solids (TDS), chlorine, and turbidity pollution.

In addition to the pollutants mentioned above, land-use changes will also affect the area's runoff conditions. Kwanchai and Koontanakulvong (2009) report that land-use changes in The Nan Basin over six years have increased runoff, negatively impacting water quality because pollutants from various land-use will spread. Bi et al. (2018) stated that land use and land cover change would affect the Total Nitrogen (TN) and Total Phosphorus (TP) content. Apart from affecting runoff, pollutants can also harm groundwater quality. Farooqui et al. (2020) explained an increase in TDS content, conductivity, total hardness, calcium, magnesium, chloride, and sulfate in groundwater.

Some efforts need to be taken to avoid water pollution due to land-use change. The bioremediation method can be used as water treatment to reduce pollutants in the water Gao et al. (2018). In addition to bioremediation, the use of constructed wetlands is also reported to manage water pollution effectively (Chen et al. 2019).

5 The Dynamics of the Nutrient and Carbon Cycle in a Watershed

Nutrient cycles can be strongly aligned to the hydrological cycle at many points. Nutrient inputs and outputs are directly related to the amount of water entering and leaving the ecosystem, as are the concepts of "washing" and "flushing." Simultaneously, the hydrological regime also dramatically influences the biogeochemical system temporally and absolutely (Bormann and Likens 1967). Moreover, the biological absorption of nutrients by plants and the release of nutrients by biological decomposition are closely related to the pattern of availability of water. Precipitation characteristics largely determine potential biomass levels within the system. Similarly, the nature and rate of weathering and soil formation are influenced by the hydrological regime, as water is essential to significant chemical weathering processes (Bormann and Likens 1967).

In terrestrial ecosystems, vegetation is a fundamental component to support the passage of many mechanisms and cycles. These mechanisms and cycles are supported by vegetation's physiological abilities (Townsend et al. 2011). Plants are crossed by nutrients, carbon, soil, and/or atmospheric water through photosynthesis and respiration mechanisms (Nieder and Benbi 2008). The mechanisms and cycles vary in each type of tropical terrestrial ecosystem, but in principle, they are related to the absorption, storage, return, and loss of elements by plants from the ecosystem (Alvarado 2016).

The photosynthetic mechanism will ideally support plant life growth when the supply of nutrients, carbon, and water is smooth, and temperature and light conditions are optimal (Lambers et al. 2008). Vegetation is a crossroads and acts as a carbon, nutrients, and water storage in terrestrial ecosystems in biomass (Lugo and Brown 1992; Townsend et al. 2011). On the other hand, vegetation's ability to produce biomass becomes a precedent in knowing the coordination of ecosystem structure and function and its interaction with the environment (Xu et al. 2015). These processes are various interactions from aspects of biological, geochemical, and atmospheric (Alvarado 2016).

Based on the elemental analysis of a plant that several studies have carried out, Alvarado (2016) states that 90–95% of the total plant dry weight consists of elements C (carbon), H (hydrogen), and O (oxygen). At the same time, 5–10% consists of growth support elements in the form of macronutrients consisting of N (nitrogen), P (phosphorus), K (potassium), Ca (calcium), Mg (magnesium), and S (sulfur), and micro-nutrients consisting of Mn (manganese), Fe (iron), Zn (zinc), Cu (copper), B (boron), Cl (chlorine), and Mo (molybdenum). The availability of nutrients in a wet tropical land is limited by biogeochemical conditions (Townsend et al. 2011). The nutrient cycle is sometimes closed (Agus et al. 2004; Townsend et al. 2011) that related to the characteristics of most of Indonesia's tropical soils, which are very weathered and leached so that the presence of vegetation plays a vital role in supporting the nutrient cycle (also the carbon cycle) through the process of absorption and return of nutrients in the ecosystem (Anderson and Spencer 1991).

Often changes and loss of forest land cover will not only disrupt nutrient cycles and mechanisms through changes in the dynamics of litter, decomposition, and nutrient mineralization (González de Andrés 2019; Agus et al. 2004) but also release and lose carbon (Noordwijk et al. 2014). Lal (1987) explained that land fertility could not be sustained when tropical natural forests are converted to agriculture or intensive forestry due to severe erosion to carry away nutrients. Meanwhile, Wang et al. (2017) also informed that the loss of nutrients in the land is also due to loss of biomass through logging and timber removal. Land clearing accompanied by burning will increase the soil nutrient content instantly but then decrease drastically. There are four mechanisms of nutrient loss in a forest ecosystem, namely through leaching, erosion, fire, and harvesting (Agus et al. 2004), which can also be used in the context of deforestation.

6 Integrated Watershed Management for Natural Resource Sustainability

Sheng (1994) argues that watershed management originated from the "torrent control" in the European Alps. This concept was then widely used in Northern America in the 1990s and was termed "watershed management" with water benefits' main objective. This notion then spread to developing countries, with the primary concerns being soil erosion and land degradation (Sheng 1994). Integrated watershed management (IWM) is an integrated effort to manage natural resources, including the utilization, arrangement, maintenance, supervision, control, recovery, and watershed development, based on preserving environmentally compatible capabilities and balanced support development improving human welfare.

Watershed management was introduced in Indonesia during the Dutch era, especially forest management, wherein the division of forests was arranged based on watershed units. Anwar (2005) reports that the first formal watershed management began in the 1970s due to massive flooding in Surakarta, Central Java. A large dam (Gajah Mungkur) was built upstream Bengawan Solo River to control the water flow and mitigate floods in the Surakarta area. Integrated watershed management was first introduced in Indonesia in the Citanduy River Basin in the eastern part of West Java, in 1981, with various cross-sectoral and interdisciplinary activities.

However, watershed management activities have emphasized the development of physical infrastructure for soil conservation activities to prevent erosion (Sudaryono 2002). It was not until 1994 that the concept of participation began to apply to Presidential Instruction on Greening and Reforestation. From 1973 to 1981, FAO and UNDP carried out various studies to find the right method for land rehabilitation and soil conservation in the Solo Watershed, both physically and socio-economically. This test's results have been implemented under the Presidential Instruction on Afforestation and Reforestation since 1976 in Indonesia's 36 watersheds (Sudaryono 2002).

Through the Ministry of Forestry, the Government of Indonesia (GoI) has issued guidance to assess Indonesia's watershed performance (Ministry of Forestry 2001). This guidance includes criteria and indicators that can be used for the sustainability of a watershed. The criteria and indicators of a healthy watershed are assessed with the following: the healthier the watershed, and the more sustainable is the natural resources. Furthermore, in 2012, the GOI issued Government Regulation No. 37 on Watershed Management to strengthen the watershed management program (Government of Indonesia 2012).

7 Conclusion

The mismatch functions of land-use change have reduced land capacity, water flow, and water quality in the watershed. Degradation of natural resources can threaten land sustainability, affecting changes in extreme rainfall (amount, duration, and intensity), greenhouse gas emissions, flooding, and landslides. Watershed management that combines biophysical and socio-economic aspects in an integrated manner can harmonize economic, environmental, and socio-cultural value benefits in encouraging the management of natural resources and live in a dignified and sustainable manner. In applying this concept to the suitability of other land uses in watershed management, we must adapt it to the area's specific biophysical and socio-cultural characteristics.

References

Adi RN (2012) Analisis neraca air kawasan hutan tanaman kayu putih di Kabupaten Gunung Kidul Daerah Istimewa Yogyakarta. Gadjah Mada University, Yogyakarta

Agus C, Karyanto O, Kita S, Haibara K, Toda H, Hardiwinoto S, Supriyo H, Na'iem M, Wardana W, Sipayung M, Khomsatun S, Wijoyo S (2004) Sustainable site productivity and nutrient management in a short rotation plantation of Gmelina arborea in East Kalimantan, Indonesia. New Forest 28(2):277–285. https://doi.org/10.1023/B:NEFO.0000040954.27630.2f

Agus C, Ilfana ZR, Azmi FF, Rachmanadi D, Widiyatno WD, Santosa PB, Harun MK, Yuwati TW, Lestari T (2020) The effect of tropical peat land-use changes on plant diversity and soil properties. Int J Environ Sci Technol 17(3):1703–1712. https://doi.org/10.1007/s13762-019-02579-x

Allen JC, Barnes DF (1985) The causes of deforestation in developing countries. Ann Assoc Am Geog. 75(2):163–184. https://doi.org/10.1111/j.1467-8306.1985.tb00079.x

Alvarado A (2016) Plant nutrition in tropical forestry In: Pancel L, Kohl M (eds) Tropical forestry handbook, 2nd edn, vol 2. Springer Reference, Germany, pp 1113–1202

Amri M, Yulianti G, Yunus R, Wiguna S, Adi A, Ichwana A, Randongkir R, Septian R (2016) Risiko bencana indonesia. Badan Nasional Penanggulangan Bencana, Jakarta

Anderson JM, Spencer T (1991) Carbon, nutrient and water balances of tropical rain forest ecosystems subject to disturbance: Unesco

Angelsen A, Kaimowitz D (1999) Rethinking the causes of deforestation: lessons from economic models. World Bank Res Obs 14:73–98

Anggana AF, Susanti PD (2020) Evaluation of water quality in the swamp river border using water quality index. J Degraded Min Lands Manage 7(4):2373

Anwar S (2005) Watershed management in Indonesia. In: Achouri M, Tennyson L, Upadhyay K, White R (eds) Preparing for the next generation of watershed management programmes and projects, Kathmandu, Nepal. ICIMOD & FAO, pp 93

Anwar M, Pawitan H, Murtilaksono K, Jaya INS (2011) Respons Hidrologi Akibat Deforestasi di DAS Barito Hulu. Kalimantan Tengah. Jurnal Manajemen Hutan Tropika 17(3):119–126

Asdak C (2002) Hidrologi dan Pengelolaan Daerah Aliran Sungai. Gadjah Mada University Press, Yogyakarta

Atmojo SW (2008) Peran agroforestri dalam menanggulangi banjir dan longsor DAS. In: Proseding Seminar Nasional Pendidikan Agroforestry Sebagai Strategi Menghadapi Pemanasan Global di Fakultas Pertanian, UNS. Solo, 2008, vol 4, pp 1–15

Bi W, Weng B, Yuan Z, Ye M, Zhang C, Zhao Y, Yan D, Xu T (2018) Evolution characteristics of surface water quality due to climate change and LUCC under scenario simulations: a case study in the Luanhe River basin. Int J Environ Res Public Health 15(8):1724

Bormann FH, Likens GE (1967) Nutrient cycling. Science 155(3761):424. https://doi.org/10.1126/science.155.3761.424

Bradshaw CJA, Sodhi NS, Peh KSH, Brook BW (2007) Global evidence that deforestation amplifies flood risk and severity in the developing world. Glob Change Biol 13(11):2379–2395. https://doi.org/10.1111/j.1365-2486.2007.01446.x

Braimoh AK, Osaki M (2009) Land-use change and environmental sustainability. Sustain Sci 5 (1):5. https://doi.org/10.1007/s11625-009-0092-2

Brath A, Montanari A, Moretti G (2006) Assessing the effect on flood frequency of land use change via hydrological simulation (with uncertainty). J Hydrol 324(1):141–153. https://doi.org/10.1016/j.jhydrol.2005.10.001

Changnon SA, Demissie M (1996) Detection of changes in streamflow and floods resulting from climate fluctuations and land use-drainage changes. Clim Change 32(4):411–421. https://doi.org/10.1007/BF00140354

Chen J, Liu Y-S, Deng W-J, Ying G-G (2019) Removal of steroid hormones and biocides from rural wastewater by an integrated constructed wetland. Sci Total Environ 660:358–365

Chiti T, Díaz-Pinés E, Butterbach-Bahl K, Marzaioli F, Valentini R (2018) Soil organic carbon changes following degradation and conversion to cypress and tea plantations in a tropical mountain forest in Kenya. Plant Soil 422(1):527–539. https://doi.org/10.1007/s11104-017-3489-1

De Roo A, Odijk M, Schmuck G, Koster E, Lucieer A (2001) Assessing the effects of land use changes on floods in the meuse and oder catchment. Phys Chem Earth Part B 26(7):593–599. https://doi.org/10.1016/S1464-1909(01)00054-5

Djaenudin D, Oktaviani R, Hartoyo S, Dwiprabowo H (2018) Analisis peluang keberhasilan penurunan laju deforestasi: pendekatan teori transisi hutan. Jurnal Penelitian Sosial Dan Ekonomi Kehutanan 15(1):15–29

Effendi H, Muslimah S, Permatasari PA (2018) Relationship between land use and water quality in Pesanggrahan River. In: IOP conf. series: earth and environmental science, vol 149, pp 012022

Farooqui KM, Kumar Sar S, Diwan V (2020) Investigation of water quality in Ambur City by water quality indexing. Holist Approach Environ 10(2):48-52

Gao H, Xie Y, Hashim S, Akhtar Khan A, Wang X, Xu H (2018) Application of microbial technology used in bioremediation of urban polluted river: a case study of Chengnan River, China. Water 10(5):643

Geng R, Sharpley AN (2019) A novel spatial optimization model for achieve the trad-offs placement of best management practices for agricultural non-point source pollution control at multi-spatial scales. J Clean Prod 234:1023–1032. https://doi.org/10.1016/j.jclepro.2019.06.277

González de Andrés E (2019) Interactions between climate and nutrient cycles on forest response to global change: the role of mixed forests. Forests 10(8), https://doi.org/10.3390/f10080609

Government of Indonesia (2012) Peraturan Pemerintah Republik Indonesia Nomor 37 Tahun 2012 Tentang Pengelolaan Daerah Aliran Sungai. In: Government of Indonesia (ed) Government of Indonesia, Jakarta

Government of Indonesia (2015) Intended Nationally Determined Contribution, Republic of Indonesia. Jakarta: Government of Indonesia

Guzha AC, Rufino MC, Okoth S, Jacobs S, Nóbrega RLB (2018) Impacts of land use and land cover change on surface runoff, discharge and low flows: evidence from East Africa. J Hydrol Reg Stud 15:49–67. https://doi.org/10.1016/j.ejrh.2017.11.005

Inyinbor Adejumoke A, Adebesin Babatunde O, Oluyori Abimbola P, Adelani Akande Tabitha A, Dada Adewumi O, Oreofe Toyin A (2018) Water pollution: effects, prevention, and climatic impact. Water Challenges Urbanizing World 33

Junaidi E, Indrajaya Y (2018) Hydrological responses of agroforestry system application which is not based on land suitability, a case study in Cimuntur watershed. Jurnal Penelitian Kehutanan Wallacea 7(1):69–81

Kodoatie RJ, Sjarief R (2008) Pengelolaan Sumber Daya Air Terpadu (Edisi Revisi). Andi, Yogyakarta

Kundu S, Khare D, Mondal A (2017) Past, present and future land use changes and their impact on water balance. J Environ Manage 197:582–596. https://doi.org/10.1016/j.jenvman.2017.04.018

Kwanchai P, Koontanakulvong M (2009) The effect of land use change on runoff in the nan basin. CATENA 69:31–35

Lal R (1987) Managing the soils of sub-Saharan Africa. Science 236(4805):1069. https://doi.org/10.1126/science.236.4805.1069

Lal R (1999) Rationale for watershed as a basis for sustainable management of soil and water resources. Integr Watershed Manage Glob Ecosyst 3–16

Lambers H, Chapin FS III, Pons TL (2008) Plant physiological ecology. Springer Science & Business Media. https://doi.org/10.1007/978-0-387-78341-3

Lambin EF, Turner BL, Geist HJ, Agbola SB, Angelsen A, Bruce JW, Coomes OT, Dirzo R, Fischer G, Folke C, George PS, Homewood K, Imbernon J, Leemans R, Li X, Moran EF, Mortimore M, Ramakrishnan PS, Richards JF, Skånes H, Steffen W, Stone GD, Svedin U, Veldkamp TA, Vogel C, Xu J (2001) The causes of land-use and land-cover change: moving beyond the myths. Glob Environ Change 11(4):261–269. https://doi.org/10.1016/S0959-3780(01)00007-3

Lawrence D, Vandecar K (2015) Effects of tropical deforestation on climate and agriculture. Nat Clim Change 5(1):27–36. https://doi.org/10.1038/nclimate2430

Li Z, Liu W-Z, Zhang X-C, Zheng F-L (2009) Impacts of land use change and climate variability on hydrology in an agricultural catchment on the Loess Plateau of China. J Hydrol 377(1):35–42. https://doi.org/10.1016/j.jhydrol.2009.08.007

Lugo AE, Brown S (1992) Tropical forests as sinks of atmospheric carbon. For Ecol Manage 54 (1):239–255. https://doi.org/10.1016/0378-1127(92)90016-3

Mawardi I (2010) Kerusakan daerah aliran sungai dan penurunan daya dukung sumberdaya air di pulau jawa serta upaya penanganannya. Jurnal Hidrosfir Indonesia 5(2)

Ministry of Environment and Forestry (2018) The State of Indonesia's Forests 2018. Ministry of Environment and Forestry Republic of Indonesia, Jakarta

Ministry of Forestry (2001) Ministry of Forestry Decree Number 52/Kpts-II/2001 about the guidance of watershed management practice. In: Forestry MO (ed) Ministry of Forestry, Jakarta

Ministry of Forestry (2009) Ministry of Forestry Decree Number 39/Kpts-II/2009 about the guidance of developing the integrated watershed management plan. In: Ministry of Forestry (ed)

Nieder R, Benbi DK (2008) Carbon and nitrogen in the terrestrial environment. Springer Science & Business Media. https://doi.org/10.1007/978-1-4020-8433-1

Ministry of Public Works and Housing (2010) Keputusan Menteri Pekerjaan Umum No. 266/KPTS/M/2010 tentang Pola Pengelolaan Sumber Daya Air Wilayah Sungai Bengawan Solo, Indonesia. In: M. o. P. W. a. Housing (ed) Jakarta

Nugroho SP (2011) Minimalisasi lahan kritis melalui pengelolaan sumberdaya lahan dan konservasi tanah dan air secara terpadu. Jurnal Teknologi Lingkungan 1(1). https://doi.org/10.29122/jtl.v1i1.165

Owa F (2013) Water pollution: sources, effects, control and management. Mediterr J Soc Sci 4 (8):65

Pawitan H (2004) Perubahan penggunaan lahan dan pengaruhnya terhadap hidrologi Daerah Aliran Sungai. Laboratorium Hidrometeorologi FMIPA IPB, Bogor

Prasena A, Shrestha DP (2013) Assessing the effects of land use change on runoff in Bedog Sub Watershed Yogyakarta. Indones J Geogr 45(1):48

Rafiei Emam A, Mishra BK, Kumar P, Masago Y, Fukushi K (2016) Impact assessment of climate and land-use changes on flooding behavior in the upper Ciliwung River, Jakarta, Indonesia. Water 8(12). https://doi.org/10.3390/w8120559

Ratnasari M, Damarraya A, Rhama DFP, Tosiani A (2020) Rekalkulasi Penutupan Lahan Indonesia Tahun 2019. In: Purwanto J, Pramono J, Limbu ES (eds) Ministry of Environment and Forestry, Jakarta

Ridwansyah I, Yulianti M, Apip Onodera S-I, Shimizu Y, Wibowo H, Fakhrudin M (2020) The impact of land use and climate change on surface runoff and groundwater in Cimanuk watershed, Indonesia. Limnology 21(3):487–498. https://doi.org/10.1007/s10201-020-00629-9

Rukmi NI, Hamidy R, Mubarak M (2013) Pengaruh luas tebangan hutan tanaman akasia terhadap karakteristik hidrograf banjir. Jurnal Ilmu Lingkungan 7(1):68–94

Schweitzer L, Noblet J (2018) Chapter 3.6—water contamination and pollution. In: Török B, Dransfield T (eds) Green chemistry. Elsevier, pp 261–290

Setyorini A, Khare D, Pingale SM (2017) Simulating the impact of land use/land cover change and climate variability on watershed hydrology in the Upper Brantas basin, Indonesia. Appl Geomatics 9(3):191–204. https://doi.org/10.1007/s12518-017-0193-z

Seymour F, Harris NL (2019) Reducing tropical deforestation. Science 365(6455):756. https://doi.org/10.1126/science.aax8546

Sheng TC (1994) Challenges and strategies in integrated watershed management in developing countries. In: 8th international soil conservation conference, New Delhi, India

Silva AC, Barbosa MS, Barral UM, Silva BPC, Fernandes JSC, Viana AJS, Filho CVM, Bispo DFA, Christófaro C, Ragonezi C, Guilherme LRG (2019) Organic matter composition and paleoclimatic changes in tropical mountain peatlands currently under grasslands and forest clusters. CATENA 180:69–82. https://doi.org/10.1016/j.catena.2019.04.017

Singh J, Yadav P, Pal AK, Mishra V (2020) Water Pollutants: Origin and Status. In: Pooja D, Kumar P, Singh P, Patil S (eds) Sensors in water pollutants monitoring: role of material. Springer, Singapore, pp 5–20

Siswamartana S, Utomo WH, Soedjoko SA, Priyono NS, Mulyana NM, Rusdiana O, Pramono IB (2002) Hutan pinus dan hasil air. BPPTPDAS, Surakarta

Sudaryono S (2002) Pengelolaan Daerah Aliran Sungai (DAS) Terpadu, Konsep Pembangunan Berkelanjutan. Jurnal Teknologi Lingkungan 3(2). https://doi.org/10.29122/jtl.v3i2.249.

Susanti PD, Miardini A (2017) The impact of land use change on water pollution index of Kali Madiun Sub-watershed. In: Forum Geografi, vol 31, pp 128–137, vol 1

Susanti PD, Wahyuningrum N (2020) Identification of the main water quality parameters for monitoring and evaluating watershed health. Indones J Geogr 52(2):239–245

Sutrisno N, Heryani N (2013) Teknologi konservasi tanah dan air untuk mencegah degradasi lahan pertanian berlereng. Jurnal Penelitian dan Pengembangan Pertanian 32(3):122–130. https://doi.org/10.21082/jp3.v32n3.2013.p122-130

Tadjoeddin MZ (2007) A future resource curse in Indonesia: the political economy of natural resources, conflict and development. Oxford University

National Statistics Bureau (2019) The critical Land in Indonesia. https://www.bps.go.id/dynamictable/2017/02/08%2000:00:00/1231/luas-dan-penyebaran-lahan-kritis-menurut-provinsi-tahun-2005-2013-ribu-ha-html

Toto Same A (2009) Transforming natural resource wealth into sustained growth and poverty reduction: a conceptual framework for Sub-Saharan African oil exporting countries: The World Bank

Townsend AR, Cleveland CC, Houlton BZ, Alden CB, White JWC (2011) Multi-element regulation of the tropical forest carbon cycle. Front Ecol Environ 9(1):9–17. https://doi.org/10.1890/100047

Tsujino R, Yumoto T, Kitamura S, Djamaluddin I, Darnaedi D (2016) History of forest loss and degradation in Indonesia. Land Use Policy 57:335–347. https://doi.org/10.1016/j.landusepol.2016.05.034

van Noordwijk M, Agus F, Dewi S, Purnomo H (2014) Reducing emissions from land use in Indonesia: motivation, policy instruments and expected funding streams. Mitig Adapt Strat Glob Change 19(6):677–692. https://doi.org/10.1007/s11027-013-9502-y

Viman OV, Oroian I, Fleşeriu A (2010) Types of water pollution: point source and non-point source. Aquaculture Aquarium Conserv Legislation 3(5):393–397

Wahyuningrum N, Putra PB (2019) Degraded land analyses of Brantas River basin to support land rehabilitation. Jurnal Penelitian Kehutanan Wallacea 8(2):135–145. https://doi.org/10.18330/jwallacea.2019.vol8iss2pp135-145

Wang X, Chu X, Liu T, Cheng X, Whittecar R (2017) Water–soil–vegetation dynamic interactions in changing climate. Water 9(10). https://doi.org/10.3390/w9100740

Xu J, Dong J, Wu L, Shao G, Yang H (2015) Land use/cover change and its impact on net primary productivity in Huangfuchuan watershed temperate grassland, China. In: Bian F, Xie Y (eds) geo-informatics in resource management and sustainable ecosystem. Springer, Berlin, Heidelberg, pp 664–683

Zhang M, Wei X (2012) The cumulative effects of forest disturbance on streamflow in a large watershed in the central interior of British Columbia, Canada. Hydrol Earth Syst Sci Discuss 9(3)

Soil and Water Conservation Planning Toward Sustainable Management of Upstream Watershed in Indonesia

Agung Budi Supangat, Cahyono Agus, Nining Wahyuningrum, Dewi Retna Indrawati, and Purwanto

1 Introduction

Indonesia has a very strategic geographical location, on the equator, between two continents, between two oceans, in the Ring of Fire, and on three geological plates (Agus 2020). Indonesia is a large and rich archipelago featuring tropical land and oceans. The land area is 1.9 million km^2 (13th-largest globally), consisting of 17,504 islands and a total sea area of 6.4 million km^2 (BPS 2020). Furthermore, Indonesia has at least 127 active volcanoes, which are part of the world's Ring of Fire, with a mountainous and hilly landscape causing steep upstream and downstream areas.

Indonesia has abundant natural resources and a variety of commodities that are required to cover basic human needs, such as water, air, food, feed, medicine, fiber, wood, energy, and industrial raw materials (Agus et al. 2017, 2019a, b, c, d, e, 2020a). The over-exploitation of natural resources has resulted in numerous environmental and humanitarian disasters (Agus 2013). An increasing number of floods, droughts, landslides, and the water crises do not support sustainable development (Agus 2020). A new paradigm of natural and human resource management is required to support a dignified and sustainable environment and life (Agus 2018).

Watershed management does not only include physical and sectoral factors but must also take on a comprehensive approach from the forestry, water resources, and agricultural sectors (DKKSDA 2020). Moreover, watershed management is constrained by the lack of integration among sectors, agencies, regions, and community participation (Pambudi 2019). Integrated watershed management is required to harmonize management between upstream and downstream, participatory

A. B. Supangat · N. Wahyuningrum · D. R. Indrawati · Purwanto
Watershed Management Technology Center, Surakarta, Indonesia

C. Agus (✉)
Faculty of Forestry UGM, Yogyakarta, Indonesia
e-mail: cahyonoagus@gadjahmada.edu

W. Leal Filho et al. (eds.), *Sustainability in Natural Resources Management and Land Planning*, World Sustainability Series,
https://doi.org/10.1007/978-3-030-76624-5_6

multi-stakeholders, and land conservation. This concept must consider the role of nature, human ecology, land use, reforestation, and sustainable forest management practices (Biswas 1990). Improved multilevel polycentric governance arrangements between the government, NGOs, the private sector, and civil society are required to improve the management of sustainable forest landscapes across Indonesia and the tropics (Riggs et al. 2018). A creative, innovative, and interdisciplinary approach is required to support community-based landscape governance.

The integrated approach also views the importance of community participation in watershed management, from planning, formulating, and implementing policies, to benefiting from the results. The forestry sector in the upstream area is well managed and sustainable. Infrastructure and facilities in the middle stream should support this by increasing downstream watershed functions and benefits (DKKSDA 2020). A collaborative planning process is essential for developing sustainable and ecosystem-based strategies. Participatory and iterative processes, including the development of "actor-based" scenarios, are required to define planning objectives, land use elements, and their relationship to ecosystem services (Karrasch et al. 2017).

The concept of landscape as a social-ecological system generates services and value to stakeholders who share specific geographic areas harmoniously and sustainably (Opdam 2018). Rural areas face various landscape changes and vulnerabilities because of financial and environmental crises. The social capacity of farming plays a role in creating viable and sustainable rural and suburban areas. Some crucial elements of the socio-ecological system in socio-agricultural practices are (1) the role of local communities and non-formal institutions, (2) the involvement of target stakeholders, and (3) a direct relationship between agroecosystems and human well-being (García-Llorente et al. 2016). Social agriculture provides a hybrid governance solution beyond the applicable market-based instruments for governance agroecosystems, including cultural ecosystem services and the well-being of rural and urban residents (Britta 2005).

The management of watershed units from upstream to downstream areas has a biophysical linkage through the hydrological cycle. Land-use changes in the upstream area will impact the downstream area through fluctuations in water flow, water quality, and transport of sediment and dissolved materials. Watershed management must consider the biophysical dimension (erosion control, prevention, control of critical land, conservative agricultural management), the institutional dimension (incentives and regulations relating to the economic sector), and the local social dimension (DKKSDA 2020).

The integrated watershed concept requires the involvement of all stakeholders, and it is planned in an integrated, comprehensive, sustainable, and environmentally friendly management unit (DKKSDA 2020). Management of land and biological resources in tropical ecosystems through genetic and environmental engineering is crucial for managing more productive, conservative, dignified, and sustainable natural resources (Agus et al. 2017, 2019a, b, c, d, 2020a).

Integrated watershed management demonstration plots have been developed in some micro-watersheds, namely Naruan, Wonosari, Pronggo, and Cilampuyang in

Indonesia (Purwanto et al. 2016; Supangat et al. 2019a, b). A participative and collaborative approach was developed in the planning process and soil and water conservation measures (Indrawati 2019; Supangat et al. 2020). Soil and water conservation activities, both vegetative and mechanical techniques, decrease soil erosion and increase community welfare (Purwanto et al. 2016; Supangat et al. 2019a, b).

This chapter discusses the participatory and collaborative planning and management of micro-watersheds (approximately 1000–5000 ha) adapted to local environmental, economic, and socio-cultural characteristics. Nature-based micro-watershed management and local wisdom are essential to ensure sustainable watershed management from upstream to downstream. A micro-watershed model was necessary and was conducted at the right location representing the upstream part, which is generally steep. This micro-based management model remedies land degradation while measuring its economic and environmental impacts.

2 Problem Facing Upper Watershed Area in Indonesia

The upstream watershed has a conservation function that must be appropriately managed to maintain watershed conditions and prevent watershed degradation. This area is usually characterized by dense vegetation cover but a steeper slope. Based on its topography, most of the upper watersheds are protected and retained as rainwater catchment areas to provide significant benefits and minimize the destructive power of the downstream (Asdak 2014; Nepal et al. 2017).

However, many people in Indonesia live in the upper watershed and farm on sloping land (Indrawati et al. 2016). The community's upstream characteristics are usually marginal; the lack of knowledge of the environment and limited land ownership, result in high community dependence on land (Kierch and Faures 2002; Jariyah 2014); therefore, intensive cultivation has overexploited natural resources (Aprisal et al. 2019; Suprayogo et al. 2019). Land management that neglects soil and water conservation aspects will result in erosion, landslides, and degraded land (Achouri 2003; Molla and Sisheber 2017). This phenomenon will impact the downstream area in the form of sedimentation, flooding, and drought. Supangat et al. (2020) stated that the upper watershed can contribute to sediment downstream, due to soil erosion in cultivated lands and settlements. Soil erosion is a severe problem for the upper watershed because it can reduce land quality and productivity, both in Indonesia (Supangat 2019) and in some other countries (Gachene et al. 2011; Li et al. 2015; Molla and Sisheber 2017). In Indonesia, some upper watersheds are in a critical condition, mostly because the existing land cover/use is no longer suitable for its function and usage (Suryatmojo et al. 2019). The average soil erosion rate of Indonesia's upper watershed is more than 100 tons/ha/year (Setyono and Prasetyo 2012; Nugraheni et al. 2013; Chaidar et al. 2017; Soplanit et al. 2018). This condition indicates that most of Indonesia's upper watershed is degraded owing to many dryland farming systems without proper and adequate soil and water conservation measures. The upper watershed with a high elevation and steep topography should have land cover in perennial trees (forests). A forest is believed to be the best land cover for maintaining the hydro-orological function of the area because it protects the land via a stratified canopy and optimal soil coverage. Thus, it can protect the soil from erosion, control the surface flow rate, and improve the microclimate (Asdak 2014; Hofer 2003; Supangat et al. 2016).

However, most of the upstream watersheds in Indonesia have insufficient forest areas to run their recharge area, which is less than 20% on average. This condition can cause land degradation (Kartika et al. 2019; Mulyadi and Jupri 2016; Wahyuningrum and Basuki 2019). Degraded land, erosion, and sedimentation have become significant concerns for reservoir management. They can only be addressed with sustainable management by implementing the best management practices, especially in erosion-prone areas (Adeogun et al. 2020).

3 Soil and Water Conservation to Combat Soil Degradation

The increase in population has created a gap between production and demand for food. Pressure on land with steep slopes forces agricultural production activities on dry land (Pawitan and Haryani 2011). It is necessary to apply appropriate soil and water conservation technology to increase land productivity sustainably and prevent negative impacts on the environment. Soil and water conservation are carried out using an agroecosystem approach, which increases farm income and land productivity while preventing environmental damage. The efforts include minimum tillage, permanent ground cover in the form of plant residues and/or cover crops, and crop rotation (FAO 2010).

Soil and water conservation are two inseparable activities. Soil conservation on sloping areas generally functions as a recharge area, controlling soil loss and functions to increase infiltration, eventually becoming a groundwater supply. The amount of rainfall absorbed into the soil will be the starting point for controlling downstream flooding. Controlling runoffs is the key to water conservation. Runoff and soil loss are the leading cause of on-site and off-site soil degradation that were subsequently causing a decrease in productivity and socio-economic problems (Zuazo and Pleguezuelo 2008). The vegetation affected runoff and sediment by its structure and functions (Zhu et al. 2015). Besides, plant species and patterns affect soil loss (Fu et al. 2011). Therefore, permanent vegetation such as shrubs and other perennials trees should be maintained on the steep slopes (Wahyuningrum 2020). The effects of vegetation on controlling runoff and soil loss are in three aspects: vertical vegetation structures, density, and pattern (Jianbo et al. 2018).

The utilization/arrangement of vegetation cover can be an economically and environmentally sound alternative in the sloping area. In the private land, farmers were reluctant to invest in soil conservation and agronomic improvements as long as they could not be confident of reaping the benefits. Compromises must be made to achieve conservation goals, which allow interaction, both ecological and economical. Agroforestry, the collective term for land-use systems and technologies, combines woody perennials (trees, shrubs, palms, bamboos, and others) and crops in some spatial arrangements or temporal sequence (Lundgren and Raintree 1982), could be the possible method. Some of the advantages of agroforestry, according to FAO (2017), are: (1) providing many ecosystem services, (2) providing permanent tree cover, (3) enhancing livelihoods by providing a variety of food, fodder, and tree products, income and relieve poverty, (4) increasing the resilience to drought & food shortages, and mitigate climate change. Their practice requires legal certainty to guarantee ownership rights to land, incentives to farmers, investment, and marketing of agroforestry products (FAO 2017). Agroforestry should be carried out by selecting plants of different ages or growing periods and have different needs for environmental factors, including water, humidity, light, and plant nutrients (Priyadi et al. 2018). Agroforestry provides higher production than the monoculture system (Amanullah et al. 2016; Rachman and Hani 2014). Agroforestry had a more significant impact when combined with mechanical efforts such as making drainage on the terraces (dead-end ditch) and mulch. According to (Dijk 2002), agroforestry combined with mechanical conservation can reduce erosion and runoff in the range of 50–70%.

Gully erosion is the development of surface erosion (Luffman et al. 2015). Gullies can be managed by controlling water flow, stabilizing channel walls and floors, and stabilizing slopes to prevent gully development. Revegetation of walls and channels is an alternative other than mechanical means (Local Land Service 2018). The impact of soil and water conservation implementation must be measurable and accountable (Manale et al. 2018). The management unit must consider an easily managed area and observed, such as in a micro catchment area.

4 Development Micro Watershed Model

The implementation of watershed management in the field incurs many obstacles because many interested parties have different tasks and functions (Supangat 2019). These difficulties in implementation make it challenging to evaluate the benefits and impacts of the watershed management program on a large scale. One of the watershed management components, soil and water conservation (SWC) planning, is often unable to solve the main problems. Therefore, a manageable watershed management unit is required on a more rational and operational scale during the planning, implementation, monitoring, and evaluation. This unit can be a micro-watershed unit upstream of a watershed. The concept of micro-watershed management has been widely used in various countries. A micro-watershed is a coherent ecosystem unit in the smallest unit of a geographic area with natural characteristics such as slope, soil, drainage, and geomorphology (Shukla 1992;

Ramakrishna 2003). The upstream micro-watershed aims to control runoff and increase groundwater supplies to overcome water scarcity problems (Gunasena 2014). In Indonesia, micro-scale watershed units can be used for the testing of planning systems, monitoring and evaluation systems, and institutional systems to produce manageable models/prototypes of watershed management practices to be replicated as examples elsewhere (Purwanto et al. 2016; Supangat 2019).

Although there is no consensus on the exact size of a micro-watershed area, one report (Purwanto et al. 2016) recommended that a micro-watershed area should be ±1,000 ha in Java; it could be up to ±5,000 ha for areas outside Java, such as Sumatra and Kalimantan. The most important principle related to this area is that hydrologically, the management impact can be measured at the micro-watershed outlet.

Micro-watersheds can be used as the most appropriate planning unit to support sustainable development (Shukla 1992). In Indonesia's hierarchy of development planning, micro-watershed management is expected to become part of the medium-term development plan in Indonesia, which lasts for five years. Annually, it can be planned to manage an area of ±200 ha (Purwanto et al. 2016). Natural resource planning at the micro-watershed scale can contribute to sustainable development because it can integrate various development programs with efficient resource use, equitable access to resources, and decentralized control. The implementation of watershed management in small (micro) units has several advantages because it is easier to implement a participatory planning process that involves village communities and implements SWC steps (Gunasena 2014; Nabi et al. 2020; Supangat et al. 2020). The village administrative area has become a planning unit related to social-community aspects. Delineation of the micro-watershed area, which only covers one to three village administrative areas, facilitates community empowerment, participatory institutional development, and collaboration among related parties (Purwanto et al. 2016). With a small hydrological unit, the impact of management activities will be easier to evaluate quantitatively (Supangat 2019).

Several models can be developed for micro-scale watershed management. Each model is based on experiences from various regions with specific biophysical characteristics and problems. However, the principles and stages of watershed management in implementing SWC measures can be replicated in other places with similar biophysical characteristics. Purwanto et al. (2016) developed a micro-watershed model that focuses on SWC measures in several locations, such as Central and East Java, to overcome land degradation problems. Supangat et al. (2019b) developed a participatory-sustainable micro-watershed management model in the Upper Solo watershed to control soil erosion and increase community welfare. In contrast, the development of "Cilampuyang Conservation Village" in West Java is more oriented toward empowering the village (community and village government) to manage the natural resources of the village by considering SWC aspects (Supangat et al. 2019a).

The essence of micro-watershed management is vegetative SWC action, mechanical SWC to control ditch erosion, and community empowerment. Participatory pattern models in collaborative planning and implementation can support sustainable natural resource management. The implementation of vegetative SWC with agroforestry patterns and mechanical SWC in the form of a gully structure can control on-site erosion that does not become sediment downstream. Albizia (*Pharaseriantes falcataria*) established that agroforestry patterns increase the soil organic matter in the O-horizon by 31.3% after three years of planting and reduce the total soil loss by 33.3% at the end of Albizia rotation (7 years) (Supangat et al. 2019b). The terrace rehabilitation in the Pronggo Micro Watershed increased maize production by 1.8 times (Purwanto et al. 2016), whereas the yield of Albizia wood from agroforestry patterns in the Naruan Micro Watershed increased farmers' income by 39% (Supangat et al. 2019b).

5 Participatory Planning of Upper Watershed Management

It must be realized that humans have a direct or indirect influence on watershed conditions (Kusbiantoro et al. 2015; Kadir et al. 2016; Aulia et al. 2018). For example, the community's behavior in land management, including the selection of vegetation and its conservation practices, will affect the resulting erosion (Suwarto et al. 2012; Kusbiantoro et al. 2015). Therefore, community participation in watershed management has an important role, especially in watersheds, where most of the area is private. Without community participation, conserving soil and water on private land will be relatively challenging to do.

Community participation is active community involvement in the decision-making process starting from planning, utilization of results, and evaluation (Akhmaddhian and Fathanudien 2015). In general, community participation in watershed management is only in the implementing activities, and the participation is mobilized by incentives (Rahadiani et al. 2014; Indrawati et al. 2016). In some cases, community participation in watershed management has started from the planning stage, but participation is only at the informing or consultation level (Dewi 2013; Indrawati et al. 2016). Participation at the informing level is passive participation, where the community only receives information about the government's planning activities. Participation at the consultation level means that the community only answers questions from the government, then the government analyzes problems and makes decisions and planning activities (Ife and Tesoriero 2008; Dewi 2013). Because of the absence of a proper participatory approach, especially in the planning stage, the community did not have a sense of belonging to the activities. The community feels that their interests have not been accommodated, and activities are considered not under community needs. The next impact is the community dependence on the government and discontinuation of activities (Indrawati et al. 2016).

Based on some experiences, community participation in watershed management should start from the planning stage. Participatory planning has been implemented in the management of the Naruan Micro Watershed. In participatory planning, the community learns to identify the existing problems in land management, find solutions, and decide activities to overcome them (Supangat et al. 2020). However, people's decisions are more influence by economic considerations to meet their needs, and less pay attention to environmental aspects (Indrawati 2019). Therefore, participatory watershed management planning cannot be carried out based on bottom-up only; it requires government guidelines based on baseline data on bio-physical and socio-economic conditions. In this case, it is necessary to compromise and integrate between community needs and environmental sustainability so that watershed management's objectives can be achieved (Indrawati 2019; Supangat et al. 2020).

Although community participation in watershed management has not been successful, it has been successful in several places. In general, this participation is carried out based on community local wisdom. Community traditions become a factor in the success of environmental management. The absence of a socio-cultural approach is one reason for the failure of forest and land rehabilitation efforts (Yeny et al. 2016).

Local wisdom in soil and water conservation are values or norms manifested in rituals (traditional ceremonies), suggestions or prohibitions, and even take the form of sanctions in the use of water and land resources (Maridi 2015). There is much local wisdom in Indonesia related to watershed management. Some of this local wisdom include (1) Considers individual trees around the spring in the Wonosadi Forest, Yogyakarta as sacred; (2) Prohibitions and sanctions to maintain the customary forest ecosystem in Karampuang, Sulawesi: (3) Prohibition of entering the prohibited forest in the Baduy tribe, Banten; and (4) Classification of the forest into the forbidden forest (*Rimbo Tuo*) which has a conservation function and *Palak* which has economic functions for the communities around Lake Singkarak (Alanindra 2012; Maridi 2015).

These facts show that community participation in watershed management can not only consider environmental factors but must also pay attention to the social, cultural, and economic factors of the community because they can influence community behavior in decision making and implementation of activities (Lovelace and Rambo 1986; Floress et al. 2015).

6 Collaborative Management to Guarantee Sustainability

A watershed is a special type of common-pool resource: an area defined by hydrological linkages, where optimal management requires the coordinated use of natural resources by all users (Savas 2000). Common-pool resources have characteristics that are difficult to exclude, even though they have individual consumptions (Savas 2000). Natural resources in the watershed cannot be consumed in parallel; therefore, the opportunities for each party are not equal. Furthermore, a report (Savas 2000) states that the allocation of common-pool resources should be managed by public organizations and not market mechanisms.

Natural resource management in the upstream of a watershed involves various parties, including (1) the people as the landowners, (2) local governments responsible for regional development and community welfare, (3) forestry agencies responsible for forest preservation and land rehabilitation, (4) agricultural agencies related to community agricultural land management, (5) agencies related to river management, and (6) private parties involved in exploiting natural resources, especially water resources (Indrawati 2019). The working areas between agencies, for example, forestry and agriculture, overlap. Sloping land should be a forest area but is still used for agricultural activities. Leaders and planning documents are required to direct all institutions and individuals to achieve similar goals. Therefore, collaborative management must be implemented.

Problems found in collaboration with a watershed in Indonesia include: (1) each institution has its objectives (Kartodihardjo 2000) and is not integrated (Raharja 2010), (2) no organization is capable of leading (Alviya et al. 2016), and (3) co-operation between institutions is weak (Alviya et al. 2016). Therefore, it is necessary to have a leader who can compile a micro-watershed management plan and direct all the abilities of stakeholders. A district planning agency has the main tasks, functions, human resource capacity, funding, and methods for performing these tasks. At the implementation level, community participation in managing land is needed. Land managers are referred to as grassroots stakeholders. Lubell (2004) stated that cooperation from grassroots stakeholders is necessary for successful collaborative watershed management. Grassroots stakeholders manage and use natural resources. In practice, not all land managers actively participate in the management.

According to Supangat et al. (2020), there are three reasons why people do not participate in watershed management: (1) they have little knowledge and understanding of the community regarding watershed management, (2) they prioritize household needs and land management must be conducted intensively, which results in land degradation, and (3) the community is uninformed about the potential to improve welfare through integrated and sustainable land management. These problems can be overcome by (1) increasing community knowledge through counseling and training, (2) following field practices of SWC, and (3) selecting cropping patterns and SWC practices that can increase community income.

Collaboration in the evaluation and monitoring process is performed under the institutions' capacity and communities managing the land. The indicators monitored are land, social, economic, institutional, and hydrological. The monitoring and evaluation results should be conveyed to regional leaders through the annual conference of the district development planning forum to provide input to the regent to plan and implement subsequent watershed management and ensure its sustainability. The collaborative upstream watershed management stages allow each micro-watershed to have one plan as the basis for each stakeholder's activities to fulfill the "one-river-one-plan" principle.. Agus (2020) developed an integrated bio-cycle farming system with natural norm-based management of land and biological resources by producing gold crops and livelihoods to develop environmental, economic, socio-cultural, and health aspects (Cahyanti et al. 2019; Agus et al. 2020a).

7 Conclusions

Watershed management in the upstream area must be conducted with conservation forests, which must also be supported by infrastructure and facilities in the middle stream to improve the functions and benefits of the downstream watershed. The planning and management unit of micro-watersheds are expected to empower local potentials in an ideal and rational manner. Participatory and collaborative planning for soil and water conservation techniques must be adapted to local environmental, economic, and socio-cultural characteristics. Nature-based micro-watershed management and local wisdom are crucial to ensure sustainable watershed management from upstream to downstream. As there are many parties involved, a leading institution is necessary to synergy each interest for implementing the management plan without obstacles and to avoid overlapping the implementation of regulations. When applying this concept to other watersheds, we must adjust to the local bio-physical and socio-cultural characteristics, which have different features.

References

Achouri M (2003) Preparing the next generation of watershed management programmes (Chapter 1). In: Achouri M, Tennyson L, Upadhyay K, White R (eds) Preparing for the next generation of watershed management programmes and projects (ASIA). Proceedings of The Asian regional workshop. Kathmandu-Nepal, 11–13 September 2003. Food and Agriculture Organization of the United Nations, Rome

Adeogun AG, Ibitoye BA, Salami AW, Ihagh GT (2020) Sustainable management of erosion prone areas of upper watershed of Kainji hydropower dam, Nigeria. J King Saud Univ Eng Sci 32:5–10. https://doi.org/10.1016/j.jksues.2018.05.001

Agus C (2013) Management of tropical bio-geo-resources through integrated bio-cycle farming system for healthy food and renewable energy sovereignty: Sustainable food, feed, fiber, fertilizer, energy, pharmacy for marginalized communities in Indonesia. In: IEEE Global Humanitarian Technology Conference (GHTC). October, San Jose, CA, USA., pp 20–23

Agus C (2018) Development of Blue Revolution Through Integrated Bio-cycles System on Tropical Natural Resources Management. In: Leal Filho W, Pociovălişteanu D, Borges de Brito P, Borges de Lima I (eds) World sustainability series: towards a sustainable bioeconomy: principles, challenges and perspectives. Springer, Cham, pp 155–172

Agus C (2020) Integrated bio-cycles system for sustainable and productive tropical natural resources management in Indonesia. In: Keswani C (ed) Bioeconomy for sustainable development. Springer Singapore, Singapore, pp 201–216

Agus C, Wulandari D, Primananda E, Hendryan A, Harianja V (2017) The role of soil amendment on tropical post tin mining area in Bangka Island Indonesia for dignified and sustainable environment and life. IOP Conf Ser Earth Environ Sci 83:012030. https://doi.org/10.1088/1755-1315/83/1/012030

Agus C, Primananda E, Faridah E, Wulandari D, Lestari T (2019) Role of arbuscular mycorrhizal fungi and Pongamia pinnata for revegetation of tropical open-pit coal mining soils. Int J Environ Sci Technol 16:3365–3374. https://doi.org/10.1007/s13762-018-1983-5

Agus C, Primananda P, Rufus M (2019d) Integrated bio-cycle system for rehabilitation of open-pit coal mining areas in tropical ecosystems. In: AG. Leal W (ed) World sustainability series. International business, trade and institutional institutional sustainability. Springer Nature Switzerland, pp 515–528

Agus C, Wulandari D, Cahyanti PAB, Bantara I, Hutahaean BP, Lestari T (2019) Environmental site engineering and integrated bio-cycles management for rehabilitation of degraded tin mining land in tropical ecosystem. IOP Conf Ser Earth Environ Sci 398:012013. https://doi.org/10.1088/1755-1315/398/1/012013

Agus C, Azmi FF, Widiyatno IZR, Wulandari D, Rachmanadi D, Harun MK, Yuwati TW (2019) The impact of forest fire on the biodiversity and the soil characteristics of tropical Peatland. In: Leal Filho W (ed) Climate change management book series: handbook of climate change and biodiversity. Springer, Zug, Switzerland, pp 287–303

Agus C, Hendryan A, Harianja V, Faridah E, Atmanto WD, Cahyanti PAB, Wulandari D, Pertiwiningrum A, Suhartanto B, Bantara I, Hutahaean BP, Suparto B, Lestari T (2019b) Role of organic soil amendment of paramagnetic humus and compost for rehabilitation of post tin-mined tropical land. Int J Smart Grid Clean Energy 8:556–561. https://doi.org/10.12720/sgce.8.5.556-561

Agus C, Ilfana ZR, Azmi FF, Rachmanadi D, Widiyatno WD, Santosa PB, Harun MK, Yuwati TW, Lestari T (2020) The effect of tropical peat land-use changes on plant diversity and soil properties. Int J Environ Sci Technol 17:1703–1712. https://doi.org/10.1007/s13762-019-02579-x

Agus C, Cahyanti PAB, Widodo B, Yulia Y, Rochmiyati S (2020a) Cultural-based education of Tamansiswa as a locomotive of Indonesian education system. In: Leal Filho W (ed) Universities as living labs for sustainable development. World Sustainable Series. Springer, Cham, pp 471–486

Akhmaddhian S, Fathanudien A (2015) Community participation in manifesting Kuningan as a conservation regency (Study in Kuningan Regency). J Ilmu Huk 2:67–90. https://doi.org/10.25134/unifikasi.v2i1.26

Alanindra S (2012) Vegetation Analysis in The Cokro and Umbul Nila Springs Catchment Area in Klaten Regency, also Mudal and Wonosadi, Gunungkidul Regency (Master-Thesis). Gadjah Mada University, Yogyakarta (ID) (Unpublished)

Alviya I, Suryandari EY, Maryani R, Muttaqin MZ (2016) Enhancing the role of stakeholders in the management of upstream Ciliwung Watershed (In Indonesian). J Penelit Sos dan Ekon Kehutan 13:121–134. https://doi.org/10.20886/jpsek.2010.7.2.93-100

Amanullah KF, Muhammad H, Jan AU, Ali G (2016) Land equivalent ratio, growth, yield and yield components response of mono-cropped vs. inter-cropped common bean and maize with and without compost application. Agric Biol J N Am 7:40–49. https://doi.org/10.5251/abjna.2016.7.2.40.49

Aprisal, Istijono B, Juniarti, Harianti M (2019) Soil quality index analysis under horticultural farming in Sumani upper watershed. Int J GEOMATE 16:191–196. https://doi.org/10.21660/2019.56.8212

Asdak C (2014) Hydrology and watershed management (in Indonesian). Gadjah Mada University Press, Yogyakarta, p 615

Aulia F, Harahap RH, Absah Y (2018) Community perception of the Deli River existence in Medan. J Pembang Perkota 6:35–39

Biswas AK (1990) Watershed management. Int J Water Resour Dev 6:240–249. https://doi.org/10.1080/07900629008722479

BPS (2020) Statistical Yearbook of Indonesia. Jakarta, Indonesia

Britta GB (2005) Predicting ecological connectivity in urbanizing landscapes. Environ Plan B Plan Des 32:763–776

Cahyanti PAB, Widiastuti K, Agus C, Noviyani P, Kurniawan KR (2019) Development of an edutainment shaft garden for integrated waste management in the UGM green campus. IOP Conf Ser Earth Environ Sci 398.: 012001. https://doi.org/10.1088/1755-1315/398/1/012001

Chaidar AN, Soekarno I, Wiyono A, Nugroho J (2017) Watershed Erosion Analysis (Case Study of Citarum Upstream Watershed). In: Taufik T, Al E (eds) ICoSI 2014. Springer, Singapore, pp 483–497

Dewi AK (2013) Community participation in land rehabilitation in the community seedling program in Plukaran Village, Gembong District, Pati Regency. J Pembang Wil Kota 9:42–52. https://doi.org/10.14710/pwk.v9i1.6525

Dijk AIJM V. (2002) Water and sediment dynamics in bench-terraced agricultural steeplands in West Java, Indonesia (Ph.D Theses). Vrije Universiteit Amsterdam, Enschede, The Netherland (Unpublished)

DKKSDA (2020) Model Pengelolaan Daerah Aliran Sungai (DAS) Terpadu. Info Kajian Bappenas 1(2):55–62

FAO (2010) Conservation agriculture. FAO, Rome, Italy. Retrieved from http://www.fao.org/ag/ca. Accessed 20 Dec 2020

FAO (2017) Agroforestry for landscape restoration: exploring the potential of agroforestry to enhance the sustainable and resilience of degraded lanscapes. FAO, Rome, Italy

Floress K, Akamani K, Halvorsen KE, Kozich AT, Davenport M (2015) The role of social science in successfully implementing watershed management strategies. J Contemp Water Res Educ 85–105. https://doi.org/10.1111/j.1936-704x.2015.03189.x

Fu BJ, Liu Y, Lu YH (2011) Assessing the soil erosion control service of ecosystems change in the Loess Plateau of China. Ecol Complex 8:284–293

Gachene CKK, Nyawade SO, Karanja NN (2011) Soil and water conservation: an overview. In: Leal Filho W et al (ed) Zero hunger. Springer Nature Switzerland, pp 115–142

García-Llorente M, Rossignoli CM, Di Iacovo F, Moruzzo R (2016) Social farming in the promotion of social-ecological sustainability in rural and periurban areas. Sustainability 1238

Gunasena C (2014) Micro catchment management and its importance in the next century. https://www.researchgate.net/publication/260268199. Accessed 20 Dec 2020

Hofer T (2003) Sustainable use and management of freshwater resources: the role of forest. State of The World's Forest 2003, Part II: selected current issues in the forest sector. FAO Forestry Department/FO-0285/T, Rome, 12 pp

Ife JW, Tesoriero F (2008) Community development: community-based alternatives in an age of globalisation. Pearson Education, Australia, 374 pp

Indrawati DR (2019) Participation and collaboration of parties in the management of the Naruan Micro Watershed (In Indonesia). In: Supangat AB, Darmawan IWS (eds) Building integration of watershed management at the implementation level. IPB-Press, Bogor, pp 45–55

Indrawati DR, Awang SA, Faida LRW, Maryudi A (2016) Community empowerment in micro watershed management: concept and implementation (in Indonesian). J Kawistara 6:113–224

Jariyah NA (2014) Characteristics of society Pengkol Sub-Watershed in relation with watershed management (Case study in Pengkol Sub-Watershed, Wonogiri Regency, Central Java). J Penelit Sos dan Ekon Kehutan 11:59–69. https://doi.org/10.20886/jsek.2014.11.1.59-69

Jianbo L, Guangyao G, Shuai W, Lei J, Xing W, Bojie F (2018) The effects of vegetation on runoff and soil loss: Multidimensional structure analysis and scale characteristics. J Geogr Sci 28:59–78. https://doi.org/10.1007/s11442-018-1459-z

Kadir S, Badaruddin, Nurlina, Ridwan I, Rianawaty F (2016) The recovery of Tabunio Watershed through enrichment planting using ecologically and economically valuable species in South Kalimantan, Indonesia. Biodiversitas 17:140–147. https://doi.org/10.13057/biodiv/d170121

Karrasch L, Maier M, Kleyer M, Klenke T (2017) Collaborative landscape planning: co-design of ecosystem-based land management scenarios. Sustainability 9:1668. https://doi.org/10.3390/su9091668

Kartika T, Dirgahayu D, Sari IL, Parsa IM, Carolita I (2019) Evaluation of critical land rehabilitation based on Lansat-8 NDVI trend (in Indonesian). J Penginderaan Jauh dan Pengolah Data Citra Digit 16:61–70. https://doi.org/10.30536/j.pjpdcd.2019.v16.a3079

Kartodihardjo H (2000) Institution analysis of watershed management and soil conservation in Indonesia (In Indonesia). Koperasi Sodaliti Press, Bogor, p 101

Kierch B, Faures JM (2002) Land-water linkages in rural watersheds. FAO L. Water Bul. - 9. Food Agric. Organ. United Nations. Rome, 90 pp

Kusbiantoro A, Awang SA, Maryudi A, Gunawan T (2015) Degradation and land-use system of Tulis Watershed. J Wana Trop 5:15–25. https://doi.org/10.1088/1751-8113/44/8/085201

Li X, Wang H, Zhang L, Wu B (2015) Soil erosion and sediment-yield prediction at basin scale in upstream watershed of Miyun Reservoir. J Hydrol Eng 20(6) C6014003: 1–7. https://doi.org/10.1061/(asce)he.1943-5584.0001098

Local Land Service (2018) Gully erosion assessment and control guide. National Landcare Program, New South Wales, Australia. 36 pp

Lovelace G, Rambo A (1986) Behavioral and social dimensions. In: Watershed resources management: an integrated framework with studies from Asia and The Pacific. East-West Center, Environment and Policy Institute, Honolulu, Hawaii, pp 81–91

Lubell M (2004) Collaborative watershed management: a view from the grassroots. Policy Stud J 32:341–361. https://doi.org/10.1111/j.1541-0072.2004.00069.x

Luffman IE, Nandi A, Spiegel T (2015) Gully morphology, hillslope erosion, and precipitation characteristics in the Appalachian Valley and Ridge province, southeastern USA. CATENA 133:221–232. https://doi.org/10.1016/j.catena.2015.05.015

Lundgren BO, Raintree JB (1982) Sustained agroforestry. In: Nestel (ed) Agricultural research for development: potentials and challenges in Asia. The Hague, the Netherlands: ISNAR., pp 37–39

Manale A, Sharpley A, DeLong C, Speidel D, Gantzer C, Peterson J, Adusumili N (2018) Principles and policies for soil and water conservation. J SOIL WATER Conserv 73:96A-99A. https://doi.org/10.2489/jswc.73.4.96A

Maridi (2015) Using culture and local wisdom in soil and water conservation. In: The XII national conference on biology education UNS, pp 20–39

Molla T, Sisheber B (2017) Estimating soil erosion risk and evaluating erosion control measures for soil conservation planning at Koga watershed in the highlands of Ethiopia. Solid Earth 8:13–25. https://doi.org/10.5194/se-8-13-2017

Mulyadi A, Jupri (2016) Study on critical land in Cikeruh Sub Watershed at Bandung Basin Area (in Indonesian). In: Prosiding Seminar Nasional Geografi UMS 2016: Upaya Pengurangan Risiko Bencana Terkait Perubahan Iklim. Universitas Muhamadiyah Surakarta-Press, Surakarta, pp 238–249

Nabi G, Hussain F, Wu RS, Nangia V, Bibi R (2020) Micro-watershed management for erosion control using soil and water conservation structures and SWAT modeling. Water (switzerland) 12:1–25. https://doi.org/10.3390/w12051439

Nepal S, Neupane N, Shrestha H, Tharu BR (2017) Upstream-downstream Linkages Catchment Level Water Use Master Plans (WUMP) in the Mid-hills of Nepal. ICIMOD Working Paper 2017/23. Kathmandu: ICIMOD, 33 pp

Nugraheni A, Sobriyah, Susilowati (2013) Comparation of erosion rate predictions using USLE, MUSLE, RUSLE methods in Keduang Watershed (in Indonesian). e-Jurnal MATRIKS Tek SIPIL 1:318–325. https://doi.org/10.20961/mateksi.v1i3.37537

Opdam P (2018) Exploring the role of science in sustainable landscape management: an introduction to the special issue. Sustainability 10:331–337

Pambudi AS (2019) Watershed management in Indonesia: a regulation, institution, and policy review (in Indonesian). Indones J Dev Plan 3:185–202. https://doi.org/10.36574/jpp.v3i2.74

Pawitan H, Haryani GS (2011) Water resources, sustainability and societal livelihoods in Indonesia. COHYDROLOGY Hydrobiol 11:231–243. https://doi.org/10.2478/v10104-011-0050-3

Priyadi R, Nuryati R, Sumarsih E, Faqihuddin (2018) Agroforestry patterns applied by farmers in South Tasikmalaya District (In Indonesia). In: Rachman E et al (eds) Prosiding seminar Nasional Agroforestry "IPTEK Agroforestry mendukung Hutan Rakyat Lestari dan Jabar Sejahtera. Balai Teknologi Agroforestry, Dinas Kehutanan Provinsi Jawa Barat, Masyarakat Agroforestri Indonesia, Jatinangor p 645–654

Purwanto, Harjadi B, Supangat AB (2016) Learning from experience: micro-watershed management (in Indonesian). UNS Press, Surakarta. p 188

Rachman E, Hani A (2014) Agroforestry pattern of Sengon (Falcataria moluccana L.) and curly red chilli in Ciamis Upland Area. West Java. J Penelit Agrofor 2:35–44

Rahadiani AAS, Dharma IG, Norken I (2014) Community participation around Lake Beratan in conservation of water resources. J Spektran 2:41–49

Raharja SJ (2010) Collaborative approach in the Citarum Watershed Management (In Indonesia). J Bumi Lestari 10:222–232

Ramakrishna N (2003) Production system planning for natural resource conservation in a micro-watershed. Electron Green J 1(18): 1–11. http://egj.lib.uidaho.edu/egj18/nallathiga1.html

Riggs RA, Langston JD, Margules C, Boedhihartono AK, Lim HS, Sari DA, Sururi Y, Sayer J (2018) Governance challenges in an Eastern Indonesian forest landscape. Sustainability 10:169:1-18

Savas E (2000) Privatization and public-private partnership. Chatam House Publisher, London, p 368

Setyono E, Prasetyo B (2012) Analysis of erosion hazard level in Lesti Sub Watershed, Malang District using Geographical Informaton System (in Indonesian). Media Tek Sipil 10:114–127

Shukla PR (1992) A Multiple objective model for sustainable micro-watershed planning with application. Indian Institute of Management Ahmedabad. Research and Publication Department in its series IIMA Working Papers with number 1058. 10 Mei 2007. http://ideas.repec.org/p/iim/iimawp/1058.html

Soplanit R, Putinella J, Siwalette ES (2018) Erosion prediction and landuse planning in Upper Wai Illa Watershed, Amahusu Village-Nusaniwe Sub District, Ambon City (in Indonesian). Agrologia 5:22–29. https://doi.org/10.30598/a.v5i1.194

Supangat AB (2019) Integration of management toward watershed sustainability: an introduction. In: Supangat AB, Dharmawan IWS (eds) Building integration of watershed management at the implementation level. IPB-Press, Bogor, pp 1–20

Supangat AB, Sudira P, Poedjirahajoe E (2016) Effect of harvesting of Eucalyptus pellita F. Muell forest stand in increasing river flow and sediment (in Indonesian). J Penelit Hutan Tanam 13:113–122

Supangat AB, Donie S, Wahyuningrum N, Purwanto, Indrawati DR (2019b) Development participatory management of micro catchment at the bengawan solo upper watershed. Technical Report No. 2017P6-INA-TR. Collaboration research: Asia-Pacific Network for Sustainable Forest Management and Rehabilitation—Watershed Management Technology Center. Surakarta, Indonesia. http://www.apfnet.cn/en/show-model6-981.html

Supangat AB, Donie S, Purwanto, Wahyuningrum N, Indrawati DR (2019a) Pengembangan Teknik Konservasi Tanah dan Air (KTA) di Desa RHL Cilampuyang (Development of Soil and Water Conservation Techniques in Cilampuyang Conservation Village). Laporan Kegiatan Pengembangan. Balai Penelitian dan Pengembangan Teknologi Pengelolaan Daerah Aliran Sungai. Surakarta, Indonesia (Unpublished), 30 pp

Supangat AB, Indrawati D, Wahyuningrum N, Purwanto, Donie S (2020) Developing a participatory planning process of micro-watershed management: a lesson learned (in Indonesian). J Penelit Pengelolaan Drh Aliran Sungai 4:17–36. https://doi.org/10.20886/jppdas.2020.4.1.17-36

Suprayogo D, Prayogo C, Saputra DD, Sari RR, Nugraha A, Hadiwijoto E, Andhika Y, Ishaq RM, Purnamasari E, Irawan DB (2019) The capacity of community on running soil and water conservation in Bangsri micro-catchment, Upper Brantas Watershed, Indonesia. IOP Conf Ser Earth Environ Sci 393:1–18. https://doi.org/10.1088/1755-1315/393/1/012054

Suryatmojo H, Putra GM, Satriagasa MC (2019) Investigation of direct runoff in the intensive farming system of upstream Merawu watershed, Indonesia. IOP Conf. Ser.: Earth Environ. Sci. 407 012013. https://doi.org/10.1088/1755-1315/407/1/012013

Suwarto S, Suwarto S, Anantanyu S (2012) Participation model of dryland farmer in land conservation. J Econ Pembang 13:218–234

Wahyuningrum N (2020) Permanent vegetation distribution: indicators of watershed management. IOP Conf. Ser.: Earth Environ. Sci. 451 012003

Wahyuningrum N, Basuki TM (2019) Analysis in degraded land for rehabilitation planning in Upper Solo Watershed (in Indonesian). J Penelit Pengelolaan Drh Aliran Sungai 3:27–44. https://doi.org/10.20886/jppdas.2019.3.1.27-44

Yeny I, Yuniati D, Khotimah H (2016) Local wisdom and practice of bamboo forest management in Bali community (in Indonesian). J Penelit Sos Dan Ekon Kehutan 13:63–72

Zhu HX, Fu BJ, Wang S (2015) Reducing soil erosion by improving community functional diversity in semi-arid grasslands. J Appl Ecol 52:1063–1072

Zuazo VHD, Pleguezuelo CRR (2008) Soil-erosion and runoff prevention by plant covers: a review. Agron Sustain Dev 28:65–86

The Integration of Energy Planning and Urban Planning: A Synergies Analysis of the Achievement of the Sustainable Development Goals

Flávia Mendes de Almeida Collaço and Rafael Barreto Castelo da Cruz

1 Introduction

The Paris Agreement (UN 2015) recognizes adaptation to climate change as a "global challenge faced by all, with local, subnational, national, regional, and international dimensions." Adding to the complexity of climate change mitigation actions is the 2030 Agenda: 17 Sustainable Development Goals (SDGs) and 169 targets that constitute the global voluntary action agenda until 2030 (United Nations Development Program 2015). In this sense, the Urban Energy Planning (UEP), as well as the Urban Energy Systems (UESs), is a topic of growing research interest, since it presents itself as a possible way to achieve the development of sustainable and/or low carbon cities under the motto, "*think globally, act locally.*"

Historically, cities have exhibited a universal trend. Since 2007, for the first time in history, more than half of the world's population has lived in urban areas (Madlener and Sunak 2011; Morlet and Keirstead 2013; Oldenberg et al. 2015). Based on the current trend, by 2050, two-thirds of the world's population will live in cities (UN-Habitat 2016). In Brazil, the residential, service, industrial, and transport (public and individual, and air and land transport) sectors demanded a total of approximately 206.9 million tons of oil equivalent in 2017 (Brazilian Ministry of Mines and Energy 2018). In other words, urban areas account for 80% of the energy demand in Brazil.

F. M. de A. Collaço (✉)
School of Arts Science and Humanities (EACH), University of São Paulo (USP), São Paulo, Brazil
e-mail: flavia.collaco@usp.br

R. B. C. Cruz
Department of Civil and Urban Construction Engineering of the Polytechnic School (POLI), University of São Paulo (USP), São Paulo, Brazil
e-mail: rafaelcastelo@usp.br

W. Leal Filho et al. (eds.), *Sustainability in Natural Resources Management and Land Planning*, World Sustainability Series,
https://doi.org/10.1007/978-3-030-76624-5_7

In addition, cities currently occupy less than 5% of the earth's surface but demand approximately 80% of the available natural resources (Rosales Carreón and Worrell 2018). Given the magnitude of cities' impact, urban regions are a potential locus for the transition of current energy systems—including electric power systems, natural gas networks, hydrogen production and transportation, district heating, refrigeration systems, electrified transportation, and the associated information and communication infrastructure (Wu et al. 2016)—to low-carbon ones. From this perspective, the integration of urban planning (UP) and energy planning (EP) strategies can help to increase access to energy, as well as provide a way to improve the well-being of the cities' inhabitants (Rutter and Keirstead 2012), mitigating global emissions, reducing energy demand (Yazdanie et al. 2017), and finally acting on both the energy demand and supply sides (Rodríguez-Rodríguez et al. 2015; Collaço et al. 2020).

According to some studies, UEP is a way to build more sustainable cities (Jovanovic et al. 2010), possibly leading to gains in terms of energy conservation (EC), energy efficiency (EE; Collaço et al. 2019a, b), and lower greenhouse gas (GHG) emissions (Keirstead and Calderon 2012). Finally, it can promote the development of local energy production (Adhikari et al. 2012).

This study assumes the results from Collaço et al. (2019a), that quantifies and implements 29 EP and UP strategies, selected through a literature review, through *the simulation model long-range energy alternatives planning* (LEAP) *system* (Heaps 2006). All the strategies were applied to the São Paulo megacity UES. This modeling exercise aimed to compare the impact of the strategies' implementation on energy savings, reductions in GHG emissions, and increase in local energy generation under four scenarios: (i) a reference scenario (C_REF), (ii) a scenario that implements EP strategies (C_EP), (iii) a scenario that implements UP strategies (C_UP), and (iv) a scenario that integrates both UP and EP strategies (C_UEP).

In the second stage, we classified the 29 EP and UP strategies into 13 categories, according to Collaço et al. (2019a). Subsequently, in addition to the modeling results, we analyzed the degree of achievement of the SDG goals based on four indices according to Cruz and Marins' methodological proposal (PMCM; Cruz 2019). The first index analyzes the comprehensiveness capacity of the strategies, as we sought to identify the strategies that achieve the largest number of SDG goals simultaneously. The second index concerned effectiveness, so as to determine which strategies have been the most effective in addressing the SDGs. The third index evaluates the joint performance of strategies for a set of problems (in this case, the 17 SDGs). Finally, the fourth index assesses the capacity of all UEP strategies to effectively achieve the 17 SDGs. Thus, it was possible to determine the overall performance of the strategies in achieving the SDGs goals, considering their level of coverage and ability to solve/address the goals.

Finally, the following four indices were used to hierarchize the strategies and to evaluate their overall performance: comprehensiveness—strategy comprehensiveness index (IAA); efficiency—strategy efficiency index ($IEA_{(e)}$); completeness—quality/completeness evaluator (AQC); and a problem service evaluator (AAP).

These indicators were summarized in a panel to analyze the strategies correlation (Cruz 2019).

This chapter is structured as follows: After this introduction, Sect. 2 describes the materials and methods used in the modeling and presents the scope of the case study, the city of São Paulo. Section 3 presents and discusses the results of the model and its relationship with the achievement of the Agenda 2030 goals by considering the four aspects of performance. Finally, Sect. 4 concludes the chapter and presents possible future lines of research.

2 Materials and Methods

The case study was conducted through the following steps: (i) a literature review of the selection of EP and UP strategies and the development of scenarios; (ii) establishment of the LEAP_SP model (modeling the São Paulo megacity UES), whose analysis period runs from 2014 (base year) to 2030 (final year of simulation); (iii) quantification of the UEP implementation potential in the context of a megacity; and (iv) critical analysis of the modeling results in light of the achievement of the SDGs goals based on the PMCM.

The PMCM consists of the application of structured panels (see an example in Fig. 1) including requirements and solutions, and the results obtained in the panels' construction are used determine an indicator, or index, that assists in hierarchizing and determining to what degree the solutions address the problems and demands (Cruz 2019). In this study, we applied this method to 13 strategies that summarize the 29 strategies modeled in the first part of the study in order to verify the individual scope of each strategy in the SDGs and the overall performance of the set of strategies.

The application of the structured panel consists of a matrix in which each row represents one of the simulated strategies and each column one of the 17 SDGs. The strategies were transcribed sequentially. The same procedure was performed for the SDGs, which were sequentially transcribed into columns.

Therefore, each column represents an SDG, and each row represents a UEP strategy. The panel was completed by experts, who classified the degree to which each of the UEP strategies addressed each of the SDGs into the following categories: (i) Addressed directly and partially (DP), when the UEP strategy directly addressed the scope of an SDG but without "achieving" it completely. (ii) Addressed indirectly and partially (IP), used when the strategy indirectly addressed an SDG without completely achieving it. (iii) Addressed directly and completely (DC), was used when the strategy directly addressed an SDG, completely achieving it. (iv) Addressed indirectly and completely (IC) was used when the strategy indirectly addressed an SDG and completely achieved it. (v) Null (N) was used when, in the opinion of the experts, the strategy did not address the given SDG.

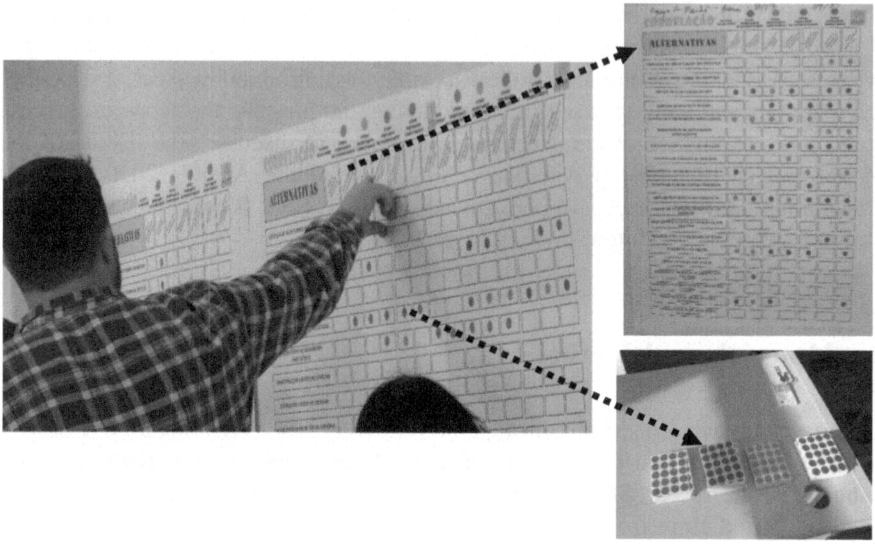

Fig. 1 Example of the application of a structured panel for the analysis and correlation of strategies

The counts of each category (DP, IP, DC, IC, and N) for each strategy were then calculated. Through the application of the PMCM, the UEP strategies were arranged hierarchically according to the IAA and the $IEA_{(e)}$ indices, which demonstrated which were the most comprehensive and effective strategies for achieving the SDGs. In turn, the AQC and the AAP indices globally evaluated the degree to which the set of UEP strategies achieved the SDGs.

The panel was validated using a response convergence technique applied by subject experts, with a Delphi approach used in applying the proposed legend. From the data obtained in the analysis, the aforementioned indices (IAA, $IEA_{(e)}$, AQC, and AAP) were calculated using the following equations:

(a) **Strategy comprehensiveness index (IAA)**: This index evaluates which strategies address the largest number of SDGs at the same time. It is calculated as the ratio of the total sum of actions of the strategy (UEP strategy) and the total sum of the problems (SDGs), according to Eq. (1):

$$IAA = \frac{\sum_{i=0}^{n} DC_i + \sum_{j=0}^{m} DP_j + \sum_{k=0}^{z} IC_k + \sum_{w=0}^{l} IP_w}{\sum p} \tag{1}$$

where

IAA Strategy comprehensiveness index;

$DC_{(i)}$ This indicates whether the *i*th strategy was classified into the "direct and complete" category;

$DP_{(j)}$ jth direct and partial marking in the strategy;
n Total number of strategies classified into the direct and complete category;
m Total number of direct and partials markings in the strategy;
$CI_{(k)}$ kth indirect and complete marking in the strategy;
z Total number of indirect and complete markings in the strategy;
$IP_{(w)}$ wth indirect and partial marking in the strategy;
l Total number of indirect and incomplete markings in the strategy;
P Total number of SDGs.

(b) **Strategy efficiency index ($IEA_{(e)}$)**: This index is used to hierarchize the most effective strategies, considering effectiveness as the ability to produce or achieve the expected objective or result. Therefore, this index assesses how effective the proposed strategies are. Its formula is

$$IEA_{(e)} = \frac{\sum_{i=0}^{n} DC_i}{\sum_{i=0}^{n} DC_i + \sum_{j=0}^{m} DP_i + \sum_{k=0}^{z} IC_k + \sum_{w=0}^{l} IP_w} \tag{2}$$

where

$IEA_{(e)}$ Strategy efficiency index;
$DC_{(i)}$ ith direct and complete marking in the strategy;
$DP_{(j)}$ jth direct and partial marking in the strategy;
n Total number of direct and complete markings in the strategy;
m Total number of direct and partial markings in the strategy;
$CI_{(k)}$ kth indirect and complete marking in the strategy;
z Total number of indirect and complete markings in the strategy;
$IP_{(w)}$ wth indirect and partial marking in the strategy;
l Total number of indirect and partial markings in the strategy.

(c) **Quality/completeness evaluator (AQC)**: This index evaluates the joint performance of a set of strategies for a set of problems (in this case the 17 SDGs). It helps to assess the ability of a combination of strategies to achieve a set of goals. It is given by the ratio of the number of goals that are completely achieved by the evaluated strategies and the total number of goals to be achieved. The value of this index is a positive real number, where the higher the number, the greater the degree of completeness in achieving the 17 SDGs. The formula for this index is

$$AQC = \frac{\left(\sum_{i=0}^{N} DC_i + \sum_{i=0}^{N} IC_i\right)}{N} \tag{3}$$

where

AQC Quality/completeness evaluator;
$DC_{(i)}$ *i*th direct and complete marking;
$CI_{(i)}$ *i*th indirect and complete marking;
N Total number of goals.

(d) **Problem service evaluator (AAP)**: This index assesses the capacity of the combination of all the UEP strategies to effectively achieve the 17 SDGs. It is calculated as the ratio of the total number of goals achieved completely/partially and directly/indirectly and the total number of goals. The index is not simply based on the number of strategies proposed but on the number of strategies that address the problems listed, as can be observed in Eq. (4).

$$AAP = \frac{\sum_{i=0}^{N} DC_i + \sum_{i=0}^{N} IC_i + \sum_{i=0}^{N} IP_i + \sum_{i=0}^{N} DP_i}{N} \tag{4}$$

where

AAP Problem service evaluator;
$DC_{(i)}$ *i*th direct and complete marking;
$DP_{(i)}$ *i*th direct and partial marking;
$CI_{(i)}$ *i*th indirect and complete marking;
$II_{(i)}$ *i*th indirect and partial marking;
N Total number of goals.

2.1 Case Study: City of São Paulo

The geographic scope of the case study is the municipality of São Paulo. The city is the main financial, corporate, and commercial center of South America (Pimenta 2010). It is the largest and most populous city in Brazil, accounting for 5.9% of the country's population, that is, approximately 12 million inhabitants (Brazilian Institute of Geography and Statistics (IBGE) 2010). São Paulo is a strategic economic hub for Brazil, and in 2011 contributed almost 12% of the national GDP. In addition, it has the largest industrial park in the country, being home to 63% of the multinationals established in Brazil (São Paulo Secretariat of International Relations 2012).

From 2007 to 2017, São Paulo was the largest consumer of electricity and natural gas in the entire state (Department of Energy of the State of SP 2007, 2008, 2010, 2012, 2013, 2014, 2015, 2016, 2017). Further, according to the 2010 census, 11% of the city's population lives in subnormal housing, with limited access to energy services (e.g., electric lighting, refrigeration, cooking, and transportation) and urban sanitation and municipal solid waste collection (IBGE 2010). This limited access to energy services is one of the challenges that the city of São Paulo needs to face to transition to a more sustainable urban energy system.

The sectors analyzed through the model are households, public infrastructure (buildings—PB and public lighting—PL), commercial and services (C&S), industry, sanitation (WT), urban mobility (UM), and air transport, and the energy sector itself (internal electricity consumption for transmission and distribution, and natural gas activities for cogeneration). In addition to the city's UES aspects, the model developed aimed to determine the aspects of EP and UP that impact the energy demand (ED) and supply.

The establishment of a UEP model for the São Paulo city aimed to measure the possible impacts of combinations of different policies that include, in addition to the traditional strategies considered within EP, strategies and drivers of UP policies that influence the ED in cities. These impacts were measured in terms of (i) energy savings through improvements in EE and EC; (ii) promotion of distributed generation and increased use of local city resources; and (iii) reduction in GHG and pollutant emissions at the local level, through the

We assumed that understanding the effects of cities' UES can promote public policies that focus on a more balanced urban environment so as to construct guidelines for the improvement of citizens' lives and promote healthier cities, as such measures would positively affect citizens' health (Saldiva 2018).

2.2 Design of Scenarios Modeled Through LEAP_SP

To analyze the impact of the adoption of different policies and strategies of UP and EP within the UES of São Paulo we use the results presented by Collaço et al. (2019a), in which four scenarios were modeled: (i) C_REF, (ii) C_EP, (iii) C_UP, and (iv) C_UEP.

The C_REF scenario considers ongoing policies (urban and energy), population's socioeconomic situation and applies historically observed rates of energy demand and supply in the city (considering the period of 2007–2016). More information on methodology used see Collaço (2019a).

The C_EP and C_UP scenarios adopted the same growth rates as in C_REF, but additional combinations of urban and energy policy strategies were simulated in each of these two scenarios. Finally, the C_UEP scenario combined UP and EP policy strategies to quantify the synergies obtained by this integration. For more information on the scenario's assumption see Collaço et al. (2019a).

We simulated 29 strategies organized as follows:

Strategies for saving energy: technological replacement, i.e.: more efficient equipment's for refrigeration, lighting, pumping, water heating etc., were simulated inside C_EP and C_UEP scenarios.

Strategies using green infrastructure/passive architecture reducing the energy needs: natural lighting, water reuse and new green areas in the city (more parks), were simulated inside C_UP and C_UEP scenarios.

Energy management strategies were simulated for buildings, industrial and water treatment sector's inside the C_EP and C_UEP scenarios.

Because transport sector is a major source of energy demand and GHG emissions, some strategies were strategically developed for the sector, such as no use of fossil fuels on public transportation (C_EP and C_UEP); all city taxis electrification; replacement of individual transportation by more public transportation; more use of bicycles and other active alternatives, were simulated inside C_UP and C_UEP.

Lastly, we simulated also strategies for the transformation side, such as: fewer losses in the grid, retrofitting old power plants, more decentralized generation (electricity from: PV rooftops, MSW, Sewage Sludge, Pruning Waste, Biomass waste and Livestock waste). A complete description of each strategy is available on Collaço, et al. (2019a).

3 Results and Discussion

In 2014, the city consumed 367 PJ, with 58% consumed by the urban mobility sector. The residential sector accounted for the second largest consumption (15%) and C&S the third (13%). The most demanded energy resources were electricity (36%), gasoline (25%), and diesel (16%). A 72% (approximately 630 PJ) growth in the city's final ED is expected by 2030.

On the supply side, in 2014, the city had an installed electricity generation capacity of 901.5 MW within its limits (BIG, ANEEL 2017), and according to LEAP_SP calculations, was able to provide 4 TWh per year. Finally, the GHG emission data show a 43% growth (from 1.8 tCO_2e./inhabitant in 2014 to 2.6 tCO_2e/inhabitant in 2030). This result goes against the city's municipal climate change policy (Prefeitura do Município de São Paulo-PMSP 2009)

Table 1 presents the evolution of FEC in the city. The largest reduction in the estimated FEC was observed in the C_UEP scenario, which decreased the city's FEC by 74 per year in 2030, that is, a 12% reduction in ED compared with the C_REF scenario for 2030.

In the C_UP and C_EP scenarios, it was observed that PU strategies resulted in a 10% reduction in ED in 2030, while PE strategies resulted in a demand reduction of only 2% compared with the C_REF scenario. C_EP strategies showed a lower contribution to the reduction in ED in São Paulo. This is because such strategies are aimed at the building sector and electricity consumption, while PU (C_UP scenario) strategies operate on the city's most energy-intensive sector (urban mobility).

In addition, within the mobility sector, the selected strategies aimed to reduce the use of individual motorized transport modes (cars). Therefore, this result agrees with the results in the scientific literature on the importance of the urban mobility

Table 1 FEC in PJ, by sector and by scenario (2014–2030)

	2014	2030	2030	2030	2030
Scenarios	BY	REF	C_EP	C_UP	C_UEP
Household	56	124	119	123	119
C&S	48	112	103	105	95
Industry	29	25	23	25	23
Public Buildings	4	5	4	5	4
Public lighting	2	4	1	4	1
Water Treatment	7	13	11	11	9
Energy Sector	0	2	2	2	2
Urban Mobility	212	340	346	290	297
Air Transportation	8	6	6	6	6
Total	367	630	615	569	556

sector in determining energy consumption patterns in cities (Marins 2014; Ruparathna et al. 2017).

Strategies promoting active mobility and public transport are important policies with great impact on energy savings in the urban environment, which are corroborated in the implementation of the 2030 Agenda. Table 2 presents the results of the implementation of the selected strategies for all the city's energy resources (except electricity). It shows that energy savings goals could be achieved by replacing individual transport to active mobility and public transport modes (Table 2). "More public transport" and "more active transport" were the second and third best-performing strategies in reducing GHG emissions and saving energy.

The "no use of fossil fuels in public transportation" strategy contributed the most to the reduction in emissions; on the other hand, it resulted in no energy savings. This strategy is expected to reduce GHG by approximately 6.5 million tCO2e by

Table 2 Energy savings and avoided GHG emissions

Strategies for energy demand	Energy savings in 2030 (PJ)	Avoided GHG emission (ktCO2e)
No fossil fuels on PT	−6	6047.52
Reduce Lighting Kerosene usage	0.0	0.03
More solar water heating	0.04	132.20
More electric cars	4	373.03
More PT	70	3630.85
More cycling	29	1518.87
Non-motor/active mobility (accessibility)	22	1136.70
Non-motor/active mobility (mix use)	101	5245.71

Table 3 Electricity and GHG emissions savings (k tCO2e.)

Strategies for electricity demand	Energy savings in 2030 (PJ)	Avoided GHG emission (ktCO2e)
More LED-Lighting	3.4	1.4
More efficient refrigeration	0.1	0.1
More efficient air conditioning	6.9	4.5
Industrial Co-generation	1.5	0.3
Industrial Energy Management	0.7	0.2
Buildings Energy Management	2.2	0.5
Water treatment Energy Management	2.0	0.5
More efficient water pumps	0.2	0.0
Water reuse- greywater	2.6	0.6
New green areas- from 100 parks to 167 parks	7.9	5.9
Natural lighting	4.1	1.0

2030; however, it is also estimated that it will increase the DE by 6 PJ, due to the replacement of diesel vehicles by vehicles powered by ethanol, a less efficient fuel. Table 3 shows the impacts of the implementation of the different strategies on the electricity economy.

The best strategy considering the buildings sector, was the creation of "new green areas," which reduced electricity demand in 8 PJ (Table 3). Important to note that this amount could be even bigger if the new green areas were located closer to the densest areas of the city. The strategy also showed a greater contribution to reducing GHG emissions. Furthermore, it is important to highlight, based on the implementation, the increase of drainage areas and surfaces and permeability in a city like São Paulo, which constantly faces floods. This fact highlights the need on the part of municipal public management to consider environmental, social, and energy goals in the PU. Once again, it demonstrates the synergistic *feedback scans* of implementing a strategy for achieving multiple goals for more than one public policy purpose (this strategy acts, for example, as a measure of adaptation and mitigation of climate change, as a policy for saving energy and resources, and as a public health policy).

Even though, in general, the C_UP scenario demonstrated greater economic capacity, the C_EP scenario performed better in terms of increasing the share of demand for renewable (*RES*) in the city. This result is also related to the urban mobility sector, since the "no use of fossil fuels in public transport" strategy exhibited the greatest impact in terms of reducing fossil consumption in the city and increasing the demand for *RES*. Consequently, one of the great added values of the integration of (synergy between) PE and PU is the possibility of achieving multiple results and objectives. Thus, in relation to *the energy consumption of RES* in the city (Table 4), the C_UEP scenario presented the highest insertion of *RES* in the

Table 4 Fossil energy consumption versus RES by scenario (2030)

Resource	C_REF (PJ %)	C_EP (PJ %)	C_UP (PJ %)	C_UEP (PJ %)
Fossil	383 (61%)	297 (48%)	351 (62%)	246 (44%)
RES	247 (39%)	318 (52%)	218 (38%)	310 (56%)
Fossil + RES	630 (100%)	615 (100%)	569 (100%)	556 (100%)

demand of the city in 2030, with 310 PJ (56%), when compared with the three other scenarios.

Regarding the evolution of power generation in the city, the results highlight an endogenous energy potential in the unexplored parts of the city that is currently neglected by the urban and energy policies. With the implementation of new biodigester plants and a conservative estimate for photovoltaic potential on roofs, it is possible to increase the city's electricity generation by approximately 8 TWh and 10 TWh in the C_EP and C_UEP scenarios, respectively. It is noteworthy that the resource that influences the increase in local generation the most is the use of roofs in the residential and commercial sectors, which allows electricity generation through photovoltaic panels.

At this point, another aspect that merits attention is the precautionary principle applied in regard to energy dependence. A megacity like São Paulo, which is so strongly influential and important for the rest of the country from an economic and social point of view, needs to be less energy-dependent and less vulnerable. When the city increases its endogenous *energy and RES*, it also increases its ability to withstand national power blackouts or the lack of other resources. In addition, a more self-sufficient city can ensure the minimum provision of essential services in the event of a national energy collapse.

Table 5 shows *emissions* per capita. Electricity imports consider the implied GHG emissions corresponding to national emissions of electricity generation (average emission factor of the Brazilian grid). It was considered that the lowest levels of GHG emissions result from the integration of C_UEP (Table 2). Thus, for cities that wish to take full advantage of their system for acting on climate change, the integration of UEP strategies is an effective approach.

While emissions in the C_EP and C_UP scenarios increased by 17% and 31%, respectively, over their respective base-year (2014) values, with C_UEP there was an increase of only 1%. In contrast, when compared with the results for C_REF 2030, C_UEP achieves a 30% reduction in emissions (approximately 9.4 million

Table 5 GHG emissions by scenario

Million metric tCO2e	2014	REF-2030	C_EP-2030	C_UP-2030	C_UEP-2030
Total emissions	20.7	31.5	25.8	29.0	22.2
tCO_2e/inhabitant	1.80	2.57	2.10	2.36	1.81

tCO2e less), while the C_UP and C_EP scenarios result in reductions of 8% and 18%, respectively. The differences in GHG emission reductions between C_UP and C_EP, although apparently conflicting with the energy-saving results achieved by these scenarios, are explained by the implementation of power generation policies in each of them. The potential of photovoltaic electricity generation was considered an EP measure/strategy and hence was simulated in the C_EP scenario, while all strategies for endogenous electricity production using biogas were simulated in the C_UP scenario (since such strategies were motivated by the need to reduce urban waste flows into the city).

Finally, the study demonstrated the importance of analyzing the quantification of synergies of UP and EP strategies. This is because it was observed that strategies that usually would not be prioritized by energy planners (due to their lower performance in terms of energy savings), when combined with PU strategies, can increase the impact of EP strategies and reduce global GHG emissions in the city of São Paulo.

3.1 UEP and the Range of the SDGs

The panel of analysis and correlation of strategies resulted in the establishment of four indicators (IAA, IEA, AQC, and AAP). Figure 2 shows the performance of UEP strategies in terms of achieving SDG results. In general, SDGs 3, SDG4, SDG6, SDG7, SDG8, SDG9, SDG11, SDG12, SDG13, SDG14, and SDG15 were all addressed by the UEP strategies modeled.

SDGs 1 (poverty eradication), 2 (zero hunger) and 10 (reduction of inequalities) were partially addressed by the strategies if the synergistic feedback effects of the UEP performance are taken into consideration. Finally, according to the consulted experts, SDGs 5 (gender equality), 16 (peace and justice), and 17 (partnerships and means of implementation) were not addressed by any strategy.

According to the IAA index (sum of all actions, whether direct or indirect, and their scope, whether complete or incomplete) of the UEP strategies analyzed using the PMCM methodology, those that contributed the most to the achievement of the SDGs were distributed generation, reuse and recycling, implementation of new green areas in the city, solar heating of water and photovoltaic energy, energy management in buildings, increases in the number of bodies of water, and replacement of fossil fuels by renewables. Table 6 presents the hierarchy of UEP strategies, starting with the most comprehensive ones (IAA > 0.5). Such strategies achieve greater simultaneous performance in terms of achieving the 17 proposed SDGs.

On the other hand, when evaluating the effectiveness of strategies through the $IEA_{(e)}$ index (which evaluates how effective the proposed strategies are), the results indicate that, *lato* sensu, such strategies are not efficient ($IEA_{(e)} < 0.5$), because on the whole there are few strategies that address an SDG directly and completely. Among the evaluated strategies, the most effective are non-motorized public

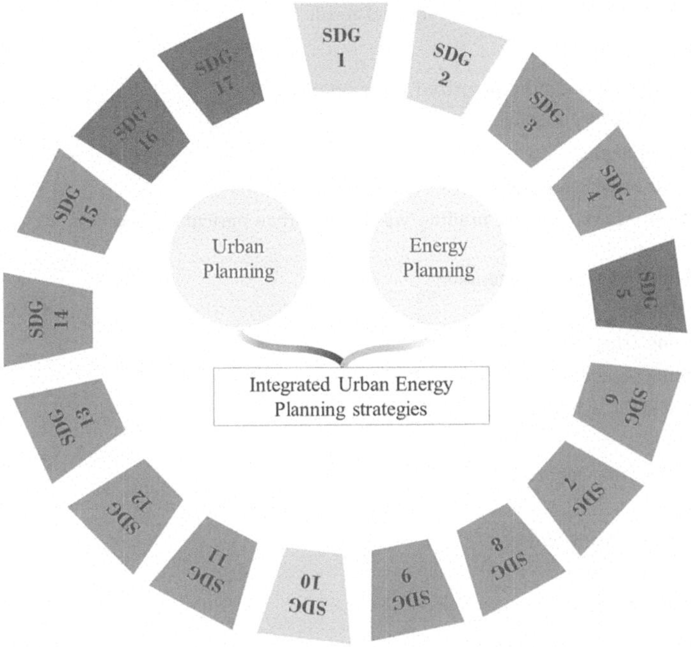

Fig. 2 Overall performance of UEP strategies in terms of achieving the SDGs

Table 6 IAA analysis and correlation panel results

Strategy	IAA
Buildings' energy management	0.58824
Energy reuse strategies	0.58824
More distributed electricity generation	0.58824
More pv and solar water heating	0.58824
New green areas	0.58824
Fuel shift	0.52941
Passive cooling (bodies of water)	0.52941
Non-motorized or active mobility	0.47059
Industrial energy management	0.47059
Technological replacement	0.47059
Public transportation and non-motorized transportation	0.41176
Water treatment energy management	0.41176
Electricity from msw, sewage, pruning waste, and urban agriculture biomass waste	0.35294
Green rooftops/walls	0.35294
Natural lighting	0.29412
Industry cogeneration	0.23529
Passive cooling (shading and wind)	0.23529

Table 7 $IEA_{(e)}$ analysis and correlation panel result

Strategy	$IEA_{(e)}$
Public transportation and non-motorized transportation	0.42857
Water treatment energy management	0.42857
Energy reuse strategies	0.40000
Non-motorized or active mobility	0.37500
Electricity from MSW, sewage, pruning waste, and urban agriculture biomass waste	0.33333
More distributed electricity generation	0.30000
New green areas	0.30000
Passive cooling (shading and wind)	0.25000
Passive cooling (bodies of water)	0.22222
More PV and solar water heating	0.20000
Natural lighting	0.20000
Industrial energy management	0.12500
Fuel shift	0.11111
Buildings' energy management	0.00000
Technological replacement	0.00000
Green rooftops/walls	0.00000
Industry cogeneration	0.00000

transport, energy management in sanitation services, and reuse and recycling, with the latter being among the most comprehensive and most effective strategies (Table 7).

It is interesting to note, however, that energy management strategies in buildings, modernization and technological replacement, incorporation of green rooftops/walls, and industrial cogeneration presented high levels of performance in the first stage of the study (in the modeling exercise, they presented the best results in terms of energy savings and avoided CO2 emissions) but a poor performance in meeting the SDGs (see "more green areas" and "more efficient air conditioners" in Table 2).

Evaluating the combined application of all strategies and its degree of effectiveness in achieving the SDGs based on the AQC index produced a value of 2.06. The scale of this index ranges from 0 (all indirect or incomplete complete markings) to 19 (only complete and either direct or indirect markings), so the process is not measured by the number of strategies proposed but by the number of strategies that act on the problems listed. This result indicates that the analyzed strategies exhibit a low rate of meeting the SDGs.

On the other hand, when evaluating the degree to which the SDGs are achieved in a generic way based on the AAP index, which measures the relationship between proposed problems and proposed solutions, the result was 10.357. Therefore, the analyzed strategies address the 17 SDGs to a large extent but lack the ability to achieve them completely.

4 Conclusions

We simulated the implementation of 29 UP and EP strategies using the LEAP_SP model to visualize São Paulo's future UES (for the period 2014–2030), in addition to quantifying the possible synergies of their implementation through UEP and relating them to the achievement of the SDGs. The implementation of these 29 strategies was simulated under four scenarios: historical rates (C_REF), energy policy strategies (C_EP), urban policy strategies (C_UP), and urban and energy policy strategies (C_UEP).

The main results of the quantification of synergies and the model indicate that UEP has greater potential when compared with individual EP and UP policies, translated into energy savings of 12% in 2030 (C_UEP scenario) compared with the C_REF scenario. Although the synergies related to the FEC of UP and EP integration are reduced, since C_UP already creates energy savings of 10% while C_UEP only creates an additional 2%, based on the other indicators, the results show the clear benefits of UEP.

This is the case, for example, for the indicator of *increased participation of RES* in the city, which increases from 11 to 17% in 2030 when a combination of EP and UP strategies is simulated, and for GHG emissions, which decrease by 30% under C_UEP when compared with C_REF. Furthermore, C_UP exhibits the largest decrease in ED compared with C_EP. Nevertheless, this is an important result, considering that energy policies do not consider UP strategies for energy saving, which demonstrates that some potential for additional improvement is wasted. Finally, a total of 11 of the 17 SDGs are addressed by C_UEP.

Moreover, when UP and EP strategies are integrated, not only is a more equitable division of efforts achieved among the different economic sectors of the city, but it is also possible to simultaneously achieve different policy objectives (e.g., mitigation GHG emission, better air quality, and better public health). We found that C_UEP exhibits the best performance among the four scenarios. Thus, through a UEP approach, it is possible to leverage ED reductions, increase the share of RES, increase the electricity generated by the city with local resources and RES, and reduce GHG emissions. However, the implementation of UEP is hampered by numerous factors; one of them is precisely the lack of pragmatic or disciplinary clarity regarding the attribution of responsibilities that every interdisciplinary or transdisciplinary theme carries with it as a strength and also as a weakness.

Thus, we conclude that achieving these goals should also be approached from both the perspective of EP and that of UP. Finally, it is expected that the implementation of UEP (which consists of integrated planning and that should dialogue with the other planning spheres) will also improve the quality of life and health of citizens, also achieving objectives of the social dimension. Future research in UEP should further verify the feasibility of its development, both in the social and economic dimensions of this type of planning.

References

Adhikari RS, Aste N, Manfren M (2012) Optimization concepts in district energy design and management—a case study. Energy Procedia 14:1386–1391. https://doi.org/10.1016/j.egypro.2011.12.1106

Brazilian Ministry of Mines and Energy (2018) National Energy Balance, 2018. Summary Report (base year 2017). MME, Brasilia

Brazilian Ministry of Mines and Energy (2013) National Inventory of Atmospheric Emissions by high-engine vehicles. Synthesis Report (base year 2012). MME, Brasilia

Brazilian Institute Of Geography And Statistics - IBGE (2010) Demographic census Atlas. IBGE, Rio de Janeiro

Collaço FMA et al (2019a) The dawn of urban energy planning—synergies between ener y and urban planning for São Paulo (Brazil) megacity. J Clean Prod 215:458–479

Collaço FMA et al (2019b) What if São Paulo (Brazil) would like to become a renewable and endogenous energy-based megacity? Renew Energy 138:416–433

Collaço FMA et al (2020) Understanding the energy system of the Paulista Macrometropolis: a first step for local action in Climate Chan-ge. Ambiente & Sociedade 23:1–23 (São Paulo)

Cruz RBC (2019) Methodological proposal for popular participation in the definition of strategies in urban planning. Thesis (Doctorate)—Polytechnic School of the University of São Paulo. São Paulo, 552 p

Department of Energy of the State of SP (2007) Statistical yearbook of energy drinks by municipality in the State of São Paulo—2008 base year 2007. São Paulo-SP-Brazil

Department of Energy of the State of SP (2008) Statistical yearbook of energy drinks by municipality in the State of São Paulo—2009 base year 2008. São Paulo-SP-Brazil

Department of Energy of the State of SP (2010) Statistical yearbook of energy drinks by municipality in the State of São Paulo—2010 base year 2009. São Paulo-SP-Brazil

Department of Energy of the State of SP (2012) Statistical yearbook of energy drinks by municipality in the State of São Paulo—2012 base year 2011. São Paulo-SP-Brazil

Department of Energy of the State of SP (2013) Statistical yearbook of energy drinks by municipality in the State of São Paulo—2013 base year 2012. São Paulo-SP-Brazil

Department of Energy of the State of SP (2014) Statistical yearbook of energy drinks by municipality in the State of São Paulo—2014 base year 2013. São Paulo-SP-Brazil

Department of Energy of the State of SP (2015) Statistical yearbook of energy drinks by municipality in the State of São Paulo—2015 base year 2014. São Paulo-SP-Brazil

Department of Energy of the State of SP (2016) Statistical yearbook of energy drinks by municipality in the State of São Paulo—2016 base year 2015. São Paulo-SP-Brazil

Department of Energy of the State of SP (2017) Statistical yearbook of energy drinks by municipality in the State of São Paulo—2017 base year 2016. São Paulo-SP-Brazil. Available from: http://dadosenergeticos.energia.sp.gov.br/Portalcev2/intranet/BiblioVirtual/menu.asp. Accessed: 11/08/2019

Heaps C (2006) LEAP data requirements for energy planning and mitigation assessment. SEI US

Jovanovic M, Afgan N, Bakic V (2010) An analytical method for the measurement of energy system sustainability in urban areas. Energy 35:3909–3920. https://doi.org/10.1016/j.energy.2010.06.010

Keirstead J, Calderon C (2012) Capturing spatial effects, technology interactions, and uncertainty in urban energy and carbon models: retrofitting newcastle as a case-study. Energy Policy 46:253–267. https://doi.org/10.1016/j.enpol.2012.03.058

Madlener R, Sunak Y (2011) Impacts of urbanization on urban structures and energy demand: what can we learn for urban energy planning and urbanization management? Sustain Cities Soc 1:45–53. https://doi.org/10.1016/j.scs.2010.08.006

Marins KRDCC (2014) A method for energy efficiency assessment during urban energy planning. Smart Sustain Built Environ 3:132–152. https://doi.org/10.1108/SASBE-12-2013-0056

Morlet C, Keirstead J (2013) A comparative analysis of urban energy governance in four European cities. Energy Policy 61:852–863. https://doi.org/10.1016/j.enpol.2013.06.085

National Eletric Energy Agency—BIG, ANEEL (2017) Management information bank. ANEEL, Brasilia. Available in: https://www.aneel.gov.br/documents/656877/14854008/Boletim+de+Informa%C3%A7%C3%B5es+Gerenciais+1%C2%BA+trimestre+de+2017/798691d2-990b-3b36-1833-c3e8c986

Oldenberg O, Murshed SM, Kremers E et al (2015) Model-based analysis of urban energy systems (On the basis of a city's energy master plan). Emerg Complex Organ 17. 10.17357.03a4747a4b28258c105148d3775522a1

Pimenta APA (2010) Legislação Básica do Setor Elétrico Brasileiro. ANEEL, Brasilia

Rodríguez-Rodríguez D, Kain JH, Haase D et al (2015) Urban self-sufficiency through optimised ecosystem service demand. A utopian perspective from European cities. Futures 70:13–23. https://doi.org/10.1016/j.futures.2015.03.007

Rosales Carreón J, Worrell E (2018) Urban energy systems within the transition to sustainable development. A research agenda for urban metabolism. Resour Conserv Recycl 132:258–266. https://doi.org/10.1016/j.resconrec.2017.08.004

Ruparathna R, Hewage K, Karunathilake H et al (2017) Climate conscious regional planning for fast-growing communities. J Clean Prod 165:81–92. https://doi.org/10.1016/j.jclepro.2017.07.092

Rutter P, Keirstead J (2012) A brief history and the possible future of urban energy systems. Energy Policy 50:72–80. https://doi.org/10.1016/j.enpol.2012.03.072

Saldiva P (2018) Vida urbana e saúde. Editora Contexto, São Paulo

São Paulo-PMSP (2009) City hall of the municipality of São Paulo climate change policy in the municipality of São Paulo. PMSP, Brazil, São Paulo

UN—United Nations Climate Change (2015) The Paris Agreement. http://bit.ly/ParisAgreementUNFCCC

UN-Habitat (2016) World Cities Report 2016—Urbanization and Development. Emerging Futures. UN-Habitat, Nairobi

United Nations Development Programme (UNDP) (2015) Transforming our world: the 2030 agenda for sustainable development. UNDP, Rio de Janeiro

Wu J, Yan J, Jia H et al (2016) Integrated energy systems. Appl Energy 167:155–157. https://doi.org/10.1016/j.apenergy.2016.02.075

Yazdanie M, Densing M, Wokaun A (2017) Cost optimal urban energy systems planning in the context of national energy policies: a case study for the city of Basel. Energy Policy (accepted) 110:176–190. https://doi.org/10.1016/j.enpol.2017.08.009

Biopesticides and Sustainability in a Land Use Context

D. Duarte, C. Gaspar, C. Galhano, and P. Castro

1 Introduction

Since the 1950s, an enormous growth in the human population has been witnessed, despite a continuous decrease in its growth rate, and worst-case scenarios predict that it could grow above 50% by 2100 (Fig. 1). This results in expanding the current economic model and consumption patterns that depend on different ecosystems' goods and services. This observation raises obvious questions about the and capacity of the Planet due to the increasing demand for food and other resources essential to life, and on meeting the targets of the Agenda 2030—Ensure sustainable consumption and production patterns (Goal 12) or for sustainable and resilient food systems (Castro et al. 2016; Pantzar et al. 2018; EEA 2019).

Presently, we are overusing the Earth's biocapacity by at least 56% to feed and fuel our lifestyle. This current consumption rate has been driving the ecosystems' degradation, which has led to 1 million species threatened with extinction (WWF 2020). One way to face this problem is the so-called "sustainable development of society", one of the most debated subjects because we need to find a way to make it work. Given that we face severe environmental menaces such as climate change, global warming, chemical pollution, loss of biodiversity or scarcity of natural resources, we urgently need to actively pursue ways to mitigate these ecological "disasters" instead of avoiding them at all (Cardinale et al. 2012; UN 2015; Silva et al. 2018; Chase et al. 2020; WWF 2020).

D. Duarte · C. Gaspar · C. Galhano · P. Castro (✉)
Centre for Functional Ecology—Science for People & the Planet, Department of Life Sciences, University of Coimbra, Calçada Martim de Freitas 3000-456, Coimbra, Portugal
e-mail: pcastro@ci.uc.pt

C. Galhano
Coimbra Agriculture School, Department of Environment, Polytechnic Institute of Coimbra, Bencanta 3045-601, Coimbra, Portugal

W. Leal Filho et al. (eds.), *Sustainability in Natural Resources Management and Land Planning*, World Sustainability Series,
https://doi.org/10.1007/978-3-030-76624-5_8

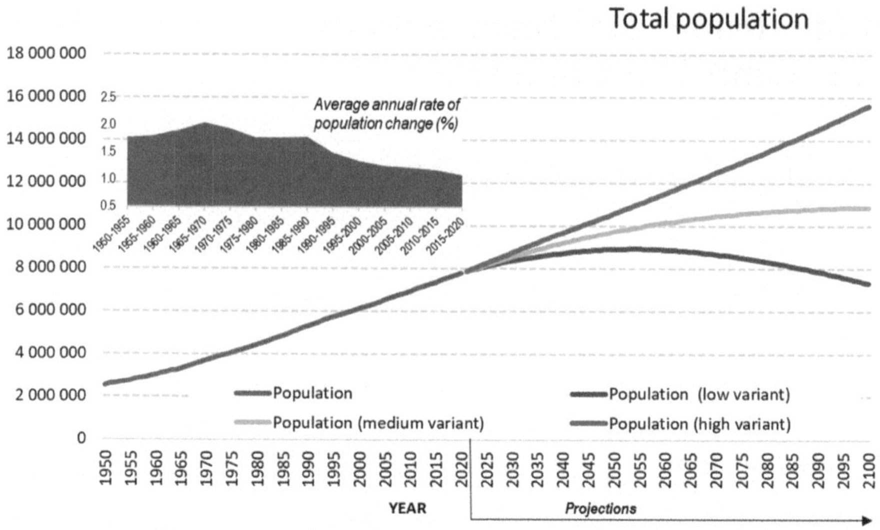

Fig. 1 Worldwide population and growth estimate up to 2100. Smaller graph: World population growth rate 1950–2020. *Data source* UN, DESA (2019). Figure elaborated by the authors with copyright permission—© August 2019 by United Nations, made available under a Creative Commons license CC BY 3.0 IGO: http://creativecommons.org/licenses/by/3.0/igo/

Land use in agriculture is a significant driver of environmental change (Godfray and Garnett 2014; Borrelli et al. 2020) and sustainable and resilient agricultural systems are needed support the global population, but more importantly, they assume even more important roles, producing several other benefits for communities, in addition to provisioning (Castro et al. 2019; Amelung et al. 2020). Linked to the sustainability problem (or solution) is the recurrent theme of pesticides (Le et al. 2018; Qadri and Bhat 2020). These substances seem to be a real environmental and human threat but necessary for agriculture and crop protection (MacLeod et al. 2010; Kim et al. 2017). In this context, this chapter pretends to address the issues related to pesticide use and reviews some recent developments regarding biopesticides, which may be one of the most appealing alternatives to synthetic chemicals commonly applied to control plant enemies. It is intended with this discussion to help to raise awareness of this specific environmental issue and make people look at this topic with different and renewed perspectives, so that consequences resulting from the use of synthetic chemicals start to be addressed more seriously in policy and decision-making, in the management of natural and semi-natural systems, as well as in everyday life.

2 Pesticides

Pesticides may be considered any substances or mixture of substances that prevents, destroys, or controls certain types of organisms reckoned to be pests (harmful organisms) and diseases. It may also protect crops or plant products during production or storage. These substances may include herbicides, fungicides, insecticides, acaricides, nematicides, molluscicides, rodenticides, growth regulators, repellents, rodenticides or even biocides transport (more on this in the European Commission website: https://ec.europa.eu/food/plant/pesticides_en; Kumar and Singh 2015).

These products were first developed to diminish agricultural yield loss and better cope with increasing population growth and the corresponding demand for food (Mcafee 2017). Today, they can be used on a broader spectrum: in addition to their wide use in agriculture, they can also be used to manage, for example, green spaces in cities (from turfgrass in sports fields and parks to housing backyards/gardens and street pavements (Gangstad 1982; Korres et al. 2018; Dhananjayan et al. 2020).

In ancient times, in Greece and Rome, some products were already used to control insects and were mainly based on arsenic and sulfur chemicals. Later, around the nineteenth century, weed control has begun to be implemented in countries like France and Germany (Zimdahl 2018). Following that first era of pesticides, it began "the modern chemical age" characterised by the development and application of an unprecedented effectiveness chemical, DDT (dichlorodiphenyltrichloroethane), pesticide, in 1939 in Switzerland (Jarman and Ballschmiter 2012). Simultaneously, in the UK, the production of phenoxy acid herbicides was also commercialised. In 1974, the most successful herbicide in history, glyphosate, was released in the market (Benbrook 2016).

Regarding the chemical fungicides, they were introduced mainly during the 1960s and 1970s (Morton and Staub 2008). From the 1970s onwards, many other synthetic chemicals (e.g. organophosphates, carbamates, pyrethroids and neonicotinoids) started to be produced and sold worldwide (Seifert 2005; Aktar et al. 2009; Costa 2018; Matsuo 2019; Dhananjayan et al. 2020). In recent decades, there has been significant technical and scientific progress in this area, leading to the application of ever-smaller dosages to control pests and diseases while guaranteeing their effectiveness (Peshin et al. 2009).

The use of these synthetic chemicals has increased over the years and has now become a general procedure. According to FAO (2020), about 4.1 million tons of pesticides are used worldwide per year. In Europe, fungicides and herbicides dominate the sales charts, while insecticides are less significant (Fig. 2).

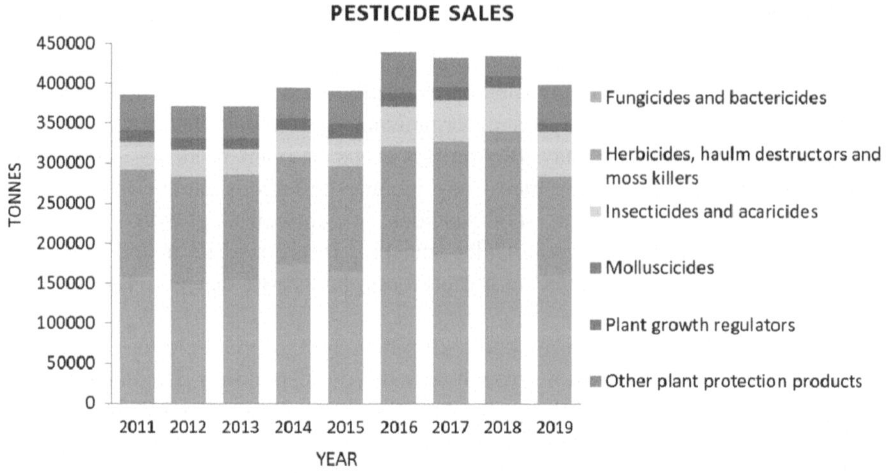

Fig. 2 Total pesticides sales, EU, from 2011 to 2019. *Data source* Eurostat—pesticide sales, [AEI_FM_SALPEST09] (02/03/2021). Figure elaborated by the authors with copyright permission of the EU under the Creative Commons Attribution 4.0 International (CC BY 4.0) licence [https://creativecommons.org/licenses/by/4.0/]

2.1 Benefits and Disadvantages of Pesticides

Notwithstanding the acknowledged negative impacts generated by these products, there were undoubtedly significant positive outcomes concerning the crop production boost, increased profits for farmers, and disease prevention (Aktar et al. 2009; Mahmood et al. 2016; Carvalho 2017; Dhananjayan et al. 2020). Although pesticides may increase food production costs, they are also perceived as relevant to increase productivity (Ahmed et al. 2011). The vast majority of farmers believe that they are indeed necessary to increase productivity and profits, decreasing the risks associated with investments in the sector justifying their use, despite their environmental and human health effects (Ríos-González et al. 2013; Wang et al. 2017). However, their long-term harmful effects on the environment and non-target organisms (including humans) have taken some time to be identified (Mahmood et al. 2016). The period after the development of DDT (and other pesticides) and the recognition of its deleterious effects led to the emergence of new movements in society (Mcafee 2017). The warning launched by Rachel Carson in (1962), through her book "Silent Spring", reinforces the launch of an environmental movement that questions and brings environmental problems to the public discussion, such as those caused by pesticides (Jarman and Ballschmiter 2012).

It is a well-known fact that pesticides are hazardous, tend to be persistent in the environment and organisms culminating in their bioaccumulation (Mackay and Fraser 2000; Chopra et al. 2011; Mahmood et al. 2016; Humphries et al. 2021). Once released into the environment, they contaminate the soil, seep into groundwater and runoff into streams. They can contaminate sprayed food directly or arrive indirectly through contaminated food through bioaccumulation (Mackay and Fraser 2000; Chopra et al. 2011; Van Bruggen et al. 2018).

Regarding public health concern, people may be exposed when working in the field and handling these products or manufacturing them. Also, the population is at risk by consuming contaminated food and contaminated water or because they may live close to sprayed fields and/or frequently maintained street pavements and other green spaces (Van Bruggen et al. 2018; Carvalho 2017; Dhananjayan et al. 2020). However, the pesticide problem goes far beyond agricultural fields and farm-workers. For example, pesticides are also widely present in other ecosystems, such as in urban areas (Md Meftaul et al. 2019; Wolfand et al. 2019). These highly artificial and human-dominated areas are continuously growing and receiving more and more people [it is expected 68% of the total population may live in cities by 2050 (UN DESA 2018)] and, in consequence, a considerable portion of the population is also being affected.

The most common adverse and short-term effects of pesticides reported on farmworkers and workers handling pesticide production include acute headaches, vomiting, stomach-aches, and diarrhoea. In the long run, constant exposure may lead to chronic health complications such as cancer, congenital disabilities, or reproductive problems (EC 2007; Koutros et al. 2016; Kim et al. 2017; Anifandis et al. 2018; Kalliora et al. 2018; Lerro et al. 2020). For instance, in 2015, glyphosate was considered “probably carcinogenic to humans” by the World Health Organization’s International Agency for Research on Cancer (IARC 2015). Some studies have also linked this substance with other health problems, such as kidney damage, mental conditions (ADHD, autism, Alzheimer’s and Parkinson’s disease) and dermatological and respiratory illnesses (Beuret et al. 2005; Ngo et al. 2012; Dardiotis et al. 2013; Zaganas et al. 2013; Jayasumana et al. 2014; Beecham and Seneff 2015; Kim et al. 2017; Van Bruggen et al. 2018).

The unsafe use of pesticides can also contaminate soil, water, air, vegetation, and interfere with non-target organisms (Mcafee 2017; Dhananjayan et al. 2020). Those products disseminated as spay can drift or volatilize from the treated area and interact with the moisture and deposit during precipitation leading to contamination of the surface water system (Van Bruggen et al. 2018; Carvalho 2017; Dhananjayan et al. 2020). Moreover, contaminated soil results in the decline of beneficial soil microorganisms (Shao and Zhang 2017; Onwona-Kwakye et al. 2020). Because of their bioaccumulation capacity, these chemicals may become toxic, interfere with food chains and cause health problems to other organisms (Dhananjayan 2013; Al-Ani et al. 2019; Kumar et al. 2019; Onwona-Kwakye et al. 2020). For example, atrazine affects the lipid peroxidation and antioxidant enzyme activity in freshwater fish (Nwani et al. 2010; Dhananjayan et al. 2020). Organophosphate-based pesticides considerably affect siderophore-producing microorganisms by inhibiting cholinesterase activity (Kumar et al. 2019).

The continued use of these highly toxic and persistent products can cause an additional negative effect: an increased resistance in the target species, i.e., once pests or diseases highly susceptible to a particular chemical can become highly resistant to its action (Curutiu et al. 2017). Therefore, only the most resistant organisms survive and transmit their genetic features to descendants. The danger associated with these "new" organisms drives the scientific community to seek more and more aggressive solutions (Dentzman et al. 2016). This vicious circle thus brings with it severe and unsustainable consequences.

Given their application background and their current rate of use, not only in agriculture but also in industry, urban areas, or domestic use, it becomes imperative to understand and control how their residues end up in the environment and food (EFSA 2020).

3 European Union—Strategies and Regulations, a Brief Review

Given all the risks associated with pesticides, the European Union has been implementing several measures to diminish the adverse consequences and because it has become more evident how much we need the environment and its services. Policies and legislation on pesticides date since the 1970s. The Council Directive 76/769/EEC was an essential step in restricting the marketing and use of certain dangerous substances and preparations. The concern of the action of plant protection products on plant species and plant production, as well as on the environment, in particular, human, animal health or on groundwater, led to the publication, on 15 July 1991, of the Council Directive regulating the authorisation, placing on the market, use and control within the Community of plant protection products in commercial form and the placing on the market and control within the Community of several active substances (Council Directive 91/414/EEC). However, the European Council continued to express concerns regarding the absence of harmonised Community provisions for biocides (which were formerly known as non-agricultural pesticides), and on 16 February 1998, a directive was published concerning the placing of these products on the market, thus complementing the existing directive (Directive 98/8/EC). This legislation was then revoked in 2012 by Regulation (EU) N.º 528/2012, concerning the making available on the market and use of biocidal products.

Since 1992 the European Commission, with other partners, had been conducting the Sustainable Use of Plant Protection Products project. These efforts culminated in the elaboration of the Communication "Thematic Strategy on the Sustainable Use of Pesticides" (COM2006, 373). This Strategy addressed the problem regarding unwanted amounts of pesticides found in the environment (in particular soil and water) and residues exceeding regulatory limits in agricultural production. Therefore, it aimed to reduce the risk to humans and the environment by minimising or eliminating exposure, promoting research and development of less harmful alternatives, and raising awareness of this thematic.

Several other EU policies and strategies compromise the use of pesticides products, such as The Common Agricultural Policy (CAP) (launched in 1962 and reformed in 2013), The Water Framework Directive (WFD) (Directive 2000/60/EC), The Waste Framework Directive (Directive 2008/98/EC) or the recent Commission Implementing Regulation (EU) 2019/533 on residue levels of pesticides.

In 2009, the Directive 2009/128/EC established a framework to achieve a sustainable use of pesticides (SUP) by reducing the risks and impacts of pesticide use on human health and the environment and promoting the use of integrated pest management and alternative approaches or techniques such as non-chemical alternatives to pesticides. This should be accomplished via national action plans in each member state. Also, Regulation (EC) N.º 1185/2009 was implemented to collect data on the use and sale of these products so it could be possible to attain a greater understanding of their use.

To complement the previous directive and update the existing legislation, the Regulation (EC) No. 1107/2009 was approved concerning the placing of plant protection products on the market, revoking Council Directives 79/117/EEC and 91/414/EEC. This regulation complemented the previous legislation, namely by updating the list of products allowed in the European Union and the standards for new product development, testing, approval, and commercialisation.

For the past 25 years, the regulating process of all chemicals in the EU is done in close cooperation with the European Food Safety Authority (ESFA), all Member States and the European Commission. ESFA (www.efsa.europa.eu) has become part of the process in 2003 as an independent source of scientific advice and risk assessment and communication associated with the food chain. It advises the EU, perform peer review of active substances and correspondent risk assessment, recommends the maximum residue levels of pesticides in products of plant or animal origin and drafts annual reports. After the active substance's approval, the product's approval containing that substance by the European Commission began in consultation with the Member States.

Overall, the measures carried out in the European Union include: the development of national plans, training of professional pesticide users and distributors, raising awareness, control or ban aerial spraying, minimise or ban pesticides in critical areas, equipment inspection, and the promotion of the principles of integrated pest management (which encourages prevention and prioritises the use of low-risk pesticides and non-chemical methods).

As European regulations have grown and are increasingly stricter, the number of approved active substances in pesticides has been reduced by more than 50% (of the previously 1000 active substances, only around 400 are currently authorised—https://ec.europa.eu/assets/sante/food/plants/pesticides/lop/index.html). For example, in 2011 the substance Acetochlor was banned from the EU; in 2013, neonicotinoids were restricted to protect bees and, most recently, a process is underway to ban Mancozeb (a fungicide) in 2021. Although the authorisation process is costly and demanding, the consumption, use and sale of pesticides have not decreased (Fig. 3).

In 2020 a special report assessed whether EU action had effectively reduced the risk related to plant protection products (PPP) use (European Court of Auditors

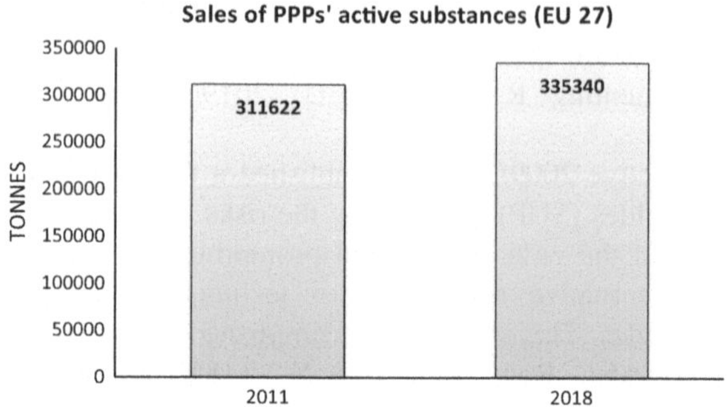

Fig. 3 Total sales of active substances used in plant protection products (PPPs) (EU 27). *Source* Eurostat (online data—retrieved on 18/02/2021). This data excludes confidential information, according to Eurostat estimate representing <3% of sales over the full dataset. Figure elaborated by the authors with copyright permission of the EU under the Creative Commons Attribution 4.0 International (CC BY 4.0) licence [https://creativecommons.org/licenses/by/4.0/]

2020). This report revealed that, despite the EU's efforts for the sustainable use of PPPs, progress had started slowly and timidly, and several weaknesses in the implementation of regulation could be identified.

Since pesticides are a need and continue to be used and sold in a large scale, the main goal defined in Europe is to continue to help producers and ecosystems/public space's managers in compliance with the 17 Sustainable Development Goals and 169 targets projected in the 2030 Agenda for Sustainable Development discussed by ONU in 2015 (https://sdgs.un.org/2030agenda). This "plan of action for people, planet and prosperity" has set numerous sustainability-related measures prepared to put in place until 2030.

It stands out, in the context of pesticide use, the SDGs No. 11 and 15, focused, respectively, on cities' sustainability (by "reducing the adverse per capita environmental impact", "providing universal access to safe, inclusive and accessible, green and public spaces") and terrestrial ecosystems conservation (main goal: "Protect, restore and promote sustainable use of terrestrial ecosystems, sustainably manage forests, combat desertification, and halt and reverse land degradation and halt biodiversity loss" (https://sdgs.un.org/2030agenda). It should also be taken into account that the new 8th EU Environment Action Programme (COM(2020) 652 final) reinforces the ambition for 2030 of a toxic free-environment, including for air, water and soil, and protecting the health and well-being of citizens from environment-related risks and impacts.

Therefore, it is essential to continue to invest in the study of sustainable alternatives to these chemicals. It is in this context that biopesticides can be an asset. In the EU, Integrated Pest Management (IPM) policies provide the incentive for new pest management strategies, especially research into the application of biopesticides.

4 Biopesticides—The Solution?

The EPA (United States Environmental Protection Agency) defines **biopesticides** as products derived from living organisms or natural materials such as animals, plants, bacteria and even minerals for controlling pests (https://www.epa.gov/ingredients-used-pesticide-products/what-are-biopesticides). At the European level, there is no accurate description, although they are generally referred to as "low risk" and "basic" substances (Villaverde et al. 2016). However, it can be found in the EU legislation the term "**biocidal** product" in the Regulation (EU) N.º 528/2012) of the European Parliament and of the Council, which must obey the following: "(i) any substance or mixture, in the form in which it is supplied to the user, consisting of, containing or generating one or more active substances, with the intention of destroying, deterring, rendering harmless, preventing the action of, or otherwise exerting a controlling effect on, any harmful organism by any means other than mere physical or mechanical action", and "(ii) any substance or mixture, generated from substances or mixtures which do not themselves fall under the first indent, to be used with the intention of destroying, deterring, rendering harmless, preventing the action of, or otherwise exerting a controlling effect on, any harmful organism by any means other than mere physical or mechanical action".

The classification of these natural control products differs amongst authors and can be based on the active substance present or according to their use or target species (Abbey et al. 2019; Liu et al. 2021). One of the simplest and most straightforward classifications is the one used by Abbey et al. (2019). These authors classify these products according to the active substance and its use. Based on the active substance, biopesticides can be divided into microbial pesticides, biochemical pesticides, and plant-incorporated protectants (PIPs). Depending on what they will be used for, they can be named bioinsecticides, biofungicides, bioherbicides and bionematicides.

Microbial pesticides consist of microorganisms or substances obtained from them, mainly referring to naturally occurring or genetically altered bacteria, fungi, oomycetes, viruses, algae and protozoa. These alternative substances are classified as ecologically safe and have target specificity. They suppress pest's development through the production of a specific toxin. Most of these toxins were identified as peptides and may vary considerably on chemical structure, toxicity, and specificity. The most known and used microbial pesticide (to control a pathogenic insect) is the bacterium *Bacillus thuringiensis* (Bt) that produces a protein crystal termed as the Bt δ-endotoxin and causes the lysis of gut cells when consumed by susceptible insects. Other examples include the *Cydia pomonella* granulovirus (CpGV) used against codling moth on apples; at least 170 different microbial biopesticides based on entomopathogenic fungi to control five insect and acarine orders in the horticulture industry (Chandler et al. 2011; Kumar and Singh 2015; Abbey et al. 2019).

Biochemical pesticides are natural substances capable of pest's control by a non-toxic mechanism. They act mainly by interfering with mating, growth and/or population build-up. They may have their origin in plants, animals or insects. Some examples include: (i) pheromones—to attract or repel unwanted organisms, such as insect sex-pheromones and the *Podisus maculiventris* hormone used against *Leptinotarsa decemlineata* already commercialized as Spined Soldier Bug Attractors™ (Aldrich and Cantelo 1999); (ii) plant secondary metabolites—which are usually plant defence mechanisms to deter herbivores [e.g. pyrethrins are a fast-acting insecticidal compound produced by *Chrysanthemum cinerariaefolium* (Silvério et al. 2009)] or reduce competition of other plants (phytotoxicity activity) (Chandler et al. 2011; Abbey et al. 2019; Liu et al. 2021).

Plant-Incorporated Protectants are compounds produced by plants from genetic material that has been added to them. For example, a specific Bt pesticidal protein gene, when introduced into the plant genetic material, induces the pesticidal protein synthesis, making it resistant to pest's infection. These proteins or genes have been incorporated into some crops such as corn (*Zea mays*), rice (*Oryza sativa*), soybean (*Glycine max*), tobacco (*Nicotiana tabacum*), sugarcane (*Saccharum officinarum*), potato (*Solanum tuberosum*) and others. These crops are the so-called genetically modified (GM) that possesses a gene that has been transferred from a different species and encodes insecticidal toxins (Fujimoto et al. 1993; Douches et al. 2004; Yu et al. 2013; Chandler et al. 2011; Abbey et al. 2019; Liu et al. 2021).

Bioinsecticides include the microbes, plants and other living forms or substances obtained from them, with insect control capacity. As mentioned before, the most used is the bacterium B. *thuringiensis*. Some other examples are the use of *Verticillium lecanii*, an approved name of an entomopathogenic fungus species that is effective against whitefly and aphids, the virus NPV for the management of butterflies and moths and the most widely used botanical compound with an insecticidal effect, neem oil, from the seeds of *Azadirachta indica* (Alavo et al. 2002; Rao et al. 2015; Chaudhary et al. 2017; Abbey et al. 2019).

When the natural active substances are able to control fungal pathogens, they are called **Biofungicides**. *Trichoderma* spp. is the most studied one and acts on different pathogens that attack different crop species (Tripathi et al. 2010; Singh and Zaidi 2017). Some plant extracts, such as *Allium* spp. and *Capsicum* spp., essential oils from *Mentha piperita* and *Ocimum basilicum*, also demonstrated this capacity to control plant diseases (Ziedan and Farrag 2008; Saeidi and Mirfakhraie 2017; Kumar et al. 2018; Abbey et al. 2019).

Bioherbicides refers to natural-based products beneficial for weed control. *For example, Puccinia chondrillina*, for the management of the rush skeleton (*Chondrilla juncea*) has been very successful in Australia and the USA (Emge et al. 1981). Bacteria such as *Xanthomonas campestris* pv. *poannua* and P. *syringae* pv. *tagetis* have been used to control the annual bluegrass (*Poa annua*) and Asteraceae weeds (Zhou and Neal 1995; Sheikh et al. 2001; Gronwald et al. 2002). Once again, essential oils show a pesticidal capacity, this time from eucalyptus, rosemary and Lawson cypress to inhibit weed growth (*Amaranthus retroflexus* and *Acroptilon repens*) (Ramezani et al. 2008; Abbey et al. 2019).

Bionematicides are the biological compounds with nematicidal properties. Some fungus (*Paecilomyces lilacinus*, *Muscodor albus*) and plant extracts

(*Peganum harmala, Raphanus raphanistrum, Taxus baccata, Ricinus communis*) demonstrate these properties (Riga et al. 2008; Hashem and Abo-Elyousr 2011; Zaidat et al. 2020). *A. indica* besides being a bioinsecticide is also a notable bionematicide as it controls root-knot nematode in soybean with 100% inhibition of egg hatch and larval mortality (Chaudhary et al. 2017; Kumar et al. 2018; Abbey et al. 2019).

The primary sources of biopesticides compounds are plants and microorganisms due to their high bioactive substances' quantity. These alternatives present themselves with an attractive list of advantages in comparison with synthetic pesticides. Being natural products, they are eco-friendly, readily biodegradable and therefore, do not contribute to long-term environmental pollution (Kumar et al. 2018; Abbey et al. 2019). Adding to that, they tend to be more target-specific, do not harm non-target organisms, and are effective in small quantities (do not promote resistance build-up among pests) (Samada and Tambunan 2020). From a consumer's point of view, and given that tastes and preferences are pending for organically produced food (in the EU, the organic area increased 70% in the last decade) (EC 2019), biopesticides appear as suitable to fulfil this requirement as they are safer to use on fresh fruits and vegetables. When it comes to production and formulation processes, these compounds are economical to attain, given their source materials' availability and time to produce (Lengai and Muthomi 2018).

Pest control will be an increasingly vital activity if agriculture is to meet the food needs of an increased human population (>9 billion) by 2050 (UN, DESA 2019). Although the development patterns of today's societies, technological development and legislation offer a promising growth area for biopesticides, it is expected that they would continue to face significant competition from synthetic chemical pesticides (Kumar et al. 2018; Liu et al. 2021). Biopesticides have a slower kill rate than conventional chemical pesticides (Liu et al. 2021), making them somewhat less competitive than synthetic chemicals. It has been hard to achieve the characteristics of the "perfect pesticide" based on natural products as they should have a high selectivity to target species but, at the same time, minimal toxicity to non-target organisms; they must be highly effective at low dosages and hold low environmental persistence (Villaverde et al. 2016).

5 Recent Works and Future Trends

The need to search for cheaper and more effective alternatives to synthetic chemical pesticides, coupled with the need in reducing adverse environmental and health impacts, will boost the opportunity for the development and massification of biopesticides (Damalas and Koutroubas 2018; Lengai and Muthomi 2018). Although it is implausible to expect biopesticides to replace chemical-based pesticides in the short term, their production and use is expected to increase, especially in developed countries (Kumar and Singh 2015; Lengai and Muthomi 2018).

The use of biopesticides increases by almost 10% every year, and there are about 175 registered biopesticides worldwide, with 700 active substance products available for use (Samada and Tambunan 2020). Worldwide, the biopesticide market is valued at around $3 billion US dollars, representing only 5% of the total crop protection market. However, it is expected to surpass the synthetic pesticides market by the early 2050s (Olson 2015; Damalas and Koutroubas 2018). In spite of the limitations pointed out by Balog et al. (2017) in the evolution of biopesticides in Europe, mainly the increased complexity of biopesticide regulations and the relatively low level of research on biopesticide in the EU as compared with the United States, India or China, its progress is expected to increase in the face of the foreseen future environmental challenges.

Current research and innovation on biopesticides aim to improve their action spectra and develop mechanisms that make them suitable to replace conventional pesticides, with new substances continuing to be reported as potential biopesticidal compounds (Table 1).

6 Final Remarks

Europe is still trying to follow-up with everything related to biopesticides. In 2021 the Biopesticides Europe 2021 conference is expected to occur on 9 and 10 June in Brussels, Belgium. In addition to the fact that it is crucial to stop or drastically reduce the use of synthetic pesticides, in recent years, a large part of European consumers has increasingly demanded products not managed with conventional pesticides (Eurobarometer 2019). Therefore, biopesticides seem to be the most desirable solution within the Integrated Pest Management (IPM) measures, one of the approaches for low-pesticide-input (Directive 2009/128/EC). Moreover, it meets the objectives set out in the Agenda 2030 and the recently Communication of the European Commission The European Green Deal (COM (2019) 640 final) aiming to make the EU a fair and thriving society with a sustainable economy, not dependent on resource use and with lower greenhouse gas emissions by 2050.

However, it is still fundamental to be aware that biopesticides are not entirely harmless and also need proper regulation (currently under the same status and regulations as synthetic pesticides). When we look into the scary data scenarios about pollution, global warming, biodiversity loss combined with the exponential population growth and chose to ignore it or relativise it, we contribute to its intensification. It is crucial to realise that these problems are real, and societies must continue to channel their efforts to avoid/mitigate further significant hazards. It is crucial to find solutions that shift completely our common future, such as biopesticides that present themselves as a promising sustainable and healthier alternative to the chemical pollution problem. The scientific community plays a fundamental

Table 1 Recent works on various substances with pest-control properties (according to its action)

Biopesticide	Product	Target pest	Reference
Bioinsecticides	Essential oils from Asteraceae plants mainly, *Santolina chamaecyparissus* and *Achillea millefolium*	*Myzus persicae*—aphids	Czerniewicz et al. (2018)
	Essential oils *of Lippia sidoides*—compound thymol	Different populations of *Sitophilus zeamais* (Coleoptera: Curculionidae).	Oliveira et al. (2018)
	Root essential oil of *Carlina acaulis* (Compositae), mainly composed by carlina oxide	*Musca domestica*	Pavela et al. (2020)
	Holy basil, *Ocimum tenuiflorum* essential oil: eugenol and caryophyllene	*Sitophilus orzyae*	Bhavya et al. (2018)
	Essential oil of *Aristolochia trilobata*: ρ-cymene and limonene	*Aedes aegypti* mosquito	Silva et al. (2018)
	Wedelia prostrata essential oil: camphene and γ-elemene	*Spodoptera litura* (Lepidoptera: Noctuidae)	Benelli et al. (2018)
	Extract of the species *Clitoria ternatea* (butterfly pea)	*Helicoverpa* spp.	Mensah et al. (2014)
	Stilbenes isolated from grapevine extracts	*Spodoptera littoralis*	Pavela et al. (2017)
	Extract from: *Suaeda maritima* AgNP (Silver nano-particules): biosynthesis from extract	*Aedes aegypti, Spodoptera litura*	Suresh et al. (2018)
	Beauveria bassiana (Balsam), orange oil, neem oil, pyriproxyfen	*Chrysoperla carnea*	Youssif and Ramadan (2020)
	Lysates of recombinant *E. coli* expressing dsRNA	*Aedes sp.*	Lopez et al. (2019)
	Cinnamon oil (*Cinnamomum zeylanicum*), Vetiver oil (*Chrysopogon zizanioides*)	*Lucilia sericata*	Khater et al. (2018)
	Orange oil (registered commercial product)	*Tuta absoluta*	Soares et al. (2019)
	Azadirachtin	*Spodoptera litura*	Shu et al. (2018)
	Bacillus thuringiensis var. *tenebrionis* strain Xd3 (Btt-Xd3).	Alder leaf beetle (*Agelastica alni*)	Eski et al. (2017)

(continued)

Table 1 (continued)

Biopesticide	Product	Target pest	Reference
Biofungicides	*Humulus lupulus*: crude extract and the essential oil	*Zymoseptoria tritici*	Bocquet et al. (2018)
	Three isolates of *Pseudomonas fluorescens*	*Botrytis cinerea* (grey mold)	Wallace et al. (2018)
	Products of the fungus *Trichoderma harzianum*	*Fusarium* root rot	Kirk and Schafer (2015)
	Three lipopeptides (fengycin—F, surfactin—S and mycosubtilin—M)	*Fusarium oxysporum* f. sp. *iridacearum*	Mihalache et al. (2018)
	n-hexane extracts from: *Eucalyptus camaldulensis (aerial parts), Vitex agnus-castus (leaves) and Matricaria chamomilla (flowers)*	*Fusarium culmorum, Penicillium chrysogenum, Rhizoctonia solani*	Salem et al. (2019)
	Cepacin	*Pythium sp.*	Mullins et al. (2019)
	Natamycin	Anthracnose crown rot of strawberry, caused by *Colletotrichum acutatum*	Haack et al. (2018)
	Pseudomonas aeruginosa	*Ganoderma boninense*	Irma et al. (2018)
Bioherbicides	Metabolites produced by *Phoma* sp.	*Bidens pilosa, Amaranthus retroflexus* and *Conyza canadensis*	Todero et al. (2018)
	Cynara cardunculus var. *altilis* leaf extracts: caffeoylquinic acids, apigenin and luteolin derivatives and lactone cynaropicrin.	*Amaranthus retroflexus, Portulaca oleracea, Stellaria media, Anagallis arvensis*	Scavo et al. (2020)
	Eucalyptus globulus (leaves extract)t	*Lactuca sativa*	Puig et al. (2018)
	Diaporthe schini	*Amaranthus viridis, Bidens pilosa, Echinocloa crusgalli, Lollium multiflorum*	Brun et al. (2020)
	Guapira graciliflora	*Cenchrus echinatus and Calotropis procera*	Rodrigues et al. (2020)
	Artemisia vulgaris (aerial biomass)	*Amaranthus retroflexus*	Pannacci et al. (2020)
	Ludwigia hyssopifolia	*Amaranthus spinosus, Dactyloctenium aegyptium, Cyperus iria*	Mangao et al. (2020)
	Aspergillus sp. and *Valsa mali*	*Parthenium hysterophorus*	Sheikh et al. (2020)

(continued)

Table 1 (continued)

Biopesticide	Product	Target pest	Reference
Bionematicides	(E)-cinnamaldehyde	*Meloidogyne incognita*	Jardim et al. (2018)
	Berberis brevissima and *Berberis parkeriana (root extracts)*	*Meloidogyne javanica*	Ali et al. (2019)
	Rhabdopeptides isolated from culture broth of *Xenorhabdus budapestensis* SN84	*Meloidogyne incognita*	Bi et al. (2018)
	verrucarin A and roridin A isolated from *Myrothecium verrucaria*	*Meloidogyne incognita*	Nguyen et al. (2018)
	Extracts (aqueous and methanolic) of *Peganum harmala, Raphanus raphanistrum, Taxus baccata, Sinapis arvensis, Ricinus communis*	*Meloidogyne incognita*	Zaidat et al. (2020)

role, in this matter, if adequately supported by institutions, investors and other sources/actors. Also, considering market growth and the promising results of some of these products, it is clear to state that there will be biopesticide solutions used worldwide in the near future!

References

Abbey L, Abbey J, Leke-Aladekoba A, Iheshiulo EMA, Ijenyo M (2019) Biopesticides and biofertilizers: types, production, benefits, and utilization. In: Simpson BK, Aryee ANA, Toldrá F (eds) Byproducts from agriculture and fisheries: adding value for food, feed, pharma, and fuels. Wiley, NY, USA, pp 479–500. https://doi.org/10.1002/9781119383956.ch20

Ahmed N, Englund J, Åhman I, Lieberg M, Johansson E (2011) Perception of pesticide use by farmers and neighbors in two periurban areas. Sci Total Environ 412–413:77–86. https://doi.org/10.1016/j.scitotenv.2011.10.022

Aktar W, Sengupta D, Chowdhury A (2009) Impact of pesticides use in agriculture: their benefits and hazards. Interdisc Toxicol 2:1–12. https://doi.org/10.2478/v10102-009-0001-7

AL-Ani M, Hmoshi R, Kanaan IA, Thanoon A (2019) Effect of pesticides on soil microorganisms. J Phys Conf Ser 1294:072007. 10.1088/1742-6596/1294/7/072007

Alavo TBC, Sermann H, Bochow H (2002) Biocontrol of aphids using *Verticillium lecanii* in greenhouse: factor reducing the effectiveness of the entomopathogenic fungus. Arch Phytopathol Plant Prot 34:407–424. https://doi.org/10.1080/713710567

Aldrich JR, Cantelo WW (1999) Suppression of Colorado potato beetle infestation by pheromone-mediated augmentation of the predatory spined soldier bug, *Podisus maculiventris* (Say) (Heteroptera: Pentatomidae). Agric For Entomol 1:209–217. https://doi.org/10.1046/j.1461-9563.1999.00026.x

Ali S, Naz I, Alamzeb M, Urrashid M (2019) Activity guided isolation of nematicidal constituents from the roots of *Berberis brevissima* Jafri and *Berberis parkeriana* Schneid. J Agric Sci 25:108–114. https://doi.org/10.15832/ankutbd.539012

Amelung W, Bossio D, de Vries W, Kögel-Knabner I, Lehmann J, Amundson R, Chabbi A (2020) Towards a global-scale soil climate mitigation strategy. Nature Commun 11:5427. 10.1038/s41467-020-18887-7

Anifandis G, Katsanaki K, Lagodonti G, Messini C, Simopoulou M, Dafopoulos K, Daponte A (2018) The effect of glyphosate on human sperm motility and sperm DNA fragmentation. Int J Environ Res Public Health 15:1117. https://doi.org/10.3390/ijerph15061117

Balog A, Hartel T, Loxdale HD, Wilson K (2017) Differences in the progress of the biopesticide revolution between the EU and other major crop-growing regions. Pest Manag Sci 73:2203–2208. https://doi.org/10.1002/ps.4596

Beecham JE, Seneff S (2015) The possible link between autism and glyphosate acting as glycine mimetic—a review of evidence from the literature with analysis. Mol Genet Med 9(4):1–16. https://doi.org/10.4172/1747-0862.1000187

Benbrook C (2016) Trends in glyphosate herbicide use in the United States and globally. Environ Sci Europe 28:3. https://doi.org/10.1186/s12302-016-0070-0

Benelli G, Govindarajan M, AlSalhi MS, Devanesan S, Maggi F (2018) High toxicity of camphene and γ-elemene from *Wedelia prostrata* essential oil against larvae of *Spodoptera litura* (Lepidoptera: Noctuidae). Environ Sci Pollut Res 25:10383–10391. https://doi.org/10.1007/s11356-017-9490-7

Beuret CJ, Zirulnik F, Giménez MS (2005) Effect of the herbicide glyphosate on liver lipoperoxidation in pregnant rats and their fetuses. Reprod Toxicol 19:501–504. https://doi.org/10.1016/j.reprotox.2004.09.009

Bhavya ML, Chandu AGS, Devi SS (2018) *Ocimum tenuiflorum* oil, a potential insecticide against rice weevil with anti-acetylcholinesterase activity. Ind Crops Prod 126:434–439. https://doi.org/10.1016/j.indcrop.2018.10.043

Bi Y, Gao C, Yu Z (2018) Rhabdopeptides from *Xenorhabdus budapestensis* SN84 and their nematicidal activities against *Meloidogyne incognita*. J Agric Food Chem 66:3833–3839. https://doi.org/10.1021/acs.jafc.8b00253

Bocquet L, Rivière C, Dermont C, Samaillie J, Hilbert JL, Halama P, Siahb A, Sahpaz S (2018) Antifungal activity of hop extracts and compounds against the wheat pathogen *Zymoseptoria tritici*. Ind Crops Prod 122:290–297. https://doi.org/10.1016/j.indcrop.2018.05.061

Borrelli P, Robinson DA, Panagos P, Lugato E, Yang JE, Alewell C, Ballabio C (2020) Land use and climate change impacts on global soil erosion by water (2015-2070). Proc Natl Acad Sci 117:21994–22001. 10.1073/pnas.2001403117

Brun T, Rabuske JE, Confortin TC, Luft L, Todero I, Fischer M, Zabot GL, Mazutti MA (2020) Weed control by metabolites produced from *Diaporthe schini*. Environ Technol 1–10. 10.1080/09593330.2020.1780477

Cardinale BJ, Duffy JE, Gonzalez A, Hooper DU, Perrings C, Venail P, Narwani A, Mace GM, Tilman D, Wardle AD, Kinzig AP, Daily GC, Loreau M, Grace JB, Larigauderie A, Srivastava DS, Naeem S (2012) Biodiversity loss and its impact on humanity. Nature 486:59–67. https://doi.org/10.1038/nature11148

Carson R (1962) Silent spring. Houghton Mifflin Harcourt

Carvalho FP (2017) Pesticides, environment, and food safety. Food Energy Secur 6:48–60. https://doi.org/10.1002/fes3.108

Castro P, Azul AM, Paiva J (2016) Conservation of biological resources: why does it matter? In: Castro P, Azeiteiro U, Bacelar-Nicolau P, Leal Filho W, Azul A (eds) Biodiversity and education for sustainable development. World sustainability series. Springer, Cham, pp 13–28. https://doi.org/10.1007/978-3-319-32318-3_2

Castro P, Azul AM, Leal Filho W, Azeiteiro UM (2019) Climate change-resilient agriculture and agroforestry: ecosystem services and sustainability. In: Leal Filho W (ed) Climate change management series. Springer. https://doi.org/10.1007/978-3-319-75004-0

Chandler D, Bailey AS, Tatchell GM, Davidson G, Greaves J, Grant WP (2011) The development, regulation and use of biopesticides for integrated pest management. Philos Trans R Soc B: Biol Sci 366:1987–1998. https://doi.org/10.1098/rstb.2010.0390

Chase JM, Blowes SA, Knight TM, Gerstner K, May F (2020) Ecosystem decay exacerbates biodiversity loss with habitat loss. Nature 584:238–243. https://doi.org/10.1038/s41586-020-2531-2

Chaudhary S, Kanwar RK, Sehgal A, Cahill DM, Barrow CJ, Sehgal R, Kanwar JR (2017) Progress on *Azadirachta indica* based biopesticides in replacing synthetic toxic pesticides. Frontiers Plant Sci 8:610. https://doi.org/10.3389/fpls.2017.00610

Chopra AK, Sharma MK, Chamoli S (2011) Bioaccumulation of organochlorine pesticides in aquatic system—an overview. Environ Monit Assess 173:905–916. https://doi.org/10.1007/s10661-010-1433-4

COM (2006) 373. Communication from the Commission to the Council, the European Parliament, the European Economic and Social Committee and the Committee of the Regions—a thematic strategy on the sustainable use of pesticides. [COM(2006) 372 final—Not published in the Official Journal]

COM (2019) 640 final. Communication from the Commission to the European Parliament, the European Council, the Council, the European Economic and Social Committee and the Committee of the Regions. The European Green Deal

COM (2020) 652 final. Proposal for a Decision of the European Parliament and of the Council on a General Union Environment Action Programme to 2030

Commission Implementing Regulation (EU) 2019/533 of 28 March 2019 concerning a coordinated multiannual control programme of the Union for 2020, 2021 and 2022 to ensure compliance with maximum residue levels of pesticides and to assess the consumer exposure to pesticide residues in and on food of plant and animal origin. Official J Eur Commun 88:28–41

Costa LG (2018) Organophosphorus compounds at 80: some old and new issues. Toxicol Sci 162:24–35. https://doi.org/10.1093/toxsci/kfx266

Council Directive 76/769/EEC of 27 July 1976 on the approximation of the laws, regulations and administrative provisions of the Member States relating to restrictions on the marketing and use of certain dangerous substances and preparations. Official J Eur Commun 262:201–203

Council Directive 91/414/EEC of 15 July 1991. Concerning the placing of plant protection products on the market of the European Communities. Official J Eur Commun 230:1–32

Curutiu C, Lazar V, Chifiriuc M (2017) Pesticides and antimicrobial resistance: from environmental compartments to animal and human infections. In: Grumezescu AM (ed) New pesticides and soil sensors. Academic Press, pp 373–392. https://doi.org/10.1016/b978-0-12-804299-1.00011-4

Czerniewicz P, Chrzanowski G, Sprawka I, Sytykiewicz H (2018) Aphicidal activity of selected Asteraceae essential oils and their effect on enzyme activities of the green peach aphid, *Myzus persicae* (Sulzer). Pestic Biochem Physiol 145:84–92. https://doi.org/10.1016/j.pestbp.2018.01.010

Damalas CA, Koutroubas SD (2018) Current status and recent developments in biopesticide use. Agriculture 8:13. https://doi.org/10.3390/agriculture8010013

Dardiotis E, Xiromerisioua G, Hadjichristodouloub C, Tsatsakisc A, Wilksd M, Hadjigeorgioua G (2013) The interplay between environmental and genetic factors in Parkinson's disease susceptibility: the evidence for pesticides. Toxicology 307:17–23. https://doi.org/10.1016/j.tox.2012.12.016

Dentzman K, Gunderson R, Jussaume R (2016) Techno-optimism as a barrier to overcoming herbicide resistance: comparing farmer perceptions of the future potential of herbicides. J Rural Stud 48:22–32. https://doi.org/10.1016/j.jrurstud.2016.09.006

Dhananjayan V (2013) Accumulation pattern of persistent organochlorine pesticides in liver tissues of various species of birds from India. Environ Sci Pollut Res 20:3149–3156. https://doi.org/10.1007/s11356-012-1225-1

Dhananjayan V, Jayakumar S, Ravichandran B (2020) Conventional methods of pesticide application in agricultural field and fate of the pesticides in the environment and human health.

In: Rakhimol KR, Volova TKJ (eds) Controlled release of pesticides for sustainable agriculture. Springer, Cham, pp 1–39. https://doi.org/10.1007/978-3-030-23396-9_1
Directive 98/8/EC of the European Parliament and of the Council of 16 February 1998 concerning the placing of biocidal products on the market. Official J Eur Commun 123:1–63
Directive 2000/60/EC of the European Parliament and of the Council of 23 October 2000 establishing a framework for Community action in the field of water policy. Official J Eur Commun 327:1–73
Directive 2008/98/EC of the European Parliament and of the Council of 19 November 2008 on waste and repealing certain Directives. Official J Eur Commun 312:3–30
Directive 2009/128/EC of the European Parliament and of the Council of 21 October 2009 establishing a framework for Community action to achieve the sustainable use of pesticides. Official J Eur Commun 309:71–86
Douches DS, Pett W, Santos F, Coombs J, Grafius E, Li W, Metry EA, Nasr el-Din T, Madkour M (2004) Field and storage testing *Bt* potatoes for resistance to potato tuberworm (Lepidoptera: Gelichiidae). J Econ Entomol 97:1425–1431. https://doi.org/10.1093/jee/97.4.1425
EC, European Commission (2007) EU Policy for a sustainable use of pesticides—the story behind the Strategy. Retrieved from: https://ec.europa.eu/environment/archives/ppps/pdf/pesticides_en.pdf
EC, European Commission (2019) Organic farming in the EU—a fast growing sector. EU Agricultural Markets Briefs, 13. Available at: https://ec.europa.eu/info/sites/info/files/food-farming-fisheries/farming/documents/market-brief-organic-farming-in-the-eu_mar2019_en.pdf
EEA, European Environment Agency (2019) The European environment—state and outlook 2020. Knowledge for transition to a sustainable Europe. European Environment Agency, Copenhagen
EFSA (European Food Safety Authority), Medina-Pastor P and Triacchini G (2020) The 2018 European Union report on pesticide residues in food. EFSA J 2020 18(4):6057, 103 pp. https://doi.org/10.2903/j.efsa.2020.6057
Emge RG, Melching JS, Kingsolver CH (1981) Epidemiology of *Puccinia chondrillina*, a rust pathogen for the biological control of rush skeleton weed in the United States. Phytopathology 71:839–843
Eski A, Demir D, Sezen K, Demirbağ ZA (2017) New biopesticide from a local *Bacillus thuringiensis* var. *tenebrionis* (Xd3) against alder leaf beetle (Coleoptera: Chrysomelidae). World J Microbiol Biotechnol 33:95. https://doi.org/10.1007/s11274-017-2263-0
Eurobarometer (2019) Special Eurobarometer Wave EB91.3. Food safety in the EU. Brussels. Retrieved from https://www.efsa.europa.eu/sites/default/files/corporate_publications/files/Eurobarometer2019_Food-safety-in-the-EU_Full-report.pdf
European Court of Auditors (2020) Special Report—sustainable use of plant protection products: limited progress in measuring and reducing risks. Special Report 5/2020. https://www.arc2020.eu/wp-content/uploads/2020/02/SR_Pesticides_EN.pdf
FAO (2020) Pesticides Use Statistics. Global, regional and country trends 1990–2018. FAOSTAT Analytical Brief Series, Rome, FAO. http://www.fao.org/3/cb0488en/CB0488EN.pdf
Fujimoto H, Itoh K, Yamamoto M, Kyozuka J, Shimamoto K (1993) Insect resistant rice generated by introduction of a modified δ-endotoxin gene of *Bacillus thuringiensis*. Bio/Technology 11:1151–1155. https://doi.org/10.1038/nbt1093-1151
Gangstad EO (ed) (1982) Weed control methods for recreation facilities management. CRC Press, Boca Raton. 10.1201/9781351077705
Godfray HCJ, Garnett T (2014) Food security and sustainable intensification. Philos Trans R Soc B 369:20120273. https://doi.org/10.1098/rstb.2012.0273
Gronwald JW, Plaisance K L, Ide DA, Wyse DL (2002) Assessment of *Pseudomonas syringae* pv. *tagetis* as a biocontrol agent for Canada thistle. Weed Sci 50:397–404. https://doi.org/10.1614/0043-1745(2002)050%5b0397:aopspt%5d2.0.co;2
Haack SE, Ivors KL, Holmes GJ, Förster H, Adaskaveg JE (2018) Natamycin, a new biofungicide for managing crown rot of strawberry caused by QoI-resistant *Colletotrichum acutatum*. Plant Dis 102:1687–1695. https://doi.org/10.1094/PDIS-12-17-2033-RE

Hashem M, Abo-Elyousr KA (2011) Management of the root-knot nematode *Meloidogyne incognita* on tomato with combinations of different biocontrol organisms. Crop Prot 30:285–292. https://doi.org/10.1016/j.cropro.2010.12.009

Humphries MS, Myburgh JG, Campbell R, Buah-Kwofie A, Combrink X (2021) Organochlorine pesticide bioaccumulation in wild Nile crocodile (*Crocodylus niloticus*) fat tissues: Environmental influences on changing residue levels and contaminant profiles. Sci Total Environ 753: https://doi.org/10.1016/j.scitotenv.2020.142068

IARC (2015) IARC Monographs Volume 112: evaluation of five organophosphate insecticides and herbicides. WHO. Retrieved from: https://www.iarc.who.int/wp-content/uploads/2018/07/MonographVolume112-1.pdf

Irma A, Meryandini A, Rupaedah B (2018) Biofungicide producing bacteria: an in vitro inhibitor of *Ganoderma boninense*. HAYATI J Biosci 25:151. https://doi.org/10.4308/hjb.25.4.151

Jardim IN, Oliveira DF, Silva GH, Campos VP, de Souza PE (2018) (E)-cinnamaldehyde from the essential oil of *Cinnamomum cassia* controls *Meloidogyne incognita* in soybean plants. J Pest Sci 91:479–487. https://doi.org/10.1007/s10340-017-0850-3

Jarman WM, Ballschmiter K (2012) From coal to DDT: the history of the development of the pesticide DDT from synthetic dyes till Silent Spring. Endeavour 36:131–142. https://doi.org/10.1016/j.endeavour.2012.10.003

Jayasumana C, Gunatilake S, Senanayake P (2014) Glyphosate, hard water and nephrotoxic metals: are they the culprits behind the epidemic of chronic kidney disease of unknown etiology in Sri Lanka? Int J Environ Res Public Health 11:2125–2147. https://doi.org/10.3390/ijerph110202125

Kalliora Ch, Mamoulakis Ch, Vasilopoulos E, Stamatiades GA, Kalafati L et al (2018) Association of pesticide exposure with human congenital abnormalities. Toxicol Appl Pharmacol 346:58–75. https://doi.org/10.1016/j.taap.2018.03.025

Khater HF, Ali AM, Abouelella GA, Marawan MA, Govindarajan M, Murugan K, Abbas RZ, Vaz NP, Benelli G (2018) Toxicity and growth inhibition potential of vetiver, cinnamon, and lavender essential oils and their blends against larvae of the sheep blowfly, *Lucilia sericata*. Int J Dermatol 57:449–457. https://doi.org/10.1111/ijd.13828

Kim K, Kabir E, Jahan S (2017) Exposure to pesticides and the associated human health effects. Sci Total Environ 575:525–535. https://doi.org/10.1016/j.scitotenv.2016.09.009

Kirk WW, Schafer RS (2015) Efficacy of new active ingredient formulations and new biopesticides for managing *Fusarium* root rot disease of gladiolus hybrids. Acta Hort 1105:55–60. https://doi.org/10.17660/ActaHortic.2015.1105.8

Korres NE, Burgos NR, Duke SO (eds) (2018) Weed control: sustainability, hazards, and risks in cropping systems worldwide. CRC Press, Boca Raton. https://doi.org/10.1201/9781315155913

Koutros S, Silverman D, Alavanja M, Andreotti G, Lerro C, Heltshe S, Lynch CF, Sandler DP, Blair A, Beane Freeman LE (2016) Occupational exposure to pesticides and bladder cancer risk. Int J Epidemiol 45:792–805. https://doi.org/10.1093/ije/dyv195

Kumar K, Sridhar J, Murali-Baskaran R, Senthil-Nathan S, Kaushal P, Dara S, Arthurs S (2018) Microbial biopesticides for insect pest management in India: current status and future prospects. J Invertebr Pathol 165:74–81. https://doi.org/10.1016/j.jip.2018.10.008

Kumar S, Singh A (2015) Biopesticides: present status and the future prospects. J Fertilizers Pesticides 6: https://doi.org/10.4172/2471-2728.1000e129

Kumar V, Singh S, Upadhyay N (2019) Effects of organophosphate pesticides on siderophore producing soils microorganisms. Biocatal Agric Biotechnol 21:101359. https://doi.org/10.1016/j.bcab.2019.101359

Le T, Chang Y, Park D (2018) Economic development and environmental sustainability: evidence from Asia. Empirical Economics 57:1129–1156. https://doi.org/10.1007/s00181-018-1494-8

Lengai G, Muthomi J (2018) Biopesticides and their role in sustainable agricultural production. J Biosci Med 6:7–41. https://doi.org/10.4236/jbm.2018.66002

Lerro CC, Freeman LEB, DellaValle CT, Andreotti G, Hoffman JN, Koutros S, Parks CH, Shrestha S, Alavanja MCR (2020) Pesticide exposure and incident thyroid cancer among male pesticide applicators in agricultural health study. Environ Int 146:106187. https://doi.org/10.1016/j.envint.2020.106187

Liu X, Cao A, Yan D, Ouyang C, Wang Q, Li Y (2021) Overview of mechanisms and uses of biopesticides. Int J Pest Manage 67:65–72. https://doi.org/10.1080/09670874.2019.1664789

Lopez SBG, Guimarães-Ribeiro V, Rodriguez JVG, Dorand FAPS, Salles TS, Sá-Guimarães TE, Alvarenga ESL, Melo ACA, Almeida RV, Moreira MF (2019) RNAi-based bioinsecticide for *Aedes mosquito* control. Sci Rep 9:4038. https://doi.org/10.1038/s41598-019-39666-5

Mackay D, Fraser A (2000) Bioaccumulation of persistent organic chemicals: mechanisms and models. Environ Pollut 110:375–391. https://doi.org/10.1016/S0269-7491(00)00162-7

MacLeod A, Pautasso M, Jeger M, Haines-Young R (2010) Evolution of the international regulation of plant pests and challenges for future plant health. Food Secur 2:49–70. https://doi.org/10.1007/s12571-010-0054-7

Mahmood I, Imadi SR, Shazadi K, Gul A, Hakeem KR (2016) Effects of pesticides on environment. In: Hakeem KR, Akhtar MS, Abdullah SNA (eds) Plant, soil and microbes. Springer, Cham, pp 253–269. https://doi.org/10.1007/978-3-319-27455-3_13

Mangao AM, Arreola SLB, San Gabriel EV, Salamanez KC (2020) Aqueous extract from leaves of *Ludwigia hyssopifolia* (G. Don) Exell as potential bioherbicide. J Sci Food Agric 100:1185–1194. https://doi.org/10.1002/jsfa.10128

Matsuo N (2019) Discovery and development of pyrethroid insecticides. Proc Jpn Acad Ser B 95:378–400. https://doi.org/10.2183/pjab.95.027

Mcafee A (2017) A brief history of pesticides. Am Bee J 157:781–783. https://doi.org/10.1101/098897.5

Md Meftaul I, Venkateswarlu K, Dharmarajan R, Annamalai P, Megharaj M (2019) Pesticides in the urban environment: a potential threat that knocks at the door. Sci Total Environ 711:134612. https://doi.org/10.1016/j.scitotenv.2019.134612

Mensah R, Moore C, Watts N, Deseo MA, Glennie P, Pitt A (2014) Discovery and development of a new semiochemical biopesticide for cotton pest management: assessment of extract effects on the cotton pest *Helicoverpa* spp. Entomol Exp Appl 152:1–15. https://doi.org/10.1111/eea.12198

Mihalache G, Balaes T, Gostin I, Stefan M, Coutte F, Krier F (2018) Lipopeptides produced by *Bacillus subtilis* as new biocontrol products against fusariosis in ornamental plants. Environ Sci Pollut Res 25:29784–29793. https://doi.org/10.1007/s11356-017-9162-7

Morton V, Staub T (2008) A short history of fungicides. APSnet features, 10, 2008–0308. https://pdfs.semanticscholar.org/5f53/60373ce2c2c7723f08c8a71606d55a4555ce.pdf

Mullins AJ, Murray JAH, Bull MJ, Jenner M, Jones C, Webster G, Green AE, Neill DR, Connor TR, Parkhill J, Challis GL, Mahenthiralingam E (2019) Genome mining identifies cepacin as a plant-protective metabolite of the biopesticidal bacterium *Burkholderia ambifaria*. Nat Microbiol 4:996–1005. https://doi.org/10.1038/s41564-019-0383-z

Ngo MA, O'Malley M, Maibach HI (2012) Pesticide-related dermatoses. In: Rustemeyer T, Elsner P, John SM, Maibach HI (eds) Kanerva's occupational dermatology. Springer, Berlin, Heidelberg. https://doi.org/10.1007/978-3-642-02035-3_68

Nguyen LTT, Jang JY, Kim TY, Yu NH, Park AR, Lee S, Bae CH, Yeo JH, Hur JS, Park HW, Kim JC (2018) Nematicidal activity of verrucarin A and roridin A isolated from *Myrothecium verrucaria* against *Meloidogyne incognita*. Pestic Biochem Physiol 148:133–143. https://doi.org/10.1016/j.pestbp.2018.04.012

Nwani CD, Wazir Singh L, Naresh SN, Kumar R, Basdeo K, Satish Kumar S (2010) Toxicity of the herbicide atrazine: effects on lipid peroxidation and activities of antioxidant enzymes in the freshwater fish *Channa punctatus* (Bloch). Int J Environ Res Public Health 7:3298–3312. https://doi.org/10.3390/ijerph7083298

Oliveira AP, Santos AA, Santana AS, Lima APS, Melo CR, Santana ED, Sampaio TS, Blank AF, Araújo APA, Cristaldo PF, Bacci L (2018) Essential oil of *Lippia sidoides* and its major compound thymol: toxicity and walking response of populations of *Sitophilus zeamais* (Coleoptera: Curculionidae). Crop Prot 112:33–38. https://doi.org/10.1016/j.cropro.2018.05.011

Olson S (2015) An analysis of the biopesticide market now and where is going. Outlooks Pest Manage 26:203–206. https://doi.org/10.1564/v26_oct_04

Onwona-Kwakye M, Plants-Paris K, Keita K, Lee J, Van den Brink P, Hogarh J, Darkoh J (2020) Pesticides decrease bacterial diversity and abundance of irrigated rice fields. Microorganisms 8:318. https://doi.org/10.3390/microorganisms8030318

Pannacci E, Masi M, Farneselli M, Tei F (2020) Evaluation of mugwort (*Artemisia vulgaris* L.) aqueous extract as a potential bioherbicide to control *Amaranthus retroflexus* L. in maize. Agriculture 10:642. https://doi.org/10.3390/agriculture10120642

Pantzar M, Strube R, Gionfra S, Modée K (2018) Sustainable consumption—policy approaches for systems change. Policy paper produced for the think 2030 conference. Institute for European Environmental Policy. Retrieved from: https://www.scp-centre.org/wp-content/uploads/2019/01/Think-2030-Sustainable-consumption.pdf

Pavela R, Waffo-Teguo P, Biais B, Richard T, Mérillon J-M (2017) *Vitis vinifera* canes, a source of stilbenoids against *Spodoptera littoralis* larvae. J Pest Sci 90:961–970. https://doi.org/10.1007/s10340-017-0836-1

Pavela R, Maggi F, Petrelli R, Cappellacci L, Buccioni M, Palmieri A, Canale A, Benelli G (2020) Outstanding insecticidal activity and sublethal effects of *Carlina acaulis* root essential oil on the housefly, *Musca domestica*, with insights on its toxicity on human cells. Food Chem Toxicol 136: https://doi.org/10.1016/j.fct.2019.111037

Peshin R, Bandral R, Zhang W, Wilson L, Dhawan A (2009) Integrated pest management: a global overview of history, programs and adoption. In: Peshin R, Dhawan AK (eds) Integrated pest management: innovation-development process. Springer, Dordrecht, pp 1–49. https://doi.org/10.1007/978-1-4020-8992-3_1

Puig CG, Gonçalves RF, Valentão P, Andrade PB, Reigosa MJ, Pedrol N (2018) The consistency between phytotoxic effects and the dynamics of allelochemicals release from *Eucalyptus globulus* leaves used as bioherbicide green manure. J Chem Ecol 44:658–670. https://doi.org/10.1007/s10886-018-0983-8

Qadri H, Bhat R (2020) The concerns for global sustainability of freshwater ecosystems. In: Qadri H, Bhat R, Mehmood M, Dar G (eds) Freshwater pollution dynamics and remediation. Springer, Singapore, pp 1–13. https://doi.org/10.1007/978-981-13-8277-2_1

Ramezani S, Saharkhiz MJ, Ramezani F, Fotokian MH (2008) Use of essential oils as bioherbicides. J Essent Oil Bearing Plants 11:319–327. https://doi.org/10.1080/0972060X.2008.10643636

Rao GR, Kumar CS, Sireesha K, Kumar PL (2015) Role of nucleopolyhedroviruses (NPVs) in the management of lepidopteran pests in Asia. In: Sree K, Varma A (eds) Biocontrol of lepidopteran pests. Soil biology, vol 43. Springer, Cham. https://doi.org/10.1007/978-3-319-14499-3_2

Regulation (EC) 1107/2009 of the European Parliament and of the Council of 21 October 2009 concerning the placing of plant protection products on the market and repealing Council Directives 79/117/EEC and 91/414/EEC. Official J Eur Commun 309:1–50

Regulation (EC) 1185/2009 of the European Parliament and of the Council of 25 November 2009 concerning statistics on pesticides. Official J Eur Commun 324:1–22

Regulation (EU) 528/2012 of the European Parliament and of the Council of 22 May 2012 Concerning the making available on the market and use of biocidal products. Official J Eur Commun 167:1–123

Riga E, Lacey LA, Guerra N (2008) *Muscodor albus*, a potential biocontrol agent against plant-parasitic nematodes of economically important vegetable crops in Washington State, USA. Biol Control 45:380–385. https://doi.org/10.1016/j.biocontrol.2008.01.002

Ríos-González A, Jansen K, Sánchez-Pérez HJ (2013) Pesticide risk perceptions and the differences between farmers and extensionists: towards a knowledge-in-context model. Environ Res 124:43–53. https://doi.org/10.1016/j.envres.2013.03.006

Rodrigues AS, Bezerra JW, Silva VB, Costa AR, Rodrigues FC, Linhares KV (2020) Phytotoxic activity of *Guapira graciliflora* (Nyctaginaceae) on weeds. J Agric Stud 8:287–301. https://doi.org/10.5296/jas.v8i1.15420

Saeidi K, Mirfakhraie S (2017) Chemical composition and insecticidal activity *Mentha piperita* L. essential oil against the cowpea seed beetle *Callosobruchus maculatus* F. (Coleoptera: Bruchidae). J Entomol Acarol Res 49:127–134. https://doi.org/10.4081/jear.2017.6769

Salem MZM, Behiry SI, EL-Hefny M (2019) Inhibition of *Fusarium culmorum*, *Penicillium chrysogenum* and *Rhizoctonia solani* by n-hexane extracts of three plant species as a wood-treated oil fungicide. J Appl Microbiol 126:1683–1699. https://doi.org/10.1111/jam.14256

Samada LH, Tambunan USF (2020) Biopesticides as promising alternatives to chemical pesticides: a review of their current and future status. OnLine J Biol Sci 20:66–76. https://doi.org/10.3844/ojbsci.2020.66.76

Scavo A, Pandino G, Restuccia A, Mauromicale G (2020) Leaf extracts of cultivated cardoon as potential bioherbicide. Sci Hortic 261: https://doi.org/10.1016/j.scienta.2019.109024

Seifert J (2005) Neonicotinoids. In: Wexler P (ed) Encyclopedia of toxicology, 2nd edn. Elsevier, pp 196–200. https://doi.org/10.1016/b0-12-369400-0/00675-x

Shao H, Zhang Y (2017) Non-target effects on soil microbial parameters of the synthetic pesticide carbendazim with the biopesticides cantharidin and norcantharidin. Sci Rep 7:5521. https://doi.org/10.1038/s41598-017-05923-8

Sheikh MH, Naher L, Karim SR, Sannasi P (2020) An assessment of in vitro herbicidal potential of fungal metabolites against parthenium weed (*Parthenium hysterophorus* L.). In: IOP conference series: earth and environmental science, 596, 012087. https://doi.org/10.1088/1755-1315/596/1/012087

Sheikh T, Wheeler TA, Dotray PA, Zak JC (2001) Biological control of Woollyleaf Bursage *(Ambrosia grayi)* with *Pseudomonas syringae* pv. *tagetis*1. Weed Technol 15:375–381. https://doi.org/10.1614/0890-037x(2001)015%5b0375:bcowba%5d2.0.co;2

Shu B, Zhang J, Cui G, Sun R, Yi X, Zhong G (2018) Azadirachtin affects the growth of *Spodoptera litura* Fabricius by inducing apoptosis in larval midgut. Frontiers Physiol 9:137. https://doi.org/10.3389/fphys.2018.00137

Silva IM, Martins GF, Melo CR, Santana AS, Faro RR, Blank AF, Alves PB, Picanço MC, Cristaldo PF, Araújo APA, Bacci L (2018) Alternative control of *Aedes aegypti* resistant to pyrethroids: lethal and sublethal effects of monoterpene bioinsecticides. Pest Manage Sci 74:1001–1012. https://doi.org/10.1002/ps.4801

Silvério FO, De Alvarenga ES, Moreno SC, Picanço MC (2009) Synthesis and insecticidal activity of new pyrethroids. Pest Manage Sci 65:900–905. https://doi.org/10.1002/ps.1771

Singh US, Zaidi NW (2017) Trichoderma—an impeccable plant health booster. In: Anwer MA (ed) Biopesticides and bioagents: novel tools for pest management. Apple Academic Press, USA

Soares MA, Campos MR, Geraldo LC, Carvalho A, Haro MM, Lavoir A, Biondi A, Zappalà L, Desneux N (2019) Botanical insecticide and natural enemies: a potential combination for pest management against *Tuta absoluta*. J Pest Sci 92:1433–1443. 10.1007/s10340-018-01074-5

Suresh U, Murugan K, Panneerselvam C, Rajaganesh R, Roni M, Aziz AT, Al-Aoh HAN, Trivedi S, Rehman H, Kumar S, Higuchi A, Canale A, Benelli G (2018) *Suaeda maritima*-based herbal coils and green nanoparticles as potential biopesticides against the dengue vector *Aedes aegypti* and the tobacco cutworm *Spodoptera litura*. Physiol Mol Plant Pathol 101:225–235. https://doi.org/10.1016/j.pmpp.2017.01.002

Todero I, Confortin TC, Luft L, Brun T, Ugalde GA, de Almeida TC, Arnemann JA, Zabot GL, Mazutti MA (2018) Formulation of a bioherbicide with metabolites from *Phoma* sp. Sci Hortic 241:285–292. https://doi.org/10.1016/j.scienta.2018.07.009

Tripathi A, Sharma N, Tripathi N (2010) Biological control of plant diseases: an overview and the *Trichoder masystem* as biocontrol agents. In: Arya A, Perelló A (eds) Management of fungal plants pathogens, pp 121–137. https://doi.org/10.1079/9781845936037.0000

UN (2015) Transforming our world: the 2030 agenda for sustainable development—A/RES/70/1. https://sdgs.un.org/2030agenda

UN DESA (2018) World urbanization prospects 2018. https://population.un.org/wup/

UN DESA (2019) The 2019 revision of world population prospects. https://population.un.org/wpp/

Van Bruggen AHC, He MM, Shin K, Mai V, Jeong KC, Finckh MR, Morris JG Jr (2018) Environmental and health effects of the herbicide glyphosate. Sci Total Environ 616–617:255–268. https://doi.org/10.1016/j.scitotenv.2017.10.309

Villaverde JJ, Sandín-España P, Sevilla-Morán B, López-Goti C, Alonso-Prados JL (2016) Biopesticides from natural products: current development, legislative framework, and future trends. BioResources 11:5618–5640. https://doi.org/10.15376/biores.11.2

Wallace RL, Hirkala DL, Nelson LM (2018) Mechanisms of action of three isolates of *Pseudomonas fluorescens* active against postharvest grey mold decay of apple during commercial storage. Biol Control 117:13–20. https://doi.org/10.1016/j.biocontrol.2017.08.019

Wang J, Deng Y, Ma Y (2017) Relationships between Safe Pesticide Practice and Perceived Benefits and Subjective Norm, and the Moderation Role of Information Acquisition: Evidence from 971 Farmers in China. Int J Environ Res Public Health 14:962. https://doi.org/10.3390/ijerph14090962

Wolfand J, Seller C, Bell C, Cho Y, Oetjen K, Hogue T, Luthy R (2019) Occurrence of urban-use pesticides and management with enhanced stormwater control measures at the watershed scale. Environ Sci Technol 53:3634–3644. https://doi.org/10.1021/acs.est.8b05833

WWF (2020) Living planet report 2020—bending the curve of biodiversity loss. https://livingplanet.panda.org/

Youssif MAI, Ramadan MM (2020) Influence of different pest control agents on some biological aspects of *Chrysoperla carnea* (Stephens). Zagazig J Agric Res 47:507–518. https://doi.org/10.21608/zjar.2020.94491

Yu H, Li Y, Li X, Romeis J, Wu K (2013) Expression of Cry1Ac in transgenic *Bt* soybean lines and their efficiency in controlling lepidopteran pests. Pest Manage Sci 69:1326–1333. https://doi.org/10.1002/ps.3508

Zaganas I, Kapetanaki S, Mastorodemos V, Kanavouras K, Colosio C, Wilks M, Tsatsakis A (2013) Linking pesticide exposure and dementia: What is the evidence? Toxicology 307:3–11. https://doi.org/10.1016/j.tox.2013.02.002

Zaidat SAE, Mouhouche F, Babaali D, Abdessemed N, De Cara M, Hammache M (2020) Nematicidal activity of aqueous and organic extracts of local plants against *Meloidogyne incognita* (Kofoid and White) Chitwood in Algeria under laboratory and greenhouse conditions. Egypt J Biol Pest Control 30:1–8. https://doi.org/10.1186/s41938-020-00242-z

Zhou T, Neal JC (1995) Annual bluegrass (*Poa annua*) control with *Xanthomonas campestris* pv. *poannua* in New York State. Weed Technol 9:173–177. https://doi.org/10.1017/S0890037X00023137

Ziedan EHE, Farrag ESH (2008) Fumigation of peach fruits with essential oils to control postharvest decay. Res J Agric Biol Sci 4:512–519

Zimdahl RL (2018) Fundamentals of weed science, 5th edn. Academic Press, London

Barriers to Sustainable Development in Agriculture

Rozélia Laurett, Arminda Paço, and Emerson Wagner Mainardes

1 Introduction

Today, promoting sustainability in agriculture is one of the most important aspects of sustainable development (SD) (Bastan et al. 2018). SD in agriculture intends agricultural activity to be profitable, contribute to improved quality of life in the rural and urban population, and to be environmentally responsible (Menalled et al. 2008). Sustainable agriculture also seeks to make best possible use of natural resources, technology (Pretty 2008) and natural fertilizers (Schaller 1993; Bagheri 2010; Van Thanh and Yapwattanaphun 2015; Zeweld et al. 2017), protect the soil (Horrigan et al. 2002), minimize the use of chemical products (Rodriguez et al. 2009; Knutson et al. 2011) and choose organic production (Rodriguez et al. 2009).

Sustainable agriculture also means reducing deforestation and air pollution, economizing on water and avoiding damage to the environment, since farming has contributed greatly to various environmental problems (Pham and Smith 2013). It must also ensure sufficient food production for a growing population. The world's population is estimated to reach 9 billion in 2030, and 10 billion in 2050 (United Nations Department of Economic and Social Affairs (UNDESA), 2017), which will require an increase of approximately 50% in food in 2030 (United Nations (UN), 2012) and 70% in 2050 (FAO, 2011; Balmford et al. 2012).

Therefore, various questions are involved in making agriculture more sustainable, but consensus has not yet been reached. On this road to making farming more sustainable, various obstacles can emerge (Milbrath 1995; Carolan 2006; Ma et al. 2009; Rodriguez et al. 2009; Sassenrath et al. 2010; Leite et al. 2014; Kata and

R. Laurett (✉) · E. W. Mainardes
FUCAPE Business School, Vitória, Espírito Santo, Brazil and NECE-UBI - Research Centre for Business Sciences, Rua Marquês d'Ávila e Bolama, 6201-001 Covilhã, Portugal

A. Paço
Universidade da Beira Interior, NECE-UBI - Research Centre for Business Sciences, Rua Marquês d'Ávila e Bolama, 6201-001 Covilhã, Portugal

W. Leal Filho et al. (eds.), *Sustainability in Natural Resources Management and Land Planning*, World Sustainability Series,
https://doi.org/10.1007/978-3-030-76624-5_9

Kusz 2015; Martin et al. 2015; Grover and Gruver 2017; Cederholm Björklund 2018). These barriers may be defined as the obstacles perceived in relation to implementation of sustainability efforts (Laurett and do Paço 2019, p. 1). However, few studies seek to identify the variables hindering SD in agriculture, from the perspective of farmers themselves, and few studies attempt to group those barriers. Therefore, this research aims to identify and group the barriers to adoption of sustainable development in agriculture, from the perception of Brazilian organic family farmers.

This study is also justified by the relevance of a country such as Brazil, considered one of the world's major food producers (Ferreira et al. 2012) and the country with the fourth largest area under agriculture (Lowder et al. 2014). Brazil has approximately 4.367.902 family farms, with 12.3 million people involved in family farming (IBGE 2009). This means that family farming has an important role in the Brazilian economy (da Veiga 1996; Guthman 2004). Knowing the barriers to sustainability can contribute to proposing actions to overcome these obstacles, thereby contributing to farmers' greater commitment to sustainability.

2 Barriers to the Implementation of Sustainable Practices in Agriculture

The literature contains various studies identifying the barriers to SD in agriculture. Standing out among them are: the lack of financial resources (Sassenrath et al. 2010; Kata and Kusz 2015; Cederholm Björklund 2018), higher production costs (Kata and Kusz 2015), the need to make high initial investment (Ma et al. 2009) when adopting certain technology, such as changing irrigation processes and/or applying for organic certification. Regarding the process of sustainable certification, Kata and Kusz (2015) mention the low number of organisations issuing such certification and the excessive rules and obligations demanded by those certifying entities as major barriers.

Other obstacles can be listed, such as: the lack of government support to help farmers become more sustainable (Carolan 2006; Cederholm Björklund 2018), the difficulty in understanding what consumers really want or need (Ma et al. 2009; Leite et al. 2014; Grover and Gruver, 2017; Cederholm Björklund 2018), farmers' lack of information about alternative methods used in the agricultural sector that can make farms more sustainable (Carolan 2006; Leite et al. 2014; Martin et al. 2015; Kata and Kusz 2015), the difficulty in innovating and introducing new ways of working (Martin et al. 2015; Cederholm Björklund 2018) and the lack of successful examples linked to sustainability, i.e., a reference (Rodriguez et al. 2009),

Besides these barriers, also identified are the lack of time (Grover and Gruver 2017), the traditional model of family management being resistant to change (Rodriguez et al. 2009; Cederholm Björklund 2018); the lack of technical support or specialist help/advice (Carolan 2006; Leite et al. 2014; Kata and Kusz 2015; Cederholm Björklund 2018), and appropriate information and technical knowledge about sustainability (Leite et al. 2014; Martin et al. 2015); the lack of understandable legislation and specific regulations (Leite et al. 2014; Czyzewski et al. 2018; Cederholm Björklund 2018), the excess of state and federal laws and regulations (Grover and Gruver 2017) and the difficulty in adopting new technology (Sassenrath et al. 2010; Martin et al. 2015), which can all be barriers to the implementation of more sustainable actions in agriculture.

Some barriers identified were related to the individual, such as: farmers' reluctance to change their behaviour (Carolan 2006; Sassenrath et al. 2010; Rodriguez et al. 2009; Cederholm Björklund 2018), risk aversion (Ma et al. 2009), and the fear of using new work methods. Others found in the literature were the lack of training directed to sustainable development (Rodriguez et al. 2009) and the difficulty in understanding what sustainable development is (Rodriguez et al. 2009). This literature review revealed that numerous barriers prevent farmers from making farming more sustainable. Therefore, identifying these barriers is the first step towards the adoption of more sustainable actions (Horhota et al. 2014), whether in agriculture or in other sectors.

3 Methods

To fulfil the aim of the research, exploratory, qualitative research was adopted. The necessary characteristics to be part of the sample were: being resident in Brazil, carrying out organic farming and belonging to family farming as set out in Law 11.326/2006 (Brazil 2006). The choice of organic agriculture was due to various authors considering this type of farming as being more sustainable that conventional farming (Sandhu et al. 2010; Seufert et al. 2012), but it should be noted that there is no consensus on this matter (Tal 2018). Family farming was chosen due to its contribution by adopting more sustainable farming practices (Tavernier and Tolomeo 2004; Ortiz et al. 2018).

The first farmers were contacted by telephone and invited to participate in the study. When they accepted, they were interviewed and asked to indicate others, leading to snowball sampling (Hair et al. 2015). A total of 13 organic farmers were interviewed. The interviewees were informed individually that the interviews would be recorded and transcribed, and that their anonymity was assured. Therefore, the farmers were identified as F1 to F13. The interviews were held face-to-face in October 2018. Regarding the characterisation of the 13 organic farmers interviewed, 10 were male and 3 female. Their age varied from 27 to 57, averaging 45.9 years. As for education, the majority had no more than average schooling (84% of the sample). Concerning marital status, they were all married with children, except for one who was single and had no children. As for the length of time in farming, the average was 31.6 years (ranging from 14 to 45 years).

The interviews were semi-structured, focusing on two questions. The first aimed to find out how the farmer operated in agriculture and gather demographic data to characterise the sample. The second focused on understanding the barriers hindering more sustainable agriculture, i.e., the factors preventing farmers from operating more sustainably. Finally, the 13 interviews were transcribed completely and subject to content analysis. The interviews were analysed using NVIVO software.

4 Data Analysis and Findings

In the course of the interviews asking farmers about the main barriers to sustainability in agriculture, 20 such barriers emerged. These were grouped in six factors: (1) Perceived semantic barriers; (2) Perceived regulation barriers; (3) Perceived individual barriers; (4) Perceveid macro-systemic barriers; (5) Perceveid capital/financial barriers; (6) Perceveid administrative/organizational barriers, as set out in Fig. 1.

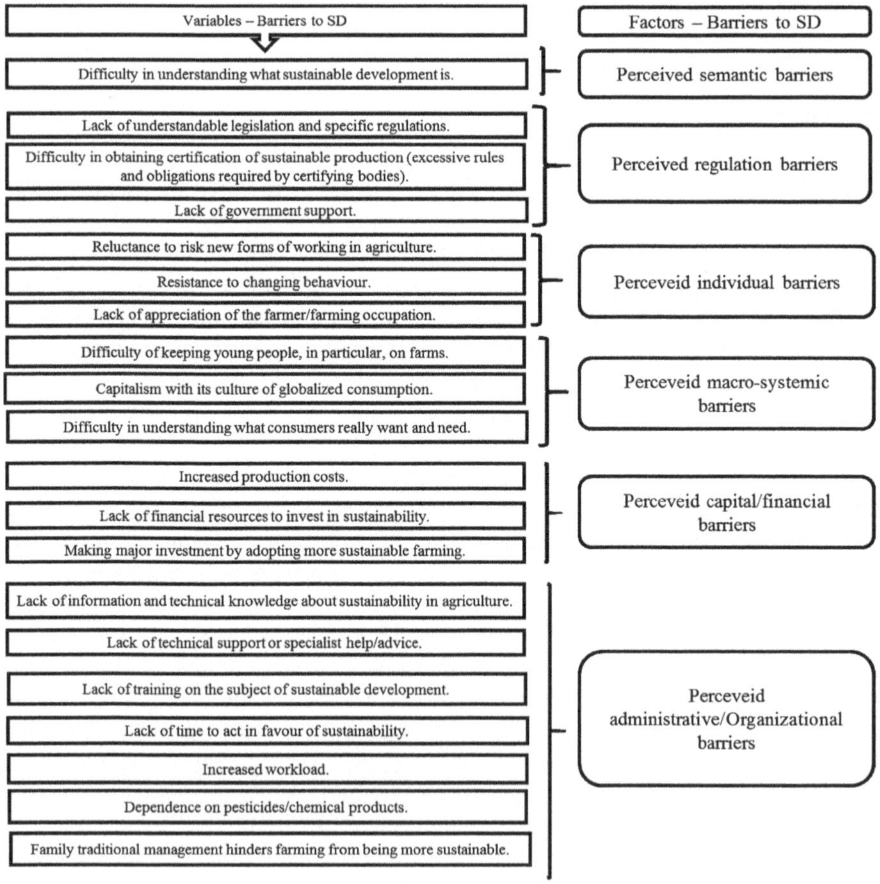

Fig. 1 Barriers to SD in agriculture. *Source* Elaborated by the authors

4.1 Perceived Semantic Barriers

The difficulty in understanding what sustainable development is was identified in various studies as one of the main barriers to the implementation of sustainability (Williams and Dair 2007; Rodriguez et al. 2009; Evans et al. 2012; Duarte 2015). Figure 1 shows that this was the first obstacle perceived by farmers (F1, F5, F6, F9, F12 and F13) in making agriculture more sustainable, as stated by F1: "*Well, it's a difficult topic to talk about, sometimes you experience it, but when it comes to speaking it's more difficult*"; F9: *"Look, as I'm going to say, a lot I don't understand either..."* and F12: *"...I just hope I've managed to clarify something, quite complicated, this business of sustainability and that...*".

This difficulty in understanding what SD is can be considered normal, since sustainable development is still a concept in evolution and construction, according to Omole and Ozoji (2014), and so is open to various interpretations. Even in the literature, there is no consensus or single definition of sustainability. To the extent that many authors consider it as a slogan (Ramsey 2015) or an oxymoron (White 2013), with it being a complex concept open to multiple interpretations (Glavič and Lukman 2007; Lozano 2008) and difficult to put into practice (Beratan et al. 2004; White 2013).

4.2 Perceived Regulation Barriers

The second factor is related to the question of regulations and is represented by 3 barriers (see Fig. 1). The first concerns the country's legislation, in this case Brazil, as a barrier to agriculture becoming more sustainable (Leite et al. 2014; Czyzewski et al. 2018; Cederholm Björklund 2018). Farmers said that laws are created without knowing the actual situation on farms. As stated by farmers F1 and F6, and in the words of F1: "*...it's my opinion, I thinks the laws are too much from the top down, we who are at the base, we have to follow what is decided at the top... even if we don't agree with it, we have to accept it, because it's already decided*". This indicates that the public sector, through its respective legislators, should make laws clearer and more accessible to farmers.

Various farmers (F2, F4, F10, F11, F12 and F13) mentioned the difficulty in obtaining certification of sustainable production, associated with the excessive rules and obligations set by the certifying entities, for example, for organic production (Kata and Kusz 2015). As shown in the following statements by F2: "*...but more generally, the question of legislation itself, because legislation for organic, to be organic, you have to prove a lot of things... with 300 norms that come to you, to do things correctly, and if you don't agree, you don't get the certificate, so working for a living today is very difficult*"; F10: "*... it's difficult, you have to comply with a lot of rules... it's very bureaucratic*"; and F12: "*The rules make things very difficult, so* *many rules, and we're not used to following rules, a lot of people give up when they see everything they've got to do…*".

The third barrier is the lack of government support to encourage more sustainable agriculture (Carolan 2006; Cederholm Björklund 2018). This was shown by five farmers (F1, F2, F4, F6 and F13). As perceived in the statements of F1: "… *because the government doesn't play its part, here the government only helps large producers, family farming doesn't get any support, nothing at all…*"; F4: "… *the government helps…but it could help more*"; and F6: "…*a great problem is public management, that doesn't give encouragement.*". So it is perceived that excessive laws and the absence of government help tend to be obstacles to farming becoming more sustainable.

4.3 Perceived Individual Barriers

This factor deals with barriers related to the individual and was formed of three obstacles (see Fig. 1). The first is risk aversion, i.e., adopting more sustainable agricultural practices requires changes and can also involve risks that farmers are not prepared to run (Ma et al. 2009), as reported by F2: "*... there is resistance to looking for new things, it's easy the way things are, so why will I take the risk of trying to do something different...* ". The second barrier is related to the resistance to change (Sassenrath et al. 2010; Carolan 2006; Rodriguez et al. 2009; Stewart et al. 2016; Cederholm Björklund 2018), and was shown in the statement by F4. Finally, disregard for the farmer, as mentioned by F1 "… *the very value of the farmer, because no value is attributed to the farmer...*". That is, the lack of regard for farmers can discourage them from innovating, or even from continuing in farming. So characteristics related to the individual are also found to hinder the implementation of more sustainable farming practices.

4.4 Perceived Macro-Systemic Barriers

Macro-systemic variables were perceived by farmers as barriers to sustainability, this factor being formed of 3 obstacles (see Fig. 1). The first barrier mentioned was keeping mainly young people on the farm, so that they can continue with farming and avoid rural depopulation, as stated by F1: "…*avoiding rural depopulation, because we have problems of this type, there are a lot of young people, I'm speaking about my children too, my children don't want to stay on the farm doing what their father does, so that's my case, and if you look around it's very common, the young people are leaving the farms to look for work outside...*". So rural depopulation is already considered as one of the challenges to be overcome by agriculture in the coming decades, especially in a situation where it will be necessary to produce more food (Godfray et al. 2010).

Besides rural depopulation, capitalism, with its culture of global consumption, previously identified in the study by Duarte (2015), was shown in the statements of F1 and F11, and F2 who argued: "*...it's a very complicated business, because today everything boils down to the financial, economic question...today we're moulded by a system that only thinks about capital, that doesn't think about well-being, the economic system is making people think more and more that there has to be a lot of production...*". This indicates that the focus on production, consumption and generating wealth can lead farmers to give less priority to the other environmental and social pillars of SD, according to Elkington (2004).

Finally, various farmers said that understanding consumers' wishes also tends to be a barrier to more sustainable agriculture (Ma et al. 2009; Grover and Gruver 2017; Cederholm Björklund 2018). This barrier was mentioned by two farmers (F12 and F13), and in the words of F12: "*...yes, because he (referring to the consumer) wants to eat everything perfectly at the time when there isn't any, it's not produced, this creates a race to achieve this, and you end up using a lot of fertilizer, a lot of poison to be able to have that...have it out of season*".

So these results show that macro-systemic questions, such as rural depopulation, the culture of globalized consumption and the difficulty in understanding consumers can hinder SD in agriculture.

4.5 Perceived Capital/Financial Barriers

Financial and capital barriers can also hinder actions in favour of sustainability in agriculture. This factor was formed of three barriers (see Fig. 1). The first is increased costs and the need for high initial investment. Farmer F5 spoke of the increased production costs of adopting organic farming, mentioning the cost of organic certification "*... this has a cost, this generates a high cost for us, we can't get our certification for less than 2000 reais a year*".

Besides increased costs, the lack of financial resources (Williams and Dair 2007; Kata and Kusz 2015; Horhota et al. 2014; Sassenrath et al. 2010; Kuppig et al. 2016; Stewart et al. 2016; Cederholm Björklund 2018) was also revealed by some farmers (F2 and F11), as reported by F2: "*All this means investment, investment of financial resources...so, they're things that we do step by step...*". Finally, adopting more sustainable production can also need high initial investment, as identified in the study by Ma et al. (2009) and mentioned by F2 and F4. So, increased production costs, the lack of financial resources and the high initial investment are perceived as financial barriers to SD in agriculture.

4.6 Perceived Administrative/Organisational Barriers

Administrative/organisational barriers can also hinder actions in favour of sustainability being put into practice in the agricultural sector. This factor was formed of 7 barriers. Among them, the lack of technical knowledge about sustainability in agriculture prevents the adoption of more sustainable practices (Leite et al. 2014; Martin et al. 2015) and the lack of technical support or specialist help/advice, as this could give guidance and information about the implementation of more sustainable actions (Carolan 2006; Leite et al. 2014; Kata and Kusz 2015; Cederholm Björklund 2018). As stated by F4: "*…technical knowledge and technical help is lacking…*"; F7: "*… technical assistance… they don't have the specialists to help us…*"; and F11: *"…the matter, for example, in fact we attend a lot of courses, training, seminars, where farmers come and participate, but the language used is generally very technical…*". There is a lack of training and courses on the subject of sustainable development (Rodriguez et al. 2009; Stewart et al. 2016), as shown by F7 and F9. This emphasizes how much knowledge and technical support are important for the introduction of more sustainable practices on farms.

Another barrier mentioned by farmers F2, F5, F9 and F12 was the lack of time (Evans et al. 2012; Stewart et al. 2016; Grover and Gruver 2017). Another farmer said that adopting more sustainable practices tends to increase the workload (Kata and Kusz 2015), as shown by F10: "*…because it's difficult the way we work, it gives a lot of work…*". Farmers' dependence on the use of pesticides in production also appeared in the statements of F6 and F4: "*In the matter of pesticides… for many things we really need agricultural pesticides…*". This highlights that farmers still believe it is difficult to produce food without using chemical products.

It was also perceived that family organisations, as in the case of family farming, may be resistant to making changes in their daily activity which could contribute to making farming more sustainable (Rodriguez et al. 2009; Cederholm Björklund 2018), as mentioned by F3. To summarise, various administrative and organisational barriers, such as the lack of information, knowledge, technical support and training related to SD, as well as the lack of time, increased workload, dependence on pesticides and family management can all prevent the implementation of actions in favour of sustainability.

5 Discussion

This research presents the results of an exploratory study based on the perspective of 13 organic family farmers in Brazil on the barriers to SD in agriculture. The interviews showed 20 barriers that tend to prevent farmers from making agriculture more sustainable. These barriers were grouped in six factors: (1) semantic barriers; (2) regulation barriers; (3) individual barriers; (4) macro-systemic barriers; (5) capital/financial barriers; as presented in summarised form in Table 1.

Table 1 Barriers to sustainability in agriculture

Factors	Variables/Barriers	Organic farmers	Citations	Total	%
Factor 1—Perceived semantic barriers	Difficulty in understanding what sustainable development is	F1, F5, F6, F9, F12, F13	6	6	12
Factor 2—Perceived regulation barriers	Lack of understandable legislation and specific regulations	F1, F6	2	13	25
	Difficulty in obtaining certification of sustainable production (excessive rules and obligations required by certifying bodies)	F2, F4, F10, F11, F12, F13	6		
	Lack of government support	F1, F2, F4, F6, F13	5		
Factor 3—Perceived individual barriers	Reluctance to risk new forms of working in agriculture	F2	1	3	6
	Resistance to changing behaviour	F4	1		
	Lack of appreciation of the farmer/farming occupation	F1	1		
Factor 4—Perceived Macro- systemic barriers	Difficulty of keeping young people, in particular, on farms	F1	1	6	12
	Capitalism with its culture of globalized consumption	F1, F2, F11	3		
	Difficulty in understanding what consumers really want and need	F12, F13	2		
Factor 5—Perceived capital/financial barriers	Increased production costs	F5	1	5	10
	Lack of financial resources to invest in sustainability	F2, F11	2		
	Making major investment by adopting more sustainable farming	F2, F4	2		
Factor 6—Perceived administrative/ organizational barriers	Lack of information and technical knowledge about sustainability in agriculture	F11	1	18	35
	Lack of technical support or specialist help/advice	F4, F6, F7, F13	4		
	Lack of training on the subject of sustainable development	F7, F9	2		
	Lack of time to act in favour of sustainability	F2, F5, F9, F12	4		
	Increased workload	F10	4		
	Dependence on pesticides/ chemical products	F4, F6	2		
	Traditional family management hinders farming from being more sustainable	F3	1		
Total				51	100

As seen in Table 1, the first factor (Perceived semantic barriers) was revealed in six interview excerpts, and therefore obtained 12% of citations. The second factor (Perceived regulation barriers), was mentioned in 13 interview statements and obtained 25% of citations. As for the third and fourth factors (Perceived individual barriers and Macro-systemic barriers), both were mentioned in 3 statements, i.e., each receiving 6% of citations. The fifth factor, Perceived capital/financial barriers, was shown in 5 statements, representing 10% of citations. Finally, the sixth factor (Perceived administrative/organizational barriers) appeared in 18 of the farmers' statements, representing 35% of citations.

From the results presented in Table 1, the 20 barriers are found in a total of 51 statements by farmers. Analysis of the barriers individually reveals that the difficulty in understanding what SD is and obtaining organic certification received the greatest number of mentions, six each. This shows that these two barriers are the ones hindering SD in agriculture most. In addition, analysing the factors, regulation barriers and administrative/organizational barriers obtained the greatest number of citations, 25% and 35% respectively. This indicates that these two barriers also tend to hinder organic farmers from making farming more sustainable.

6 Conclusion

This study aimed to identify and group the barriers to adoption of Sustainable Development in agriculture. Analysis of the interviews showed 20 barriers that tend to prevent farmers from making agriculture more sustainable. These barriers were grouped in five factors (semantic barriers; regulation barriers; individual barriers; macro-systemic barriers; capital/financial barriers).

Various implications for theory and practice were presented. Regarding theory, it contributes to advancing research related to sustainable development in agriculture, principally by showing farmers' perceptions. It identified and grouped the barriers, something which had not yet been done from the agricultural perspective and with a sample of organic family farmers. Also in relation to theory, it shows the need for more studies devoted mainly to understanding and explaining what sustainable development in agriculture is, since this was one of the barriers most frequently mentioned by farmers. Without understanding what sustainability is, it is difficult to practise it. The study also identified the factors that tend to be the biggest obstacles to sustainability in agriculture, which were regulation barriers and administrative/organisational barriers. Finally, it is highlighted that the study can contribute to the 17 objectives of SD proposed by the United Nations.

As practical implications, public managers, researchers and farmers themselves can use the results of the research, as identifying the barriers and grouping them facilitated understanding of the obstacles to SD in agriculture and can mean actions being proposed to mitigate them.

The study can also have several limitations. The first is related to the research method adopted and the size of the sample, which do not allow generalization of the results obtained. The second limitation was farmers' difficulty in speaking about the barriers due to their problems in understanding what SD is. Another limitation is that there may be other barriers not mentioned by the farmers interviewed. As a suggestion for future research, this study could be replicated in other contexts and countries. The results obtained here can be replicated in quantitative studies, in order to identify the factors that tend to be most relevant from the perspective of family farmers.

Acknowledgements This research was supported by Brazilian National Council for Scientific and Technological Development (CNPq/Brazil), project 304209/2018-0, by Foundation for Research Support of Espírito Santo (FAPES/Brazil), projects 84513772 (599/2018) and 85395650 (228/2019), by Portuguese Science Foundation (FCT/Portugal) through NECE (Núcleo de Estudos em Ciências Empresariais), project UID/GES/04630/2020, and by IFTS (Instituto Fucape de Tecnologias Sociais), project 2018–2021 and project Post-Doctoral fellowship 2020–2021.

References

Bagheri A (2010) Potato farmers' perceptions of sustainable agriculture: the case of Ardabil province of Iran. Procedia-Soc Behav Sci 5(1):1977–1981

Balmford A, Green R, Phalan B (2012) What conservationists need to know about farming. Proc R Soc B: Biol Sci 279(1739):2714–2724

Bastan M, Ramazani Khorshid-Doust R, Delshad Sisi S, Ahmadvand A (2018) Sustainable development of agriculture: a system dynamics model. Kybernetes 47(1):142–162

Beratan KK, Kabala SJ, Loveless SM, Martin PJ, Spyke NP (2004) Sustainability indicators as a communicative tool: building bridges in Pennsylvania. Environ Monit Assess 94(1–3):179–191

Brazil L (2006) 11.326, de 24 de Julho de 2006. Estabelece as diretrizes para formulação da política nacional da agricultura familiar e empreendimento familiares rurais, Diário Oficial da União-Seção, 1–25. http://www.planalto.gov.br/ccivil_03/_ato2004-006/2006/lei/l11326.htm

Carolan MS (2006) Do you see what I see? Examining the epistemic barriers to sustainable agriculture. Rural Sociol 71(2):232–260

Cederholm Björklund J (2018) Barriers to sustainable business model innovation in Swedish agriculture. J Entrepreneurship Manage Innov 14(1):65–90

Czyzewski B, Matuszczak A, Muntean A (2018) Influence of agricultural policy on the environmental sustainability of European farming. J Environ Prot Ecol 19(1):426–434

da Veiga JE (1996) Agricultura familiar e sustentabilidade. Cadernos de Ciência & Tecnologia 13 (3):383–404

Duarte FP (2015) Barriers to sustainability: an exploratory study on perspectives from Brazilian organizations. Sustain Dev 23(6):425–434

Elkington J (2004) Enter the triple bottom line. In: Henriques A, Richardson J (eds) The triple bottom line: does it all add up? Earthscan, London, pp 1–16. http://www.johnelkington.com/archive/TBL-elkington-chapter.pdf

Evans N, Whitehouse H, Gooch M (2012) Barriers, successes and enabling practices of education for sustainability in far North Queensland schools: a case study. J Environ Edu 43(2):121–138

FAO (2011). The state of world's land and water resources for food and agriculture (SOLAW): managing systems at risk. Rome, FAO, London, Earthscan. www.fao.org/docrep/017/i1688e/i1688e.pdf

Ferreira J, Pardini R, Metzger JP, Fonseca CR, Pompeu PS, Sparovek G, Louzada J (2012) Towards environmentally sustainable agriculture in Brazil: challenges and opportunities for applied ecological research. J Appl Ecol 49(3):535–541

Glavič P, Lukman R (2007) Review of sustainability terms and their definitions. J Clean Prod 15(18):1875–1885

Godfray HCJ, Beddington JR, Crute IR, Haddad L, Lawrence D, Muir JF, Toulmin C (2010) Food security: the challenge of feeding 9 billion people. Science 327(5967):812–818

Grover S, Gruver J (2017) 'Slow to change': Farmers' perceptions of place-based barriers to sustainable agriculture. Renew Agric Food Syst 32(6):511–523

Guthman J (2004) Agrarian dreams: the paradox of organic farming in California. University of California Press, Berkeley, CA

Hair JF, Celsi M, Money A, Samouel P, Page M (2015) The essentials of business research methods, 3rd edn. Routledge, New York, US

Horhota M, Asman J, Stratton JP, Halfacre AC (2014) Identifying behavioral barriers to campus sustainability: a multi-method approach. Int J Sustain High Educ 15(3):343–358

Horrigan L, Lawrence RS, Walker P (2002) How sustainable agriculture can address the environmental and human health harms of industrial agriculture. Environ Health Perspect 110(5):445

https://biblioteca.ibge.gov.br/visualizacao/periodicos/50/agro_2006_agricultura_familiar.pdf. Accessed on: 15/12/2017

IBGE - Instituto Brasileiro de Geografia e Estatística (2009). Censo Agropecuário: Agricultura Familiar: Primeiros Resultados 2006. Rio de Janeiro: Instituto Brasileiro de Geografia e Estatística –IBGE, 2006. Available at: https://biblioteca.ibge.gov.br/visualizacao/periodicos/50/agro_2006_agricultura_familiar.pdf

Kata R, Kusz D (2015) Barriers to the implementation of instruments assisting sustainable development of agriculture. Sci Pap Ser Manage Econ Eng Agric Rural Dev 15(1):239–248

Knutson CL, Haigh T, Hayes MJ, Widhalm M, Nothwehr J, Kleinschmidt M, Graf L (2011) Farmer perceptions of sustainable agriculture practices and drought risk reduction in Nebraska, USA. Renew Agric Food Syst 26(3):255–266

Kuppig VD, Cook YC, Carter DA, Larson NJ, Williams RE, Dvorak BI (2016) Implementation of sustainability improvements at the facility level: motivations and barriers. J Clean Prod 139:1529–1538

Laurett R, do Paço A (2019) Sustainability barriers. Encycl Sustain High Educ 1608–1614

Leite AE, De Castro R, Jabbour CJC, Batalha MO, Govindan K (2014) Agricultural production and sustainable development in a Brazilian region (Southwest, São Paulo State): motivations and barriers to adopting sustainable and ecologically friendly practices. Int J Sustain Dev World Ecol 21(5):422–429

Lowder SK, Skoet J, Singh S (2014) What do we really know about the number and distribution of farms and family farms in the world. Background paper for the State of Food and Agriculture, 8

Lozano R (2008) Envisioning sustainability three-dimensionally. J Clean Prod 16(17):1838–1846

Ma Y, Chen L, Zhao X, Zheng H, Lü Y (2009) What motivates farmers to participate in sustainable agriculture? Evidence and policy implications. Int J Sustain Dev World Ecol 16(6):374–380

Martin S, Rieple A, Chang J, Boniface B, Ahmed A (2015) Small farmers and sustainability: institutional barriers to investment and innovation in the Malaysian palm oil industry in Sabah. J Rural Stud 40:46–58

Menalled F, Bass T, Buschena D, Cash D, Malone M, Maxwell B, Weaver10 D (2008) An introduction to the principles and practices of sustainable farming. Montana State University, pp 1–4

Milbrath LW (1995) Psychological, cultural, and informational barriers to sustainability. J Soc Issues 51(4):101–120

Omole CO, Ozoji BE (2014) Science education and sustainable development in Nigeria. Am J Educ Res 2(8):595–599

Ortiz W, Vilsmaier U, Osorio ÁA (2018) The diffusion of sustainable family farming practices in Colombia: an emerging sociotechnical niche? Sustain Sci 13(3):829–847

Pham LV, Smith C (2013) Agricultural sustainability in developing countries: an assessment of the relationships between drivers and indicators in Hoa Binh Province, Vietnam. Agroecol Sustain Food Syst 37(10):1144–1186

Pretty J (2008) Agricultural sustainability: concepts, principles and evidence. Philos Trans R Soc Lond B: Biol Sci 363(1491):447–465

Ramsey JL (2015) On not defining sustainability. J Agric Environ Ethics 28(6):1075–1087

Rodriguez JM, Molnar JJ, Fazio RA, Sydnor E, Lowe MJ (2009) Barriers to adoption of sustainable agriculture practices: change agent perspectives. Renew Agric Food Syst 24(1): 60–71

Sandhu HS, Wratten SD, Cullen R (2010) Organic agriculture and ecosystem services. Environ Sci Policy 13(1):1–7

Sassenrath GF, Halloran JM, Archer D, Raper RL, Hendrickson J, Vadas P, Hanson J (2010) Drivers impacting the adoption of sustainable agricultural management practices and production systems of the Northeast and Southeast United States. J Sustain Agric 34(6): 680–702

Schaller N (1993) The concept of agricultural sustainability. Agr Ecosyst Environ 46(1–4):89–97

Seufert V, Ramankutty N, Foley JA (2012) Comparing the yields of organic and conventional agriculture. Nature 485(7397):229–232

Stewart R, Bey N, Boks C (2016) Exploration of the barriers to implementing different types of sustainability approaches. Procedia CIRP 48:22–27

Tal A (2018) Making Conventional agriculture environmentally friendly: moving beyond the glorification of organic agriculture and the demonization of conventional agriculture. Sustainability 10(4):1078

Tavernier EM, Tolomeo V (2004) Farm typology and sustainable agriculture: does size matter? J Sustain Agric 24(2):33–46

UNDESA—United Nations DESA Population Division (2017) Probabilistic population projections based on the world population prospects: the 2017 revision. United Nations, New York

UN—United Nations (2012). Resilient people, resilient planet: a future worth choosing, United Nations Secretary General's High-Level Panel on Global Sustainability, New York

Van Thanh N, Yapwattanaphun C (2015) Banana farmers' adoption of sustainable agriculture practices in the Vietnam Uplands: the case of Quang Tri Province. Agric Agric Sci Procedia 5(1):67–74

White MA (2013) Sustainability: I know it when I see it. Ecol Econ 86:213–217

Williams K, Dair C (2007) What is stopping sustainable building in England? Barriers experienced by stakeholders in delivering sustainable developments. Sustain Dev 15(3):135–147

Zeweld W, Van Huylenbroeck G, Tesfay G, Speelman S (2017) Smallholder farmers' behavioural intentions towards sustainable agricultural practices. J Environ Manage 187(1):71–81

Sustainability Practices and Motivation for Adopting Organic Pineapple Farming in the Ekumfi District of the Central Region of Ghana

Shine Francis Gbedemah, Sandra Swatson-Oppong, and Selase Kofi Adanu

1 Introduction

Even though Organic agriculture has been in existence since time immemorial, it has emerged recently as an important priority area globally in view of the growing consciousness for safe and healthy food, long term sustainability and environmental concerns (Wani et al. 2017). Organic agriculture has undergone drastic and remarkable growth, particularly since the mid-1980s, when it gained the attention of policy-makers, international organizations, non-governmental organizations (NGOs), farmers, environmentalists and consumers across the globe (Kristiansen et al. 2006; Stolze and Lampkin 2009). Paull (2011) points out that there has been about 8.9% per annum compound growth in organic land over the last decade. Thus, organic agriculture is now practiced in about 178 countries around the world accounting for 57.8 million hectares of land (Willer and Lernoud 2017).

The remarkable growth seen in organic farming over the years has also been attributed to growing global concerns about issues such as environmental and health implications of high-input agriculture, food scarcity, growing demand for organically grown foods and the introduction of policy supports for environmental initiatives (Kristiansen et al. 2006; Stolze and Lampkin 2009).

According Willer and Lernoud (2019), Oceania, covers about 47% of organic land distribution on earth. This is followed by Europe (23%) Latin America (12%) and Asia (9%). Africa occupies just 3% of organic land distribution on earth.

S. F. Gbedemah (✉)
Department of Geography and Earth Science, University of Environment and Sustainable Development, PMB, Somanya, Ghana

S. Swatson-Oppong
Department of Development and Environmental Studies, Wisconsin International University College, Accra, Ghana

S. K. Adanu
Ho Technical University, P.O. Box HP 217, Ho, Ghana

W. Leal Filho et al. (eds.), *Sustainability in Natural Resources Management and Land Planning*, World Sustainability Series,
https://doi.org/10.1007/978-3-030-76624-5_10

Despite the impressive statistics and progress, the organic sector in most African countries is considered to be underdeveloped, and yet to optimize its potential. The reasons for this low performance of African countries in the organic market can be attributed to political and logistical constraints like unfavourable national policy environments that promote high-input agriculture, inadequate government support, inadequate organic extension agents, and underdeveloped local and regional markets (Azam and Shaheen 2019; Opoku et al. 2020; Parrott et al. 2006; Walaga 2005).

The International Federation of Organic Agriculture Movements (IFOAM) defines *o*rganic agriculture as "production system that sustains the health of soils, ecosystems and people (IFOAM 2017: 1). Organic agriculture relies on ecological processes, biodiversity conservation and cycles that adapts to local conditions, rather than the use of inputs with adverse effects. Under organic agriculture, traditional systems of farming are combined with innovation and science which benefits the environment and the farmers (IFOAM 2017).

The exciting growth that organic agriculture has witnessed since the mid-1980s is said to have largely taken place in developing and emerging markets of Oceania, Latin America and Asia (Kristiansen et al. 2006). However, more than one-third of the world's organically managed land (12.5 million hectares) and 80% of the world's organic producers are reported to be in developing countries (Willer and Kilcher 2012).

Ghana is the sixth country in the world that has large organic area for the cultivation of fruits in hectares apart from Italy, Spain, China, Mexico and USA. The main fruits produced are mangoes, bananas and pineapples. The export pineapple industry provides tremendous benefits to a large number of smallholder farmers and now, large scale farmers in the southern sectors of the country since 1983. The organic sector in Ghana, like many African countries is underdeveloped, and yet to optimize its potential. All regions in the country practice some form of organic farming but the Ekumfi district of the Central region of Ghana is home to many organic pineapple producers and has the greatest share of organic certified pineapple farmers in the country (GSS 2014). The objective of this paper is to assess the sustainability practices and motivations driving small scale pineapple farmers in the study area to engage in organic pineapple production by small scale farmers in the study area.

2 Literature Review

2.1 Organic Farming

Organic farming is governed by rules established by IFOAM. The accepted IFOAM standard for a product to be acknowledged as organic consists of rules governing its cultivation, pest and weed control (IFOAM, n.d.). According to Giovannucci

(2006: 1), "the term 'organic' is referring not to the type of inputs used. The farm is seen as an organism, in which all the components - the soil minerals, organic matter, microorganisms, insects, plants, animals and humans - interact to create a coherent, self-regulating and stable whole". Reliance on external inputs, whether chemical or organic, is very low. Rao (2014: 699) the other hand points out that, organic agriculture is a "production system which avoids or largely excludes the use of synthetic compounded fertilizers, pesticides, growth regulators and livestock feed additives". It relies on crop rotation, crop residues, animal manure, legumes, green manure, off farming organic waste and aspects of biological pest control.

Organic agriculture is being promoted throughout the world and Ghana in particular because it is good for the environment; it is a socially and economically sound production of food and fibers; and its aim is to achieve sustainable development. Organic agriculture utilizes both traditional and scientific knowledge; it relies on the management of ecological systems and cycles using agronomic and biological methods, rather than synthetic materials (Giovannucci 2006). Table 1 shows a comparison between organic and conventional agriculture.

Like other organic products, organic pineapple earns a premium price on the market compared to conventional varieties (Kleemann 2016). Studies have found that, certified organic products like pineapples leave farmers better off in developing countries because of the higher prices farmers are given for their products (Maertens and Swinnen 2009). Pineapple production and export is one of the most important non-traditional export crops in the study area and it is a leading supplier of organic pineapple to the European market (Kleemann 2011). Are the farmers practicing organic agriculture as being promised for sustainability? This is the main question this study is aimed at answering.

Table 1 Comparison between conventional and organic farming systems

Conventional	Organic
Large markets	Smaller markets
Easy market access	Less accessible markets
Modest growth rates	Robust growth rates
Intense competition	Moderate competition
Rewarded for quality, quantity and low price	Rewarded for quality and processes
Government support: subsidy, R&D, extension	Limited government support
Capital intensive	Labour and knowledge intensive
Short learning curve	Longer learning curve
May face trade barriers	Incorporates traceability, maximum residue levels, less trade barriers
Strong downward market prices	Enjoys premium price in the marketplace

Source Adapted from Giovannucci (2006: 12)

2.2 Sustainability Processes Under Organic Pineapple Farming

Sustainable farming can be said to be farming that requires no burning of the land by farmers. In terms of clearing the field, little or limited use of equipment that use fossil fuel is recommended. As much as possible, the vegetation that was available before planting is to be used as mulch and organic manure for the soil. There is the need to classify suckers for planting and the best suckers are to be used. Farmers are expected to use planting material from their own farms or obtain it from their neighbours or other organic certified seedling farms. Plants are to be planted in a regular spacing to improve productivity and maximize yield (Agro Eco 2008).

Crop rotation system is encouraged especially by using green manure (green fallow land as a first pre-crop). Farmers are to make compost an essential part of their fertilization programme. Three (3) main types of fertilizers recognized under organic agriculture are: composting; manure (poultry, cow, sheep, goats etc.) and pruning waste and other mulching material from the farm. The process under organic farming is shown in Fig. 1. Under organic cultivation, the rule is for farmers not to use conventional chemicals on their farms. Pest and diseases are to be treated with natural methods like the use of plants that repel insects. In terms of flower induction, farmers are expected to allow the fruit to ripe naturally or use recommended flower induction methods. The harvesting and selling of the ripe fruit are expected to follow the recommended rules. In Fig. 1, any deviation on the part of farmers nullifies the term organic pineapple. The fruit is thus expected to be treated as none certified conventional produce.

2.3 Farmer's Motivations for Engaging in Organic Farming

Organic agriculture can contribute to meaningful socio-economic and ecologically sustainable development, especially in less developed countries like Ghana. This is due to the application of organic principles, which means efficient management of local resources like local seed varieties, manure and therefore cost effectiveness (Kilcher 2007). Many farmers see organic agriculture as providing solutions to some of the problems associated with conventional agricultural practices, such as environmental degradation, depletion of non-renewable resources, and food safety issues (Tilman et al. 2002). Pechrová (2014), suggested that by avoiding the use of agrochemicals, organic agriculture will help make food relatively 'free' of synthetic chemicals and thus healthier in comparison to food produced based on conventional agricultural practices.

Prices for organic produce are usually higher than conventional produce due to the higher production costs, limited supply and small volumes of production. However, the prices of such organic products may only partially compensate for its

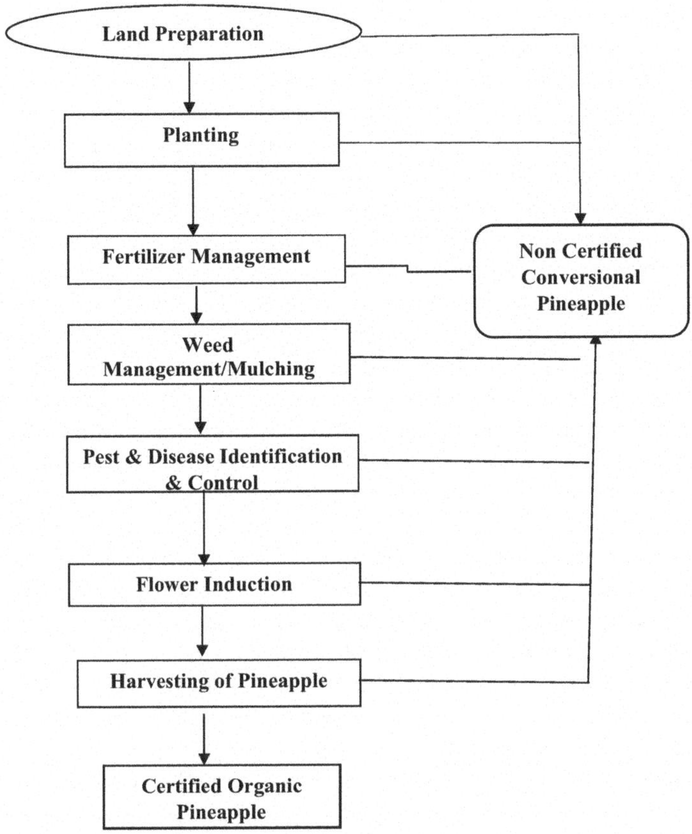

Fig. 1 Processes towards organic farming and certification. *Source* Elaborated by the authors (2021)

relatively high production costs (Inkoom 2017). Research also suggests that organic farmers in developing countries can increase their income by 30–200% after the organic conversion (Onduru et al. 2002).

3 Study Area, Materials and Methods

The Ekumfi District is located along the Atlantic Coastline of the Central Region of Ghana. The district was purposively selected for being the leader in pineapple production in the country (GSS 2014). This makes the selection of the Ekumfi district a good representation of the country. The district is bounded to the West by Mfantsiman Municipality, to the North by Ajumako–Enyan–Essiam District, to the East by Gomoa West District and to the South by the Gulf of Guinea.

According to the 2010 census of Ghana, the district had a population of 52,231 (GSS 2014). "Of the employed population, about 52.3% are engaged as skilled agriculture, forestry and fishing, 17.5% in service and sales, 16.6% in craft and related trade, and 6.1% are engaged as managers, professionals, and technicians" (GSS 2014: x). The Trans ECOWAS highway divides the district and offers cheap and easy proximity to the two main ports of Accra and Takoradi, and the only international airport in the country, Kotoka International Airport.

The research design for this study is descriptive. Descriptive research design is aimed at explaining a happening, an event or giving an accurate or factual description of the population under consideration (Boru 2018). In this study, the opinions of smallholder pineapple farmers and other stakeholders in the pineapple industry in relation to motivation and organic practices are explored through questionnaires, interviews and focus group discussions (FGD). Primary data were generated from a cross-sectional survey of farmers belonging to producer organizations like the Aponopono Biakoye Organic Cocoa Farmers Association, Coalition for the Advancement of Organic Farming (CAOF), Citrus Producers of Ghana, Adwendaho Organic Pineapple Producers and Suppliers (ACCOPPS) and Adwumadzen ma Mpontu Organic Pineapple Producers Association (AMOPPA). Other farming units in the sector that data was gathered from are Bomart farms, Eden tree, Eloc farms and Grow West Africa. These groups form cooperatives and they obtained their certification in 2018 with support from GIZ. Most of the members were part of an organic pineapple supply chain managed by WAD Ghana limited since 2004. They are thus very experienced in the organic pineapple production and certification business.

Representatives of the organic export companies to which farmers are affiliated were also interviewed in relation to organic pineapple production standards and certification. A structured questionnaire was used to gather data from the farmers. Kothari (2004) points out that one of the major criteria to use to select a sample size is the extent to which the sample is distributed within the population. There are a lot of certified organic pineapple farmers in the area but they constitute less than one percent of pineapple farmers in the district. Farmers among the nine identified organizations were identified from the list of members. The lists provided a relatively complete inventory of some certified farms in Ekumfi District. The lists of pineapple farmers within the producer organizations were used to randomly select 81 farmer respondents for the study. Out of the 81 farmers identified, random sampling was used to gather data from 78 certified organic pineapple farmers who were willing to provide data of pineapple production in the district.

This sampling technique reduced systematic errors and minimize the chance of sampling biases. Inferences drawn from the sample were therefore generalisable to the population (Alvi 2016). The 78 questionnaires were administered to the farmers with the help of agricultural extension officers in the Ekumfi District. To Morgan et al. (1998), focus group discussion (FGD) consists of four major steps: (1) research design, (2) data collection, (3) analysis and (4) reporting of results. According to these authors, FGD is a "cost-effective" and "promising alternative" in participatory research (Morgan et al. 1998). This study used FDG to gather qualitative data from the famers where the researchers adopted the role of "facilitators" or a "moderators". Only 10 farmers belonging to two producer organizations —Adwendaho Organic Pineapple Producers and Suppliers (ACCOPPS), and Adwumadzen ma Mpontu Organic Pineapple Producers Association (AMOPPA) were used for the FDG. Descriptive statistics which mainly used percentages and frequencies were used to analyse the socioeconomic and organic pineapple production features of the study, as well as farmers' reasons for undertaking organic practices. The next section will be used to present the findings from the field.

4 Results

4.1 Background of Respondents

The data in Table 2 explains the age range, educational background and years of experience of the respondents. From this table, it can be seen that, few of the respondents (1.3%) are within 18–25 years, this is followed by 15.4% who are within the age range of 26–35 years, 28.2% are between 36 and 45 years and the majority of the remaining respondent (55.1%) are within the age range of 46 years and above. This indicates that young people are shying away from organic farming, while older folks are more drawn to farming in the area, a contradiction to the findings of McCann et al. (1997), who stated that farmers in Michigan, USA undertaking organic farming are relatively young and well-educated people.

The findings in Table 2 also shows that most youth see agriculture as backward and a task for rural settlers and the elderly. Majority of them move to the urban centers like Accra, Cape Coast and Kumasi in search of none existing white colour jobs. This is wrong because, agriculture is now a lucrative venture leading to the emergence of large fruit farms in the area. Instead of seeing agriculture as a hard skilled labour activity, the youth have to embrace it because agri-business is now a big time business activity.

This finding is however in line with that of a multi-country study (five sites in four African countries) that concluded that "farming is not a favoured option for the younger generation in rural areas of developing countries, even those in which agriculture remains the mainstay of livelihoods and the rural economy" (Leavy and Hossain 2014: 38). This can be attributed to the rising aspirations of the youth and the low social status of small-scale farming in the country.

The findings from this study indicates that, the percentage of men (66.7%) are twice that of women (33.3%) showing that there are more men involved in organic pineapple farming than women and this can be attributed to the tedious nature of pineapple cultivation and management. It is also important to note that, a higher proportion of females (20.1%) than males (11.9%) are, however, engaged in craft and related trade. This shows that there is gender stereotyping of occupation in the District (GSS 2014). This lends credence to the view by Doss (2001) that, gender

Table 2 Demographic background of the respondents

Age range	Frequency	Percent (%)
18–25 yrs.	1	1.3
26–35 yrs.	12	15.4
36–45 yrs.	22	28.2
Above 46 years	43	55.1
Educational background of respondents		
No formal education	7	9.0
Elementary	9	11.5
Junior High School	43	55.1
Senior High School/'O' Level	16	20.5
Tertiary	3	3.9
Years of experience in cultivating pineapples		
1–10 yrs.	37	47.4
11–20 yrs.	31	39.8
21–30 yrs.	10	12.8
Farm size (acreage) of respondents		
1–3	49	62.8
4–6	12	15.4
7–10	8	10.3
11–5	4	5.1
16+	5	6.4
Total	**78**	**100.0**

Source Elaborated by the authors. Fieldwork (2019)

disparities in the agricultural sector were heightened by the introduction of a cash crop economy during colonial days where men played the central role in the production of cash crops for export whilst women on the other hand were relegated to the production of food crops. A situation that can be alluded to in this study. The number of women who are into production for export in the pineapple sector is very low compared to men.

The gender aspect of this study is very important. This is because, it will enable us to show the gender relations with organic farming especially of small scale pineapple farmers. In most areas, especially in Africa, men are perceived as heads of households who are expected to provide for members of their household, However, in recent years, some women have played a similar role in their homes, either in support of their male partners or as head of the family and have also taken farming as an occupation. This traditional view may be incorrect, since women have recently played roles that were originally played by men (Lease 2003).

The study investigates the respondents' level of education in order to ascertain if they are well equipped or will be able to easily acquire the necessary knowledge and skills in the ever changing export sector of organic farming. The data presented in Table 2 also shows that all the respondents have some form of formal educational background. Table 2 clearly shows that, 11.5% have elementary education and majority of the respondents constituting 55.1% have JHS qualification. This is followed by 20.5% of the respondents who have Senior High School (SHS/'O'/A) Level qualification, whilst 3.8% of the respondents have tertiary qualification and the remaining respondents do not have any formal education. The implication of this findings is that, majority of the farmers will find it easy to learn new things (ideas and technology) and standards of certification. Indeed, each farmer can read instructions and carry out ecological agriculture with ease. This findings erases the stereotypical thoughts of farmers being hillbillies, uneducated, having a big straw hat with overalls, missing teeth and even being unhygienic.

Table 2 also shows that, 47.4% of the respondents stated that they have been farming pineapples between 1 and 10 years, 39.7% have 11–20 years' experience and only 12.8% of the remaining respondents have been in pineapple production for more than 21 years. Most of the farmers have been in this enterprise for a long time because, pineapple production is the main farming activity in the District (GSS 2014). According to the respondents, farming offers satisfactions not often found in other tasks, echoing the sentiments of the Secretary of Agriculture Claude R. Wickard, himself a farmer who told a meeting of the Future Farmers of America on October 10, 1944 that: "*I like farming. It is a good life. After more than 30 years of it, I still would rather farm than do anything else. I envy you future farmers who have such great things ahead*" (Wickard 1944: 1). During the FGD with farmers, it was revealed to the study by farmers who have been in the business for more than 21 years that, they have more knowledge about the factors that could affect yields, best planting seasons and disease symptoms than the younger farmers. That, older farmers have more farming experience hence less inefficiency in the sector. The next section of the study will focus on the motivations driving farmers in the study area to engage in organic pineapple agriculture.

4.2 Motivations to Engage in Organic Farming

Several papers have addressed motivational issues for converting to organic farming (Adebiyi et al. 2019; Lahmar 2010; Soane et al. 2012). The choice of the type of farming depends on many variables, such as age, education, experience in agriculture, farm size, type of crop, economic conditions, etc. The literature shows, however, that these dependencies are not clear-cut (Assouline 1997).

Figure 2 shows the mixed motivations that prevails among farmers for engaging in organic pineapple cultivation in the area. The figure shows that 42% of the respondents are motivated by health and environmental factors for engaging in organic farming. During the FGD, it was revealed that, the main environmental factor for adopting organic agriculture was the availability of fertile land. Some points out that, the land belongs to them and they only cut the trees on it and plant the suckers. They indicated that the ownership status of the land is very important in deciding whether to go organic or not. Residential status of the farmer is thus very important in deciding whether to engage in organic or inorganic. It was revealed to this study that, the likelihood of a native farmer to produce organic pineapples is very high compared to migrant or non-native farmers. The reason for native farmers engaging in organic farming is not farfetched. They chose the more fertile part of the land to till and give the over used portions to migrant farmers. Also, they are in a more favourable position to invest more time and resources on the land since they will not be ejected compared to those tilling the land on share cropping basis (Anaafo 2015). This leads to the observation that farmers in Ekumfi district of Ghana temper their profit motive with a small amount of self-sacrifice to meet social and/or environmental goals (Peterson et al. 2012). Some farmers do not till the land with machines but use cutlass and hoe on the farm. The farmers pointed out that, no-tillage of the land is in conformity to their traditional ways of farming as such they are more comfortable applying it on their pineapple farms. One can also say the use of no-tillage on the farms stems from the high prices mechanical tillage cost the peasant farmer compared with using the hoe and cutlass. The cost of ploughing one hectare of a field is beyond some of the peasant farmers pushing them to use no-tillage which is ideal for organic farming. It should be noted here that, majority of the farmers during the FDG points out that, most farmers are encroaching into virgin forests in order to take advantage of the fertile land. These people are mostly the owners of the land. Should the land exhaust its fertility, they leave this piece of land to migrant farmers to be used in cultivating conventional pineapples. Others said they themselves use it to cultivate conventional inorganic pineapples. This finding shows organic farming is reducing the vegetative cover of the land which is not good for the environment.

The findings also shows that good pricing is the second motivation for many respondents (30%) followed by market availability 28%. The price that is offered to the products of organic farmers is high compared to conventional pineapples. Some farmers are swayed by this economic reason for engaging in organic farming. Most Ghanaian farmers are peasant poor farmers as such any activity that will alleviate their poverty will be welcome leading to about 30% of the respondents pointing out that they engage in organic farming due to the price. Indeed, Padel (2010) stated that financial reasons are one of the prominent factors for the adoption of organic

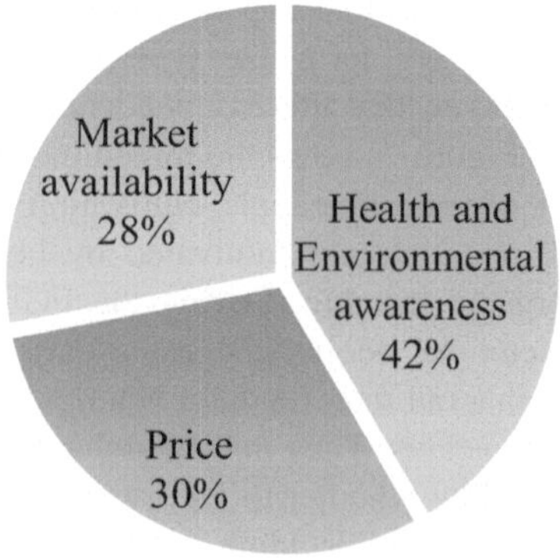

Fig. 2 Farmers motivation to engage in organic farming. *Source* Elaborated by the authors. Fieldwork (2019)

production and this was found to be true to some extent in our study. Other literature established that, proper market demand and adequate price premium paid for organic foods will attract the poor farmer since he will be assured of reasonable profit for his endeavor (Argilés and Brown 2010). We can therefore point out that, farmers in the Ekumfi district of the Central region of Ghana will increase production if there is a better pricing and market for their products.

Market availability is the last but not the least response that the farmers of our study gave as the reason why they engage in organic farming. In the developed markets of Europe and North America where the organic market sector is, market niche opportunities are available (Kleemann 2016). Even in Ghana these days, the ever expanding supermarkets are creating opportunities for organic farmers to market their produce easily and at affordable cost. Even though, local market demand is typically modest, specialised stores for organic products are growing in the country leading to some supermarkets to start offering organic food to their customers. The export market in Ghana is dominated by different actors making it difficult for farmers to identify buyers. To take advantage of the export market, farmers formed cooperatives to help sell the products. Some of the exporters of pineapple in the area are WAD (Weija Agric Development), SPEG (Sea-Freight Pineapple Exporters of Ghana), FAGE (Federation of Association of Ghanaian Exporters), Golden Exotics Ltd, Yayra Glover, to mention a few. Some of these firms are producers as well as marketers or agents in the supply chain of pineapples in Ghana. The farmers can either market their produce to the consumer directly in the market by selling the produce on the road side or in the market. This option does not come with high premium price like those for export. However, these options are for the left overs that have been rejected by exporters or the supermarkets. They can also market their produce to the processes like Weija Agricultural Development and Golden Exotics Ltd. Due to the high profit margin these producers/agents/marketers enjoys, their numbers are on the ascendancy in the country.

These findings re-enforce Porter's (2016) opinion that organic farmers hold non-economic motivations in very high regard. This is not surprising as organic agriculture has historically been considered a socially and environmentally guided countermovement to what some consider the "rationalization" of agriculture. It also yield credence to Rogers (1983) view that, both environmental and economic factors are at play in motivating farmers to engage in their organic farming activities. Interestingly, during the FGD, it was alluded to by the farmers that, their motivation tends to change with time. It started from purely environmental point and ends with financial motives. This is the main reason why some farmers who are engaged in conventional farming are now converting their farms to organic farming (Flaten et al. 2006). These findings are in line with Sheeder and Lynne (2011: 433) who concluded that, "……while it is undeniable that profits do play a role, the assumption that they play the only role in economic decision making is highly contentious". Additionally, this goes to show that organic farmers may be more likely to restructure their farms in ways that maintain its ability to persist over time than their conventional counterparts. Our discussion in the next section will now centre on sustainability practices under pineapple cultivation in the Ekunfi district of the Central region of Ghana.

4.3 Sustainability Practices Undertaken by Organic Pineapple Farmers

The farmers in the Ekumfi district mentioned that land preparation for planting on newly acquired lands involve clearing with cutlass by peasant farmers while large scale farmers use heavy equipment to clear the land for initial planting. The large mass of plant residues produced as a result of land clearing are normally used as mulch and organic manure for the soil. On previously cropped lands however, land preparation involves weeding and reshaping of the land by incorporating compost into the soil. Burning the land, as is done by most conventional peasant farmers in Ghana is strictly not an acceptable practice under organic land preparation. All farmers during the focus group session attest to the fact that they have never burnt their vegetation to plant suckers. These findings shows that land preparation by organic farmers in the area is sustainable.

Under indigenous farming, farmers do not plant the crops in rows but under organic farming, farmers were trained in planting their suckers in a manner that will help increase yield. Special materials are required to be used for planting. The field is expected to be design for planting of the pineapple suckers. For instance, farmers mentioned that they classify suckers for planting under organic pineapple cultivation. In other words, there are categories of planting materials for planting. Farmers said they were advised to select the big size suckers (500–550 g) for early fruit bearing (18 months) which are from their own farms or obtain it from neighbour farmers or other organic certified farms. Small size suckers are thus used on conventional farms.

In terms of field design, farmers said they divides their fields into four parts. After doing this, they again sub-divide each plot into four for easy planting. They use a distance of 30 cm (using alternating planting holes) between plants. This spacing is believed to be sufficient in double-row systems that yield best results. The farmers said they plant their pineapple suckers on flat land in a regular spacing to improve productivity and maximize yield. They also try to avoid water-logged land or completely dry land since both situations can lead to low yield or they attracts pest and diseases. In planting the suckers, the farmers have to press the shoots into the slightly loosened soil. After inserting the suckers into the soil, the loosened soil is used to cover or pressed back on the sucker to prevent its root from being exposed. Exposure of the sucker's roots will lead to withering of the roots. One farmer during the FGD lamented '*I employed an inexperience farm hand who planted the suckers without pressing the soil back. After a few days, I lost most of the suckers he planted while those that I planted myself germinated*' (Farmer 2, FGD 1).

The negative side of conventional agricultural practices are said to stem mainly from the use of chemicals to enrich the soil. Some of these negative practices leads to reduced soil fertility, water pollution, destruction of natural habitat and production of greenhouse gases among others. Under organic farming, the nutrient requirements of the crop rotation system are mainly provided by green manure. The farmers points out that the remnants from the pineapple production are spread over the soil which they called mulching. The decomposed mulch is a very vital nutrient for the soil. Compost, according to participants in the FGD is prepared by all of them. The compost is mainly spread before the suckers are planted and another before flower formation. The rule is that it should not be close to the suckers but rather at least few inches away from the suckers. All the three main fertilizer varieties recognized under organic agriculture; compost; animal manure (poultry, cow, sheep, goats etc.) and pruning waste from the farm are being used by participants of this study. The participants stated that, they prefer composting because it increases yields as they are readily available in the area. Another merit of compost leading to its preference is that it increases organic matter content of the soil and suppresses weed and disease growth.

Today, most conventional farmers are using chemicals to suppress weed development on their fields but organic farmers do not apply chemicals. Managing weeds on organic pineapple farms is a difficult task because of the thorny nature of the plant. Indeed, the whole field is almost occupied with pineapple plants with little room for the feet as such, weeding is a difficult and herculean task to be performed. The hand, cutlass or hoe are used by the farmers to remove the weeds. The uprooted or cut weeds are then used as mulch. Farmers at the focused group discussion observed that when mulching is done well, it prevents weed growth and help to conserve moisture for the growth of pineapple plants. All farmers responded negatively when asked whether some people clandestinely use chemicals on their organic farms. This suggests that, the pineapple farmers are truthful to the tenets of organic farming.

Another area where organic pineapple farmers practice sustainable production is the control of pests and diseases. Under organic pineapple cultivation, one of the rules is 'no use of conventional chemicals'. Most of the participants explained that they practice integrated pest management (IPM), also known as integrated pest control (IPC). This is a broad-based approach that integrates different practices for economic sustainable control of pests (Miller and Spoolman 2015). IPM is aimed at suppressing pest populations. This in turn implies reliance on local knowledge and participatory input into the development of agricultural techniques. The farmers said they use biological methods of pest control that involves the use of natural pest predators, parasites and disease organisms. The most important predator that they claim to use are spiders. Spiders are noted to kill more insects every year than we humans do (Nyffeler and Birkhofer 2017). Biological control of pest may at times have a beneficial effect in slowing down the population growth of hemipteran pests (Birkhofer et al. 2016). It should be noted that, the presence of spiders and at times in combination with other predators is very beneficial to organic farming. Farmers said they were advised to announce to their internal control officers before they use biological methods of pest control and this must be documented. The farmers in this study stated that they usually observe the chemicals that are being used by their neighbours who are into conventional farming. Some also said they do not want to farm organic crops close to conventional farms just to avoid spill-over into the organic farms.

Flower formation under organic pineapple cultivation is enhanced by the use ethylene, which all farmers said they use. This is a growth regulator that is used to force the plant to produce more flowers or to ripen the fruit. The phytohormone ethylene controls growth and senescence of plants (Nazar et al. 2014). The farmers pointed out that special preparations they used to induce flower formation in conventional pineapple farming are not permitted on organic farms. They have to source their ethylene from recommended shops. The farmers said they do not use calcium carbide on their conventional farms let alone on their organic farms. The use of calcium carbide in farming is not recommended worldwide because of the health hazards it is associated with. Calcium carbide treatment of fruits is very hazardous because it contains traces of arsenic and phosphorous, and once dissolved in water, it produces acetylene gas, which affects different body organs and causes various health problems like headache, dizziness, sleepiness, mental and memory loss, cerebral edema, seizures and prolonged hypoxia (Asif 2012). These findings shows that the farmers are using sustainable means to cultivate their pineapples which are sustainable.

On processes used to harvest their pineapples, the respondents points out that, the ripen fruits are removed from its stem by cutting it with a sharp clean knife or cutlass. They explained that, they try as much as possible not to damage the fruits during harvest. If the fruits are damaged as a result of harvesting, the wounds will leave behind ideal spots and fungus easily develops on them which shortens the life of the harvested pineapple on the shelf. The participants points out that, the plant is harvested when green-ripe or half ripe. Full ripped fruits gets spoilt within days as such they do not leave them to get to that stage. Crates are used to convey the fruits to minimize bruising and ensure quality fruits. During harvesting, it is the responsibility of the farmer to ensure that, the organic fruits are not mixed with any other fruits.

5 Discussion

It can be seen that, the sustainability measures used by participants of this study are in line with most practices (Lori et al. 2017; Mehrabi and Ramankutty 2019; Smith et al. 2019). This is important because of the large number of conventional farms in the world that are damaging the environment. Since organic farms are more reliant on ecosystem services for production of high yielding crops, it is important that they are encouraged for posterity sake.

Majority of the respondents were males compared to very few female farmers. This is not new in most parts of Ghana. Literature shows that agriculture in Ghana is dominated by men (Doss 2001; GSS 2014). The main reason is not farfetched. The main reason is that women do not own land in the southern parts of Ghana due to discrimination against them in terms of rights to land (Bugri 2008). The domination of men over women in farming in the country is also due to the type of crop in question. When it comes to cash crops, Ghanaian men are seen but when it is about domestic backyard gardens which does not help them to earn income, women predominates and here is where their knowledge counts.

For agricultural production to be successful in most parts of Ghana, traditional knowledge is relied upon especially by the peasant farmers. Indeed, traditional knowledge serves as the life wire to rural farmers as scientific information is either absent or very limited. Observation of the local environment for many generations has empowered indigenous people to develop skills that enabled them increase their resilience to changing weather and climate events like rainfall, drought and floods (Handmer and Iveson 2017; Muguti and Maposa 2012). Now, these age tested knowledge is being used in organic farming.

Organic agriculture according to Lampkin (1994) is a production system in which the use of synthetic compounded fertilizers are avoided, or largely excludes the use of pesticides, growth regulators and livestock feed additives. The findings of study shows that organic farming is actually indigenous to the people leading to its easy adoption in the communities. The momentum with which indigenous knowledge is gaining grounds in both developed and developing countries as a result of failure of scientific knowledge to adequately address some human problems gives credence to the viability of indigenous knowledge in promoting agriculture (Mchombu 2002). For the sake of progress in agriculture and ensuring food security and sustainable development of the world, local people especially the rural farmers in developing countries like Ghana should collectively be educated on ways of using simple to use farming technologies. These technologies were labelled as backward, peasant and rural in nature but this is what is being taught our farmers called sustainability practices. Because of the labels given these technologies, there is a risk of losing valuable indigenous knowledge when the world is controlled by science to the extent of making indigenous knowledge outmoded due to rapid urbanization (King et al. 2008; Masinde 2015).

As has been seen in Fig. 2, most of the process under organic farming are indigenous in nature apart from the rule of none burning of the bush and the application of growth enhancers. In Africa, bush burning is seen as a means of clearing large tracts of land but this has been seen to be a bad practice. Planting in rows is also not practiced as an indigenous method of farming. Fertilizer application, weed management, pest and disease identification and control, and harvesting are all indigenous means that are being transferred to them as organic farming.

The question that has to be answered is this: is sustainability practice observed under pineapple farming in the Ekumfi District of the Central Region of Ghana good? For instance, organic farming has many positive attributes of food production and distribution. Some are superior food quality free from pesticides, low emissions of greenhouse gases, efficient energy use, insignificant nutrient leaching, high biodiversity, and sufficient and sustainable food production (Kirchmann 2019). Are these attributes correct? It is true that organic yields are lower than conventional farming methods but its cultivation requires more land than conventional farming. If we have to value organic farming as sustainable and worthy of adoption on a large scale, yield level and land used should be considered and not only environmental benefits like none burning and no chemical usage. This is because, reliance on only organic farming will lead to suffering, loss of life, political strife and war in the world and Africa in particular. This is because, many peasant farmers in Ghana are using organic farming practices but are not enjoying the premium price being paid to the certified farms leading to them being negatively impacted in social terms of sustainability. Certification is then being used to cheat them of their hard work and if care is not taken, will lead to an entrenchment of their poverty.

6 Conclusion

Organic farming and certification have a range of strengths in various areas including sustainable food production, good health and process auditing which assures consumers of the safety of their food. Fuelled by strong global consumer demand for ethically produced goods, organic farming and certification is expected to grow and diversify. More needs to be done to ensure that organic agriculture remains an attractive option for Ghanaian farmers. To begin with, policies need to be put in place to provide organic smallholders with much needed structural support. Heeding to this call will enable Ghana build a commanding global position in organic agriculture. Developing and regulating agricultural input supply and distribution chains will make it very easy for small scale farmers to gain access to organic inputs in order to help them address disease incidence and soil fertility issues on their fields.

It is also recommended that more research is done on organic farming to develop techniques that will help control diseases, maintain soil fertility and increase yields for small scale farmers. This is because for organic farming to be successful, agricultural innovations should be based on interactive process between farming communities and agricultural research institutions. Applied and adaptive agricultural research could be successfully implemented through location-specific and on-farm research, jointly with rural and tribal families, where partnerships, extension, and communication are integrative parts of the process of research and knowledge sharing. This on-farm research component can support rural communities and generate new knowledge that will benefit all farmers.

References

Adebiyi JA, Olabisi LS, Richardson R, Saweda L, Liverpool-Tasie O, Delate K (2019) Drivers and constraints to the adoption of organic leafy vegetable production in Nigeria: a livelihood approach. Sustainability 12(1):1–21. https://doi.org/10.3390/su12010096

Agro Eco (2008) Improve your sugarloaf pineapple production illustrated manual for organic farmers in Ghana. FAO, Rome, Italy

Argilés JM, Brown ND (2010) A comparison of the economic and environmental performances of conventional and organic farming: evidence from financial statements. Agric Econ Rev 11(1): 69–83

Asif M (2012) Physico-chemical properties and toxic effect of fruit-ripening agent calcium carbide. Ann Trop Med Public Health 5(3):150

Assouline G (1997) Conditions et obstacles a la difference de modeles d'agriculture durable en Europe. Report for Ministere de l'Environment. QAP Decision, Theys, Paris

Anaafo D (2015) Land reforms and land rights change: a case study of land stressed groups in the Nkoranza South Municipality, Ghana. Land Use Policy 42(1):538–546. https://doi.org/10.1016/j.landusepol.2014.09.011

Azam MS, Shaheen M (2019) Decisional factors driving farmers to adopt organic farming in India: a cross-sectional study. Int J Social Econ 46(4):562–580

Birkhofer K, Arvidsson F, Ehlers D, Mader VL, Bengtsson J, Smith HG (2016) Organic farming affects the biological control of hemipteran pests and yields in spring barley independent of landscape complexity. Landscape Ecol 31:567–579

Boru T (2018) Chapter five research design and methodology 5.1. Introduction https://doi.org/10.13140/rg.2.2.21467.62242

Alvi M (2016) A manual for selecting sampling techniques in research. https://mpra.ub.uni-muenchen.de/70218/

Bugri JT (2008) Gender issues in Africa land tenure: findings from a study of North-East Ghana. J Ghana Inst Surveyors 1(1):21–35

Doss CR (2001) Men's crops? Women's crops? The gender patterns of cropping in Ghana. World Dev 30(11):1987–2000

Flaten O, Lien G, Ebbesvik M, Koesling M, Valle P (2006) Do the new organic producers differ from the 'old guard?' Empirical results from Norwegian dairy farming. Renew Agric Food Syst 21:174–182. https://doi.org/10.1079/RAF2005140

Giovannucci D (2006) Salient trends in organic standards: opportunities and challenges for developing countries. World Bank/USAID Trade and Standards E-learning Course, Jan–March, 2006

GSS (2014) 2010 population and housing census. District analytical report. Ghana Statistical Service, Accra, Ekumfi district

Handmer J, Iveson H (2017) Cyclone pam in Vanuatu: learning from the low death toll. Aust J Emerg Manage 32:60–65

IFOAM (2017) Guidelines for public support to organic agriculture. IFOAM, Bonn, Germany. https://www.ifoam.bio. 15 Jan 20221

Inkoom SA (2017) Encouraging organic agriculture: the effects of conversion subsidies. Electronic Theses and Dissertations. No. 1214. http://openprairie.sdstate.edu/etd/1214/

Kilcher L (2007) How organic agriculture contributes to sustainable development. J Agric Rural Dev Tropics Subtropics 89:31–49. University of Kassel, Witzenhausen

Kirchmann H (2019) Why organic farming is not the way forward. Outlook Agric 48(1):22–27

King DNT, Skipper A, Tawhai WB (2008) Māori environmental knowledge of local weather and climate change in Aotearoa—New Zealand. Clim Change 90:385–409. https://doi.org/10.1007/s10584-007-9372-y

Kleemann L (2016) Organic pineapple farming in Ghana—a good choice for smallholders? J Developing Areas 50(3):109–130

Kleemann L (2011) Organic pineapple farming in Ghana—a good choice for smallholders? Kiel Working Paper No. 1671| Jan 2011
Kothari CS (2004) Research methodology, methods and techniques. New Age International Publishers, New Delhi
Kristiansen P, Taji A, Reganold J (eds) (2006) Organic agriculture: a global perspective. CSIRO Publishing, Collingwood, VIC, Australia, pp 484
Lahmar R (2010) Adoption of conservation agriculture in Europe. Land Use Policy 27:4–10. https://doi.org/10.1016/j.landusepol.2008.02.001
Lampkin NH (1994) Organic farming: sustainable agriculture in practice. In: Lampkin NH, Padel S (eds) The economics of organic farming—an international perspective. CAB International Publishers, Oxon (UK)
Lease SH (2003) Testing a model of men's non-traditional occupational choices. Career Dev Q 51(13):244–258. https://doi.org/10.1002/j.2161-0045.2003.tb00605.x
Leavy J, Hossain N (2014) Who wants to farm? youth aspirations, opportunities and rising food prices. IDS working paper 439. Institute of Development Studies, Brighton
Lori M, Symnaczik S, Mäder P, De Deyn G, Gattinger A (2017) Organic farming enhances soil microbial abundance and activity—a meta-analysis and meta-regression. PLoS One 12: https://doi.org/10.1371/journal.pone.0180442
McCann E, Sullivan S, Erickson D, De Young R (1997) Environmental awareness, economic orientation, and farming practices: a comparison of organic and conventional farmers. Environ Manage 21:747–758
Maertens M, Swinnen JFM (2009) Food standards, trade and development. Rev Bus Econ 54(3): 313–326
Masinde M (2015) An innovative drought early warning system for sub-Saharan Africa: integrating modern and indigenous approaches. Afr J Sci, Technol, Innovations Dev 7:8–25. https://doi.org/10.1080/20421338.2014.971558
Mchombu KJ (2002) Sharing knowledge for community development and transformation: a handbook. Oxfam Canada, Ottawa. http://www.schoolnetafrica.net/fileadmin/resources/SharingKnowledge.pdf. 22 Mar 2007
Mehrabi Z, Ramankutty N (2019) Synchronized failure of global crop production. Nat Ecol Evol 3:780–786. https://doi.org/10.1038/s41559-019-0862-x
Miller GT, Spoolman S (2015) Environmental science, 15th edn. Cengage Learning, Boston, MA
Morgan DL, Krueger RA, King JA (1998) The focus group kit, vols 1–6. Sage, Thousand Oaks, CA
Muguti T, Maposa RS (2012) Indigenous weather forecasting: a phenomenological study engaging the Shona of Zimbabwe. J Pan African Stud 4(9):102–113
Nazar R, Khan MIR, Iqbal N, Masood A, Khan NA (2014) Involvement of ethylene in reversal of salt-inhibited photosynthesis by sulfur in mustard. Physiol Plant 152:331–344. https://doi.org/10.1111/ppl.12173
Nyffeler M, Birkhofer K (2017) An estimated 400–800 million tons of prey are annually killed by the global spider community. Sci Nat 104(30):1–12. https://doi.org/10.1007/s00114-017-1440-1
Onduru DD, Diop JM, Van Der Werf E, de Jager A (2002) Participatory on-farm comparative assessment of organic and conventional farmers' practices in Kenya. Biol Agric Hortic 19:295–314
Opoku PD, Bannor RK, Oppong-Kyeremeh H (2020) Examining the willingness to produce organic vegetables in the Bono and Ahafo regions of Ghana. Int J Soc Econ 47(5):619–641
Padel S (2010) Conversion to organic farming: a typical example of the diffusion of an innovation? Sociologia Ruralis 41(1):40–61
Parrott N, Olesen JE, Høgh-Jensen H (2006). Certified and non-certified organic farming in the developing world. Global Dev Org Agric: Challenges Prospects:153–179. https://doi.org/10.1079/9781845930783.0153
Paull J (2011) The uptake of organic agriculture: a decade of worldwide development. J Soc Dev Sci 2(3):111–120

Pechrová M (2014) Determinants of the farmers' conversion to organic and biodynamic agriculture. AGRIS On-line Papers Econ Inform 6(4):113. http://online.agris.cz/files/2013/agris_on-line

Peterson HH, Barkley A, Chacon-Cascante A, Kastens TL (2012) The motivation for organic grain farming in the United States: profits, lifestyle, or the environment? J Agric Appl Econ 44(2): 137–155

Porter M (2016) Farm household motivations and diversification strategies of organic farmers at the rural urban interface. Ohio State University, Ohio

Rao LKM (2014) Organic farming in India: problems and prospects. Indian Econ J 62(1):697–720

Rogers EM (1983) The diffusion of innovations, 3rd edn. Free Press, New York

Sheeder RJ, Lynne GD (2011) Empathy conditioned conservation: 'Walking-in-the shoes-of-others' as a conservation farmer. Land Econ 87:433–452

Smith OM, Cohen AL, Rieser CJ, Davis AG, Taylor JM, Adesanya AW, Jones MS, Meier AR, Reganold JP, Orpet RJ, Northfield TD, Crowder DW (2019) Organic farming provides reliable environmental benefits but increases variability in crop yields: a global meta-analysis. Front Sustain Food Syst 3:82. https://doi.org/10.3389/fsufs.2019.00082

Soane BD, Ball BC, Arvidsson J, Basch G, Moreno F, Roger-Estrade J (2012) No-till in northern, western and southwestern Europe: a review of problems and opportunities for crop production and the environment. Soil Till Res 118:66–87. https://doi.org/10.1016/j.still.2011.10.015

Stolze M, Lampkin N (2009) Policy for organic farming: rationale and concepts. Food Policy 34(3):237–244. https://doi.org/10.1016/j.foodpol.2009.03.005

Tilman D, Cassman G, Matson PA, Naylor R, Polasky S (2002) Agricultural sustainability and intensive production practices. Nature 418:671–677. https://doi.org/10.1038/nature01014

Walaga C (2005) Organic agriculture in Kenya and Uganda: study visit report. CTA Working Document, 8033. The Netherlands, Wageningen

Wani SA, Wani MA, Mehraj S, Padder BA, Chand S (2017) Organic farming: present status, Scope and prospects in Northern India. J Appl Nat Sci 2272–2279. https://journals.ansfoundation.org/index.php/jans/article/view/1523

Willer H, Lernoud J (2017) The world of organic agriculture - Statistics and emerging trends 2017. FiBL and IFOAM, Bonn; Frick

Willer H, Lernoud J (eds) (2019) The world of organic agriculture. statistics and emerging trends 2018, 19th ed. Research Institute of Organic Agriculture FiBL and IFOAM Organics International, Frick and Bonn

Willer H, Kilcher L (eds) (2012) The world of organic agriculture—statistics and emerging trends 2012. IFOAM & Frick, Switzerland Research Institute of Organic Agriculture, FiBL, Bonn, Germany

Wickard CR (1944) Address by the Secretary of Agriculture. Annual convention of the Future Farmers of America, United States Department of Agriculture, Kansa City

The State of Agricultural Land Transformations in Asia: Drivers, Trade-Offs, and Solutions

Sivapuram Venkata Rama Krishna Prabhakar

1 Introduction

The land is a primary natural resource in Asian countries, and agriculture is still the sector that makes the most use of land in Asia (Food and Agriculture Organization 2016). Agricultural land plays a vital role in delivering the food, climate, and human security goals of countries in Asia, and any unsustainable land use could undermine the ability to achieve these goals. Agricultural land in Asia has been undergoing significant transformation due to drivers such as growing population, urbanisation, changing income levels, and lifestyle changes, including modified food consumption habits. These drivers are putting pressure on land resources leading to land degradation with impacts on the environment and economy, as well as on the livelihoods of communities that are dependent on land resources. With Asia's population projected to grow soon, land resources in Asia will likely come under additional pressure with negative consequences. This trend calls for a greater understanding of the emerging drivers of land use and land-use changes in Asia so that appropriate policy solutions can be put in place.

Understanding the impacts of land-use changes on agricultural land can help to manage the land sustainably. Efforts have been made to present land-use changes in published literature. Several of these studies utilise remote-sensing applications, mainly at the sub-regional level (e.g. Leisz et al. 2009) or at the national level (e.g. Hinz et al. 2020; Meiyappan et al. 2017). While these studies are very informative, they tend to require complex understanding and are often not accessible to the policy community. Besides, they do not usually have sufficient time-series data (Saah et al. 2019). Regional environmental synthesis exercises, including that of Environmental Outlook (GEO) for Asia-Pacific, provide a good overview of environmental changes. However, it can take time to produce such synthesis

S. V. R. K. Prabhakar (✉)
Institute for Global Environmental Strategies, Hayama, Kanagawa, Japan
e-mail: prabhakar@iges.or.jp

W. Leal Filho et al. (eds.), *Sustainability in Natural Resources Management and Land Planning*, World Sustainability Series,
https://doi.org/10.1007/978-3-030-76624-5_11

reports, and so there are often delays in communicating significant environmental changes to the relevant stakeholders. Although there have been calls to produce environmental status synthesis reports focusing on specific sectors over short intervals, efforts have not yet created a sectoral synthesis for land-use transformations in Asia with frequent updates (King 2019).

For the above reasons, this chapter presents a hybrid approach to analysing the secondary data on land-use changes available from various sources and presents some trends in combination with a review of literature on the state of major agricultural land transformations in Asia. The main objective of the chapter is to provide an easily accessible analysis of land transformations in Asia and to provide available evidence on the drivers behind these land transformations. The chapter identifies important drivers and trade-offs countries are facing due to land-use changes and identifies policy measures to halt any negative trends. As official consultations for the UNEP GEO-7 have already begun, the results presented here could provide meaningful input into the GEO process.

The hybrid approach set out in this chapter consists of a literature review supplemented by original analysis of data from secondary sources. Such a hybrid approach is increasingly being adopted in several environmental assessment processes, including that of the UNEP GEO reports. Data were largely drawn from the Food and Agriculture Organization Corporate Statistical Database (FAOSTAT) for the years 2002–2015 when data on agriculture production and land use are available for most Asian countries.

Wherever necessary, a simple statistical correlation analysis was performed to show the relationship between two variables in the discussion. For example, a statistical correlation between agricultural land and the urban population was performed for major Asian countries to visualise how changes in urban population have resulted in changes in agricultural land. Wherever relevant, the trends were presented using simple statistical measures such as the percentage change from the base year 2002, using bar graphs. Selecting these methods ensures that the analysis is accessible to as wide a range of readers as possible, including their application in limited data and skill environments.

2 The State of Agricultural Land Transformations in Asia

Two kinds of agricultural land transformations can be observed in Asia—agricultural land-use change and agricultural intensification. The former refers to changes in land use either by conversion into agriculture or away from agriculture. The latter refers to how agricultural land is managed.

2.1 Agricultural Land-Use Change

Countries in Asia have witnessed significant transformations in agricultural land over the past few decades. During the period 1961–2015, the percentage of agricultural land compared to total land areas remained constant in eight countries, declined in 10, and increased in 29 (Fig. 1, based on the data from FAOSTAT (Food and Agriculture Organization 2018a). The countries that have lost a significant amount of agricultural land include Brunei Darussalam (66.7% decline compared to 1961), Japan (48%), Republic of Korea (35.3%), and Mongolia (24.6%). The five countries that have experienced the highest rate of growth in agricultural land are Malaysia (49.5%), Vietnam (47.8%), United Arab Emirates (44.5%), and Bhutan (40.7%).

China (153 Mha), Indonesia (18.6 Mha), and Saudi Arabia (87.2 Mha) have converted vast areas of land for agricultural use. In comparison, Mongolia (27.8 Mha) and Iran (13.7 Mha) have lost significant agricultural land during the

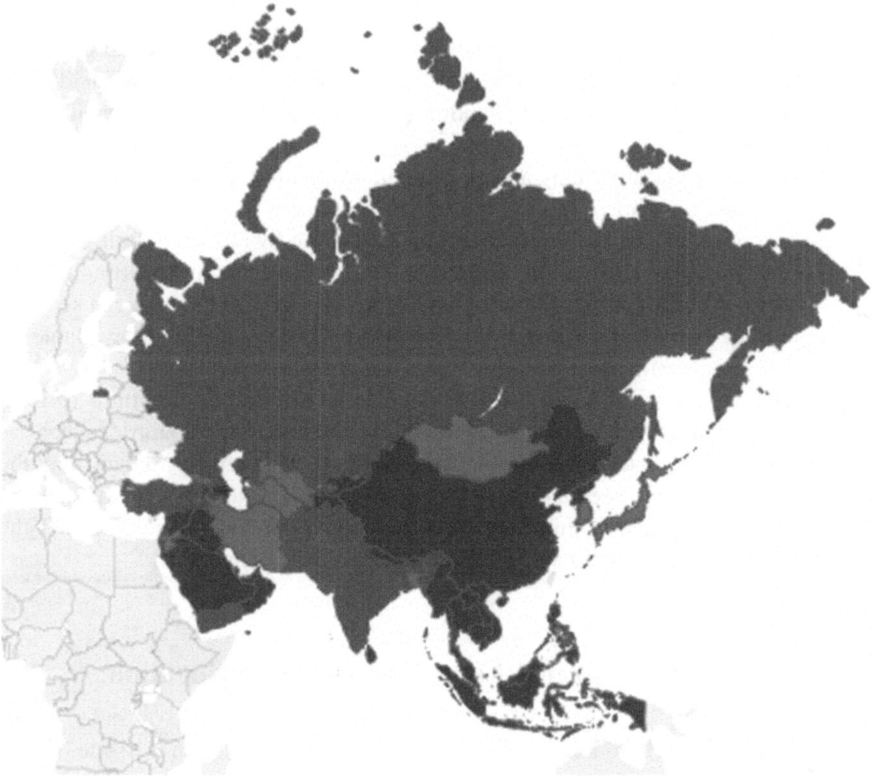

Fig. 1 Change in agricultural land in Asia (Green: no change, Red: Declining, Blue: Increasing). *Source* Author

same period. The source of land differed for countries that gained agricultural land. A substantial amount of agricultural land came from 'other land' categories in China and Saudi Arabia. In Indonesia, the gain in agricultural land came from forests. Agricultural land in Iran and Mongolia was lost to afforestation or other land-use categories. Various factors can explain why agricultural land was lost or gained. In countries where agricultural land was lost to other land-use types, factors behind the loss may include the way agricultural land was managed for its productivity, the role played by the global agricultural trade that has normalised supplies of certain food products from the least costly sources, and the fact farmers are shifting from full-time farming to non-farming income sources. In countries where agricultural land was gained, factors such as national food security needs, subsidies, food export to niche markets, and employment laws may explain such changes. Some of these factors are discussed in the following section on agricultural intensification.

2.2 *Agricultural Intensification*

Agricultural intensification has been defined as the degree of yield increase caused by production choices made (Dietrich et al. 2012). Although crop yields can indicate agricultural intensification, caution is needed in interpreting intensification based only on yield since it is not only human factors that contribute to better crop yields—favourable weather may also play a part. Agricultural intensification can be measured by the amount of land employed to produce a unit amount of food per unit amount of input. Irrigation can drive intensification by allowing multiple crops in a year, ensuring the cultivation of high yielding varieties, and permitting the use of fertilisers, pesticides, and mechanisation. Therefore, irrigation alone can serve as the single most important indicator of intensification when data is not available on other forms of intensification.

The data on chemical inputs and percentage of irrigated agriculture (Fig. 2) show that chemical inputs have significantly increased in countries such as Bangladesh, China, India, Indonesia, and Vietnam. Conversely, Japan and the Republic of Korea have seen a drop in the use of chemicals largely due to a decline in agricultural land, and a shrinking farming population. Indonesia recorded the largest increase in chemical use, followed by India and Bangladesh.

3 Drivers of Agricultural Land Use and Intensification

The term 'driver' refers to a factor that contributes to the change in the state of land from one land-use category to another one. There are multiple drivers concerning agricultural land in the Asian region, and characterising these drivers is a challenge for three main reasons.

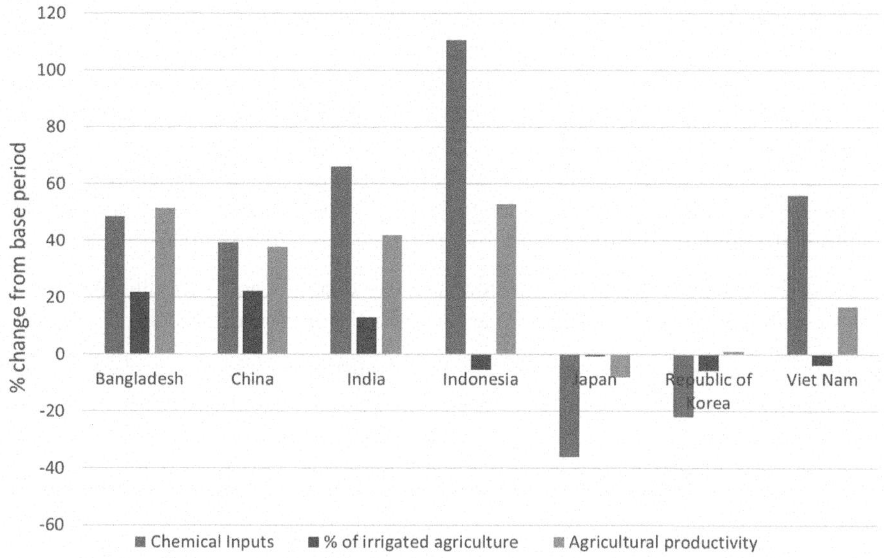

Fig. 2 Trends in agricultural inputs in selected Asian countries, values are for 2015 expressed as % change from 2002. *Source* Author, based on data from FAOSTAT

(a) There are feedback connections between the various drivers. Drivers seldom operate independently, and interlinkages make it a challenge to pinpoint which driving factor has contributed to the land-use changes and to what extent compared to others. For example, rapid urbanisation can provide a better quality of life and therefore pull the population out of agriculture as a livelihood. Hence, urbanisation is a visible driver here. However, other common drivers, such as education and growing income levels also affect both urbanisation and agricultural land-use changes. Furthermore, drivers such as food exports, niche food markets in urban areas, and agricultural subsidies may act as pull factors towards agriculture. This complex interaction among drivers makes analysis quite difficult, so it is necessary to treat them as cascading factors (primary and secondary drivers) or a matrix of driving factors rather than taking them in isolation.

(b) Extrinsic drivers are equally important as drivers intrinsic to the agriculture sector. Several drivers are outside the influence of the decisions made within the agriculture sector. For example, the growing need to build new infrastructure and increased energy demands mean that a significant amount of land has been taken out of agriculture and other land-use categories. Decisions made by infrastructure and energy ministries do not often take into consideration the impact of these decisions on agriculture and food needs and may disregard the additional pressure put on agricultural land. This lack of coordinated action can make it difficult to account for major drivers of land-use changes.

(c) Different drivers operate at various levels with complex ways connections between different levels. For example, drivers operating at the local level are not often captured at the macro level, such as at the national and regional levels. This could lead to a masking effect when only macro drivers are considered. Because of this, macro drivers do not often explain land-use changes at the local level in a specific context. Hence, there is a need to take into consideration drivers at the local and macro levels, and the interaction between both should be understood.

Table 1 presents a set of drivers for land-use changes and agricultural intensification. The following conclusions can be drawn from these drivers:

(a) There is a common set of drivers that contribute to land-use changes and intensification in Asia. For example, high economic growth can turn

Table 1 Drivers reported in literature for land-use change and agricultural intensification in agriculture

Land-use change drivers	References	Agricultural intensification drivers	References
• Population growth	Food and Agriculture Organization (2016)	• Economic development • Commercial agriculture	Alauddin and Quiggin (2008)
• Economic development • Rural enterprise development • Infrastructure	Azadi et al. (2011)	• Crop diversification • Improved varieties	Gunasena (2001)
• Commercial agriculture • Private sector investments • FDI in agriculture • Agricultural trade • Food security priorities	Hosonuma et al. (2012)	• Irrigation • Groundwater use	Alauddin and Quiggin (2008)
• Urbanisation	Azadi et al. (2011)	• Fertilizers	Gunasena (2001)
• Land productivity • Capital-labour ratio • Land-tenure security • Governance of land-use change	Azadi et al. (2011)	• Non-farm income • Family labor	Hao et al. (2015)
• Climate change	Oliver and Morecraft (2014)	• Mechanization • Controlled environments	Nani et al. (2011)

Source Author from references cited

populations away from agriculture. Conversely, economic growth can also contribute to farm mechanisation, high productivity, and access to best management practices.

(b) The majority of agricultural intensification drivers are intrinsic to the agriculture sector, but factors external to agriculture also played an important role. Agricultural intensification is largely explained by the expansion in irrigation facilities, adoption of high yielding varieties, and the spread of best management practices through farmer field schools and other innovations. However, factors other than agriculture such as non-farm income can also contribute to agricultural intensification as farmers with non-farm income may invest in improving their agricultural practices. Similarly, high land prices may drive efficient use of agricultural land by allocating the land to commercial crops and high-value food crops near urban centres.

3.1 Drivers of Land-Use Changes

Commercial agriculture has been one of the most important drivers of land conversion to agriculture (Food and Agriculture Organization 2016). The increase in commercial agriculture can be explained by the growing trend in urbanisation as well as an expanding food supply chain. Furthermore, commercial agriculture and agricultural futures markets are providing farmers with advantages such as economy of scale, product specialisation and income security (Wilkinson 2019). Mechanisation has had a positive impact on land conversion to agriculture as it has reduced labour dependency. Private sector investments and foreign direct investments in agriculture are emerging as new drivers sustaining the recent demand for agricultural land (Ravanera and Gorra 2011). Food trade opportunities due to regional economic integration in the ASEAN region have also sustained demand for agricultural land.

Driving factors that resulted in agricultural land being lost include urbanisation, industrialisation, and infrastructure development. High land prices for housing in peri-urban areas have led to the conversion of agricultural land to housing (Jiang and Zhang 2016; Rondhi et al. 2018). Declining rural populations in developed Asian countries have driven agricultural land to be converted for non-agricultural purposes. Declining farm profits, market imperfections, land degradation, poor farm price policies, natural hazards, and rising education levels have driven people out of farming, predisposing land to be converted away from agriculture.

Urbanisation is an important and emerging driver in several countries. Countries with the highest urban population growth rates are Lao PDR, Thailand, China, and Bhutan. Countries in West Asia are slowly de-urbanising (e.g. the urban population in Lebanon declined by 12% over the past 15 years). Countries with a negative correlation between agricultural land and urban population have recorded a significant decline in agricultural land, and countries with a positive correlation have shown a significant increase in agricultural land in Asia (Fig. 3). Countries in red in the top half and bottom half of the figure show significant positive and negative correlation respectively between urban population and land-use change.

Biofuels signify a substantial food-fuel-water nexus in Asia. It was feared that biofuels played a major role in the conversion of land into biofuel production in several Asian countries. The demand for biofuels has resulted in the allocation of agricultural land for biofuels in India (Schaldach et al. 2011), China, Indonesia, and Malaysia (Valin et al. 2015).

Rapid economic development, increasing middle-class society, changing lifestyles, and increased trade are also affecting land use in Asia. A traditional grower of cereals in 1970, by 2010, Asia was producing more fruits and vegetables than cereals, with livestock products nearly doubling in the same period (Asian Development Bank 2013). Livestock production and feed are competing for agricultural land and water, which has serious land and water pollution implications (Ahuja 2012).

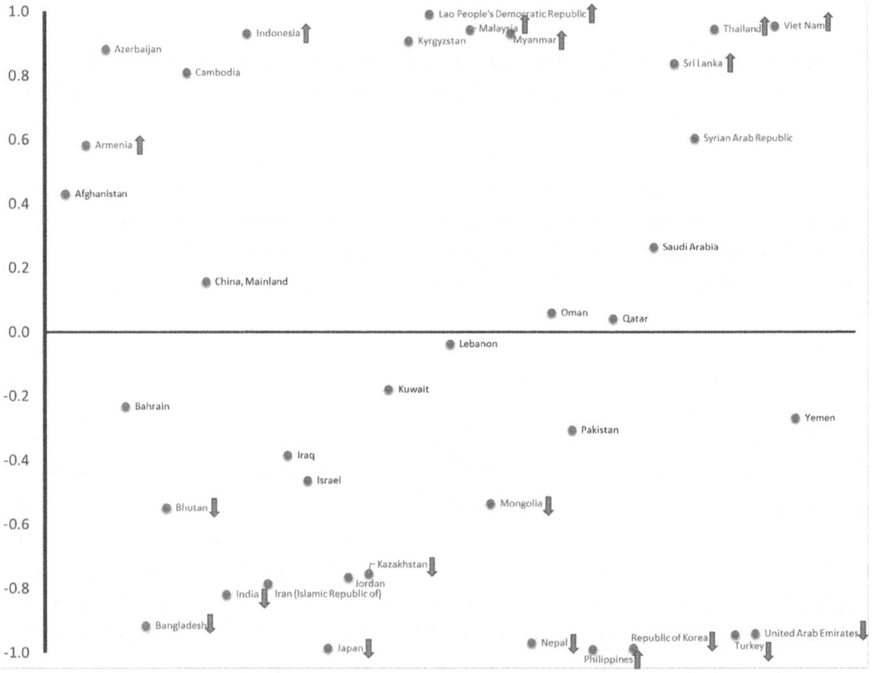

Fig. 3 Relationship between urbanisation and loss of agricultural land (Countries in red show significant correlation at a probability level of 0.05). *Source* Author, based on data from FAOSTAT

3.2 Drivers of Agricultural Intensification

Trends in agricultural intensification as a function of water, fertilizers, and pesticides are presented in Fig. 4 for three groups of countries: (a) countries that gained agricultural land; (b) countries that lost agricultural land; and (c) countries with stable agricultural land expressed as a percentage change from 2002 values (see Fig. 4).

The use of pesticides and fertilizers played a major role in agricultural intensification in countries where agricultural land expanded. However, countries do not differ much in terms of other characteristics such as irrigated areas and agricultural productivity. Countries with agricultural land loss had marginally higher productivity gains over the countries where agricultural land expanded. Observations indicate that marginal gains in irrigated areas and agricultural productivity may have contributed to the decline in agricultural land in some countries, while other countries gained agricultural land due to the use of pesticides and fertilizers.

Minimum support prices and subsidies on agricultural inputs are important for most developing countries in Asia, while developed countries are slowly moving away from input subsidies to direct cash payments (Hudson et al. 2011). These policies are justified to support economic and agricultural growth, poverty reduction, farm profitability, and to meet food security goals (Lopez et al. 2017). The policies also help to retain farmers, assist in the expansion of certain agricultural crops, and enable farmers to sustain a rate of agricultural input use that may not be possible in the absence of subsidies contributing to negative environmental and social impacts.

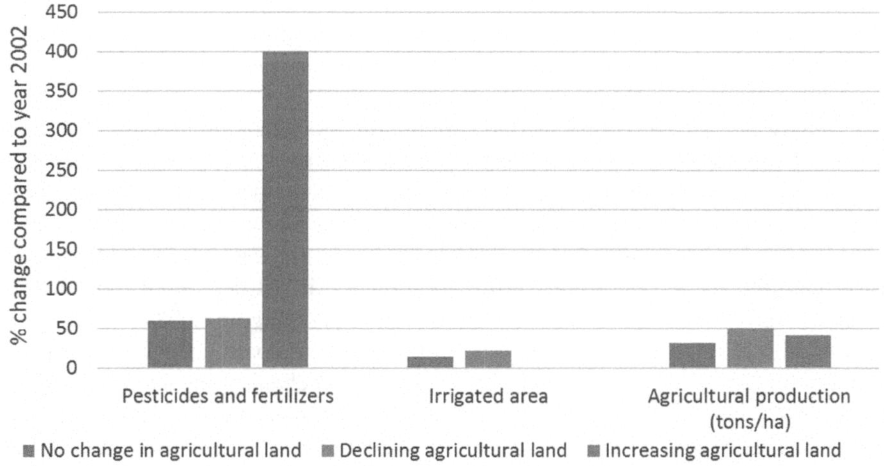

Fig. 4 Percentage change in drivers of agricultural intensification a) countries grouped as per the land-use change categories. *Source* Author, based on data from FAOSTAT

Agricultural exports are another significant driver of intensification in Asia as they constituted 13.2, 17, 15.2, and 12.2% of the GDP of India, Thailand, Vietnam and Malaysia respectively in 2016 (World Trade Organization 2016). Agricultural export policies have contributed to agricultural intensification. This trend is supported by premium prices in international markets, and an expansion in agribusiness, finance, and technologies for commercial crops (Asian Development Bank 2013).

4 Socio-Economic and Environmental Trade-Offs

The previously discussed agricultural land transformations have socio-economic and environmental trade-offs in Asia. There has been a steady rise in agricultural areas with declining total factor productivity in the Gangetic Basin (Prabhakar and Elder 2009). Food subsidies such as providing minimum support prices to food crops have led to a disproportionate expansion of crops, which in turn has led to increased water and chemical consumption (Sharma 2015). The intensive use of chemicals has led to pollution of sub-surface and surface water sources, increased runoff, and silting of freshwater bodies resulting in eutrophication affecting freshwater (Wasantha et al. 2015). There have also been reports on the negative impact of land-use changes in soil and water on ecosystems (Vlek et al. 2017).

Biofuel production consumes a significant amount of water (e.g. soy feedstock can require 47–215 litres of water per MJ energy produced (O'connor 2009). Expansion of biofuels resulted in net positive greenhouse gas emissions since the land was converted from forests, grasslands, and peatlands in Indonesia and Malaysia (Fulton et al. 2017).

Evidence indicates the negative impact of subsidised rural electrification programmes on the overuse of groundwater in some parts of India (Sharma 2015). Studies have shown that growing energy needs in the future could further exacerbate the water deficit and increase water demand for energy production by five times by 2050 in Asia (Mitra and Bhattacharya 2012).

Groundwater use has contributed to agricultural intensification in South Asia and is making agriculture environmentally-intensive (Acharya 2014). This entails significant damage to the physical environment and is a threat to agricultural production in the long run. The increasing use of groundwater has a greater environmental cost and it has also been argued that the price paid for groundwater use has often been lower than the social opportunity cost. Additionally, it has been used more intensively in regions such as southern Asia (Alauddin and Quiggin 2008).

Agricultural intensification has also contributed to climate change through the promotion of fertiliser made from fossil fuels, farm mechanisation, and the burning of crop residues in parts of Asia (Nani et al. 2011). Expansion of agriculture in the tropics has been reported to have a warming effect on the atmosphere due to surface brightening and the consequent reduction of net radiation not balancing the increase in temperature associated with reduced transpiration (Duveiller et al. 2017).

5 Measures for Easing the Pressure on Land

Any recommendations for addressing agricultural land transformations should consider the food security and developmental needs of Asia. In addition, climate change mitigation, adaptation, and resilience considerations need to be juxtaposed with these needs. Some relevant recommendations are discussed in this section to reduce the demand for agricultural land and address the perils of intensification. Some of the available options for reducing the pressure on agricultural land include improving the productivity of rain-fed agriculture, promoting organic agriculture, promoting renewable energy use in agriculture to reduce the environmental footprint of agriculture while meeting the necessary energy demands, promoting water-efficient technologies, reducing food waste, boosting urban agriculture with necessary safeguards, and promoting integrated governance of land resources. These solutions have differing potentials in reducing the demand for land and easing the pressure of agricultural intensification, as shown in Fig. 5a, b. These figures indicate that there is a high potential for unconventional productivity gains such as in rain-fed areas, while solutions such as food labelling may reduce the environmental footprint of the food with the least effect on the demand for agricultural land. Similarly, there are solutions that need policy coordination between several ministries which is not happening at the moment. Solutions that demand such policy coordination include promoting renewable energy in agriculture, changes in food consumption patterns towards sustainable ones, and integrated land resource management.

Improving agricultural productivity can be a low-hanging fruit that can reduce the demand for new agricultural land. Agricultural productivity gains in rain-fed areas remained sub-par, and efforts to improve rain-fed agriculture have largely been unsuccessful. This provides immense opportunity for the region to invest more in research and development in rain-fed agriculture. The private sector puts a very limited focus on rain-fed areas, and agricultural innovation from national research and development efforts is far from being satisfactory. In order to enhance productivity in rain-fed areas, there is a need to promote public-private partnerships that can promote innovation and scale as complementarities.

Organic agriculture occupies a significant share of sustainable agriculture in Asia, and it has great potential to expand further (Sano and Prabhakar 2009). Climate-smart agriculture (CSA) can bring climate change mitigation benefits while reducing the impact of climate change on agriculture (Lipper et al. 2018). The 'smartness' in CSA involves utilising the opportunities provided by climate change while tackling the challenges associated with it. CSA seeks to integrate global environmental and developmental issues in a single approach. CSA practices aim to reduce the demand for off-farm inputs, energy, and water and can increase production and productivity within sustainability bounds while protecting food production and rural livelihoods from climatic vagaries. CSA is attracting growing attention from policymakers in many Asian developing countries.

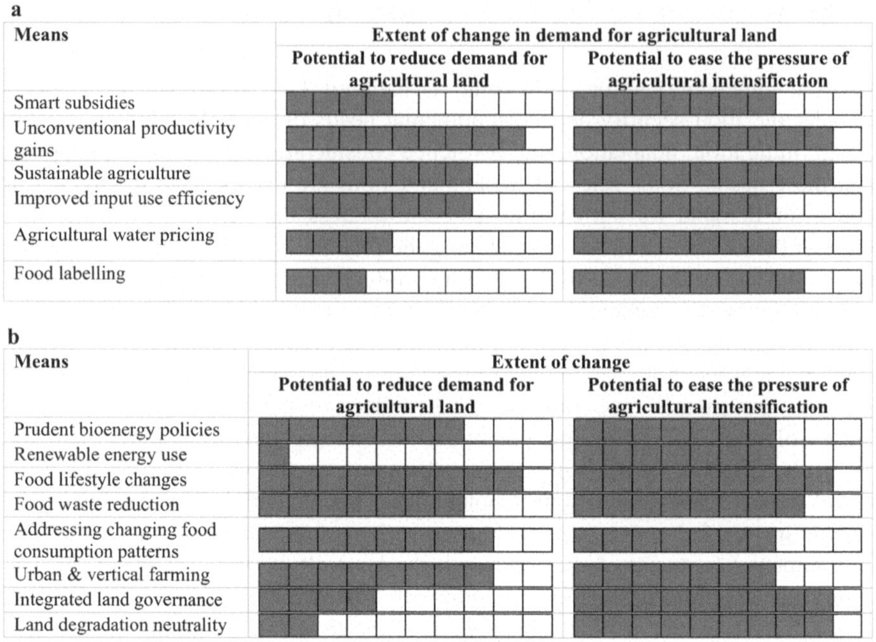

Fig. 5 **a** Potential of various agricultural approaches for reducing demand for agricultural land and easing agricultural intensification pressure. **b** Potential of various solutions that need policy coordination across different ministries of governments. *Source* Author; Note: Bars indicate the extent of change brought by the means listed on the left

Food labelling can mitigate the negative impacts by informing consumers about the environmental burden of food products such as the use of water, chemical and pesticide load, and carbon emissions. In Asia, agricultural input subsidies play a critical role in agricultural intensification. Subsidies can help meet certain national objectives, such as poverty reduction and food security, but they can also result in inefficiencies and excessive use of inputs. Subsidies are providing perverse incentives in agriculture. There needs to be a better way of targeting poor farmers other than blanket subsidies (Dorward 2009). Commodity price support has a distorting effect on markets and needs to be reformed, thereby incentivising sustainably-produced food.

Food-fuel conflicts are a concern for the region and were prominent during the biofuel boom of the early 2000s. With declining fossil fuel prices globally, the biofuel fervour has also been on the decline with several major private sectors plans to invest in biofuels in the region either being put on hold or scrapped. Biofuel production in India has either stabilised (as in the cases of ethanol) or has continuously risen at a low pace (as in the case of biodiesel) (United States Department of Agriculture 2014).

Food-fuel conflicts need to be addressed through policies such as the National Biofuel Policy by the Government of India, which limited biofuels to non-food feedstock on degraded lands (Prabhakar and Elder 2010). It is essential to combine such policies with support from farmer groups such as cooperatives like pressure groups that discourage farmers from growing biofuel crops on fertile lands.

Renewable energy use in agriculture in India and East Asian countries has reduced the demand for grid electricity, reduced the demand for coal, reduced the water for thermal power production, and contributed to GHG mitigation (Ministry of Finance 2011; Ministry of Natural Resources and Environment 2014). Biogas has reduced the demand for fossil fuels, motivated farmers to keep cattle, generated manure for crops, and helped improve agricultural yields and soil health while increasing the energy security of rural households.

Water-efficient technologies can reduce pumping needs and make water available for other priority sectors. These technologies can be scaled up by building capacity among farmers, providing positive incentives, addressing electricity subsidies (perverse subsidies), implementing water pricing, engaging water user associations, and rationing irrigation (Hasanain et al. 2012).

On the demand side, reducing food waste is an important policy area for both developed and developing countries. Food wastage can be reduced by strengthening storage and transportation facilities, enforcing food handling standards, changing food consumption patterns, developing the food processing industry, implementing food pricing, and addressing the seasonal agricultural glut (World Food Program 2009). Food waste policies have to cover the entire lifecycle of food, and must also be a multi-sectoral and multi-stakeholder collaboration.

Urban farming can help to reduce the pressure on agricultural soils. Urban farming can minimise carbon emissions from food transport and water miles, and can also reduce the ecological footprint of cities (Food and Agriculture Organization 2018b). There is a need to assess the food production potential of cities and the associated environmental costs such as space, electricity, and water shortages prevalent in Asian cities. Urban farming needs to be coupled with renewable energy, water harvesting, organic inputs, and ensuring waste reduction and recycling approaches. There is also a need for standardising agricultural practices for urban microclimates, infrastructure, and institutional structures.

Land must be governed in an integrated manner to avoid environmental harm, social conflict, and suboptimal outcomes. Integrated land governance consists of structures and processes that bring different stakeholders from different sectors together to agree on how land should be used and managed sustainably. Currently, land governance is highly fragmented and tends to ignore the interlinkages between various forms of land and their impacts on the overall wellbeing of dependant populations. Integrated land governance can help agencies (e.g. the Ministry of Housing) to understand the impacts of their decisions on the land under the purview of other agencies (e.g. the Ministry of Agriculture).

6 Conclusions

This chapter identified trends in agricultural land-use changes and provided an understanding of various drivers behind these changes. The food production and productivity improvements in the region so far have come from significant intensification with associated damage to the agricultural ecosystems even though these production gains have lifted millions of people out of poverty, hunger, and malnutrition. Achieving the SDGs while safeguarding the sustainability of agricultural lands and the health of agricultural soils presents a paradoxical problem for Asian countries. Any additional gains in food production and productivity in the future has to come without any extra cost to the environment and society. Our ability to address issues such as unhealthy consumption patterns, food loss, and inefficient food production practices provide us with the necessary leverage to ensure that ecosystems recover and rejuvenate. In the long term, the growing population is likely to demand other forms of achieving food security goals. Looking further into the future, this will require new and innovative means of achieving food security without harming the environment. This could include innovative means of food production including urban and vertical farming, and most importantly a change in our food habits towards healthy choices and introduction of new sources of food that may demand significant cultural adjustments but do not burden the environment.

It is evident that the impacts of agricultural land-use decisions do not remain in the agriculture sector alone but affect several other sectors. The fact that the land is still governed in a fragmented manner by various ministries, each responsible for a different economic sector, indicates the need for a paradigm change in the way the land is governed. Such a change could include greater policy coordination between ministries that govern land so that decisions made in one ministry or sector do not have negative consequences on other ministries. Hence, there is a need for a system that visualises the impacts of land-use decisions in an integrated fashion. This calls for developing integrated land-use decision support systems that are currently not well researched and developed for policy decisions in Asia.

References

Acharya A (2014) Groundwater, climate change and sustainable well being of the poor: policy options for South Asia, China and Africa. Procedia - Soc Behav Sci 157:226–235

Ahuja V (2012) Asian livestock: challenges, opportunities and response. International Livestock Research Institute, Bangkok, Thailand

Alauddin M, Quiggin J (2008) Agricultural intensification, irrigation and the environment in South Asia: issues and policy options. Ecol Econ 65(1):111–124

Asian Development Bank (2013) Agriculture and structural transformation in developing Asia: review and outlook. ADB, Manil, The Philippines

Azadi H, Ho P, Hasfiati L (2011) Agricultural land conversion drivers: a comparison between less developed, developing and developed countries. Land Degrad Dev 22(6):596–604

Dietrich JP, Schmitz C, Müller C, Fader M, Lotze-Campen H, Poppa A (2012) Measuring agricultural land use intensity—a global analysis using a model-assisted approach. Ecol Model 232(10):109–118

Dorward A (2009) Rethinking agricultural input subsidy programs in a changing world. Centre for Development, Environment and Policy, London, UK

Duveiller G, Hooker J, Cescatti A (2017) The mark of vegetation change on Earth's surface energy balance. Nat Commun 9:679. https://doi.org/10.1038/s41467-017-02810-8

Food and Agriculture Organization (2016) State of the world's forests: forests and agriculture. Land use challenges and opportunities. FAO, Rome

Food and Agriculture Organization (2018a) FAOSTAT. http://www.fao.org/faostat. Accessed 06 Jan 2020

Food and Agriculture Organization (2018b) Food for the cities programme: building food secure and resilient city regions. http://www.fao.org. Accessed 18 Dec 2019

Fulton L, Mejia A, Arioli M, Dematera K, Lah O (2017) Climate change mitigation pathways for Southeast Asia: CO_2 emissions reduction policies for the energy and transport sectors. Sustainability 9(1160):1–16. https://doi.org/10.3390/su9071160

Gunasena H (2001) Intensification of crop diversification in the Asia-Pacific region. In: Papademetriou ML, Dent F (eds) Crop diversification in the Asia-Pacific region. FAO, Bangkok, pp 156–165

Hao H, Li X, Tan M, Zhang J, Zhang H (2015) Agricultural land use intensity and its determinants: a case study in Taibus Banner, Inner Mongolia, China. Front Earth Sci 9(2):308–318

Hasanain A, Ahmad S, Mehmood M, Majeed S, Zinabou G. (2012) Irrigation and water use efficiency in South Asia. GDN Agriculture Policy Series Briefing Paper No 9: 1–8

Hinz R., Sulser TB, Huefner R, Mason-D'Croz D, Dunston S, Nautiyal S, Ringler C, Schuengel J, Tikhile P, Wimmer F, Schaldach R (2020) Agricultural development and land use change in India: a scenario analysis of trade-offs between UN Sustainable Development Goals (SDGs). Earth's Future 8(2). https://doi.org/10.1029/2019ef001287

Hosonuma N, Herold M, Sy V, Fries R, Brauckhaus M, Verchot L, Romijn E (2012) An assessment of deforestation and forest degradation drivers in developing countries. Environ Res Lett 7:1–12

Hudson D, Pan S, Mutuc M, Yates S, Ethridge D (2011) Crop subsidies in foreign countries. Cotton Economic Research Institute, Lubbock, TX

Jiang L, Zhang Y (2016) Modeling urban expansion and agricultural land conversion in Henan Province, China: an integration of land use and socioeconomic data. Sustainability 8:920. https://doi.org/10.3390/su8090920

King P (2019) Sixth Global Environment Outlook (GEO6): Outcome of United Nations Environment Assembly 4 (Mar 2019). https://archive.iges.or.jp/en/sdgs/unea4/geo6.html. Accessed 24 Sep 2020

Leisz SJ, Yasuyuki K, Fox J, Masayuki Y, Rambo TA (2009) Land use changes in the uplands of Southeast Asia: proximate and distant causes. Southeast Asian Stud 47(3):237–243

Lipper L, McCarthy N, Zilberman D, Asfaw S, Braca G (2018) Climate smart agriculture: building resilience to climate change. Springer, Cham, Switzerland

Lopez R, He X, Falcis E (2017) What drives China's new agricultural subsidies? World Dev 93:279–292

Meiyappan P, Roy PS, Sharma Y, Ramachandran RM, Joshi PK, Defries RS, Jain AK (2017) Dynamics and determinants of land change in India: integrating satellite data with village socioeconomics. Reg Environ Change 17:753–766

Ministry of Finance (2011) Power and energy sector road map: an update. Finance Division, Ministry of Finance. Government of the People's Republic of Bangladesh, Dhaka, Bangladesh

Ministry of Natural Resources and Environment (2014) Biogas. Ministry of New and Renewable Energy, New Delhi, India. http://mnre.gov.in. Accessed 06 Mar 2020

Mitra B, Bhattacharya A (2012) Long term energy scenario and water use—a case study on India. IGES Policy Brief, 8. Institute for Global Environmental Strategies, Hayama, Japan

Nani B, Sitaula B, Bajracharya R (2011) Agricultural intensification in South Asia and its contribution to greenhouse gas emission: a review. Asian J Water Environ Pollut 8(3):11–17

O'connor T (2009) Water usage in biodiesel production. In: Paper presented at the water energy sustainability symposium. Groundwater Protection Council, Salt Lake City, Utah, pp 13–16

Oliver T, Morecraft M (2014) Interactions between climate change and land use change on biodiversity: attribution, problems, risks and opportunities. Wiley Interdisc Rev: Clim Change 5:317–335

Prabhakar SVRK, Elder M (2009) Biofuels and resource use efficiency in developing Asia: back to basics. Appl Energy 86(1):S30–S36

Prabhakar SVRK, Elder M (2010). Not all biofuels are bad: a rural India case. http://ourworld.unu.edu. Accessed 06 Mar 2020

Ravanera R, Gorra V (2011) Commercial pressures on land in Asia: an overview. International Land Coalition, Rome, Italy

Rondhi M, Pratiwi PA, Handini VT, Sunartomo AF, Budiman SA (2018) Agricultural land conversion, land economic value, and sustainable agriculture: a case study in East Java, Indonesia. Land 7:148. https://doi.org/10.3390/land7040148

Saah D, Tenneson K, Matin M, Uddin K, Cutter P, Poortinga A et al (2019) Land cover mapping in data scarce environments: challenges and opportunities. Front Environ Sci. https://doi.org/10.3389/fenvs.2019.00150

Sano D, Prabhakar SVRK (2009) Some policy suggestions for promoting organic agriculture in Asia. J Sustain Agric 34(1):80–98

Schaldach R, Priess JA, Alcamo J (2011) Simulating the impact of biofuel development on country-wide land use change in India. Biomass Bioenerg 35(6):2401–2410

Sharma S (2015) The political economy of the water-energy nexus in Indian irrigation. http://blogs.lse.ac.uk/southasia/2015/09/09. Accessed 06 Mar 2020

United States Department of Agriculture (2014) India biofuels annual 2014. USDA, Washington, DC, USA

Valin H, Peters D, Berg M, Frank S, Havlik P, Forsell N, Hamelinck C (2015) The land use change impact of biofuels consumed in the EU: quantification of area and greenhouse gas impacts. Ecofys Netherlands, Utrecht, Netherlands

Vlek P, Khamzina A, Azadi H, Bhaduri A, Bharati L, Braimoh A, Taheri F (2017) Trade-offs in multi-purpose land use under land degradation. Sustainability 9(2196):1–19

Wasantha A, Hoang V, Wilson C (2015) Impact of agricultural chemicals on the environment and human health in Asia: are there solutions? In: Managi S (ed) Handbook of environmental economics in Asia. Routledge, New York, pp 321–348

Wilkinson J (2019) Large-scale forces, global tendencies and rural actors in the light of the SDG goals. Food and Agricultural Organization, Rome, Italy. http://www.fao.org/3/ca5058en/ca5058en.pdf. Accessed 18 Nov 2020

World Food Program (2009) Hunger and markets. Earthscan and World Food Program, London, UK

World Trade Organization (2016) International trade and market access data. http://stat.wto.org/CountryProfile. Accessed 23 Dec 2019

Sustainable Land Management Paradigm: Harnessing Technologies for Nutrient and Water Management in the Great Lakes Region of Africa

Patrick Musinguzi, Peter Ebanyat, Twaha Ali Basamba, John Baptist Tumuhairwe, Emmanuel Opolot, Giregon Olupot, John Stephen Tenywa, and Jackson Gilbert Majaliwa Mwanjalolo

1 Introduction

Land remains a critical resource in the provision of goods and services for livelihoods. Land is defined as a terrestrial system that comprises of the natural resources (such as soil, vegetation, climate, biota, geology, topography, water and air), the ecological processes, human settlement and the infrastructure, all operating within a system (FAO 2008). Despite its critical role, land resources are currently highly endangered due to rapid degradation that is orchestrating poverty, food and energy insecurity; and malnutrition particularly in the fragile ecosystems and sociocultural systems of the world. The human population, which is predicted to increase from 7.7 billion to about 9.7 billion by 2050, is one of the major threats to the integrity of land (UN DESA 2019). Currently, about 5 billion hectares of land (43% of earth's vegetative surface) is degraded, with a raging degradation rate as high as 10–12 million ha yr^{-1} (Thomas et al. 2018). In Africa, about 75% of the total arable land is degraded, with about 4–7% in Sub Saharan Africa (SSA) severely degraded (Orchard et al. 2017).

Degraded land is a major limitation to livelihoods for millions of people in SSA as land is their daily work table for income. The problem has been aggravated by climate change and rapid changes in technologies, which have led to increased floods, drought and community conflicts. It is now a global agenda that action oriented interventions be effected to counter land degradation and boost productivity. Sustainable technology applications that are characteristic of the right

P. Musinguzi (✉) · P. Ebanyat · T. A. Basamba · J. B. Tumuhairwe · E. Opolot · G. Olupot · J. S. Tenywa
Department of Agricultural Production, Makerere University, Kampala, Uganda

J. G. M. Mwanjalolo
Department of Geography, Geo-Informatics and Climatic Sciences, Makerere University, Kampala, Uganda

W. Leal Filho et al. (eds.), *Sustainability in Natural Resources Management and Land Planning*, World Sustainability Series,
https://doi.org/10.1007/978-3-030-76624-5_12

methods, techniques, skills and management approaches are vital for inclusion in the planning processes.

Sustainable Land Management (SLM) is an approach that can respond to degradation challenges with support of verified relevant technologies. It considers integration of soils, crops, livestock, water, energy, finance and human resources (WOCAT 2016). WOCAT (2016) clearly describes SLM as the use of the land resources, which include soils, water, animals and plants, for the production of goods and services to meet changing human needs, while simultaneously ensuring long-term productive potential of these resources, without compromising environmental health. The adoption of SLM has proved impactful on all ecosystem functions and services (Terefe et al. 2020). The benefits of SLM range from high and stable crop yields, regulated surface runoff and soil loss rates, reduced nutrient losses, which consequently result in increased landscape resilience, reduced production risks, improved food security and livelihoods (Branca et al. 2013; Martínez-Mena et al. 2020). The drivers of effective and efficient SLM processes are appropriate soil and water management processes, and these (codenamed indigenous technologies) have been practiced by millions of farmers for many years, though largely in the survival mode (Liniger 2011). Against this backdrop are emerging (improved) technologies with ability to catalyze meaningful sustainable land management. Limited empirical assessment has been made of these indigenous and improved technologies in the context of their adoption for sustainable land management within the sociocultural settings of the target great lakes communities. This paper reviews the existing soil water and nutrient management measures as a guide for out scaling of high potential SLM technologies and practices to achieve sustainable livelihoods in the great lake's region of Africa.

2 Sustainable Land Management: Definition and Pillars

There are several definitions of SLM that are tailored to sustainable use of natural resources to secure improved livelihoods. FAO Terr Africa defined SLM as the 'adoption of land use systems through the use of sound management measures or practices, and enabling land users to achieve maximum social and economic benefits from the land; while maintaining the ecological functions of the land resources (FAO 2008). World Bank (2006) defines SLM as a knowledge based procedure that helps to integrate land, water, biodiversity and environmental management, with the goal of meeting human needs while sustaining ecosystem services and livelihoods. Smyth and Dumanski (1995) earlier defined SLM as a combination of technologies, policies, and activities that are aimed at integrating social economic principles with environmental considerations but maintaining and enhancing productivity, reducing risks and enhancing soils buffer capacity against degradation; protecting natural resources such as water and soil from degradation; and aiming at socially acceptable and economically viable measures easily accessed by communities. Hurni (2000) defined SLM as a system of technologies and/or planning that aims at integrating ecological, socio-economic and political principles in land management for agricultural and other purposes to achieve intra- and inter-generational equity. WOCAT (2016) more recently defined SLM as the 'use of land resources which includes soil, water, animals and plants, for the production of goods to meet changing human needs while simultaneously ensuring the long-term productive potential of these resources and maintaining their environmental functions. The element of integration is emphasized across the spectrum of definitions, aiming at preserving land resources including nutrient and water resources. These resources remain the drivers of food, fibre, forage and fuel production, all of which are critical in sustaining livelihoods for millions of people.

Sustainable Land Management mitigates possible depletion of nutrients, moisture stress, soil loss and overall decline in land productivity. Water productivity is enhanced at field and catchment levels by ensuring water availability, storage capacity and quality. The mitigation interventions are in line with the widely defined pillars of SLM (Gessesse and Zerihun 2017). According to Gessesse and Zerihun (2017), the pillars of SLM include both agro-ecological and socio-economic factors; namely (i) income (when farm household income is evident), (ii) land productivity (when high yields are evident), (iii) environment protection/conservation (when resources are preserved in right quantities and quality), (iv) viability (when management options are locally adoptable and implementable), and (v) acceptability (when social acceptance is evident). The application of SLM pillars has other impacts on land productivity including reduction in agricultural related risks when social response, environment and economic factors are considered (Emerton and Snyder 2018). The SLM approach enhances nutrient and soil carbon re-stocking, consequently minimizing soil fertility decline but also increasing water availability, especially in moisture stressed environment.

3 Land Degradation Impacts and Solutions Relevant in Water and Nutrient Management for Sustainable Land Management

Several processes of land degradation are widely reported and include nutrient decline, loss of soil organic matter, water and wind erosion, soil acidification, soil salinization, soil sealing, heavy metal contamination and many others (Henry et al. 2018). These processes affect water and nutrient cycles thereby affecting soil, water and plant relationships. Processes (causes and effects of degradation) that disrupt nutrient supply and water availability can lead to decline in soil health, loss in land productivity and degraded environment. Table 1 highlights the causes and effects of degradation commonly reported in the great lakes region; potential impacts of degradation on nutrient and water resources and suggestions for possible solutions. The analysis is based on the causes and effects of degradation widely reported

(Olupot et al. 2019; Kiage 2013; Terefe et al. 2020). Some of the interventions are being promoted to improve water and nutrient management although challenges still exist in adoption due to the differences in the type of landscape and social-cultural establishment.

Although some of the interventions have been promoted in the Great Lakes region (Table 1), there is increasing evidence of less adoption by several farmers which is a threat to SLM. The low adoption of SLM techniques is a function of many factors including the social-cultural factors. The approach remains less adopted among land users in the region not because of lack of scientific technologies but because of lack of incentives and measures to sustain the indigenous approaches. The success of SLM demands that land users remain mindful of competing investment options that result in higher returns. If the returns from SLM are not profitable, there is always a shift in plans to choose other quick trade and businesses alternatives (Tenge et al. 2004). Other land users resort to rapid shift from one land use to another when there is decline far below the sustainability threshold for production. Therefore, SLM options relevant to water and nutrient management require an approach that considers a range of factors that influence land management decisions. A people centered management approach backed by existing local and national (government) institutions is critical. Emerton and Snyder (2018) denoted that interventions with less regard of land users can never address land degradation. The inclusion of the end-users is a critical step among the pillars of SLM in ensuring sustainable application of technologies.

4 Practices and Technologies for Sustainable Nutrient Management and Soil-Water Conservation

Land degradation can be addressed by applying practices and technologies that are in line with the pillars of SLM so as to achieve high land productivity, environmental conservation and income. The World Overview of Conservation Approaches and Technologies (WOCAT) has proposed various principles that support practices critical for soil water and nutrient management (Liniger 2011). The principles promoted by WOCAT include water-use efficiency and productivity, soil fertility, plants and their management, and micro-climate management (WOCAT 2016). Most of these principles resonate with practices and technologies vital for SLM.

Several SLM measures are being promoted in the region, all targeting the conservation of soil and water for increased land productivity. These practices include resource conserving measures such as organic farming, eco-agriculture, conservation agriculture, agroforestry, integrated nutrient management, integrated pest management, integrated livestock systems, water harvesting and aquaculture. These practices have demonstrated impact on soil water and nutrient management by reducing on unproductive losses of water, replenishing nutrient stocks,

Table 1 Land degradation, impacts and potential solutions relevant to water and nutrient management in the great lakes region of Africa

Selected causes (with example of effects) of land degradation	Impacts on water and nutrients	Potential solution(s)
Limited planning (misuse of land)	Loss in water quality	Land use planning
Substandard agro-inputs (nutrient depletion)	Limited supply of nutrients	Testing fertilizer quality
Inadequate organic material use (SOC depletion)	Limited capacity to restock organic carbon	Promote organic material regeneration
High surface runoff (soil loss/siltation)	Bare land and shallow soils with low water retention	Soil loss control
Rapid land use changes (high greenhouse gas emissions)	Nutrients lost through volatilization	Proper land use
Surface vegetative cover depletion (evapotranspiration)	Water loss from crop field	Use of conservation agriculture
Over-cultivation (degradation of soil structure and carbon)	Low nutrient retention capacity	Conservation tillage
Flooding (water logging)	Excess water and leaching of nutrients	Improved water management e.g. Ridging
Desertification (biodiversity loss)	Permanent loss of water and biota	Restoration with desert tolerant vegetation
High soil temperature (loss of soil biota and moisture)	High evapotranspiration	Use of soil covers
Inherent salinity/mineral weathering (salinization)	Limited nutrient availability	Adopt tolerant varieties
Uncertain precipitation (decline in ground water level)	Soil moisture stress	Apply soil water conserving measures and supplementary irrigation
Leaching (fertility decline)	Loss of mobile nutrients	Apply slow releasing fertilizers with soil conservation measures
Animal overstocking/ overgrazing (soil compaction)	Low water infiltration capacity	Optimize livestock numbers per land area
Volatilization (nutrient loss)	High loss of volatile nutrients	Minimize soil disturbances
Non-judicious use of agro-chemical (toxicity)	Decline in soil biota which subsequently affect nutrient release processes	Proper use of agro-inputs
Acidity (non-responsive soils)	Low availability of nutrients	Restoration of soil quality through liming
Alkalization (non-responsive soils)	Unavailability of some nutrients	Using tolerant crop varieties

(continued)

Table 1 (continued)

Selected causes (with example of effects) of land degradation	Impacts on water and nutrients	Potential solution(s)
Massive export of plant materials (loss in nutrient stocks)	Negative nutrient balance from farmers' fields	Increase residue retention
Burning (loss of nutrients)	Permanent loss of nutrient stocks	Stop burning of residues
Pollutant depositions (soil contamination, microbial death)	Unavailability of nutrients	Improved policies on soil protection
Wind erosion (depositions and enrichment)	High pollution of soil water	Monitor depositions and effects
Deforestation (bare land and soil loss)	Loss of nutrient stocks and high surface water runoff	Encourage tree planting
Social economic challenges		
High price of inputs (limited use)	Inability to purchase fertilizers to replenish nutrients	Fertilizer subsidies
High cost of water equipment (inability to access water resources)	Improper use of available water	Reduce taxes on water storage and supply infrastructure
Weak policies (poor soil management)	Misuse of agrochemicals	Policy guide on chemicals use
Poverty (lack of capital)	Inability to apply fertilizers or water conservation materials	Revolving funds
Limited land ownership (low confidence in land use and management)	Limits use of soil water conservation measures	Secure land ownership
Political instability (misuse of land)	Affects plans for sustainable soil water conservation measures	Promote political dialogues
Rapid urbanization (contamination and pollution)	Pressure on water resources	Effective urban planning

Source Authors based upon Olupot et al. (2019), Tully et al. (2015), Kiage (2013), Sanginga and Woomer (2009), Musinguzi (2017)

increasing storage and reducing pollution (Motavalli et al. 2013; Bossio et al. 2010). The majority of these SLM practices have been refined with the new approach for effective nutrient and water optimization, in a system defined as precision farming. In this approach, the land use system is steered by site specific management interventions. The land use measures are controlled by real-time data on soil and landscape variability to guide precision farming processes for increased efficiency in water and nutrient application. Some farming systems apply drones supported by GIS mapping and geo-statistics to accurately diagnose the constraints and provide interventions. The precision management approach is still at infancy in many African countries but it is a critical tool for optimizing resource use for high yields and water productivity, at farm scale and landscape levels. Besides, site specific measures with precision, are central in closing the yield gap in areas that produce below their potential, so as to make substantial impact on world food security (Tilman et al. 2011). The yield gap can be tackled with some of the technologies or practices highlighted in soil water conservation and nutrient management.

5 Technologies for Sustainable Nutrient Management

The implementation of SLM requires that nutrient management options are evaluated to guide land users establish affordable, acceptable, accessible and result-oriented technologies and practices that would result in highest yields. In this section, different management and technology options are categorized based on expertise and available literature in the context of material availability, application skills and sustainability (Table 2). The component of material availability and application skills is herewith represented as not available, moderately available and most available. The sustainability component is described in the context of resource constrained land users (mainly famers) in the great lakes region of Africa. All nutrient management measures are categorized basing on the pillars of SLM with emphasis on the social-cultural factors that influence adoption of a technology. Those that are common, affordable and acceptable (are represented as ++++); those that are less common, modestly affordable and acceptable (are represented as +++); those that are not common, not affordable and may not be acceptable (are represented as +). The analysis provides information on existing capacity in the use of nutrient management options and the use of strategic interventions in terms of increasing material production, skilling strategies and policy measures to enhance adoption. The projections are tailored to smallholder farming systems that are usually resource constrained with limited capacity to establish some of the best-bet land management technologies.

Table 2 reveals that most smallholder farming systems (that are resource constrained) have major challenges in accessing most of the sustainable nutrient management measures. Some nutrient management interventions are limited by skills and materials in an effort to achieve sustainable land management. This poses a threat to the current state of soil and land management, food, forage, fiber and fuel security and livelihoods in the region. There is a limited use of a range of agro-inputs that replenish depleted nutrients although organic related inputs remain locally available and easily accessible. Soil fertility management technologies and practices that are recommended are rarely practiced by majority of land users. The main limitations to the use of inputs is the lack of skills and the capacity to purchase inputs that are vital in advancing scientific innovation in SLM. As such, there is

Table 2 Sustainable nutrient management options commonly applied in the Great Lakes Region of Africa

Nutrient management options	Availability of materials and skills	Likelihood for sustainability (on adoption)
Inorganic fertilizers		
Mineral (macro) fertilizers (blended fertilizers)	Not available	+
Mineral (macro) fertilizers (Granular/compound fertilizers) e.g. Urea etc.	Moderately available	+++
Micronutrient fertilizers	Not available	+
Foliar fertilizers	Moderately available	+++
Agro-minerals (rock phosphates, limestone, potash, gypsum, pyrite)	Not available	+
Pelleted fertilizers/slow nutrient release fertilizer	Not available	+
Soil conditioners (humates, vermiculite etc.)	Not available	+
Fertigation	Not available	+
Organic fertilizers		
Green manure (incorporating legumes, grass, shrub litter)	Available	++++
Organic residues (e.g. mulch materials)	Available	+++
Compost	Moderately available	+++
Liquid manure/plant tea	Not available	+++
Animal manure (from cattle, poultry, farmyard, goats, pigs)	Available	++++
Mycorrhyza inoculants	Not available	+
Factory-made organic fertilizers e.g. Neem fertilizers	Not available	+
Beneficial soil organisms (Bio-fertilizers e.g. inoculants, free living fixers)	Moderately available	+
Agro-industrial product waste	Not available	+
Nutrient fortified compost (e.g. Rock phosphate fortified compost)	Not available	+
Vermicompost	Not available	+
Nutrient management approaches and tools		
Soil analysis—laboratory tests	Moderately available	+
Plant tissue tests	Not available	+
4R fertilizer management protocol	Moderately available	+
Micro-dozing	Not available	+
Organic and inorganic inputs (Integrated Soil Fertility Management Packaging)	Moderately available	+++
Organic matter management protocol	Not available	+
Slow/synchronized nutrient release measures	Not available	+

(continued)

Table 2 (continued)

Nutrient management options	Availability of materials and skills	Likelihood for sustainability (on adoption)
Crop residual composting and management	Available	+++
Crop nutrient requirements computations	Not available	+
Local indigenous diagnostics	Available	++++
Use of nutrient balance calculators	Not available	+
Instant soil testing kits	Moderately Available	+++
The fertilizer optimizer tool (fertilizer calculator)	Not available	+
Decision support systems applications	Not available	+

Source Authors based upon Sanginga and Woomer (2009), Bekunda et al. (2010)

need for a deliberate policy and intervention to promote management measures including the right use of inputs, adopting the right technologies, and practices that can promote improved land management.

6 Physical, Biological and Agronomic Technologies for Soil and Water Conservation

Optimal soil and water management is one of the critical components for boosting land productivity for high income and improved livelihoods. Global projections show that there is need to optimize water use, if smallholder farming systems are to overcome drought related constraints and food insecurity. It is projected that by 2050, more fresh water will be needed for agro-related services as the population increases and extreme events such as propagated drought increase (De Fraiture et al. 2007). Targeting technologies that enhance water productivity (more crop per drop) are critical in the planning process. This, however, requires identifying farmer friendly soil and water conservation measures that are in tandem with the pillars of sustainable land management.

Soil and water conservation options that reduce water loss by evaporation and runoff while increasing soil water storage were evaluated. Soil and water conservation measures common across the great lakes region were identified although these can vary with country depending on the inherent nature of the landscape and social-cultural factors (Mati 2006). For example, in Ethiopia, there are evidences of graded bench terraces, hill terracing, cut-off drain, grass strips, and contour bunds. Water harvesting measures such as underground tanks, ponds, open pans, spate irrigation are also common (Biazin et al. 2012). In Kenya and Uganda, common practices that retain water include conservation tillage, runoff harvesting, vegetative barriers, terracing, mulching, and trenches (Zhang et al. 2020). In Tanzania, some of the common soil and water conservation measures include ridging, bunded basins, pitting systems, surface runoff diverted from footpaths, and conservation tillage (Tenge and Hella 2005). Most of these countries in the great lakes promote increased water supply for use in various land use systems by promoting rainwater harvesting. Rainwater and runoff harvesting is important and there is notable link between water harvesting and soil water conservation since water can be conserved in situ and runoff can be stored in large area or storage unit for reuse.

A number of selected soil and water conservation measures were assessed while evaluating the likelihood for their adoption in resource constrained farming systems in Africa. It is important to note that in this evaluation, soil and water conservation measures are better implemented when: (i) the land ownership is assured; (ii) there is supply of labor; (iii) there are tools and materials for implementation; (iv) there are skills to support establishment; and (v) there are minimum funds to support the purchase and maintenance processes. Using a similar evaluation criteria for nutrient management (Table 2), social-economic aspect that form part of the pillars for sustainable land management (acceptability, affordability and viability) were considered (Emerton and Snyder 2018). The common, affordable and acceptable soil and water conservation measures (are represented as ++++), those that are less common, modestly affordable and acceptable (are represented as +++). Those that are not common, not affordable and may not be acceptable (are represented as +). For the component of availability of soil and water conservation support factors such as materials, labor and skills, these were categorized as available, moderately available and not available (Table 3). These were projected within the common setting of small-scale farming systems in Africa although research in SLM application remain inadequate.

The adoption of soil and water conservation technologies cannot be underrated in all efforts to ensure sustainable land management. These practices and technologies not only increase water infiltration capacity of the soil but also improve land and water productivity especially when soil fertility and rainfall distribution pattern are favorable (Sanginga and Woomer 2009). From the list of physical, biological and agronomic management measures, there is variability in how the communities in Africa are likely to adopt to the technologies. It is clearly evident that agronomic and biological soil and water conservation measures are more likely to be adopted than the physical structures that can be demanding (financially and otherwise) for the resource constrained farmers. However, physical soil and water conservation have always been very effective in minimizing soil loss and runoff compared to other soil conservation measures (Mati 2006). However, these require the use of more labor, specialized skills and resources.

Table 3 Physical, biological and agronomic measures common for soil and water conservation in the great lakes region of Africa

Soil and water conservation measures	Availability of materials, labour and skills	Likelihood for sustainability (on adoption
Physical measures		
Trench farming	Not available	+
Fanya Juu terraces	Available	++++
Fanya chini terraces	Available	++++
Stone lines/Rock bunds	Moderately available	+
Gully plugs	Not available	+
Bench terraces	Not available	+
Sloping terraces	Moderately available	+
Permanent planting basins	Not available	+
Hedge rows	Available	+++
Contour furrows and ridges	Not available	+
Diversion ditches/Cut-off drains	Available	++++
Road water harvesting drains	Moderately available	+++
Tied ridges	Not available	+
Zai pits	Not available	+
Trash lines along the contours	Moderately available	+++
Double dug beds	Not available	+
Half-moon furrows	Not available	+
Trenches/Anti-erosion ditches	Available	++++
Water tapping pits	Moderately available	+++
Stubble mulch tillage	Available	++++
Gulley control	Not available	+
Fish scale pits	Not available	+
Soil moisture retainers (Humates, hydrogel, biochar etc.)	Not available	+
Biological measures		
Surface mulching	Available	++++
Non-contour grass bunds/strips e.g. *Pennisetum purpureum*, *Tripsacum andersonii* etc.)	Available	++++
Contour grass bunds (planted)	Moderately available	+++
Natural grass strips	Available	++++
Crop residue retention in field	Available	++++

(continued)

Table 3 (continued)

Soil and water conservation measures	Availability of materials, labour and skills	Likelihood for sustainability (on adoption
Boundary tree planting/wind breakers	Available	++++
Agro-forestry/Multi-storeyed agro-forestry	Moderately available	+++
Silvo-pasture farming	Not available	+
Grass water-ways	Moderately available	+
Under-grazing (Paddocking)	Moderately available	+++
Surface vegetation cover	Available	++++
Maintaining of natural forest and trees	Available	++++
Right choice of land use type (Minimal soil disturbance)	Available	++++
Agronomic measures		
Deep tillage	Available	++++
Ripping	Not available	+
Dual purpose crop rotation	Moderately available	+++
Intercropping	Available	+
Zero/minimum tillage	Available	++++
Contour cultivation	Available	++++
Legume based cropping systems/cover cropping	Available	++++
Mixed cropping (various crops)	Available	++++
Alley cropping (trees and crops)	Available	++++
Relay cropping (double cropping)	Moderately available	+++
Contour planting	Available	++++
Early planting	Available	++++

Source Authors based upon Mati (2006), Liniger (2011), Emerton and Snyder (2018)

7 Integrating Soil Water and Nutrient Management Technologies

Soil water and nutrient management remain the central factors in increasing land productivity. The promotion of SLM technologies require that both nutrients and water are optimized. Recent advancements in agricultural land productivity have been pressing for integration science with the Integrated Soil Fertility Management concept widely promoted to respond to the principles of SLM (Vanlauwe et al. 2010; Musinguzi 2017). The combined use of organic, inorganic inputs, improved germplasm, soil water conservation and local adaptation, is emphasized with the ultimate goal of achieving high nutrient use efficiency and land productivity with minimal environmental effects. In SLM, the benefits of both nutrient and water use efficiency are fronted, in addition to the landscape approach in packaging the interventions. The impacts of such intervention even stretch to achieving the land degradation neutrality concept which is promoted to achieve stable and quality landscapes in space and with time. The adoption of combined approaches (physical, biological, agronomic and nutrient management packages) avail more water and nutrients although these can be laborious, time consuming, costly and demanding in skills. These biophysical benefits supersede the economic benefits since these integrated measures control soil loss, reduce run-off and minimize evaporation; replenish nutrients, and optimize production of food, forage, fuel and income (Martínez-Mena et al. 2020). Soil and water conservation measures such as physical structures, often result in long-term benefits (not immediate returns) compared to crop improvement technologies such as nutrient application. In order to overcome such limitations, it is advisable that emphasis in land management be on immediate intervention (moisture conservation and soil fertility) while the physical structures are simultaneously established (Ellis-Jones et al. 2001). This approach is highly commendable especially among smallholder farmers who cannot afford the complete combined package.

Use of a combination of technologies can be a success when other social-cultural factors are considered. It is critical to establish the landscape environment, the farmers' conditions in terms of their aspirations, effectiveness, preference and capacities but also understand the various needs that influence capacity to use some technologies (Emerton and Snyder 2018). Most smallholder farming communities are habitually stationed in sloping landscapes (steep, moderate and gentle slopes) in the region with varying needs for SLM interventions. It is important to observe the characteristic nature of farmers and farmlands while promoting SLM principles. The governance systems at catchment/landscape level ought to be considered since most SLM practices are not effective at farm level alone but at landscape (catchment) level. An economic analysis of the implications of using combined technologies is vital in adoption. The computation of net returns on costing the interventions can guide on possible investment options to restore degraded land. Physical structural measures remain costly for most poor smallholder farmers (Ellis-Jones et al. 2001). Recent findings indicate that there is need for simple practices for SLM with support of policy since costly practices take longer for higher returns to be realized (Dallimer et al. 2018). Technologies that require less resources (e.g., intercropping, mixed cropping, manure etc.) might be more dependable with less implementation costs compared to constructed soil conservation measures (e.g. physical terraces, trenches etc.) though the need might be guided by the setting in the landscape ecology. Some of these measures require collective action at catchment level. In all, combined packages of water and nutrient management are envisaged to have a higher cost-benefit ratio both in the short run (short-term establishment phase) but also in the long run (long-term maintenance phase). Therefore, a combination of technologies remains the most dependable and impactful management approach to avert the challenges of low productivity, soil loss and nutrient depletion. However, on-site evaluation with people-centered choice of options for application would promote sustainable land management and transform livelihoods. Research to explore various practices and technologies is vital since there are limited studies on the efficiency of various SLM measures in the great lakes region.

8 Conclusion

Nutrient application and soil water conservation are critical measures in achieving sustainable land management. The application of integrated approach for optimizing soil fertility and moisture is the most promising since it is effective and impactful. However, the success of nutrient and soil water management measures require the involvement of stakeholders especially the lead land users who are the farmers. Additionally, the computation of the cost effectiveness in the use of integrated technologies is a major incentive and a lead driver for adoption. However, the most affordable technologies can be implemented with the farmers' full consent in terms of technology acceptability and viability both in the short and long-run. Deciphering farmers' perceptions and choice in decisions for adoption of any SLM practice is important. The decisions may vary from farm unit level to village or landscape level if SLM is to provide higher societal and environmental benefits. Innovations from evidence-based research processes in the communities combined with capacity building in integrated technologies, policy and all social economic pillars of SLM are fundamental in stimulating land and water productivity.

References

Bekunda M, Sanginga N, Woomer PL (2010) Restoring soil fertility in Sub-Sahara Africa. Adv Agron 108:183–236

Biazin B, Sterk G, Temesgen M, Abdulkedir A, Stroosnijder L (2012) Rainwater harvesting and management in rainfed agricultural systems in sub Saharan Africa—a review. Phys Chem Earth 47–48:139–151. https://doi.org/10.1016/j.pce.2011.08.015

Bossio D, Geheb K, Critchley W (2010) Managing water by managing land: addressing land degradation to improve water productivity and rural livelihoods. Agric Water Manag 97(4): 536–542

Branca G, Lipper L, McCarthy N, Jolejole MC (2013) Food security, climate change, and sustainable land management: a review. Agron Sustain Dev 33(4):635–650

Dallimer M, Stringer LC, Orchard SE, Osano P, Njoroge G, Wen C, Gicheru P (2018) Who uses sustainable land management practices and what are the costs and benefits? Insights from Kenya. Land Degrad Dev 29(9):2822–2835

De Fraiture C, Wichelns D, Rockstrom J, Kemp-Benedict E, Eriyagama N, Gordon LJ, Karlberg L (2007) Looking ahead to 2050: scenarios of alternative investment approaches. In: Molden D (ed) Comprehensive assessment of water management in agriculture, water for food, water for life: a comprehensive assessment of water management in agriculture. In: International water management institute. Earthscan, Colombo, London, pp 91–145 (Chapter 3)

Ellis-Jones J, Miiro D, Lwakuba A, Critchley W (2001) Scaling up pilot research approaches and technologies: a case study from Kabale, Uganda. In: Natural resources systems programme, hillsides systems. Silsoe research Institute, UK

Emerton L, Snyder KA (2018) Rethinking sustainable land management planning: understanding the social and economic drivers of farmer decision-making in Africa. Land Use Policy 79:684–694

FAO (2008) TerrAfrica: a vision paper for sustainable land management in Sub-saharan Africa. Food and Agriculture Organisation of the United Nations, Rome

Gessesse D, Zerihun M (2017) Sustainable land management for agricultural risk management: agricultural risk management in Africa. In: Obiri JAF, Driver MF, Onyekwelu JC, Akpoko JG, Ramasawmy B, Dramé-Yayé A (eds) Agricultural risk management in Africa: a contextualized manual for tertiary institutions and development practitioners. African Network for Agriculture, Agroforesty and Natural Resources Education (ANAFE), Nairobi, Kenya, pp 30–47

Liniger HP, Mekdaschi Studer R, Hauert C, Gurtner M (2011) Sustainable land management in practice: guidelines and best practices for Sub-Saharan Africa. TerrAfrica, World Overview of Conservation Approaches and Technologies (WOCAT) and Food and Agriculture Organization of the United Nations (FAO)

Henry B, Murphy B, Cowie A (2018) Sustainable land management for environmental benefits and food security. A Synthesis Report for the GEF (Global Environmental Facility). GEF, Washington, DC, USA

Hurni H (2000) Assessing sustainable land management (SLM). Agr Ecosyst Environ 81(2):83–92

Kiage LM (2013) Perspectives on the assumed causes of land degradation in the rangelands of Sub-Saharan Africa. Prog Phys Geogr 37(5):664–684

Martínez-Mena M, Carrillo-López E, Boix-Fayos C, Almagro M, Franco NG, Díaz-Pereira E, de Vente J (2020) Long-term effectiveness of sustainable land management practices to control runoff, soil erosion, and nutrient loss and the role of rainfall intensity in Mediterranean rainfed agroecosystems. CATENA 187:

Mati BM (2006) Overview of water and soil nutrient management under smallholder rain-fed agriculture in East Africa, vol 105. International Water Management Institute, Working Paper 105

Motavalli P, Nelson K, Udawatta R, Jose S, Bardhan S (2013) Global achievements in sustainable land management. Int Soil Water Conserv Res 1(1):1–10

Musinguzi (2017) Critical soil organic carbon range for optimal maize response to nitrogen in a Ferralsol. A PhD Thesis, Makerere University, 220pp

Orchard SE, Stringer LC, Manyatsi AM (2017) Farmer perceptions and responses to soil degradation in Swaziland. Land Degrad Dev 28(1):46–56

Olupot G, Basamba TA, Ebanyat P, Musinguzi P, Opolot E, Katusabe AA, Bekunda MA, Singh BR, Coyne D, Abberton M, Adetonah S (2019) Making sense out of soil nutrient mining and depletion in sub-Saharan Africa. In: Soil degradation and restoration in Africa. CRC Press, pp 38–60

Sanginga N, Woomer PL (eds) (2009) Integrated soil fertility management in Africa: principles, practices, and developmental process. CIAT

Smyth AJ, Dumanski J (1995) A framework for evaluating sustainable land management. Can J Soil Sci 75(4):401–406

Tenge A, de Graaff J, Hella JJ (2004) Social and economic factors for adoption of soil and water conservation in west Usambara highlands, Tanzania. Land Degrad Dev 15:99–114

Tenge AJ, Hella JP (2005) Financial efficiency of major soil and water conservation measures in West Usambara highlands, Tanzania. Appl Geogr 25:348–366

Terefe H, Argaw M, Tamene L, Mekonnen K, Recha J, Solomon D (2020) Effects of sustainable land management interventions on selected soil properties in Geda watershed, central highlands of Ethiopia. Ecol Proc 9(1):1–11

Thomas R, Reed M, Clifton K, Appadurai N, Mills A, Zucca C, Mapedza E (2018) A framework for scaling sustainable land management options. Land Degrad Dev 29(10):3272–3284

Tully K, Sullivan C, Weil R, Sanchez P (2015) The state of soil degradation in Sub-Saharan Africa: baselines, trajectories, and solutions. Sustainability 7(6):6523–6552

Tilman David et al (2011) Global food demand and the sustainable intensification of agriculture. PNAS 108:20260–20264

UN Department of Economic and Social Affairs (UN DESA) (2019) Growing at a Slower Pace, World Population Is Expected to Reach 9.7 Billion in 2050 and Could Peak at Nearly 11 Billion Around 2100

Vanlauwe B, Bationo A, Chianu J, Giller K, Merckx R, Mokwunye U, Ohiokpehai O, Pypers P, Tabo R, Shepherd KD, Smaling EMA, Woomer PL, Sanginga N (2010) Integrated soil fertility management operational definition and consequences for implementation and dissemination. Outlook on Agriculture 39(1):17–24

WOCAT (2016) What is SLM for WOCAT? https://www.wocat.net/en/slm

World Bank (2006) Sustainable land management: challenges and opportunities and tradeoff. The International Bank of Reconstruction and Development/The World bank, Washington, DC

Zhang G, Majaliwa MJ, Xie J (2020) Leveraging the landscape: case study of erosion control through land management in the lake Victoria Basin

Novel Climate Smart Water and Nutrient Conservation Technologies for Optimizing Productivity of Marginal Coarse-Textured Soils

Giregon Olupot, A. J. M. Smucker, S. Kalyango, Emmanuel Opolot, Orum Boniface, Musinguzi Patrick, A. B. Twaha, and B. R. Singh

1 Introduction

Water and nutrients are the two most limiting factors to productivity especially under rain-fed agriculture (RFA) in moist, semi-arid and arid regions. The dramatic doubling in yields and production from the 1960s to the 1990s, were through rapid expansion into irrigated prime land (FAO 1986; Higgins et al. 1988; Fischer et al. 2010; FAO 2011a). The prospects of further expanding agriculture into uncultivated lands and, into lakes, rivers, swamps and marshes (as alternative sources of water) are bleak. Withdrawal and consumption of these resources have already surpassed the Earth's supply and regeneration capacity (Fischer et al. 2010). Higgins et al. (1988) estimated water requirements for food intake (in cereal equivalents) of 300 kg cap^{-1} yr^{-1} at 600 to 3,000 m^3 cap^{-1} yr^{-1} (1 m^3 = 1,000 L). Out of the Earth's renewable water resources totaling 42,000 km^3 annually, 3,900 km^{-3} is already being abstracted from rivers and aquifers distributed as 70% (2,730 km^3) irrigation, 19% (741 km^3) industrial and 11% (429 km^3) municipal uses (Fischer et al. 2010; FAO 2011a). Water withdrawals exceeding 20% exert substantial pressure on renewable freshwater resources and those exceeding 40% are considered critical (Fischer et al. 2010; FAO 2011a). Withdrawals in the Middle East, North Africa and Central Asia already exceed critical thresholds with demand outstripping supply, posing serious threats to ecological functions of the water

G. Olupot (✉) · S. Kalyango · E. Opolot · O. Boniface · M. Patrick · A. B. Twaha
School of Agricultural Sciences, College of Agricultural and Environmental Sciences, Makerere University, P. O. Box 7062, Kampala, Uganda

A. J. M.Smucker
Subsurface Water Retention Technology (SWRT) Solutions, LLC 2095 Riverwood Dr, Okemos, MI 48864, USA

B. R. Singh
Faculty of Environmental Sciences and Natural Resource Management, Norwegian University of Life Sciences, P.O. Box 5003, 1432 Ås, Norway

W. Leal Filho et al. (eds.), *Sustainability in Natural Resources Management and Land Planning*, World Sustainability Series,
https://doi.org/10.1007/978-3-030-76624-5_13

resources (FAO 2011a; Fischer et al. 2010). China, India, USA, Russia, Germany and Pakistan dominate irrigated area globally (Higgins et al. 1988; Fischer et al. 2010; FAO 2011a).

Countries leading in water abstractions also top in irrigated lands some of which, are already developed to full capacity. In contrast, countries and regions with least abstractions also lag behind in irrigation infrastructure development: sub-Saharan Africa (SSA); Central America, Caribbean, and Southern America; and the Pacific Island countries (PICs) (Fischer et al. 2010). SSA's problems are unique: few large rivers; smaller alluvial valleys; dispersed and more remotely located production units and dependence on smaller, seasonal streams (Higgins et al. 1988). Only 5.8% of SSA's irrigation potential is tapped on barely 3% of its cultivated area (FAO 2011a) contrasting miserably with the 32.8% irrigation developed by North Africa on 80–90% of cultivated land (Fischer et al. 2010). Consequently, the yield gap for crops is widest in SSA (76%) and narrowest in East Asia (11%) which accounts for 66% of total area irrigated globally (FAO 2011a). There is a high potential to more than double output in SSA and all regions where the yield gap is 50% or wider: North Africa, Central America and the Caribbean, Southern America, Western Asia, Central Asia, South Asia, Eastern Europe and Russia, and the PICs (Fischer et al. 2010; FAO 2011a).

Traditionally, the focus has been to expand irrigation into rainfed agriculture (RFA) on arid and hyper arid lands (FAO 2011a, b). Yet, most irrigation systems operate at very low water conveyance, supply, distribution and use efficiencies (Higgins et al. 1988). Consequently, surface and groundwater abstractions to furnish inefficient irrigation systems are intensifying and key aquifers are getting depleted (Higgins et al. 1988; Fischer et al. 2010; FAO 2011a). Water shortages are already severe especially in Western, Central and South Asia and in North Africa where water abstractions for irrigation now exceed renewable sources and water tables dropping (Fischer et al. 2010; Shrestha et al. 2018). Even where water resources may appear abundant and poorly developed with high potential for expansion into uncultivated land such as SSA, Central America and Latin America, much of the water is not accessible, distant and development costs for irrigated agriculture prohibitive (FAO 2011a). Besides, diversion of water for health, industrial and other urban uses is increasingly assuming greater importance as are upstream considerations including watershed hydrology and protection. About 66% of the >9.2 billion people projected to inhabit the Earth by 2050 (Cribb 2010; UN 2019) will dwell in cities (Shrestha et al. 2018). This population will be consuming 50% of freshwater abstracted globally, diminishing the amount of water available for agriculture at a time when production should further double to meet the food and fiber needs of this population. Besides, water pollution in urban settings and city slums is expected to increasingly constrain freshwater availability for agriculture. According to UNWWAP (2003), two million tons of sewage, industrial and agricultural wastes were already being dumped into the world's water bodies daily by 2000, with about 1,500 km^3 of wastewater produced annually. This wastewater exceeds 70% freshwater in all the world's rivers (Shiklomanov 1993) and is equivalent to 38.5% of total freshwater annual withdrawals globally.

Rain-fed agriculture (RFA) is therefore, expected to increasingly assume more importance as competition for freshwater from other equally important sectors intensifies. RFA is possible where rainfall exceeds 300 mm per annum, depending on temperature, soil conditions, and rainfall distribution (Higgins et al. 1988; Fischer et al. 2010; FAO 2011a). About 80% (1,300 million hectares) of total cultivated land globally (1,600 million hectares) is under RFA, accounting for 60% of crop output globally and as much as 97% of staple production in SSA (FAO 2011a, b). Erratic rains and rainfall distribution patterns, prolonged intense and more frequent droughts and flash floods (FAO 2011a; IPCC 2014) could cut crop yields from RFA by 50%, exacerbating food insecurity (FAO 2011b; Shrestha et al. 2018). Yet, less than 30% of rainfall is used by plants in the process of biomass production (Molden 2007) in a drying planet. Even where rainfall is adequate and soil moisture not limiting, inherent poor soil fertility and decades of unabated soil nutrient mining, limit the potential of RFA (Olupot et al. 2020). Interventions to restore degraded landscapes and bring marginal lands into production exist, scattered in various sources. This chapter consolidates novel climate smart innovations for optimizing productivity of marginal coarse-textured soils by coupling soil water and nutrient conservation within plant root zones.

2 Climate Smart Soil Water and Nutrient Conservation Technologies for Rainfed Agriculture

Soil moisture and rainfall are the two most important, reliable and secure direct sources of water for RFA both now and in the future. Novel climate smart technologies exist for more than tripling productivity of marginally coarse-textured soils by coupling soil water and nutrient retention within plant root zones. We propose 10 considerations (modified from FAO 2003) prior to and during implementation of these technologies:

1. Maintaining integrity of soil pore space is more important than protecting soil particles. Consider reducing soil disturbance to the bare minimum needed to optimize germination, plant root growth and function, and sustainable restoration of damaged soil pores architecture (Figs. 1, 2, and 3).
2. Creating conditions on the soil surface that maximize infiltration and storage of rainwater makes more sense than lamenting high runoff, evaporation and deep percolation.
3. Keeping soil permanently under cover is more cost-effective than erecting physical barriers downslope (Fig. 1).
4. Building and capitalizing on complementary interactions (both above- and belowground) to optimize recovery and use efficiencies of growth resources (Fig. 1) can deliver both agricultural and environmental benefits.

Fig. 1 Different mulching materials top row: self-mulching bananas (extreme left), soy (middle) and piles of mowed grass (extreme right). Middle row: *Lablab* fallow preceding rice (extreme left), maize-*Calliandra* intercrop (middle) and *Calliandra* perennial cover (extreme right). Bottom-right: banana-finger millet-maize intercrop with yams on the foreground (extreme left), maize intercropped with *Desmodium sp* and surrounded by *Pennisetum purpureum* (Bottom row, center) and typical complementary multiple canopy-root interactions (extreme right). *Source* Compiled by the authors

5. Step-wise co-promotion of a basket of complementary, interdependent and mutually reinforcing technologies in preference to one-size-fits-it-all approaches (Figs. 2 and 3).
6. Reducing the risks of crop failure from droughts is more plausible than bemoaning increasing severity, duration and frequency of occurrence of droughts (FAO 2011b; IPCC 2014). More localized portable weather stations especially in the high potential RFA areas of SSA, Latin America, Central America, Central and Southeast Asia, the PICs and New Zealand could provide more localized site-specific information about rainfall episodes to aid climate smart precision cropping especially with drought-tolerant and drought-escaping germplasm adapted to such sites with less supplemental irrigation.
7. Building soil from the surface downwards rather than waiting for it to deepen from the bottom upwards through natural processes of soil formation (FAO 2003) is a sound approach to restoration especially of tropical soils at terminal stages of development.

Fig. 2 Physical water and soil conservation techniques from left to right Top row: *zai* pits (planting pits), planting basins and half-moons; Middle row: tied ridges, stone bunds and rip lines; Bottom row: earth bunds stabilized with grass and planting across the slope, fanya juu terraces and stabilized contour with tie bunds. *Source* Compiled by the authors

8. Shifting focus away from 'soil and water conservation' to 'water and soil conservation ensures preventive tackling the root causes rather than symptoms of soil degradation.
9. Protecting and stabilizing landscapes through broader goals of environmental conservation can achieve better results than lamenting flash floods and soil loss from poorly farmed arable lands (Fig. 1).
10. Proper coordination at research, policy, institutional and advocacy level can streamline water, soil, atmospheric and environmental conservation efforts and tradeoffs.

2.1 Biological Techniques

Ensuring permanent presence of cover on land is the cheapest and most effective practice for conserving water and protecting the soil against erosion (Wischmeier

Fig. 3 Integrating agroforestry with livestock and fisheries farming Top row: banana-coffee-tree planting in Uganda (extreme left); a naturally established windbreak between soy fields (center) and a typical agro-silvo-pastoral system (extreme right). Center row: a zero-grazing unit fitted with manure collection mechanisms (extreme left), a bio-digester ready for re-filling with fresh cattle dung (center) for biogas generation and the slurry oozing out from the bio-digester (extreme right). Bottom row: a poultry house above a fish pond (left) where the bird droppings can first be collected and sterilized under the sun before they are poured in the pond to feed the fish (Right). The droppings not consumed by the fish together with fish wastewater fertilize the rice, which in turn filters out the nutrients and purifies the water for the fish

and Smith 1978). Cover can be maintained through stubble retention, mulching and incorporation of crop residues in soils. Some crops like banana (Fig. 1: Top row extreme left) and soy (top row center), coffee and sugar cane are self-mulching. Mulching is particularly suited to regions with limited rainfall, shallow-rooted crops and soils with shallow water tables (<1–2 m) where crops could benefit from capillary rise of the surface water (FAO 2003). Grass mowed from non-arable lands (Fig. 1: Top row extreme right), pruned biomass, litter fall from deciduous vegetation or on forest floors are potential mulching materials. Mulching increases water use efficiency by suppressing weeds, cooling the soil, absorbing and dissipating energy of destructive raindrops, boosting soil biological activities, nutrient recycling, increasing soil organic matter, stabilizing soil aggregates and maintaining integrity of soil pores thus, increasing infiltration of rainfall. About 4–5 Mg ha^{-1} of mulch is sufficient to minimize runoff and aid rainfall infiltration into the soil. A 0.05–0.1 m thickness of mulch (equivalent to 30–40 Mg ha^{-1} mulch) increased banana yields from 4.3 to 10.8 Mg ha^{-1} in Uganda (Speijer et al. 1998). Guarding against catching fire and invasion by destructive termites can optimize benefits of mulching.

Live biomass from improved fallows, forage legumes and agroforestry (Fig. 1: Middle row) or stubble can offset the challenge of mulches. Effectiveness of live cover depends on crop type (annual vs. perennial), plant density (high vs. low), growth habits (erect, horizontal, spreading, climbing, broad- vs. narrow-leaved) and vigor of stands. Closely spaced, low-lying, spreading and vigorous vegetation is more effective in water and soil conservation than scattered, stunted or tall vegetation. Management: intercropping versus sole cropping (spatial) or crop rotation versus continuous cropping (temporal) factors also influence effectiveness of cover in conserving water and soil. Improved fallows such as *Lablab* (Fig. 1: Middle row extreme left), *Mucuna*, *Tephrosia*, etc., are better established towards onset of prolonged dry season. Presence of cover on soil is crucial at the onset of turbulent and destructive rains typical after a long dry spell when the soils are often bare. Where land is scarce, intercropping or alley cropping cereals like maize with fodder legumes like *Calliandra* (Fig. 1: Middle row, center) is advisable. After harvesting maize, the legume remains protecting the soil during onset of turbulent and intense rains especially in the tropics. Niches can be switched when the legume occupies the space previously under the cereal and vice versa for the cereal. Where land is not limiting, a perennial cover of multi-purpose forage legumes such as *Calliandra* (Fig. 1: Middle row extreme right) can be established for fodder for livestock, biological nitrogen fixation, stabilizing soil, shade, fuel wood, lopping as green manure, windbreaks, etc. Maize, finger millet and bananas with yams on the fore where the soil is permanently moist (Fig. 1: Bottom row extreme left) is a common multiple cropping practice in eastern Uganda.

The 'push-pull' technology controls destruction of maize and cotton by moth caterpillar (*Helicoverpa amygera*). It encompasses intercropping maize with a wild legume either green leaf desmodium (*Desmodium intortum)* or silver leaf desmodium (*Desmodium unicinatum*) surrounded by Napier/elephant grass (*Pennisetum purpureum*) (Fig. 1: Bottom row center). *Desmodium* produces volatile organic compounds which repel the pest whereas *P. purpureum* produces pheromone-like volatile organic compounds which attract the caterpillars to the refugee crop where they mature as important pollinators. *Desmodium* also fixes atmospheric nitrogen, covers the soil with live mulch, drops litter which increases SOM on decomposition and boosts soil biology and soil health (Singh et al. 2011). Multiple cropping should be planned to optimize complementary aboveground (canopy) and belowground (root) interactions among component crops (Fig. 1 Bottom row, extreme right).

2.2 Physical Techniques for Conserving Water and Nutrients Endemic to Sub-Saharan Africa

Africa has had a lasting tradition of innovating in mitigating against the damaging effects of droughts on crops. FAO (2003, 2011a) capture some of these revolutionary technological innovations (Fig. 2). *Zai* pits or simply planting pits (Fig. 2: Top row extreme left) are traceable to Niger. Circular holes of diameter 0.15–0.3 m and depth 0.3 m are excavated on flat or gently sloping land during the dry season. Soil may be heaped downslope and reinforced with stone lines to control and redirect runoff into the pits (FAO 2003). The pits may be stuffed to attract termites to feed on the trash and construct galleries which aid infiltration. Two weeks to planting, animal dung is applied in the pits at 1–2.5 Mg ha^{-1} and covered with earth and millet sown at the onset of rains. Planting basins 0.15 m wide, 0.3 m long and 0.3 m deep (Fig. 2: Top row center) are also manually established in a staggered manner on fields. For half-moons, plots of area 10 to 20 m^2 are constructed in the form of half-moons 2–6 m wide and 0.2 m deep at 4 m intervals along contour lines in a staggered arrangement (Fig. 2 Top row extreme right). Half-moons are designed to direct runoff to the center where they are deepest and aid free exit of excess water through their tapering ends. Contour tied earth ridges are generally 0.15–0.2 m high along the contour at 1.5–3.0 m intervals (Fig. 2: Middle row extreme left) constructed by throwing the soil downslope. Prior to construction, base of the ridge must first be disturbed by prior tillage to increase binding of the ridge to soil beneath (FAO 2003). Cross ties (bunds) no more than one third of the height of the ridge are constructed in the furrow every 4–5 m. Bunds slow runoff, increase infiltration and prevent runoff from accumulating at the lowest point and overtopping or breaking through the ridge. Tied ridges are suitable for a range of crops and multipurpose trees (planted on either side of the ridge). The land between successive ridges is left bare to generate runoff (FAO 2003). Stone bunds are popular in Burkina Faso and Ethiopia where sorghum yields have been doubled and runoff reduced by 33% (Zougmore et al. 2000, 2018). They are built up of stones to a height of 0.25 m and about 0.35–0.4 m wide by positioning smaller stones upslope and large stones downslope at 25 m intervals on 1–3% slopes. Stones may be held together with a cloth wire mesh (Fig. 2: Middle row center). The small stones let runoff through but trap some sediment thus, initiating slow development of terraces. The base may be set in a shallow trench 0.05–0.1 m deep to prevent the stones from being swept downhill by runoff (FAO 2003). Rip lines (Fig. 2: Middle row extreme right) are best established using ox ploughs on flat or gently sloping land (1–3% slopes) preferably across the slope for water conservation (FAO 2011a). In West Nile Uganda, farmers on steep slopes of Erussi adopted earth bunding stabilized with grass and planting across the slope to reduce soil erosion (Fig. 2: Bottom row extreme left). Bottom center: farmers in Thailand constructing bench terraces (left) and fully established terraces (right) (FAO 2011a). Figure 2 Bottom extreme right: a contour with tie bunds stabilized with *P. purpureum* and *Calliandra* on a degraded foot slope of Mount Wanale under the World

Bank-supported Sustainable Land Management Project in Mbale, southeastern Uganda.

These techniques aid precision tillage and minimum soil disturbance by limiting cultivation to planting spots. Manure and water management are concentrated in these spots resulting in more than doubling of yields compared to conventional planting (FAO 1996; Zougmore et al. 2018). Planting pits are being modified and customized to specific crops of interest, including seedlings establishment for agroforestry (Ouedraogo and Sawadogo 2000). In Uganda, planting coffee in pits of sides 0.6 m and depth 0.6 m is now standard practice where the black soil mixed with manure is filled first followed by the subsoil (also mixed with manure) then coffee is planted.

2.3 *Novel Double-Dug Beds for Coupling Soil Subsurface Water and Nutrient Retention*

Double-digging and sub-soiling improve receptiveness of soils with compacted subsurface horizons to water and plant roots (FAO 2003). It involves 'scrapping off' the topsoil to expose and break up the compacted layer with sub-soilers before the topsoil is placed back alone or mixed with manure before planting. Double-digging marginal coarse-textured soils and installing with subsurface water retention membranes intercepts and retains water and nutrients within plant root zones (Smucker et al. 2018). Beds of width ranging from 0.3 m and depth 0.75 m are installed with U-shaped subsurface water retention membranes (SWRM) (Smucker et al. 2018) and filled back. The subsoil is mixed with manure and placed back followed by topsoil (also mixed with manure and mineral fertilizers), taking care to maintain original soil bulk density. The beds are staggered to allow excess water to freely maneuver and re-infiltrate soil free of SWRM (Erickson et al. 1968).

Once properly installed, SWRM can intercept and store 8.2 m^3 of water which would otherwise be lost m^{-2} day^{-1} without the SWRM (Smucker and Basso 2014), translating into >54,600 m^3 ha^{-1} day^{-1} of water (if 67% of the field is installed with the SWRM). A maize plant weighing 0.8 kg is about 0.7 kg (0.7 L) water at tussling and will have transpired 20–50 L of water by this stage (Boyer 1995). At 0.75 m × 0.3 m spacing, theorized population is about 44,444 such plants ha^{-1}. Total water transpired by anthesis exceeds 1,555 m^3 ha^{-1} and about the same amount must be transpired by physiological maturity. The >3,000 m^3 of water transpired by maize ha^{-1} by harvest time is within the 600 to 3,000 m^3 water equivalent in food consumed cap^{-1} yr^{-1} (Higgins et al. 1988) and well within the >54,600 m^3 water savings ha^{-1} day^{-1} with installation of SWRM. The water and nutrients intercepted by the SWRM are kept within the plant root zone to cushion against intervening dry spells between rainfall events during the rainy season and during the dry season (Smucker and Basso 2014). Where SWRM have been installed, up to 60% of supplemental irrigation requirements have been dispensed with, with 50–503% increases in production from 1 to 3 cropping cycles yr^{-1} in various cropping systems (Smucker 2016). The technology has successfully been implemented in the USA (Smucker 2016); Iraq (Aodi et al. 2021); Japan (Smucker et al. 2013); China (Yang et al. 2012).

2.3.1 Case Study in Uganda

We conducted a field experiment to evaluate tomato yield responses to subsurface water and nutrient retention technology (SWRT) at Makerere University Agricultural Research Institute, Kabanyolo (MUARIK) to test the hypothesis that tomato yields would be higher under coupled soil subsurface water and nutrient retention than under either factor singly. Fertilizer amendments at four levels: control, mineral NPK, vermicompost and vermicompost enriched with mineral NKP were applied to each of the three tillage levels: conventional tillage, double-digging and double-digging with installed polythene sheets. This field experimental design enabled us to disaggregate and aggregate the effects of both water and nutrient stresses to quantify their relative importance to tomato yields. Double-dug beds each 1.2 m wide, 1.0 m deep and 3 m long were established following the FAO (2003) procedures. Water impermeable black polythene sheets were carefully installed to eliminate wrinkling, before filling with soil as already described. Mineral NPK was applied at the recommended rates of 78, 50 and 100 kg N, P, K, respectively as ammonium sulphate, triple superphosphate and muriate of potash. Nitrogen was applied in three splits of 20, 40 and 40% at planting, anthesis and at fruiting whereas vermicompost 26 Mg ha^{-1} (full rate) and 13 Mg ha^{-1} (half rate combined with mineral NPK) and PK were basal applied once. Hardened off tomato seedlings were transplanted on January 31, 2019 at a recommended spacing of 0.6 m $\times$ 0.5 m, giving a total population of 12 tomato plants per bed. Seedlings were adequately watered and kept weed free manually to ensure good survival and establishment, as the rains set in and stabilized. Ready fruits were sequentially harvested as they got ready and initially weighed per plant before the aggregate weight was pooled together after the last harvest. Tomato fruit yield data were subjected to ANOVA after checking for conformity to and compliance with the assumptions for ANOVA using the R statistical package and means separated using Tukey's HSD.

Highest mean fruit yield (68.7 $\pm$ 7.7 t ha^{-1}) was from double-digging with polythene sheets and NPK + vermicompost. Mean fruit yield for this same treatment was 46.6 $\pm$ 4.9 t ha^{-1} under conventional tillage and 46.3 $\pm$ 5.2 t ha^{-1} under double-digging without polythene sheets (Table 1). From these results, there was clearly, no benefit of double-digging even with installation of SWRM, without fertilizer application, in line with our hypothesis. It was also clear that combined application of vermicompost and mineral NPK at half the full rates was superior to full rate of either amendment alone. This was a predominantly sandy loam (45% sand, 11% silt and 44% clay) Oxisol with low plant available P (9.6 mg kg^{-1}). Water did not limit tomato yields significantly because the rainfall was above normal throughout this particular experiment. With climate change escalating the longevity, severity and frequency of droughts, SWRT must be rapidly disseminated to the >2 billion ha of marginal coarse-textured soils globally to their increase productivity (Fischer et al. 2010; Smucker 2016).

2.4 Integration of Methods

2.4.1 Agroforestry

Agroforestry is a novel practice for combating land degradation, restoring degraded lands, bringing marginal unproductive lands into production and minimizing greenhouse gas emissions from forest conversion into croplands (Olupot et al. 2017). El Tahir et al. (2013) observed highest net positive soil nitrogen (N), phosphorus (P) and potassium (K) balances of 280, 11 and 74 kg ha^{-1} yr^{-1}, respectively under pasture with high tree density (HTD) in Sudan. Although net NPK balances were negative under sole crops of sorghum and roselle as well as grass (for N), introduction of agroforestry offset the net negative NPK balances in all tree-crop or tree-pasture (silvo-pasture) combinations. The results highlight the importance of integrating trees with crops especially at high tree density (HTD). In Uganda, inter-planting bananas with coffee and shade trees is a traditional practice in all coffee-growing regions (Fig. 3: Top row extreme left). Even where sole cropping is practiced, trees and grass are commonly left to establish along field boundaries as natural windbreaks and source of fuel wood and fodder for livestock (Fig. 3: Top row center).

Table 1 Tomato yield responses to soil water conservation with subsurface water retention membranes (SWRM) and fertilizers (Mg ha^{-1}) at MUARIK from end of January to April 2019

Soil amendment	Soil water conservation technology		
	Conventional tillage	Double-digging	Double-digging + SWRM
Control	2.24 ± 0.45a	2.68 ± 0.41a	3.84 ± 0.65a
Mineral NPK	17.80 ± 3.56b	16.60 ± 2.31b	28.10 ± 7.02b
Vermicompost	29.60 ± 5.37b	22.60 ± 5.73b	32.60 ± 5.36b
Vermicompost + mineral NPK	46.60 ± 4.90c	46.30 ± 5.20c	68.70 ± 7.75d

Means with the same letter down each column and across each row did not differ significantly ($P < 0.05$)

2.4.2 Integration of Cropping with Livestock and Fisheries

The only profitable way to keep arable land under pasture is by integrating crops with livestock (FAO 2006a). The trees provide shade for livestock and together with pasture ensure maximum cover on the surface of land (Fig. 3: Top row extreme left), resulting in zero soil erosion. The dung from livestock, rapid turnover of fine roots of both trees and pastures as well as litter fall from trees and pasture ensure production of high soil organic matter and abundance of a rich diversity of soil organisms (Singh et al. 2011). These conditions improve soil physical, chemical and biological properties and increase its receptiveness to water, with excellent water retention capacity. Where the livestock are confined, it is possible to collect high quality manure (Fig. 3: Middle row extreme left) for energy generation (Fig. 3: Middle row, center) and use the slurry (liquid by-product) as fertilizer (Fig. 3: Middle row, extreme right). Focus is also increasingly on integrating cropping with fish farming (FAO 2006b). Rice farming with poultry and fish (Fig. 3: Bottom row): a poultry house is established above a fish pond and the bird droppings either fall directly on the water to feed the fish or are first sterilized under the sun before dropping on the water to feed the fish. Rice filters out the nutrients from the water (Fig. 3: Bottom right) before it is recycled for fish farming. This practice is common in Asia and is fast spreading to SSA (FAO 2011a).

3 Conclusion

Increasing the usefulness of soil moisture and nutrients within plant root zones is crucial for fostering resilience to droughts and climate change. There is no one-size-that-fits-it-all technology that can address this challenge universally but a package of options already exist which can be customized to specific contexts. Keeping land under cover is the cheapest and most effective strategy for conserving soil moisture and nutrients and increasing their usefulness to crops. A combination of covers have been presented for consideration and customization to specific contexts of farming communities. For live cover, complementarity among component vegetation types in multiple cropping systems must be enlisted involving legume-cereal intercrops and agroforestry to optimize aboveground and below-ground interactions for increased utilization efficiency for growth resources. For potentially problematic landscapes, the physical soil moisture and nutrient conservation structures presented can be introduced step-wise to reinforce cover without straining farmers' resources beyond their containment capacity. For coarse-textured soils, installation of water retention membranes is a novel technology to intercept and retain water and nutrients within plant root zones. The best results for this SWRT are obtained when it is coupled with combined application of mineral and organic fertilizers. As novel technologies continue to evolve, diversification to integrate cropping with perennial deep rooters (trees), livestock, poultry and fisheries can increase productivity and diversify livelihood options. When carefully integrated, these technologies and practices can ease pressure on scarce water and land resources, increase productivity while maintaining environmental quality and integrity.

4 Research Imperatives

Although the technologies already presented and discussed are not new, there is limited empirical evidence about interrogation of their costs and benefits. Rigorous democratized research involving specific technology options under defined management practices accompanying these technologies must be jointly conducted and evaluated with farming communities to generate evidence for their promotion and to inform conducive policy for their dissemination. Long-term studies are needed to identify and refine technologies with evidence of increasing productivity as well as optimizing environmental services such as sequestration of greenhouse gases, water purification, improvement and sustenance of soil biodiversity and soil health.

References

Aodi IM, Shatha SM, Hussein AM, Fadhel HA, Smucker AJM, Robertson GP (2021) Agron J

Boyer JS (ed) (1995) Measuring the water status of plants and soils. Academic Press, New York, p 178p

Cribb J (2010) The coming famine: the global food crisis and what can be done to avoid it. CSIRO Publishing, Collingwood, VIC 3066, Australia. 247pp

El Tahir BA, Daldoum MA, Ardö J (2013) Nutrient balances as indicators of sustainability in *acacia Senegal* land use systems in the Semi-arid Zone of North Kordofan, Sudan. Stand Sci Res Essays 5:93–112

Erickson AE, Hansen CM, Smucker AJM (1968) The influence of subsurface asphalt barriers on the water properties and the productivity of sand soils. In: Transactions of the 9th international congress of soil science transactions, vol 1. Adelaide, pp 331–337

FAO (1986) Consultation on irrigation in Africa. Irrigation and drainage Paper 42, Proc. Consultation on Irrigation in Africa Lome, Togo

FAO (1996) Land husbandry—components and strategy. By E. Roose. FAO Soils Bulletin No 70. FAO, Rome. ISBN 95-5-203451-6

FAO (2003) Optimizing soil moisture for plant production: the significance of soil porosity. FAO Soils Bulletin 79, Rome, Italy

FAO (2006a) Livestock's long shadow. Rome, FAO. http://www.fao.org/docrep/010/a0701e/a0701e00.HTM

FAO (2006b) Integrated agriculture-aquaculture. FAO Fisheries Technical Paper 407, Rome

FAO (2011a) The state of the world's land and water resources for food and agriculture (SOLAW)—managing systems at risk. Food and Agriculture Organization of the United Nations, Rome and Earthscan, London

FAO (2011b) Climate change, water and food security. FAO Water Reports 36. Rome, FAO. http://www.fao.org/nr/water/jsp/publications/search.htm

Fischer G, Hizsnyik E, Prieler S, Wiberg D (2010) Scarcity and abundance of land resources: competing uses and the shrinking land resource base. SOLAW Background Thematic Report—TR02. Rome, FAO. http://www.fao.org/nr/solaw/

Higgins GM, Dieleman PJ, Abernethy CL (1988) Trends in irrigation development, and their implications for hydrologists and water resources engineers. Hydrol Sci J 33(1):43–59. https://doi.org/10.1080/02626668809491222

IPCC (2014) Climate change 2014: mitigation of climate change. In: Contribution of Working Group III to the Fifth Assessment Report of the Intergovernmental Panel on Climate Change [Edenhofer, Pichs-Madruga OR, Sokona Y, Farahani E, Kadner S, Seyboth K, Adler A, Baum I, Brunner S, Eickemeier P, Kriemann B, Savolainen J, Schlömer S, von Stechow C, Zwickel T, Minx JC (eds)]. Cambridge University Press, Cambridge, United Kingdom and New York, NY, USA

Molden D (ed) (2007) Water for food, water for life. Comprehensive assessment of water management in agriculture. Colombo, IWMI and Earthscan, London. http://www.iwmi.cgiar.org/assessment/

Olupot G, Otukei JR, Muwanika VB, Esaete J, Tabuti JRS (2017) Conversion of native vegetation in protected areas fuels CO_2-equivalent losses in Uganda. In: Mukul SA, Manzoor Rashid AZM (eds) Protected areas: policies, management and future directions. Nova Publishers, New York, USA, pp 179–202

Olupot G, Twaha AAB, Ebanyat P, Musinguzi P, Opolot E, Amoding AK, Bekunda MA, Singh BR (2020) Making sense out of soil nutrient mining and depletion in Sub-Saharan Africa. In: Stewart BA, Lal R (eds) Advances in soil science: soil degradation and restoration in Africa. CRC Press, USA, pp 39–60

Ouedraogo A, Sawadogo H (2000) Three models of extension by farmer innovators in Burkina Faso. LEISA (ILEIA Newsletter for Low External Input and Sustainable Agriculture) 16(2): 21–22

Shiklomanov I (1993) World freshwater resources. In: Gleick PH (ed) Water in crisis: a guide to the world's freshwater resources. Oxford University Press, New York, pp 13–24

Shrestha S, Hoan NAT, Shrestha PK, Bhatta B (2018) Climate change impact on groundwater recharge and suggested adaptation strategies for selected Asian cities. Asia Pacific Network (APN) for Global Change Research. https://doi.org/10.30852/sb.2018.499

Singh BP, Cowie AL, Chan KY (eds) (2011) Soil health and climate change. Springer, London, pp 397

Smucker AJM, Yang Z, He X, Lai F, Basso B (2013) SWRT is a New technology for converting arid sands into oases of food and fiber production. Int J Bio-Based Mater Bioenergy

Smucker AJM, Basso B (2014) Global potential for a new subsurface water retention technology—converting marginal soil into sustainable plant production. In: Churchman GJ, Landa ER (ed) The soil underfoot: infinite possibilities for a finite resource, Chapter 24. CRC Press, pp 315–324

Smucker AJM (2016) SWRT White Paper: Subsurface Water Retention Technologies (SWRT) Converts Marginal Sandy and Other Excessively Permeable Soils into Highly Sustainable Food and Fiber Production Systems while Improving Soil Fertility, Soil Carbon Sequestration, Many Ecosystem Services and biodegradation of Toxins Resulting in Sustainable Economic Growth and Environmental Protection. https://www.google.com/?gws_rd=ssl#q=SWRTDroughtSmucker

Smucker AJM, Levene BC, Ngouajio M (2018) Increasing vegetable production transformed sand to retain twice the soil water holding capacity in plant root zone. J Hortic 5:246. https://doi.org/10.4172/2376-0354.1000246

Speijer PR, Mudiope J, Ssango F, Adiapala E (1998) Comparison of plant stages for the evaluation of nematode damage to East African Highland banana (Musa AAA-EA). Afr Plant Prot 4:1–7

UN-Environment (2019) Climate change adaptation technologies for water: a practitioner's guide to adaptation technologies for increased water sector resilience. Accessed October 2019

UNWWAP (United Nations World Water Assessment Programmed) (2003) The World Water Development Report 1: Water for People, Water for Life. UNESCO, Paris, France

Wischmeier WH, Smith DD (1978) Predicting rainfall-erosion losses-a guide to conservation planning. Agric. Handbook No. 537. USDA, Washington, DC

Yang Z, Smucker AJM, Jiang G, Ma X (2012) Influence of the membranes on water retention in saturated homogeneous sand columns. In: International symposium on water resource and environmental protection (ISWREP), pp1590–1593. 978-1-61284-340-7/111©2011IEEE

Zougmore R, Guillobez S, Kambou NF, Son G (2000) Runoff and sorghum performance as affected by the spacing of stone lines in the semi-arid Sahelian zone. Soil Tillage Res 56(3–4): 175–183

Zougmore RB, Partey ST, Ouedraogo M, Torquebiau E, Campbell BM (2018) Facing climate variability in sub-Saharan Africa: analysis of climate-smart agriculture opportunities to manage climate-related risks. Cahiers Agric 27:34001. https://doi.org/10.1051/cagri/2018019

Rural Land-Use Planning and Livelihood Dynamics in Post-2000 Zimbabwe

Innocent Chirisa and Verna Nel

1 Introduction

Zimbabwe has experienced remarkable evolutionary processes concerning rural land-use and livelihoods. The Fast Track Land Reform Programme (FTLRP) of the early 2000s is unique for the speed in executing the programme (James 2015). Such programmes are purported to be poverty reduction strategies for 'ordinary' Zimbabweans. Land reform programmes were assumed to cushion rural populations from natural disasters through the improved livelihoods that would inherently accompany the FTLRP. Unfortunately, the envisaged benefits of the land reform programme were distorted by climate change and economic meltdown post 2000 (ZIMVAC 2010, 2011, 2013; Zvinorova et al. 2013).

Despite these environmental and economic adversities, it is argued that the Zimbabwean rural populations are resilient (United Nations Development Programme 2016). Unlike typical descriptions of simple agrarian livelihoods, this chapter contends that Zimbabwean rural livelihoods are complex and diverse influenced by household capabilities, local environmental characteristics and national socio-economic conditions (Maphosa 2005; Mano 2007). Mutami and Chazovachii (2012) observe that the risks to livelihoods viability are compounded by the failure of macro-economic policies implemented in rural areas together with environmental disruptions due to climatic changes. World Bank (2019: 6) observes that "With climatic variability increasing, natural disasters are likely to occur more frequently and have the potential to hit the most vulnerable parts of the population,

I. Chirisa
Department of Demography Settlement and Development, University of Zimbabwe, Harare, Zimbabwe

I. Chirisa (✉) · V. Nel
Department of Urban and Regional Planning, University of the Free State, Bloemfontein, South Africa
e-mail: NelVJ@ufs.ac.za

W. Leal Filho et al. (eds.), *Sustainability in Natural Resources Management and Land Planning*, World Sustainability Series,
https://doi.org/10.1007/978-3-030-76624-5_14

the poor, in a disproportionate way due to their hazard exposure and relatively weak coping mechanisms."

This chapter explores the sustainability of the complex livelihood strategies employed by the rural populations in Zimbabwe. Using archival data and case studies, the chapter examines the concepts and realities of the interface of climate change and urbanisation. The research employed a desk-top review of published documents and governmental papers. The chapter then engages three case studies of rural districts in Zimbabwe—Chipinge in Manicaland, Nkayi in Matabeleland North and Gokwe in Midlands.

2 Conceptual Framework

Livelihoods comprise various ways of making a living to meet individual and household needs (Chambers and Conway 1992) within a broader context of economic development and well-being including freedom, autonomy and empowerment. Rural livelihoods are dynamic as captured in the Sustainable Livelihoods Framework (SLF) (Thennakoon 2004). This approach considers the combination of assets and activities to do with threats and avert disasters (Mutami and Chazovachii 2012). Assets are a crucial element of the SLF that enables rural populations to survive using their innate skills (Moser 1998). The SLF considers five forms of capital assets (natural, physical, human, social, and financial) as essential resources facilitating livelihood strategies (Thennakoon 2004). Such coping or adaptive measures can include intensifying or extending agriculture, livelihood diversification and migration. Livelihood diversification (on or off-farm) pertains to activities undertaken to reduce vulnerability arising from dependence on one activity for survival and welfare (Scoones 2000). Thus, the SLF views vulnerability as a push factor towards invention and adoption of better means of attaining the wellbeing of people through the reduction of insecurity and uncertainties (Chambers 1997; Scoones 2000). A sustainable livelihood can cope with, and recover from, both environmental and economic stress and shocks in such a way that it provides opportunities even for the next generation (see Fig. 1).

3 Literature Review

Livelihood resilience highlights the role of human agency and the collective capacity to respond to stressors (Chitongo and Casadevall 2019). Governmental institutions regulating the rural environment are key for enhancing livelihoods and sustainability as they provide resources to develop the essential coping strategies. Public-private partnerships and the pooling of resources among the stakeholders are thus, present in most resilient rural areas.

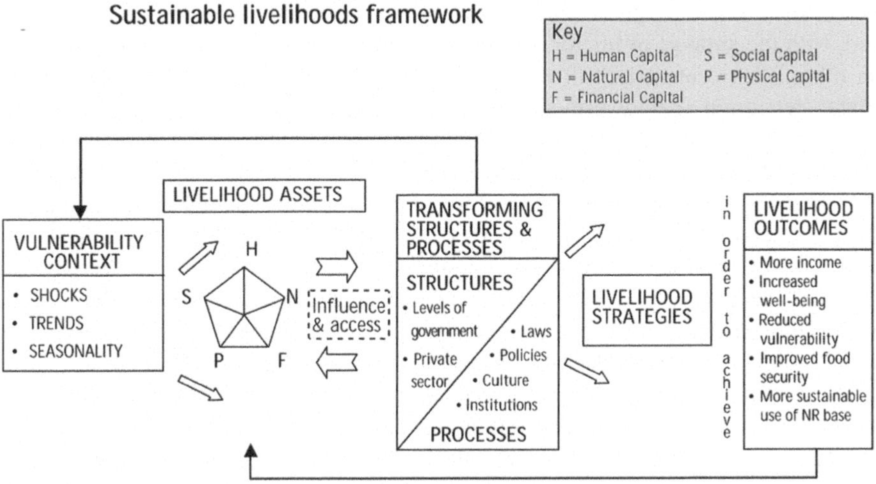

Fig. 1 Sustainable livelihoods framework. *Adopted from* DFID (2001: 23)

Local economic development (LED) strategies can encourage rural livelihood resilience LED is a participatory development process encouraging partnerships between the private and public stakeholders to promote socio-economic development (ILO 2012). Strong private sector involvement in local issues can generate employment and transfer essential skills. Hence, LED decentralises development, emphasising local socio-economic development. In Indonesia, the government partnered with the ILO to facilitate decent jobs for the rural people of East Java. A programme "Job Opportunities for Young women and men", implemented from 2007–2010 developed the resilience of beneficiaries. It focussed on sustainable enterprise development, employment-intensive investment programmes, skills development and financial access and led to a bio-gas initiative. This endeavour addressed fuel price hikes that negatively impacted on the livelihoods of the self-employed, traders and landless farmers (ILO 2012). Therefore, outsourcing is a viable option for governments lacking the resources or capacity to assist local communities.

Humanitarian assistance as part of government policies is an important input and a transformative force in sustainable livelihoods. Nonetheless, for effective humanitarian assistance, accountability, transparency and shared ownership of aid programmes among the government, donor agencies, and the rural populations as required by the Paris Declaration is needed. People in the remote districts of Ramanathapuram and Tuticorin in India were left vulnerable to poverty after the 2004 devastating tsunami. The non-governmental organization (NGO), Terre des Hommes Suisse, partnered with People's Action for Development to provide livelihood support within these two districts (Terre des Hommes 2007). This partnership concentrated on fishing micro-entrepreneurship initiatives, reinforcing existing community self-help groups (SHG). The enterprises included fishermen and women assuming post-fishing activities, such as processing, drying and selling and, with the support of microcredit facilities from the donors, the SHGs embarked on other micro enterprises (Terre des Hommes 2007). These enterprises demonstrate economic empowerment, diversity, and sustainability of livelihoods. Consequently, humanitarian assistance may enable resilient and sustainable livelihoods if sufficient institutions are enabled to monitor the fungibility of donor funds.

Agriculture is fundamental to rural livelihoods in Sub-Saharan Africa and a key driver of economic development in Africa (WFP 2015). The colonial era in Africa bequeathed a dual economy where the agricultural economy was a buffer for the industrial economy, and vice versa (Maxwell 1988; Mander 1998; Khan 2004; McSweeny 2004). These backwards and forward linkages are reflected in the apportionment of land. As such, access to land will improve food security and reduce poverty, creating sustainable livelihoods (World Bank 2008: 2–3). However, most African governments fall short in the equitable distribution of the resources needed for the realisation of sustainable livelihoods. Corruption and partisanship are so rampant in Africa negating efforts to improve livelihoods

One livelihood strategy that is common in Africa is artisanal small-scale mining (ASM). ASM refers to low-tech, labour-intensive mineral extraction and processing (Hilson and McQuilken 2014). Mining is a resilient livelihood strategy during environmental and economic shocks in rural areas (Karaki 2018). ASM involves around 10 to 15 million gold miners worldwide, including 4.5 million women and 600,000 children, ASM contributes to the livelihood of about 200,000 mining household in the Eastern provinces of DRC (Karaki 2018). Most African governments, however, relegate local communities in support of large-scale mining and this has often led to mining-induced displacements and subsequent conflicts and disruption of rural livelihoods (World Bank 2019).

The local communities perceive ASM as an easier way to generate a higher income than agriculture enables Mkodzongi and Spiegel (2019). It is contended that ASM and agricultural activities are often inseparable, citing that in Mozambique, ASM income was used to purchase agricultural inputs. In Ghana mining earnings provide start-up capital for small business growth and agricultural expansion (Wilson et al. 2015). These examples reveal the complexity of African rural livelihoods.

4 The Impact of the Triple Challenges of Climate Change, Economic Stress and Urbanisation

Global climate change is a force behind the increase of incidences of natural hazards, including the variability of rainfall, temperature and occurrences of climatic shocks (Chitongo and Casadevall 2019). In this regard, climate-related issues and farmers' livelihood strategies are different the world over (Ellis 2000). This way, over-generalisation tendencies when it comes to the issues of climate change and their impact to rural livelihoods are excluded.

There is a causal relationship between climate change and water-related disasters (WRDs). Liua (2012) observe that WRDs increase with changes in temperature, precipitation, and other climate variables. In this regard, the direct impact of climate change on rural livelihoods are increased risks and uncertainties from negative shocks, such as intermittent rainfalls and unpredictable weather patterns (Ellis 2000; Henderson 2004). As such, high temperatures and lower than anticipated precipitation result in drought leading to famines and deaths if the disaster is not mitigated. Meanwhile, the loss of vegetation due to overgrazing, deforestation, and droughts increases the risk of landslides and the probability of downstream flood disasters. Floods in vulnerable communities may result in an increase of water-borne diseases, such as cholera (Liua 2012; UNICEF 2018). There is need of close monitoring of urbanisation factors leading to emigration and enhancing sustainable livelihoods through an integration of urban and rural development planning (Dube and Chirisa 2012).

Climate change has compounded upon an already fragile livelihood situation in most parts of rural Sub-Saharan Africa. It was estimated that by 2020 between 75 and 250 million people would likely be exposed to increased water stress and that rain-fed agricultural yields could reduce by 50% if practices remain unchanged in some African countries (Angelsen and Wunder 2003; Bryceson 2004, 2009; Carswell 2002; Dold and Cocks 2002; Bilal 2017).

The key population most vulnerable to climate change and its impact are women and young people. FAO (2011) put forward that about 65% of the active West African and Sahelian population eke out their living from the agricultural sector increasing their vulnerability to climate hazards and ensuing disasters in ill-managed environments. The agricultural activities in the region are mainly rain-fed subsistence farming and extensive livestock husbandry. It is evident that climate change will result in the reduction of the quantity and quality of pastures in this region, resulting in new forms of transhumance in the sub-region, thus increasing the risk of weakening of livestock and the spread of animal diseases and the risk of territorial conflicts among the local farmers (FAO 2011). In this regard, the two-pronged vicious cycles of poverty and climate change are proven, and call for developmental interventions to enhance sustainable livelihoods among the vulnerable populations in this region.

The hazard occurrence trend in the SADC region is on the rise. UNECA (2015) reveals that between 1900 and 2013, occurrence of hydro-meteorological hazards had increasing frequency owing to climate change. The frequency of biological hazards that was on the increase during the same period may also be explained by the increased occurrences of environmental disasters, such as floods and cyclones, whose frequency owed to climate change. The biological hazards in this region have been compounded by urbanisation and unplanned sprawling of urban areas. UNECA (2015) observes that, in the SADC region, the top five countries vulnerable to disasters are Mozambique, Madagascar, Zimbabwe, the United Republic of

Tanzania and Malawi, whereby Mozambique has the lowest adaptive capacity, while Zimbabwe has the lowest coping capacity.

The common hydro-meteorological hazards in the SADC region are droughts, floods, and tropical cyclones. Of all the environmental hazards, drought is the most common one in the region. However, to "the credit of SADC member States, while droughts tend to trigger food insecurity due to reductions in sub-regional food production, these droughts have not led to famine, suggesting some level of resilience" (UNECA 2015: 9). On the other hand, the vulnerability of the SADC member states to flooding are a result of the following: "...community exposure to rivers, especially river confluences, terrain configuration, and building types" (UNECA 2015: 9). Such factors are related to housing stewardship in both rural and urban areas in the region. Overpopulation may result in the relinquishing of land rights from traditional custodians whereby the latter may succumb to the influence of land-barons, resulting in resettlement in wetlands. Concerning tropical cyclones, the SADC region is vulnerable by nature due to its proximity to the Indian Ocean where the tropical cyclones build from. It is of concern to note that "Vulnerability to disasters and poverty are intricately linked in SADC" (UNECA 2015: 15). As such, it is suggested that economic hardships, that describe most of the SADC member countries, are a hindrance to disaster risk reduction progress in the SADC region since resources, are channelled towards superficial poverty reduction, discounting the root causes of such poverty prevalence as valorised by the impact of natural disasters.

5 Rural Development and Practice in Zimbabwe Since 2000

The land issue has always been contentious in Zimbabwe. Cliffe et al. (2011: 907) observe that the debate on the FLRP has been 'highly polarised... between welcoming a reversal of a racial distribution of land - some of them bewailing the manner of implementation and its distorting of the state - and those who condemn the end, in principle, and the means'. Such a myriad of views in the motives and processes of the land reform programme has resulted in different perspectives on whether the land redistribution programme had been a success or a failure. Moyo (2009: 174) point out that "there is scant evidence to support most of the commonly held assertions regarding the outcome of the fast track land reform process". This perspective is stressed by Scoones et al. (2010: 236) who reveal that "the reality on the ground often does not match the myths perpetuated in many quarters". In this regard, the success or failure of the FLRP may not be comparatively ascertained since there are no established standards for comparison.

The land tenure for the beneficiaries of the FLRP was one set-back for the poor Zimbabweans. The "ordinary" farmers could not access lines of credit or loan from any financial institution for lack of collateral—their land ownership rights were unclear as they had only Offer Letters to show for it. The government in a bid to address the security of tenure introduced the 99-Year Lease, to formalise occupancy of re-distributed farms to beneficiaries who would have paid lease rentals (GoZ 2006). However, the issuance of these leases slows down enhancing the livelihood of the new farmers as farm valuations need to be carried out before issuance (Cusworth 1990; Chimhowu et al. 2011; Homann-KeeTui et al. 2013; Jombo 2016).

The productivity of the FLRP has been hindered by lack of inputs. With regards to the production of maize, "Zimbabwe's smallholders have had to deal with hyperinflation, recurring drought, shortages of all agricultural inputs, lack of credit, and limited public investment in agriculture…" (James 2015: 100). The government devised a facility to overcome such economic constraints through a Special Agriculture Production Programme popularly coined "Command Agriculture", that is a partnership of the government and the private sector to facilitate agricultural inputs for the beneficiaries of the FLRP. However, such initiatives have been marred by corruption.

The Zimbabwe Agenda for Sustainable Socio-Economic Transformation (ZimAsset) was an economic blue-print for Zimbabwe from 2013–2018. This Results-Based Agenda was built around four strategic clusters: "food security and nutrition; social services and poverty eradication, infrastructure and utilities, and, value addition and beneficiation". As such, agriculture was to be among the key strategic drivers for the blue-print. For rural development, ZimAsset had the following package: The Presidential Agricultural Input Support Scheme; and recapitalizing and capacitating AgriBank and the Grain Marketing Board (GMB), the Agricultural Marketing Authority. The main thrust of the Plan was to ensure that the Presidential Input Support Scheme focused on supporting the vulnerable groups at household and community level and ensuring that other farmers timeously accessed affordable inputs. Moreover, the economic blue-print paved way for the governmental promotion of contract farming initiatives.

The clusters for the economic blue-print included "infrastructure and utilities" and focused on the rehabilitation of infrastructural assets and the recovery of utility services in Zimbabwe (Kinsey 1999; PLRC 2003; Tevera 2008; TIMB 2014). As such, in the rural areas, the emphasis was on water and sanitation infrastructure and the construction of dams and conveyance systems were realised, as exemplified by the construction of Tokwe-Mukosi Dam in Masvingo Province, and the Gwayi-Tshangani Dam in Matebeleland North Province. Access to clean water was improved through the rehabilitation of broken-down pumps and drilling of new boreholes in most parts of rural Zimbabwe.

The other key result areas in the infrastructure and utilities cluster of ZimAsset was the construction and maintenance of government buildings and transport infrastructure and development. These key areas envisioned the construction of government buildings, such as Hwedza district composite office and Mutoko and Siyakobvu district composite offices. In addition, district registries were to be constructed in Hwedza, Goromonzi, Guruve and Nkayi districts. On the other hand, concerning the transport sector, the blue-print envisioned "improved road-networks". As such, some roads were expected to be constructed while others were to be dualised, rehabilitated, or resealed. For example, the following roads were rehabilitated: Nyazura-Dorowa (50 km); Golden Valley-Sanyati (100 km); and Gokwe-Siabuwa (100 km). Such investment on road-networks enhanced the development of rural areas in Zimbabwe as they eased accessibility and linkage between rural areas and urban areas.

Zimbabwe is agro-based with around 85% of its 390,757 km^2 being agricultural land (GoZ 1982; ILO 2012). Prior to 2000, the agricultural sector contributed to about 40% of Zimbabwe's gross national product. However, since the inception of the land redistribution programme, the contribution of the agricultural sector to the national economy has been reduced. For example, the GDP growth rate was estimated at under 3% in the period 2014–2017, whereby the agricultural sector contributed only about 12.5% (CIA World Factbook, October 2018). The reduced performance of the agricultural sector, that had an adverse effect on the whole economy, could be explained by many factors, chief among them being the following: economic distress and climate change.

The downfall of the Zimbabwean economy may be traced back to the late 1990s whereby the government of the day engaged in appeasement policies, such as the unbudgeted payment of gratuities to war veterans of the liberation struggle. Subsequently, socio-economic indicators, such as poverty prevalence, unemployment rate and human development indices reflect that sustainable socio-economic development is yet to be reached (Moyo 2002; Parliament of Zimbabwe 2003; Matsa and Muringaniza 2011; Zamchiya 2013; Zembe et al. 2014).

The impact of climate change is mostly tackled at national level as stressed by the adoption of the 2014 National Climate Change Response Strategy in response to the challenges posed by climate, and the need for adaptation and mitigation strategies. By and large, the effect of climate change is felt in the disruption of agricultural activities. For example, it is projected that the cereal production has been forecast to be 24% lower on an annual basis in the 2018/2019 growing season owing to climate change. The climate change exacerbated the hardships brought on by corruption and poorly planned policies and policy inconsistencies. In 2017 Zimbabwe ranked 157 out of 180 on Transparency International's Corruptions Perceptions Index (Transparency International, 21 February 2018).

5.1 Evidence from Gokwe Rural District

Regional livelihoods mapping defines Gokwe under Zone 6, that is described as a "Cereal and High Cotton Communal Zone", ZimVac (2010). Musevenzi (2012) reveals that the major crops grown in the area are cotton and maize. However, after the FTLRP in 2003, the small-scale cotton farmers in Gokwe revealed that inputs were now scarce on the open market and accessing cotton inputs became a challenge for most rural people due to price hikes. However, social differentiation in this rural area is evident in that, tribalism is believed to have played a hand also in determining the major economic activities in the area. Musevenzi (2012) points out that the Derukas (immigrants) dominated crop production whilst local tribes (Shangwe and Tongas) were largely engaged in less conventional agricultural activities and other non-farm livelihood activities, such as poaching, hunting, fishing and harvesting wild fruit.

Climate change is evident in the two districts of Gokwe (North and South) through sporadic rainfall and high temperatures. According to ZimVac (2019), in the 2018/2019 growing season, Gokwe received less than average rainfall yet the area depends on rain-fed agriculture. Resultantly, the agricultural produce was minimal and this strained the rural livelihoods of the people as evidenced by food insecurity that was 56.3% and 39.7% for Gokwe North and Gokwe South respectively. Due to chronic droughts and adverse climatic conditions, livestock condition in Gokwe South was described as "poor". This was compounded by the fact that there were inadequate and poor-quality pastures due to low rainfalls in the areas. Of concern is that the area is prone to attacks by livestock diseases, such as anthrax and foot and mouth diseases (ZmVac 2019), thus constraining the livelihoods of the people.

Rural livelihood diversification the major coping strategies of the rural people of Gokwe. It is pointed out that for the planting season 2018/2019, the cultivated area for cotton in Gokwe was less than the previous season (ZimVac 2019). As such, on-farm casual labour in Gokwe South is usually scarce, although ploughing and weeding in Gokwe North are sometimes available. Climate change has forced some of the people in Gokwe to indulge in poaching to supplement their diet and as a source of income (Musevenzi 2012). Poaching in the Chirisa game reserve to the west and Nyaminyami game reserve to the north has become a livelihood coping strategy for some of the people.

5.2 *Evidence from Nkayi Rural District*

Nkayi district falls under Zone 10 of the ZimVac livelihood mapping, summarised as the Eastern Kalahari Sandveld Communal. The area is in Natural Region IV and has an average annual precipitation of less than 620 mm. Farmers in this area are always vulnerable to climate change that is usually manifested through high temperatures and occurrence of droughts every two to five years. It is revealed that, for Nkayi, drought proneness was mostly severe and this was reached using both the Standardised Precipitation Index SPI and the Water Requirement Satisfaction Index (WRSI) for maize as a proxy for drought. Despite these chronic hazards, farmers mainly grow cash crops, such as cotton and groundnuts. Accordingly, in the 2018/2019 growing season, cotton was grown in the eastern parts of the district, although less area was planted as compared to the previous agricultural year (ZimVac 2019).

Poverty prevalence in the district is among the highest in the country whereby more than 76% of the rural population lives below the poverty line (<1.5 USD per day), and more than 22% is extremely poor (living <1 USD per day). The population that was food insecure in 2019 was 54.8% (ibid.). As such, the livelihood coping strategies for Nkayi are classified under the emergency category that means that populations use drastic measures, such as selling assets to alleviate food insecurity.

As a coping strategy, the farming systems in Nkayi are integrated and interlinked as a cattle-maize enterprise. It is pointed out that the crop residue is mostly used to feed the livestock during lean seasons while, the livestock are used as draught power to cultivate large pieces of land (Homann-Kee Tui et al. 2015). AgMIP (undated) points out that social heterogeneity in the farming community of Nkayi is evident whereby the three types of community farmers were distinguishable: extremely poor households with no cattle, cultivating 1.5 ha on average; poor households with 0–8 cattle, cultivating 2 ha; and better-off households with more than 8 cattle, cultivating 2.7 ha. In addition, the other differentiating factor, that is also a significant form of livelihood in the district, is the amount of remittances (cash and food) sent by the relatives who are in the diaspora, mostly in South Africa and Botswana (ZimVac 2010). The most vulnerable people in the district collect wild fruits and mopane worms as supplementary food. In addition, casual labour in the fields from the district's well-off people also cushion the welfare of the poor people in the area. Subsequently, with the chronic effects of climate in the district, food aid has become a significant source of food security in the district.

5.3 Evidence from Chipinge Rural District

The district may broadly be subdivided into the Highveld and the Lowveld regions. The Highveld region is depicted as Zone 8, that is described as the Eastern Highlands Commercial (ZimVac 2010). In this vein, the Highveld harbours agricultural estates that grow cash crops, such as tea, coffee, exotic tree plantations, bananas, macadamia nuts, and avocado peas, and these estates have not been affected by the FLRP (James 2015). Although in general, the precipitation in this area is reliable and is usually above average, and there are low incidences of flooding, farmers in some areas fail to harness these geographical features to their economic advantage due to input constraints, upon that the new farmers have no financial muscle to effectively practise commercial farming (ZimVac 2010).

As a consequence of high population growth rate in Chipinge district, land holdings shrink in size and there is also a decline in land productivity due to the over-use of the land without the application of adequate fertilisers (Chifamba and Mashavira 2011). Therefore, in some parts of the district, poverty is on the rise, and outmigration into South Africa perceived as a mitigation measure whereby remittances are a significant source of money and food in the district. In addition, other poor people in the district opt to work as farm labourers in the agro-industrial estates dotted in the area. Agroforestry is practised in Chipinge district whereby farmers practise mixed land-use by integrating livestock rearing with the growth of crops and trees, mostly fruit trees.

On the other hand, the Lowveld parts of Chipinge are susceptible to frequent droughts and high temperatures, conditions that do not favour the growth of staple cereal crops, such as maize. These natural geographical features increase the vulnerability of the area to the adverse effects of climate change. The farmers in these areas grow cotton as a cash crop that favours the dry conditions. The relatively new ARDA/Sabot Green Fuel sugar estate is a source of employment for many of the rural people in the area. The Green Fuel Ethanol Project was awarded National Project Status by the Government of Zimbabwe due to the significant benefits it brings to the country, such as being a source of employment for over 4,500 people. Such a venture covers 48,000 ha, hence is a hindrance for some irrigation schemes, such as Zuwarabuda Irrigation Scheme that is located downstream of Save River since most of the water upstream was dammed for the Ethanol Project that significantly reduced the water available for the irrigation schemes (Dengu et al. 2010). Some irrigation schemes along the Save River in areas, such as Mutema, Chibuwe, and Mutandahwe have been success stories in enhancing the livelihoods of the people in the Chipinge Lowveld. As such, intensive gardening of onions, tomatoes, cabbages, and maize have become a source of food and cash for most of the people in these areas.

6 Discussion

The FLRP offered rural Zimbabweans opportunities to improve their livelihoods. However, the way it was carried out disrupted the livelihoods of many people. For example, it has been pointed out that some farm workers lost their livelihoods due to the FLRP. In addition, most of the beneficiaries were "ordinary and poor people", who lacked the know-how and capital to catapult their agronomical skills towards the envisaged economic growth. Unfortunately, the recurrent droughts due to climate change did not help, and such natural phenomena have contributed to tarnish the FLRP as a "failure". However, in some locations, households, the land redistribution was a success. It is echoed that agriculture is the mainstay of the Zimbabwean economy. As such, there are both forward and backwards linkages between agriculture and the industrial sector in Zimbabwe. In other words, the failure of one factor would have negative, ripple effects to the other. In this regard, in the face of climate change, the agrochemical sector has the mandate to innovate drought resistant seed hybrids for the staple cereal crops, and making effective fertilisers and chemicals to use in such harsh environments. This observation extends to the control of pests and animal diseases. However, it was reflected that the agro-chemical industrial sector in Zimbabwe has not been performing to the expected levels since the inputs for the industrial sector are sometimes scarce, and if they are available, they demand high prices. This may be indicative of the deindustrialisation wave in the manufacturing sector. Resultantly, agricultural produce has been affected, and this has affected national food security and rural livelihoods have been put in jeopardy.

Climate change has become an issue of concern in Zimbabwe. The direct effect is that the trend of agricultural produce has been spiralling downwards, together with that of rainfall, and that of temperatures going upwards. This proves that rain-fed agriculture in this scenario is a courtship towards both economic and livelihoods disaster (Shackleton and Shackleton 2004; Sunderlin et al. 2005; Peng 2011; Meteorological Office UK 2012). In this regard, it was revealed that some rural areas engage in irrigation schemes while others diverse from agricultural activities altogether, a phenomenon known as deagrarianisation. However, deagrarianisation without viable options for a livelihood may also result towards doom. The young generations in the rural areas have resorted to emigration into cities and the Diaspora, particularly into South Africa. However, with global economic slowdown affecting most economies, the envisaged remittances have been reduced. By and large, most of the rural people in the country have employed diverse coping strategies, as determined by their localised environments.

7 Conclusion and Recommendations

When confronted with adversity, it has been proven that people innovate ways to come through. In this regard, the rural people suffer from environmental hazards and economic hardships. Therefore, diversifying and introducing conservative agriculture is recommended. On the other hand, outsourcing agricultural inputs from donor organizations may also boost agronomical activities. Notwithstanding that, social dialogue between the rural people and contract agencies in contract farming may also reduce the envisaged "exploitation" of the farmers. The SI79 of 2017 that regulates "Command Agriculture" has also the dictates to protect the rural people from unruly results of contract farming (GoZ 2017). In this regard, awareness of "rights" for the rural people concerning contract farming is called for. The FLRP, has been a blunt instrument towards poverty reduction in the rural areas. Besides lacking capital inputs, human, and social, capital is also a constraint in agricultural production. In other words, social differentiation and heterogeneity that defines the rural areas in Zimbabwe result in disunity of purpose that culminates in loss of the needed social capital in cooperatives. Climate change calls for innovative ways of agricultural production. In this regard, there is need for vocational training tailor-made towards combatting the effects of climate change. This may be done through public-private engagements, between the government and donor organisations.

References

Angelsen A, Wunder S (2003) Exploring the forest–poverty link: key concepts, issues and research implications. CIFOR, Bogor. https://www.cifor.org/publications/pdf_files/occpapers/op-40.pdf . Accessed 26 Oct 2020

Bilal S (2017) On the political economy of reforms in the extractive sector. GREAT Insights Magazine, 6(3). July/August 2017. ECDPM, Maastricht

Bryceson D (2004) Agrarian vista or cortex: African rural livelihood policies. Rev Afr Polit Econ 37(102):617–629

Bryceson D (2009) Sub-Saharan Africa's vanishing peasantries and the spectre of a global food crisis. Mon Rev 61(3):1–14

Carswell G (2002) Livelihood diversification: increasing in importance or increasingly recognized? Evidence from Southern Ethiopia. J Int Dev 14:789–804

Chambers R, Conway G (1992) Sustainable rural livelihoods: practical concepts for the 21st century. Institute of Development Studies. Discussion Paper 296, 1992

Chambers R (1997) Whose reality counts: putting the first last. Intermediate Technology Publications, London

Chifamba E, Mashavira N (2011) Adaption and mitigation strategies in sustainable land resource management to combat the effects of climate change in Chipinge, Zimbabwe. J Sustain Dev Afr 13(2):15–27

Chimhowu A, Manjengwa J, Feresu S (eds) (2011) Moving forward in Zimbabwe: reducing poverty and promoting growth. IES/BWPI, Harare

Chitongo L, Casadevall SR (eds) (2019) Rural livelihood resilience strategies in the face of harsh climatic conditions: the case of Ward 11 Gwanda, South, Zimbabwe. Cogent Soc Sci 5(1):1–19

CIA World Factbook (2018) Africa: Zimbabwe. https://www.cia.gov/library/publications/the-world-factbook/geos/print_zi.html

Cliffe L, Alexander J, Cousins B, Gaidzanwa R (2011) An overview of fast track land reform in Zimbabwe: editorial introduction. J Peasant Stud 38(5):907–938

Cusworth J (1990) Land Resettlement Issues, Background Paper: Zimbabwe Agriculture Sector Memorandum. A Paper prepared for World Bank, Harare

Dengu E, Mutambara S, Matembudze N, Mugore A (2010) Draft report: end of project evaluation on food security and livelihood independence program. Mercy Corps, Zim-ahead and SAFIRE, Harare

DFID (Department for International Development) (2001) Sustainable Livelihoods Guidance Sheets. London. http://www.efls.ca/webresources/DFID_Sustainable_livelihoods_guidance_sseet.pdf. Accessed 27 Oct 2020

Dold AP, Cocks ML (2002) The trade in medicinal plants in the Eastern Cape Province, South Africa. S Afr J Sci 98:589–597

Dube D, Chirisa I (2012) The informal city: assessing its scope, variants and direction in Harare, Zimbabwe. Glo Adv Res J Geogr Reg Plann 1(1):016–025

Ellis F (2000) Rural livelihood diversity in developing countries: analysis, policy, methods. Oxford University Press, Oxfod, Oxford

FAO (2011) Disaster risk management strategy in West Africa and the Sahel (2011–2013). FAO, Rome

GoZ (2017) SI 2017-079 Agriculture Marketing Authority (Command Agriculture Scheme for Domestic Crop, Livestock and Fisheries Production) Regulations. http://www.veritaszim.net/node/2128. Accessed 24 Oct 2020

GoZ (1982) The Transitional National Development Plan 1982/83-1984/85, vol 1. Government Printers, Harare

GoZ (2006) National A2 Land Audit Report. Ministry of Lands, Land Reform and Resettlement, Harare

Henderson LJ (2004) Emergency and disaster: pervasive risk and public bureaucracy in developing nations. Public Organ Rev 4(2):103–119

Hilson G, McQuilken J (2014) Four decades of support for artisanal and small-scale mining in Sub-Saharan Africa: a critical review. Faculty of Business, Economics and Law, University of Surrey, Guildford

Homann-Kee Tui S, Valbuena V, Masikati P, Descheemaeker K, Nyamangara J, Claessens L, Erenstein O, van Rooyen A, Nkomboni D (2015) Economic trade-offs of biomass use in crop-livestock systems: exploring more sustainable options in semi-arid Zimbabwe. Agric Syst 134:48–60

Homann-KeeTui S, Bandason E, Maute F, Nkomboni D, Mpofu N, Tanganyika J, Van Rooyen AF, Gondwe T, Dias P, Ncube S, Moyo S, Hendricks S, Nisrane F (2013) Optimizing livelihood and environmental benefits from crop residues in smallholder crop-livestock systems in Southern Africa. Socio-economics Discussion Paper Series. Series Paper Number 11. http://oar.icrisat.org/7277/1/S_Homann-Kee_Tui_et_al_2013_ISEDPS_11.pdf. Accessed 26 Oct 2020

ILO (2012) Rural Policy Briefs: Boosting local economies in rural areas, Geneva

James GD (2015) Transforming rural livelihoods in Zimbabwe: experiences of fast track land reform, 2000–2012, PhD Thesis, University of Edinburgh

Jombo SS (2016) Quantifying the impact of the Land Reform Programme on land-use and land cover changes in Chipinge District, Zimbabwe, based on Landsat observations, Masters Dissertation, University of Witwatersrand, Johannesburg

Karaki K (2018) Artisanal Gold Mining in DRC: time to get down to earth, European Centre for Development Policy Management (ECDPM) Discussion Paper No. 223. https://ecdpm.org/wp-content/uploads/DP-223-Artisanal-gold-mining-in-DRC.pdf. Accessed 1 Nov 2020

Khan MH (2004) Power, property rights and the issue of land reform: a general case illustrated with Reference to Bangladesh. J Agrarian Chang 4(1 & 2):17–44

Kinsey L (1999) Report of the study on the Determinants of Rural Household Incomes and their Impact on Poverty and Food Security in the Resettlement and Communal Areas of Zimbabwe, Report submitted to FAO, Rome

Liua C (2012) Susceptibility evaluation and mapping of China's landslide disaster based on multi-temporal ground and remote sensing satellite data. In: International Archives of the Photogrammetry, Remote Sensing and Spatial Information Sciences, XXXIX-B8. XXII ISPRS Congress, 25 Aug–1 Sep 2012. Melbourne, Australia

Mander M (1998) Marketing of indigenous medicinal plants in South Africa: a case study of KwaZulu-Natal. FAO, Rome, p 151

Mano R (2007) Zimbabwe's smallholder agriculture and livelihood performance and the challenges of recurrent food security crisis: a causes and consequences think piece. Paper prepared for CASS, University of Zimbabwe, Harare

Maphosa F (2005) The impact of remittances from Zimbabweans working in South Africa on rural livelihoods in the Southern districts of Zimbabwe. Forced Migration Working Paper Series No 14, Forced Migration Studies Programme, University of the Witwatersrand: 1–23

Matsa M, Muringaniza K (2011) An assessment of the land-use and land cover changes in Shurugwi District, Midlands Province, Zimbabwe. Ethiop J Environ Stud Manage 4(2):88–100

Maxwell S (1988) National food security planning: first thoughts from Sudan. Paper presented to workshop on food security in the Sudan, IDS, Sussex, 3–5 Oct 1988. In: Smith M, Ponting J, Maxwell S (1993) Household food security: concepts and definitions. An annotated bibliography. IDS Development Bibliography 8, pp 28

McSweeny K (2004) Tropical forests product sale as natural insurance: the effects of household characteristics and the nature of shock in Eastern Honduras. Soc Nat Resour 17:39–56

Meteorological Office, UK (2012) Climate: observations, projections and impacts, fact sheet. Case Study. http://www.metoffice.gov.uk/media/pdf/7/9/COP18_Factsheet_-_China.pdf. Accessed 26 Oct 2020

Mkodzongi G, Spiegel S (2019) Artisanal gold mining and farming: livelihood linkages and labour dynamics after land reforms in Zimbabwe. J Dev Stud 55(10):2145–2161

Moser C (1998) The asset vulnerability framework: reassessing urban poverty reduction strategies. World Devel 26(1):1–19

Moyo S (2002) Peasant organisations and rural civil society in Africa: an introduction. In: Moyo S, Romdhane B (eds) Peasant organisations and democratisation in Africa. CODESIRA, Dakar, pp 1–26

Moyo S (2009) Fast track land reform baseline survey in Zimbabwe: trends and tendencies, 2005/06. AIAS, Harare

Musevenzi J (2012) Rural livelihood diversification in semi-arid districts of Zimbabwe: an analysis of Muzarabani, Gokwe and Mwenezi districts. PhD Thesis in Development Studies, Nelson Mandela Metropolitan University

Mutami C, Chazovachii B (2012) Effective livelihood strategies in distressed environments: the case of Mudzi District of Zimbabwe. Curr Res J Soc Sci 4(5):362–372

Parliament of Zimbabwe (2003) A report on the plight of farm-workers and newly resettled farmers. Report prepared by the Public Service Labour and Social Welfare Committee, Third Session, Fifth Parliament, Parliament of Zimbabwe, Harare

Peng J (2011) WaterWorld. Market Report: Developing Desalination in China. https://www.waterworld.com/international/desalination/article/16201861/market-report-developing-desalination-in-china

PLRC (2003) The Presidential Land Review Committee on the Implementation of the Fast Track Land Reform Program, 2000–2002 Government Publications, Harare

Scoones I, Marongwe N, Mavedzenge B, Mahenehene J, Murimbarimba F, Sukume C (2010) Zimbabwe's land reform: myths & realities. James Currey, Oxford

Scoones I (2000) Rural livelihoods and diversity in developing countries. Frank Allis, New York

Shackleton CM, Shackleton SE (2004) The importance of nontimber forest products in rural livelihood security and as safety nets: evidence from South Africa. S Afr J Sci 100:658–664

Sunderlin WD, Angelsen A, Belcher B, Burgers P, Nasi R, Santoso L, Wunder S (2005) Livelihoods, forests and conservation in developing countries: an overview. World Dev 33:1383–1402

Terre des Hommes (2007) From Post-Tsunami Emergency Assistance to Livelihood Recovery in South India: Exploring the Contribution of Micro Entrepreneurship Initiatives in the Gulf of Mannar, Tamil Nadu. http://www.terredeshommes.org/pdf/publication/working_paper_regnier.pdf. Accessed 30 Oct 2020

Tevera TG (2008) Gender, migration and remittances in Southern Africa. SAMP Migration Policy Series No. 49. The International Development Center, Cape Town

Thennakoon S (2004) Rural livelihood strategies and the five capitals: a comparative study in selected villages of Sri Lanka. In: Proceedings of the 18th European conference on modern South Asian Studies. Lund, Sweden, July, pp 6–9

TIMB (2014) National tobacco workshop proceedings. TIMB, Harare

UNECA (2015) Assessment report on mainstreaming and implementing disaster risk reduction in Southern Africa. UNECA, Addis Ababa

UNICEF (2018) Zimbabwe Cholera Outbreak Report. UNICEF, Harare

United Nations Development Programme (2016) Mapping of Selected Hazards Affecting Rural Livelihoods in Zimbabwe: A District and Ward Analysis. Zimbabwe Resilience Building Fund (ZRBF) Management Unit, Harare

Wilson ML, Renne E, Roncoli C, Agyei-Baffour P, Tenkorang EY (2015) Integrated assessment of artisanal and small-scale gold mining in Ghana—Part 3: social sciences and economics. International J Environ Res Public Health 12(7):8133–8156

WFP (2015) Technical guidance note. WFP, Rome

World Bank (2008) World development report: agriculture for development. World Bank, Washington, DC

World Bank (2019) Zimbabwe IDAI Recovery Project (P171114), 31 May 2019, World Bank Harare

Zamchiya P (2013) The role of politics and state practices in shaping rural differentiation: a study of resettled small-scale farmers in South-Eastern Zimbabwe. J South Afr Stud 39(4):937–953

Zembe N, Mbokochena E, Mudzengerere F, Chikwiri E (2014) An assessment of the impact of the fast track land reform programme on the environment: the case of the Eastdale Farm in Gutu District, Masvingo. J Geogr Reg Planning 7(8):160–175

ZimVAC (2010) Rural livelihoods assessment. Government of Zimbabwe, Harare. http://fnc.org.zw/wp-content/uploads/2019/01/2010-ZimVac-Rural-Livelihoods-Assessment.pdf

ZIMVAC (2011) Rural livelihoods assessment. Government of Zimbabwe, Harare. http://fnc.org.zw/wp-content/uploads/2018/10/ZimVac-Rural-LivelihoodsAssessment-2011.pdf. Accessed 30 Oct 2020

ZimVAC (2013) Rural livelihoods assessment. Government of Zimbabwe, Harare. https://reliefweb.int/sites/reliefweb.int/files/resources/2013%20ZimVAC%20DRAFT%20REPORT.pdf. Accessed 30 Oct 2020

ZimVAC (2019) Rural livelihoods assessment. Government of Zimbabwe, Harare. https://reliefweb.int/report/zimbabwe/zimbabwe-vulnerability-assessment-committee-zimvac-2019-rural-livelihoods-assessment. Accessed 30 Oct 2020

Zvinorova PI, Halimani TE, Mano RT (2013) Viability of smallholder dairying in Wedza, Zimbabwe. Trop Anim Health Prod 45(4):1007–1015

Identity and Environment: Historical Trajectories of 'Traditional' Communities in the Protection of the Brazilian Amazon

Euzimar G. Silva, Hannah M. Reardon, Amadeu M. V. M. Soares, and Ulisses Miranda Azeiteiro

1 Introduction

"Environmental emergency" is the phrase used by nine former ministers of the environment to describe the current situation in Brazil. In an unprecedented act of mobilisation, the ministers asked that the leaders of the National Congress and Senate intervene on the basis of this emergency, one provoked by "setbacks in Brazilian socio-environmental policy and the ostensive campaign by representatives of the federal executive branch in favour of a totally outdated development model for the Amazon and other biomes in the country" (OAB Nacional 2019).

This act of political resistance took place at the height of the Amazon fires in August 2019, when the Institute for Space Research (INPE) demonstrated a 140% increase in the number of hectares of forest cover lost, compared to the same period the previous year (INPE 2020). According to experts, the increase was due to a rise in deforestation rates, tracked but ignored by the administration over the preceding months (Escobar 2019; Ferrante and Fearnside 2019).

Environmental protection agencies and the scientific community are alarmed by the possible consequences of environmental policies. They are especially concerned for indigenous peoples and other traditional communities, who are the segment of society most vulnerable to the consequences of current changes (Begotti and Peres 2019; Ferrante and Fearnside 2019).

This article focuses on the region referred to today as the Brazilian Legal Amazon, a political-economic-administrative territory that comprises nine states

E. G. Silva (✉) · A. M. V. M. Soares · U. M. Azeiteiro
Department of Biology and Centre for Environmental and Marine Studies-CESAM, University of Aveiro, Campus de Santiago, Aveiro 3810-193, Portugal
e-mail: euzimar@ua.pt

H. M. Reardon
Department of Anthropology, McGill University, Montreal, QC, Canada

W. Leal Filho et al. (eds.), *Sustainability in Natural Resources Management and Land Planning*, World Sustainability Series,
https://doi.org/10.1007/978-3-030-76624-5_15

and corresponds to 61% of the country's national territory. The Amazon holds the largest megadiverse tropical rainforest in the world. It also plays a crucial role buffering against global climate change and regulating local and regional weather patterns (Nobre et al. 2016). In addition, the region presents a remarkable ethnic diversity.

Much has been written about the role of traditional peoples in environmental conservation efforts, especially in the Amazon region. Some more recent work in the field of political ecology has sought to deconstruct received ideas about the "natural" propensity of traditional groups to protect the environment (Almeida 2004; Dove 2006; Nolte et al. 2013; Silva and Simonian 2015). These contributions complexify the categories of "indigenous" and "traditional", and reject the widespread myth of the "noble savage", frozen in an era of subsistence agriculture and simplistic livelihoods. As these authors, and many of their contemporaries, argue, the identity categories "indigenous" and "traditional" englobe a wide variety of communities and collectives with diverse political views and dynamic relationships to the environment, each other, and external forces.

The intellectual work that goes into complexifying these simplistic categories is valuable and important. It has contributed to expanding academic understandings of the complex and varying roles of traditional peoples in environmentalist struggles throughout the Amazon (Little 2006; Gerhardt 2007; Vadjunec and Rocheleau 2009). One potential downside to this work, however, is that particular examples may sometimes be instrumentalised to discredit the political claims of traditional communities. Vadjunec and Rocheleau, for example, describe how cattle farming is a valuable source of income for some communities in Extractive Reserve protected areas in Acre, Brazil (2009). They argue that the manner in which the cattle farming is carried out has a negligible environmental impact. However, in other contexts, cattle farming is a practice that is frowned upon by environmental conservation authorities, and practices related to subsistence hunting or cattle raising are strictly forbidden within the boundaries of conservation zones (Antunes et al. 2019).

The contradictions between state-led environmental conservation policies and local needs can produce tensions. In the context of political conflict, discourse around environmental "stewardship" is often instrumentalised to legitimise or discredit the land claims of indigenous or traditional peoples. In this article, we offer a brief survey of the historical relationships which connect "traditional" groups in the Brazilian Amazon—broadly divided into indigenous peoples, *ribeirinho* communities, and *quilombola* collectives—to the territories they inhabit. In so doing, we seek to reinforce the legitimacy of traditional peoples to present themselves as environmental stewards, since this political identity is deeply rooted in the region's history. Despite the dynamic nature of environmental stewardship, and the varying application of its principles among different communities, traditional peoples have a right to assert themselves as environmental stewards in order to defend their claims to the land and the resources upon which their sociocultural survival depends. Recognition of this fact in academic and environmentalist circles is urgent and necessary.

2 Unpacking the Language: "Traditional Communities"

Brazil is home to a wide range of groups considered 'traditional' by law, including indigenous peoples, riverine communities, *quilombolas*, extractivists, and rubber tappers (Maretti and Sanches 2003). However, only two social groups have their rights guaranteed by the current Federal Constitution (1988): indigenous peoples and *quilombolas*. Art. 231 of the constitution guarantees originary land rights to indigenous peoples, who are granted exclusive use of the lands they occupy. Art. 68 of the Transitory Provisions Act recognises *quilombola* communities as an ethnic group with definitive property rights over the lands that they inhabit.

Other social groups in Brazil are considered part of the vague category of "rural inhabitants" and often occupy the forests or under-used land as squatters. In rural contexts, "squatters" are occupants of a particular space that do not hold proprietary title to the land they live on. Many poor, rural families have lived and continue to live in these precarious conditions. The many legal and bureaucratic hurdles that prevent them from obtaining title is testament to their reality of daily resistance and the ongoing struggle that many people face to obtain land tenure or use rights appropriate to their economic, environmental and cultural traditions (Allegretti 2008). Regardless of the results of debates on whether these groups do or do not have constitutional rights to traditional land tenure, the traditional knowledge accumulated by these people over generations has led to recognition of their important role in protecting environmental resources (Joa et al. 2018; Reyes-García and Benyei 2019).

Of course, there are well-known difficulties in the use of the term "traditional". Often used as a catch-all, "traditional" may encompass all forms of autochthonous or originary cultures, including *quilombolas*, indigenous groups and *ribeirinhos* (riverine communities). It may also be used as an umbrella term for ambiguous peasant or rural groups not fitting the constitutionally-defined categories of indigenous or *quilombola*. In both cases, the term carries implications of race, geography, and class, meaning that "traditional" groups are often imagined as racialised, rural and poor.

Filho (2009) argues that such associations may be dangerous in the sense that they freeze so-called "traditional" groups into an "enforced primitivism". In other words, the term "traditional" is a loaded term, a product of colonial power dynamics and an indicator of their ongoing contemporary impacts. Indeed, the category "indigenous" itself is a product of colonial race relations, and its political and cultural reappropriation is evidence of the persistent neo-colonial power dynamics which continue to operate in many parts of the world today (Bolaños 2011). Many scholars have studied the political operationalisation of identity categories in the context of the Amazon (Cunha and Almeida 2000; Leff 2010; Bolaños 2011; Cardozo and Junior 2012). They argue that such categories are problematic since they play on stereotyped essentialisations of what it means to be "indigenous", "*quilombola*", or simply "traditional". Often, fitting into stereotyped imagery becomes essential to claiming political rights in the case of land claims and other fundamental struggles.

Despite the objections and warnings raised by different authors, this article works with generalising terms (traditional, *quilombola* and indigenous) as *political* categories, seeking to emphasise the agency of the people who reclaim and inhabit these identities as forms of resistance, political, social and/or cultural. These identities are the products of historical struggles, and as such they should also carry an association with resistance and political agency.

3 Indigenous Peoples of the Amazon

The now-antiquated term “Indian” was a misnomer from the era of Spanish conquest, when the conquistadors docked in America, assuming they had arrived in India. It would be the first of many errors and impositions wrought upon the indigenous peoples of the Americas through the painful process of colonisation which would ensue. During the first centuries after European contact, the native population of the Americas was drastically reduced, with disease, slavery and abuse resulting in the death of up to 90–95% of the indigenous population (Dobyns 1966).

It is estimated that in the year 1500, the first inhabitants of Brazil were organised into over a thousand distinct indigenous nations, with a total population of between 2 and 5 million people (Azevedo 2008). The process of colonisation in Brazil began with the ethnocide of its indigenous population, through slavery, widespread massacres and contagion of diseases brought by the Europeans. In the historical imaginary of the Brazilian nation, the figure of the “Indian” would come to represent cheap labour, rebels or savages to be dominated, civilised, enslaved, or killed (Almeida 2017).

Particularly nefarious were the policies designed to concentrate indigenous populations into centralised settlements—often implemented by religious missionaries and official agencies—since the high population density of these planned settlements favoured the spread of epidemic diseases (Cunha 2012). In the Amazon, settlement practices also served as a strategy for regional control and were widely implemented from the eighteenth century onward. All structures of the extractive-mercantile economy were formed and regulated by the missions (Alexiades and Peluso 2015). Religious missions were divided between the different Catholic orders—Carmelites, Franciscans, Mercedarians, and Jesuits—although all shared a common goal: to convert and “civilise” the indigenous inhabitants of the Amazon region. The religious missionaries promoted “descent” expeditions, venturing up the tributaries of the Amazon rivers to convince the various indigenous groups to abandon their villages, and move into the Christian settlements established by the religious orders, thereby surrendering their traditional culture and language (Alencastro 2000; Tavares 2011).

Under Pombaline policy, missions served to establish territorial control. Later on, the missions would become villages, named after Portuguese cities, gradually replacing indigenous place names over time (Tavares 2011). The first missions were established between 1750 and 1777 (Tavares 2011), although the practice of entering the more isolated forest regions and capturing indigenous groups for colonising projects persisted into the late nineteenth century.

At the beginning of the twentieth century, in 1910, the Indian Protection Service was established. It was the first political initiative for indigenous "protection" and "assistance". Soon after its establishment, however, it was revealed to be plagued by cases of corruption and aggression against indigenous people, and was eventually replaced by the current National Indigenous Foundation (FUNAI) in 1967. In the 1970s, the military government's development plan and the policy of occupying the supposed "demographic void" in the northern region strongly impacted indigenous communities. Partly in response to the effects of these policies, indigenous groups began to organise themselves formally in social movements, bolstered by the support of national and international NGOs (Gonçalves and Valle 2014).

The imperatives established by the "representatives of progress", as the military government liked to portray itself, impacted several groups such as Parakanã, in the state of Pará and the Waimiri Atroari (Rodrigues 1885; Cunha 2012). The National Truth Commission recently investigated human rights violations that occurred during the military dictatorship, and the partial report estimated 8,350 cases of murder of indigenous people, committed by agents of the Brazilian State (Brasil 2014). After Brazil's military dictatorship came to an end in the 1980s, and with the gradual return to democracy throughout the following decades, there has been a resurgence of activism. This new wave of activism is marked by the uprising of forest peoples throughout the Amazon, and their demands for the right to remain on the land they inhabit (Gonçalves and Valle 2014). It is a well-known fact that the indigenous social movements which emerged under the violent oppression of the dictatorship were the primary precursors of this process.

An important legal provision for guaranteeing the rights of indigenous peoples was the *Indian Statute of 1973* (DOU 1973), which dictated the nature of political relations between the state, civil society and the indigenous peoples of Brazil. The law stipulated that indigenous groups fall under the tutelage of the FUNAI. Currently, the revision of this outdated statute is one of the main demands of contemporary indigenous movements, alongside demands for streamlined demarcation procedures.

The seven paragraphs which make up article 231 of the Federal Constitution of 1988, may be one of the greatest legal advances conquered by indigenous peoples in the history of Brazil. This is because the article replaces the former political model based on notions of tutelage and welfare. According to the article: "Indigenous people are recognised for their social organisation, customs, languages, beliefs and traditions, and their originary rights over the lands they have traditionally occupied. It is the responsibility of the Union to demarcate these lands, protect them and to respect all of their material assets" (DOU 1988). In addition to the recognition of indigenous social organisation, customs, languages and beliefs, the constitutional rights of these groups also include their right to traditionally occupied land, as well as permanent possession and, hence, exclusive use of the ground, the rivers and the lakes therein.

In order to respond to the need to establish administrative demarcation procedures, Decree no. 1775/96 was promulgated in the 1990s (DOU 1996). Other notable recent advances include the establishment of the National Commission on Indigenous Policies (2006) and the National Council for Indigenous Policies (2015), established under Decree no. 8.593/15 (DOU 2015). This council provides an opportunity for indigenous peoples to expand their participation in the development of public policies for indigenous groups.

Only a decade ago, in 2010, Brazil conducted a demographic census to tally the number of self-identified "indigenous people" in the nation. The census was only the second of its kind in the history of Brazil. The results of the 2010 census suggested that 0.4% of the population self-identifies as indigenous—approximately 886,000 individuals present in all states and in both rural and urban areas. The northern region of the country has the most significant rates of population growth in this category, with the state of Amazonas accounting for the largest concentration of self-identifying indigenous peoples—168,700 people. The most populous ethnic group recorded by the census are the Tikuna—46,000 people—also largely residing in the state of Amazonas. They are followed by the Guarani Kaiowá, Kaingang, Makuxi and Terena (FUNAI 2020).

Currently, there are 723 Indigenous Territories (ITs) in Brazil. They cover approximately 13.8% of the national territory (Instituto Socioambiental 2020), and are home to over 305 different ethnic groups. According to the National Indian Foundation there are approximately 107 indigenous groups living in isolation throughout the Amazon region. Thirty of these groups are considered "isolated" or "un-contacted" groups, although their presence has been confirmed by anthropological reports and aerial photographs (FUNAI 2020).

Given the long history of abuse and oppression to which indigenous groups have been subjected, the nature of indigenous identity and identification with ethnic categories is fluid. Despite the difficulty of framing indigenous culture within the refractory categorisation of the colonial gaze, it is evident that there is invaluable cultural diversity among the various indigenous peoples of Brazil. In the last national survey, the FUNAI identified 274 indigenous languages spoken across the national territory (FUNAI 2020).

Indigenous territories which have been homologated and decreed in law cover approximately 23% of the Legal Amazon. There are 424 ITs in the region, totalling 1,153,444 km^2, and housing 173 different ethnic groups (Veríssimo et al. 2012; Instituto Socioambiental 2020). In addition to safeguarding the socio-cultural richness of the indigenous peoples established there, ITs play a key role in biodiversity conservation, and often have higher rates of conservation success than other protected areas. Recent studies have shown that indigenous territories in the Brazilian Amazon are more effective at preventing deforestation than any other type of conservation area (Toledo 2001; Nepstad et al. 2006; Nolte et al. 2013). This research demonstrates that indigenous knowledge and traditional practices of conservation can be highly effective for protecting the Amazon biome against deforestation. Furthermore, the long history of indigenous peoples struggle for recognition and political legitimacy justifies a reconceptualisation of conservation as a mechanism for social justice as well as environmental protection.

By way of conclusion to this brief overview of the history of indigenous peoples in Brazil, it may be pertinent to discuss some other significant advances gained by indigenous movements in recent years. Between 2002 and 2015, under some of the public inclusion policies implemented by the Brazilian government, Law No. 12.711/2012, also known as the Quota Law (DOU 2012), obligated universities and federal educational institutes to reserve 50% of admissions for indigenous, low-income, brown and black students. According to data posted by the last Post-Secondary Education Census in 2016, 49,026 indigenous students were enrolled in Brazilian public universities.

In 2019, however, the question of indigenous rights is once again attracting national and international attention. Attacks on indigenous territories, reports of invasion, abuse and death are once again increasing (Andreoni and Casado 2019; Begotti and Peres 2019; Bledsoe 2019; Ferrante and Fearnside 2019; Terra 2019). After attempting to transfer the responsibility for demarcating indigenous lands from the National Indian Foundation (FUNAI) to the Ministry of Agriculture, in February 2020 the federal government signed a bill that aims to open up indigenous territories to mining, oil and gas exploration, and hydroelectric ventures, without any form of veto for the indigenous groups concerned.

4 Amazonian Riverine Communities

According to the Brazilian Institute of Geography and Statistics (IBGE), Brazil's northern region is home to approximately 16 million inhabitants, accounting for about 8% of the country's total population. Of these inhabitants, approximately 11.5 million reside in urban areas and 4.5 million in rural areas (IBGE 2010). The rural population of the Amazon includes many diverse riverine communities, known colloquially as *ribeirinhos*.

The term *ribeirinho* refers to individuals, families, and communities that reside along the banks of rivers and lakes, areas which are often subject to seasonal flooding. As a social group, they were constituted during the process of colonisation, as indigenous people were pressured to "integrate" into colonial society (Parker 1989; Pereira 2014). Indigenous *ribeirinhos* trace their genealogical history to traditions of indigenous resistance. As individuals and families deserted their villages and adapted to colonial society in various ways, a mixing of cultures and ethnicities contributed to the emergence of the *ribeirinho* identity (Coelho 2005). Throughout the history of development in the Amazon, *ribeirinho* communities would also come to integrate migrants from the Northeast, who sought out wealth and land under various government resettlement programs. In a process of migration, desertion and escape, riverine communities emerged organically, evolving through the interaction of various cultural matrices over centuries of widespread miscegenation.

Ribeirinho communities frequently settle along floodplains, supporting their livelihoods by using both the floodplain and the mainland. As the name would suggest, floodplains are areas of land which experience seasonal flooding. One of the most striking features of the Amazon biome is its water regime. Between the dry and wet seasons, the water levels can vary up to 10 m (Lamotte 1990). Typically, the dry period runs from November to January, while flooding typically occurs between May and July.

The survival of *ribeirinho* groups is intricately linked to this variability in the water regime. For those living along the region's riverbanks, the change of seasons means an important fluctuation in the availability of resources for sustenance. During the period of flooding, for example, there is a relative shortage of plant products and an increased dispersal of aquatic fauna (Fraxe et al. 2007). Therefore, seasonal variation results in changing strategies for the use and exploration of natural resources (Harris 2005; Lira and Chaves 2016). For instance, during the flooded period fish stocks may render difficult the practice of fishing for family subsistence. It is a well-documented fact that the lifestyles and livelihoods of most riverine communities have a negligible environmental impact in the region; some argue that this is the result of a cultural inheritance of traditional indigenous practices (Parker 1989; Neto and Furtado 2015).

Some current research appears to show that extractive reserves and other sustainable areas that are home to *ribeirinho* communities are less effective at preventing deforestation than indigenous territories (Nolte et al. 2013). However, other studies suggest that the conclusions drawn by Nolte and others may be partially explained through mitigating factors. For example, indigenous territories tend to be located in isolated areas that are difficult to access, while extractive reserves and other sustainable use areas are frequently located closer to urban areas (Barreto-Filho 2004). While this does not entirely explain the remarkable effectiveness of indigenous territories in preventing deforestation, it should serve as a reminder that the conservationist habits of other traditional peoples should not be dismissed forthwith. Important qualitative research shows that *ribeirinho* communities play a significant role in protecting and restoring areas which have already experienced a high degree of human impact (Almeida 2004; Lima and Pozzobon 2005; Benatti 2011). Both indigenous and *ribeirinho* communities throughout the region are proven to contribute in significant ways to the preservation and restoration of the Amazon biome. Research demonstrates that where local actors play a leading role in official conservation programs, these programs demonstrate higher levels of success (Lima and Pozzobon 2005; Schweickardt 2012; Silva and Simonian 2015).

Due to the fluctuations of water levels, Amazonian *ribeirinhos* are often simultaneously farmers, fishermen, hunters, and extractivists, alternating their activities according to the river regime. Whether on the banks or in the water, riverine communities organise their lives and their ways of working according to the movement of the river, which is the source of subsistence and family income

(Neto and Furtado 2015). *Ribeirinho* communities are important repositories of traditional knowledge about the Amazonian environment, as Freitas reports in his work on the composition, successional dynamics and ecological functions of islands formed in the floodplain forests during seasonal water regimes (Freitas et al. 2015).

Recognition of the value of this knowledge has only occurred relatively recently. In the 1980s, a burgeoning *ribeirinho* social movement began to emerge contemporaneously with other socio-environmental grassroots movements. However, the recognition of the riverine populations as traditional people only occurred in 2007, under the National Policy of Sustainable Development of Traditional Peoples and Communities (DOU 2007a).

Since *ribeirinho* communities have remarkably diverse origins and livelihood practices, there is significant debate within the academic community about the "natural" propensity of these communities to safeguard and preserve natural resources. This hesitance is justified by the fact that some communities are simply more likely to protect natural resources than others. Nevertheless, the habits of some communities should not serve to delegitimise the claims of others. The vast majority of *ribeirinho* communities depend on low-impact forms of agriculture, fishing, and the extraction of non-timber forest products to feed their families and obtain a modest income. The knowledge required to carry out these activities and the sustainable nature of *ribeirinho* lifestyles should be respected and protected. Just like indigenous peoples, *ribeirinho* communities have a historically and scientifically justified right to use arguments of environmental stewardship to support their political claims for land and livelihood.

5 *Quilombolas* Communities

The word *quilombo* originates from *kilombo* or *quimbundo*, one of its meanings being "warrior camp in the forest". The *quilombo* was the most common form of collective resistance to slavery, and the presence of *quilombos* (under different names) has been documented in various Latin American countries (Hooker 2005). After the abolition of slavery, *quilombos* persisted and remained necessary, since there was no effort to elaborate any policy to include the Black population in mainstream society (Costa 1999; Silva 2018).

Quilombola communities, which, today, are also sometimes called "remnant *quilombo* communities", were formed by groups of African slaves as early as the 17th century. *Quilombos* represented the opportunity to reconstruct community life, based on shared memories of the African homeland. In most *quilombos*, daily subsistence relied on the efforts of all community-members, the presence of outside goods or products being relatively rare. Cultural traditions related to cooking, worship, dance and daily life were carefully guarded and passed on to younger generations. In these enclaves, a new model of society was made possible, beyond the barbarism of slavery. In several *quilombos*, the population consisted not only of escaped slaves and their descendants, but also of freemen and indigenous peoples (Reis and Gomes 1996). The political organisation of *quilombos* and their determined resistance in the face of a violent and oppressive colonial society, was key in the historical process which led to the abolition of slavery in Brazil.

The *Quilombo dos Palmares* was the largest *quilombola* community of the 17th century, with an estimated population of six thousand people in the main camp in 1644 (Funari 1996). In 1676, the Portuguese government launched its first major military offensive to try to dismantle the historic quilombola settlement, without success. It was only 19 years later, on November 20th, 1695, that Zumbi, the military and political leader of Palmares, was captured and beheaded, his remains gruesomely exposed in the capital. The history of Palmares is an enduring symbol of political resistance, and its traditions were upheld in *quilombos* which persisted throughout the country, varying in size and notoriety, but all standing as oases of courage, freedom and resistance for Brazil's Black population.

From the seventeenth century onward, with the intensification of coffee farming, the demand for African slaves in Brazil increased, thus fostering a profitable international market. However, under English pressure in 1831, the Brazilian parliament forbade the importation of slaves. Despite this decision, trade continued openly and increased over the following decades (Costa 1999). In 1850, the *Eusébio de Queiroz* law once again attempted to prohibit the slave trade, outlining punishments for those caught buying or selling slaves. Combined with the immense pressure exerted by Black communities in Brazil, and their repeated revolts, the decades that followed the *Eusebio de Queiroz* law saw several advances that would bring Black Brazilians closer to freedom. For example, the Free Womb Law of 1871 declared that children born to slaves, from the date of its publication onwards, should be considered freemen. The Sexagenarian Law of 1885 decreed that all slaves over the age of 60 should be set free. Nevertheless, these laws remained widely unenforced for many years.

In 1888, the Golden Law (DOU 1888) put an official "end" to slavery in Brazil, and began a new chapter in Brazilian history, marked by a search for space and for meaning in the new role of Black Brazilians as freemen. As Costa (1999) describes, abolition did not mean the immediate inclusion of Black people in society. Since the Brazilian elite refused to treat the newly freed slaves as salaried workers, there was little room for them in the changing landscape of the country. The land access policy, *the Land Law*, was vigorously exclusionary, undermining the possibilities for integration. Marginalised and scorned, many Black Brazilians opted to create their own communities, fleeing once more to remote areas to live close to the land and to support themselves and their families with subsistence-based agriculture and the sustainable extraction of forest products. In this way, through solidarity and resilience, these communities were able to survive for centuries in the face of a deeply racist and elitist social order.

From a legal viewpoint, the *quilombola* question was a point of concern for the State, both during the colonial and imperial periods, leading to the criminalisation and persecution of *quilombola* communities, which were considered symbols of subversion (Gonçalves 2017). Nevertheless, textual references to "*quilombos*" or "*quilombo* remnant communities" only appear in the legal texts from the twentieth century onward. More specifically, in the Constitution of 1988, and, in policy documents responding to the Black civil rights movement that began in the 1930s and was reinvented and reappropriated throughout the 1970s (Gonçalves 2017).

Article 68 of the Act for Transitional Provisions of the Federal Constitution (DOU 1988) states the following: "Remnant *quilombo* communities, living on their traditional land, are recognised as having the right to permanent titles and possession over said lands, and the State must emit the respective and appropriate titles". As could be expected, the legal language used in Article 68 generated considerable debate and controversy. Theoretical discussion about who and what might be considered a "remnant *quilombo* community" sent ripples through government and civil society stakeholders, concerned with the *quilombola* cause (Silva 2018).

In response to the controversies and impasses resulting from the language used in Article 68, Decree no. 4887 was promulgated in 2003 in order to outline regulations and procedures for identifying, recognising, delimiting, demarcating, and titling lands occupied by remnant *quilombo* communities, who self-declare themselves as such (DOU 2003b). In order to avoid further confusion, this Decree also establishes a legal definition for the term "remnant *quilombo* communities".

As a result of this legal definition, *quilombola* communities finally found themselves included in processes of environmental policy negotiation. This situation contributed to the "re-appropriation" of values, practices, and to the construction of a quilombola ethnic identity which had, in most cases, not been previously defined (Penna-Firme and Brondizio 2007). Once the legal bases had been established, the government moved to implement public policies aimed at *quilombola* communities, such as the National Policy for the Promotion of Racial Equality (DOU 2003a), The Brazil *Quilombola* Program (DOU 2007b) and the Policy for the Sustainable Development of Traditional Peoples and Communities (DOU 2007a).

Currently, in Brazil there are 1,715 remnant *quilombola* territories involved in the legal process of land titling (INCRA 2020b), although only 124 have been officially titled since the legal mechanisms were put in place (INCRA 2020a). In 2020, registered *quilombola* territories covered approximately 16,000 km^2 in the Amazon (0.3% of the territory) and included around 81 communities with an estimated population of 9,993 households (INCRA 2020a).

In the context of traditional peoples, it is important to remember that the notion of community is fluid, and many 'communities' may not yet be recognised as such, as they remain without strictly defined spatial boundaries—this is especially the case in the eastern portion of the Amazon, where territorial spaces are vast and communities are often geographically isolated (Veríssimo et al. 2012).

The desire for swift and effective demarcation procedures is common to all traditional populations. The slow progress of *quilombo* land titling demonstrates that the sluggishness of public bureaucracy continues to threaten the cautious gains made over the past few decades, following centuries of sustained political struggles to guarantee the rights of Brazil's most marginalised and oppressed.

The problems that emerge from the lack of basic social services, often put both *quilombolas* and indigenous populations in a situation of complete social exclusion. Furthermore, an aggravating factor for Afro-Latin communities is that, compared to other traditional populations, the practices of social exclusion based on skin colour may be even more severe (Hooker 2005). Much the same as indigenous peoples and *ribeirinho* communities, *quilombola* communities use environmental stewardship arguments to legitimise their political claims, and protect their cultural and territorial rights. History shows that *quilombola* communities have close connections to the territories they inhabit. Linking local political struggles to environmental protection through sustainable livelihoods should be a priority of conservation policy in the region.

6 Concluding Remarks

Inserting itself within academic debates that question whether traditional communities should be regarded as "natural stewards" of the environment, this chapter has argued that conservation efforts in the Brazilian Amazon simply cannot succeed without the active participation and leadership of traditional peoples. The current environmental and political context in Brazil has encouraged a climate of scepticism around the validity of the land claims and livelihoods of many traditional communities. While nuanced debate about the essentialisation of traditional identities is necessary for casting off Eurocentric notions of the "noble savage" (Dove 2006), we argue that the environmental and political urgency of the situation in Brazil demands a recognition of the crucial role of traditional communities in environmental preservation. The environmental stewardship of many traditional communities is rooted in a well-document historical reality of political, social and cultural resistance. In this chapter we have worked to draw connections between these histories of resistance, and the new politics of conservation and land claims.

The traditional knowledge required to sustain the low-impact livelihoods should be cherished, and in response to those who argue against the natural environmentalism of traditional groups, this article has discussed the recurrent culture of silencing and erasure of traditional populations, as well as the tremendous social vulnerability and political difficulties that these groups have faced over centuries of destructive and imperialistic occupation in the region.

In the contemporary context, 2019/2020 are undoubtedly years which should make us pause and reflect upon the history of resistance of social movements and traditional populations in the Amazon. This resistance continues, without respite, and recognition and support for the social movements that defend the forest is increasingly urgent. Those who stand in support of a progressive revaluation of conservation and defend the recognition of traditional people as the main protagonists in the defence of the Amazon's future, are concerned with the very existence of the Brazilian Amazon. If we are to break the cycle of economic developmentalism, we must be willing to think radically, to deconstruct our ways of knowing, and learn new ways of valuing and conserving the natural environment from the traditional guardians of the forest. The fate of the Amazon, and perhaps of the entire world, depends on it.

Acknowledgment This work was supported by the *Fundação para a Ciência e a Tecnologia* (FCT- PhD grantPD/BD/142835/2018) and *Fundo Social Europeu* (FSE).

References

Alencastro LF (2000) O trato dos viventes: formação do Brasil no Atlântico Sul séculos XVI e XVII. Companhia das Letras, São Paulo, p 525

Alexiades MN, Peluso DM (2015) Introduction: indigenous urbanization in lowland South America. J Latin Am Carib Anthropol 20(1):1–12

Allegretti MH (2008) A construção social de políticas públicas. Chico Mendes e o movimento dos seringueiros. Desenvolvimento e Meio Ambiente 18:39–59

Almeida MRC (2017) A atuação dos indígenas na história do Brasil: Revisões historiográficas. Revista Brasileira de Historia 37(75):17–38

Almeida MWB de (2004) Direitos à floresta e ambientalismo: seringueiros e suas lutas. Revista Brasileira de Ciências Sociais 19(55):33–52

Andreoni M, Casado L (2019) "Guardian" of the Amazon Killed in Brazil by Illegal Loggers. *The New York Times*. https://www.nytimes.com/2019/11/04/world/americas/brazil-indigenous-amazon.html

Antunes AP, Rebêlo GH, Pezzuti JCB, Vieira MARM, Constantino PAL, Campos-Silva JV, Fonseca R, Durigan CC, Ramos RM, Amaral JV, Pimenta NC, Ranzi TJD, Lima NAS, Shepard GH (2019) A conspiracy of silence: subsistence hunting rights in the Brazilian Amazon. Land Use Policy 84:1–11

Azevedo MM do A (2008) Diagnóstico da população indígena no Brasil. Ciência e cultura 60 (4):19–22

Barreto-Filho HT (2004) Notas para uma história social das áreas de proteção integral no Brasil. Terras Indígenas e Unidades de Conservação: o desafio das sobreposições. São Paulo, Instituto Socioambiental, pp 53–64

Begotti RA, Peres CA (2019) Brazil's indigenous lands under threat. Science 363(6427):592

Benatti JH (2011) Propriedade comum na Amazônia: Acesso e uso dos recursos naturais pelas populações tradicionais. Terras e territórios na Amazônia: demandas, desafios e perspectivas. Editora Universidade de Brasilia, Brasilia, pp 93–114

Bledsoe A (2019) Racial antagonism and the 2018 Brazilian presidential. J Latin Am Geogr 18 (2):165–170

Bolaños O (2011) Redefining identities, redefining landscapes: indigenous identity and land rights struggles in the Brazilian Amazon. J Cult Geogr 28(1):45–72

Brasil (2014) Relatório/Comissão Nacional da Verdade. Brasília. http://www.memoriasreveladas.gov.br/administrator/components/com_simplefilemanager/uploads/CNV/relatóriocnvvolume_1_digital.pdf

Cardozo IB, Junior IC do V (2012) Etnozoneamento da porção paraense das Terras Indígenas Trombetas-Mapuera e Nhamundá-Mapuera. EDUFRO, Porto Velho, pp 200

Cunha MC (2012) Índios no Brasil: história, direitos e cidadania. Claro Enigma, São Paulo, pp 54–97

Cunha MC da, Almeida MWB (2000) Indigenous people, traditional people and conservation in the Amazon. Daedalus 129(2):315–338

Coelho MC (2005) Do sertão para o mar: um estudo sobre a experiência portuguesa na América, a partir da colônia: o caso do Diretório dos Índios (1751–1798). Universidade de São Paulo, pp 433

Costa EV da (1999) Da monarquia à república: momentos decisivos. UNESP, São Paulo, p 492

Dobyns HF (1966) An appraisal of techniques with a new hemisferic estimate. Curr Anthropol 7 (4):395–416

DOU (1888) Lei n° 3.353 Declara extinta a escravidão no Brasil. Brasília. http://www.planalto.gov.br/ccivil_03/leis/lim/LIM3353.htm

DOU (1973) Lei n° 6.001 dispõe sobre o Estatuto do Índio. Brasília. http://www.planalto.gov.br/ccivil_03/Leis/L6001.htm

DOU (1988) Constituição da República Federativa do Brasil. Brasília. http://www.planalto.gov.br/ccivil_03/constituicao/constituicao.htm

DOU (1996) Decreto n° 1.775 dispõe sobre o procedimento administrativo de demarcação das terras indígenas e dá outras providências. Brasília. http://www.planalto.gov.br/ccivil_03/decreto/d1775.htm

DOU (2003a) Decreto n° 4.886 institui a Política Nacional de Promoção da Igualdade Racial-PNPIR e dá outras providências. Brasília. http://www.planalto.gov.br/ccivil_03/decreto/2003/D4886.htm

DOU (2003b) Decreto n° 4.887 regulamenta o procedimento para identificação, reconhecimento, delimitação, demarcação e titulação das terras ocupadas por remanescentes das comunidades dos quilombos de que trata o art. 68 do Ato das Disposições Constitucionais Transitór. Brasília. http://www.planalto.gov.br/ccivil_03/decreto/2003/d4887.htm

DOU (2007a) Decreto n° 6.040 institui a política nacional de desenvolvimento sustentável dos povos e comunidades tradicionais. Brasília. http://www.planalto.gov.br/ccivil_03/_ato2007-2010/2007/decreto/d6040.htm

DOU (2007b) Decreto n° 6.261 dispõe sobre a gestão integrada para o desenvolvimento da agenda social quilombola no âmbito do programa Brasil quilombola, e dá outras providências. Brasília. http://www.planalto.gov.br/ccivil_03/_ato2007-2010/2007/decreto/d6261.htm

DOU (2012) Lei n° 12.711 dispõe sobre o ingresso nas universidades federais e nas instituições federais de ensino técnico de nível médio e dá outras providências. Brasília. http://www.planalto.gov.br/ccivil_03/_ato2011-2014/2012/lei/l12711.htm

DOU (2015) Decreto n° 8.593 dispõe sobre a criação do Conselho Nacional de Política Indigenista-CNPI e dá outras providências. Brasília. http://www.planalto.gov.br/ccivil_03/_Ato2015-2018/2015/Decreto/D8593.htm

Dove MR (2006) Indigenous people and environmental politics. Annu Rev Anthropol 35:191–208

Escobar H (2019) Amazon fires clearly linked to deforestation, scientists say. Science 365 (6456):853

Ferrante L, Fearnside PM (2019) Brazil's new president and "ruralists" threaten Amazonia's environment, traditional peoples and the global climate. Environ Conserv 46:261–263

Filho HB (2009) Traditional peoples: introduction to the political ecology critique of a notion. Amazon peasant societies in a changing environment: political ecology, invisibility and modernity in the rainforest. Springer, Dordrecht, pp 95–129

Fraxe TJP de, Pereira HS dos, Witkoski AC (2007) Comunidades ribeirinhas amazônicas: modos de vida e uso dos recursos naturais. EDUA, Manaus, pp 224

Freitas CT, Shepard GH, Piedade MTF (2015) The floating forest: Traditional knowledge and use of matupá vegetation islands by riverine peoples of the central Amazon. PLoS One 10(4):1–15

FUNAI (2020) Índios no Brasil. http://www.funai.gov.br. Accessed 1 June 2020

Funari PPA (1996) A República de Palmares e a arqueologia da Serra da Barriga. Revista USP 28:6–13

Gerhardt C (2007) A invisibilização do outro nos discursos científicos sobre áreas naturais protegidas. Estudos Sociedade e Agricultura 15(2):268–309

Gonçalves AC (2017) Políticas públicas para quilombolas: a construção da cidadania na comunidade remanescente de quilombo do baú. Universidade Federal de Juiz de For a, 151

Gonçalves MA, Valle RST do (2014) Advances and Setbacks in territorial rights in Brazil. Rights and Resources Initiative and Instituto Socioambiental, pp 32

Harris M (2005) Nature makes them lazy: contested perceptions of place and knowledge in the lower Amazon floodplain of Brazil. Conserv Soc 3(2):461–478

Hooker J (2005) Indigenous inclusion/black exclusion: race, ethnicity and multicultural citizenship in Latin America. J Latin Am Stud 37(2):285–310

IBGE (2010) Censo demográfico. https://censo2010.ibge.gov.br/. Accessed 20 Aug 2020

INCRA (2020a) Andamento dos processos quilombolas. http://www.incra.gov.br/sites/default/files/incra-andamentoprocessos-quilombolas_quadrogeral.pdf. Accessed 30 Sep 2020

INCRA (2020b) Processos abertos quilombolas. http://www.incra.gov.br/media/docs/quilombolas/processos_abertos.pdf. Accessed 30 Sep 2020

INPE (2020). Banco de dados de queimadas. http://queimadas.dgi.inpe.br/queimadas/bdqueimadas/. Accessed 1 July 2020

Instituto Socioambiental (2020) Localização e extensão das TIs. https://pib.socioambiental.org/pt/Localização_e_extensão_das_TIs. Accessed 8 June 2020

Joa B, Winkel G, Primmer E (2018) The unknown known—a review of local ecological knowledge in relation to forest biodiversity conservation. Land Use Policy 79:520–530

Lamotte S (1990) Fluvial dynamics and succession in the lower Ucayali River basin, Peruvian Amazonia. For Ecol Manage 33–34:141–156. https://doi.org/10.1016/0378-1127(90)90189-I

Leff E (2010) Latin American environmental thought: a heritage of knowledge for sustainability. ISEE, Publicacion Ocasional 9:1–16

Lima D, Pozzobon J (2005) Amazônia socioambiental: sustentabilidade ecológica e diversidade social. Estudos Avançados 19(54):45–76

Lira TDM, Chaves MDPSR (2016) Comunidades ribeirinhas na Amazônia: organização sociocultural e política. Interações (Campo Grande) 17(1):66–76

Little PE (2006) Ecologia política como etnografia: um guia teórico e metodológico. Horizontes Antropológicos 12(25):85–103

Maretti C, Sanches RA (2003) Protected areas and indigenous and local communities in Brazil: lessons learned in the establishment and management of protected areas by indigenous and local communities. IUCN, V World Parks Congress, Durban, p 72

Nepstad D, Schwartzman S, Bamberger B, Santilli M, Ray D, Schlesinger P, Lefebvre P, Alencar A, Prinz E, Fiske G, Rolla A (2006) Inhibition of Amazon deforestation and fire by parks and indigenous lands. Conserv Biol 20(1):65–73

Neto FR, Furtado LG (2015) A ribeirinidade amazônica: algumas reflexões. Cadernos de campo 24:158–182

Nobre CA, Sampaio G, Borma LS, Castilla-Rubio JC, Silva JS, Cardoso M (2016) Land-use and climate change risks in the amazon and the need of a novel sustainable development paradigm. Proc Natl Acad Sci USA 113(39):10759–10768

Nolte C, Agrawal A, Silvius KM, Soares-Filho BS (2013) Governance regime and location influence avoided deforestation success of protected areas in the Brazilian Amazon. Proc Natl Acad Sci USA 110(13):4956–4961

OAB Nacional (2019) OAB e ex-ministros pedem apoio da Câmara para o enfrentamento da crise na Amazônia. https://www.oab.org.br/noticia/57499/oab-e-ex-ministros-pedem-apoio-da-camara-para-o-enfrentamento-da-crise-na-amazonia. Accessed 19 Aug 2020

Parker EP (1989) A neglected human resource in Amazonia: The Amazon "caboclo". Adv Econ Botany 7:249–259

Penna-Firme R, Brondizio E (2007) The risks of commodifying poverty: rural communities, quilombola identity, and nature conservation in Brazil. Habitus 5(2):355–373

Pereira EAD (2014) As encruzilhadas das territorialidades ribeirinhas: transformações no exercício espacial do poder em comunidades ribeirinhas da Amazônia Tocantina Paraense. Universidade Federal Fluminense, pp 434

Reis JJ, Gomes FS (1996) Liberdade por um fio: historia dos quilombos no Brasil. Companhia das Letras, pp 512
Reyes-García V, Benyei P (2019) Indigenous knowledge for conservation. Nat Sustain 2(8):657–658
Rodrigues JB (1885) Rio Jauaperi: pacificação dos Crichanás. Imprensa Nacional, pp 271
Schweickardt KHSC (2012) Faces do estado na Amazônia: entre as curvas do rio Juruá. São Paulo, Annablume, p 384
Silva ARF da (2018) Políticas públicas para comunidades quilombolas: uma luta em construção. Revista de Ciências Sociais 48:115–128
Silva JB, Simonian LTL (2015) Traditional population, Extractive Reserves and state rationality in the Brazilian Amazon. Desenvolvimento e Meio Ambiente 33:163–175
Tavares MG da C (2011) A Amazônia brasileira: formação histórico-territorial e perspectivas para o século XXI. GEOUSP-Espaço e Tempo, 29:107–121
Terra CP da (2019) Nota das Pastorais do Campo sobre o assassinato de liderança na Terra Indígena Wajãpi (AP). Comissão Pastoral da Terra. http://www.cppnacional.org.br/noticia/nota-das-pastorais-do-campo-sobre-o-assassinato-de-liderança-na-terra-indígena-wajãpi-ap. Accessed 20 July 2020
Toledo Victor M (2001) Indigenous peoples and biodiversity. Encycl Biodivers 3:451–463
Vadjunec JM, Rocheleau D (2009) Beyond forest cover: land use and biodiversity in rubber trail forests of the Chico Mendes Extractive Reserve. Ecol Soc 14(2)
Veríssimo A, Rolla A, Vedoveto M, Futada SM (2012) Protected areas in the Brazilian Amazon: challenges and opportunities. Imazon, Belém, p 72

Understanding and Defining Pathways for Ecosystems Services Decision Making, Sustainability and Livelihoods of Rural Communities of the Mozambique

Serafino Afonso Rui Mucova, Ulisses Miranda Azeiteiro, Antonio García Vinuesa, Walter Leal Filho, and Mário Jorge Verde Pereira

1 Introduction

The Millennium Ecosystem Assessment (MEA) defined ecosystem services as "the benefits people obtain from ecosystems" (MEA 2005). However, the Economics of Ecosystems and Biodiversity Ecological and Economic Foundations (TEEB) presented a more holistic definition: "ecosystem services are the direct and indirect contributions of ecosystems to human well-being" (TEEB 2010). Nevertheless, the Intergovernmental Science-Policy Platform on Biodiversity and Ecosystem Services (IPBES) reinforced the understanding stating that ecosystem services are "nature's contributions to people", and can be divided into supporting, regulating, provisioning and cultural (IPBES 2019a, b).

Ecosystem services have been widely discussed and addressed at different latitudes and scales (Hugé et al. 2020; Müller et al. 2020; Grima and Singh 2020; Riegels et al. 2020; Reynolds et al. 2020; Martin et al. 2020). Its context varies from

S. A. R. Mucova (✉)
Faculty of Natural Sciences, Lúrio University, Pemba 958, Mozambique

S. A. R. Mucova · U. M. Azeiteiro · M. J. V. Pereira
CESAM (Center for Environmental and Marine Studies) and Department of Biology, University of Aveiro, Aveiro, Portugal

A. G. Vinuesa
Facultade de Ciencias da Educación, Universidade de Santiago de Compostela, Santiago, Spain

W. L. Filho
Hamburg University of Applied Sciences, Hamburg, Germany

W. L. Filho
School of Science and the Environment, Manchester Metropolitan University, Manchester, UK

W. Leal Filho et al. (eds.), *Sustainability in Natural Resources Management and Land Planning*, World Sustainability Series,
https://doi.org/10.1007/978-3-030-76624-5_16

community regions (Chakraborty and Gasparatos 2019; Obeng et al. 2019; Quevedo et al. 2020), rural regions (Hardaker et al. 2020; Li et al. 2020; Zhou et al. 2020), urban areas (Filho et al. 2020; Shi et al. 2020; Campbell et al. 2020), islands (Havinga et al. 2020; Xu et al. 2020a, b), coastal areas (Barbier et al. 2011; Yang et al. 2019; Cabana et al. 2020), estuarine zones (Boerema and Meire 2017; Martin et al. 2020), marine environments (Müller et al. 2020; Nahuelhual et al. 2020; Depellegrin et al. 2020), mountain regions (Liu et al. 2019; Schirpke et al. 2020), as well as integrated approaches in a global context (Filho et al. 2020; Hasan et al. 2020; Xu et al. 2020a, b; Blythe et al. 2020). Its importance and contribution to the development and livelihood of populations is undeniable, being quite critical and fundamental for the least developed countries (TEEB 2010; Blasiak et al. 2015; IPBES 2019a).

In this context, ecosystem services are sources of food, medicine, water purification, disease regulation and can improve mental health through exposure to well-protected natural areas (Pogue et al. 2017; Ellis et al. 2019; IPBES 2019a). Thus, about 2 billion people rely on wood fuel to meet their primary energy needs. Yet more than 4 billion people rely primarily on natural medicines for their health care, and about 70% of drugs used for cancer are natural products inspired by nature. Additionally, more than 75% of global food crop types rely on animal pollination (IPBES 2019a). For example, in Africa about 62% of the population depend directly on these services in rural areas (IPBES 2018). This is a clear and evident demonstration of how indispensable ecosystem services are for people's lives and survival. Moreover, the intensive, predatory, and indiscriminate search for ecosystem services to satisfy current needs has led to one of the greatest degradations and massive reduction of these services in the century.

Estimates suggest that about 75% of the land surface is significantly altered, and more than 66% of the ocean area is experiencing increasing cumulative impacts. Yet approximately 85% of wetlands (area) have been lost. Additionally, about 32 million hectares of primary or recovering forest were lost between 2010 and 2015 (IPBES 2019a).

Another assessment shows that natural ecosystems have declined by 47% on average, relative to their earliest estimated states. Yet, about 25% of species are already threatened with extinction. Additionally, about 72 per cent of indicators developed by indigenous peoples and local communities show ongoing deterioration of elements of nature important to them (IPBES 2019a).

The events and massive deterioration of ecosystem services suggest a global, multisectoral, coordinated response approach and policies capable of face the imbalance between current needs, capacity for restoration and the provision of ecosystem services.

However, a considerable part of the global strategies and policies in progress involves declaring most ecosystems as conservation or protected areas and introducing specific measures for the sustainability of ecosystem services. For instance, in Africa, 14% of the continent's land mass and 2.6% of the seas have declared protected areas, while some sites have been designated as wetlands of international importance, important bird, and biodiversity areas. Still, other ecosystems have been declared as community conserved areas, World Heritage Sites; and Biosphere reserves (IPBES 2018).

However, despite considerable advances in protocols for the effective implementation of protected areas to safeguard ecosystem services, gaps and barriers remain, such as criteria, mechanisms, and the way in which decision-making is integrated into the rural context (Martinez-Harms et al. 2015; Kabisch 2015; Thompson et al. 2016; Verburg et al. 2016; Keenan et al. 2019; Morea 2019). At the same time, the context of decision-making for the protection of ecosystem services in which it integrates spatial planning, formulation of conservation policies, monitoring criteria, sustainability and assessments of trade-offs, systematically ignores the specific needs and characteristics of local or rural communities where ecosystem services are mostly concentrated (e.g. Tusznio et al. 2020).

Thus, in addition to being limited or scarce, the scientific literature available in the region does not clarify how community attributes can be framed within the decision-making process for the protection and sustainability of ecosystem services, as well as how their results (scientific research) are used to inform decisions affecting ecosystem services and rural communities. It is in this perspective that we are predisposed to address the gaps and discuss possible ways to raise the level of involvement of rural communities in the decision-making process about ecosystem services and their sustainability. In addition, our research examines the perception of the different ways of use, exploitation, importance, and attitude of the rural communities of northern Mozambique on the local ecosystems services as a way of contributing to sustainable strategies.

2 Materials and Methods

2.1 The Case Study Area

The northern region of Mozambique is made up of three provinces, namely: Nampula, Cabo Delgado and Niassa (Fig. 1). However, our research focused on Cabo Delgado province, one of the country's 11 provinces. Cabo Delgado is divided into 17 districts, namely: Ancuabe, Balama, Chiúre, Ibo, Macomia, Mecúfi, Meluco, Metuge, Mocímboa da Praia, Montepuez, Mueda, Muidumbe, Namuno, Nangade, Palma, Pemba, and Quissanga. It also has about 56 administrative posts and 134 villages.

This province is simultaneously the fourth most populous (2,333,278 inhabitants, about 8.36% of the Mozambique population) and extensive (82,625 km^2, corresponding to about 10.30% of the national territory) in Mozambique. Four extremes limit the province to other regions, namely: the Rovuma River and the Republic of Tanzania border it on the north. To the west are the Lugenda River and the Niassa province. To the south are the Lúrio River and the Nampula province. To the east is the Indian Ocean.

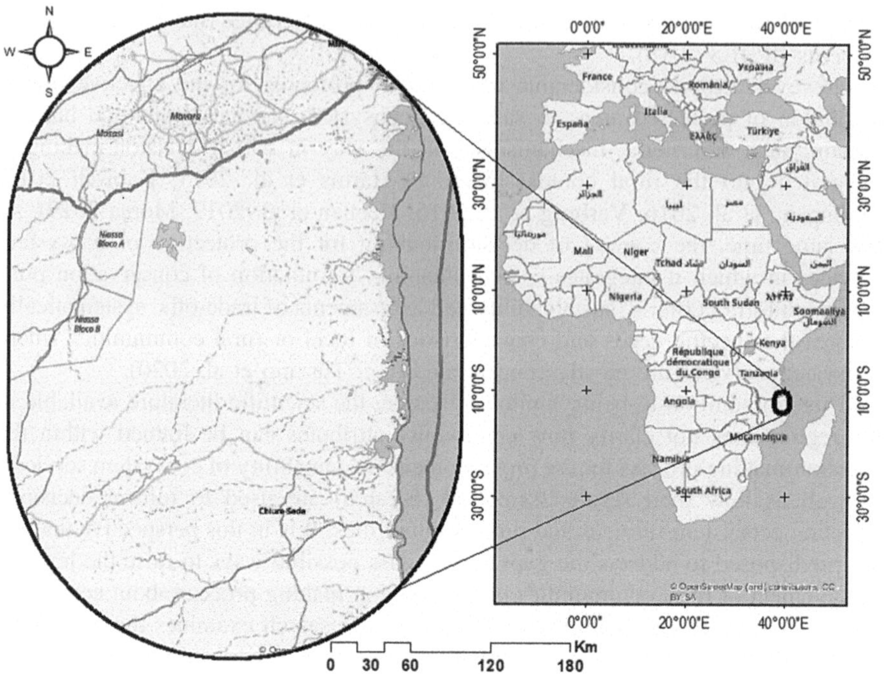

Fig. 1 Plotting the research study area (left side), northern Mozambique. *Source* Elaborated by the authors

A tropical climate characterises this province. The hot and rainy season is from November to April. While the dry and cool season is from May to October. The average annual temperature is 25 °C, while the average annual precipitation is 900 mm.

The majority of the population of Cabo Delgado has Emakhuwa as their native language, with 66.8%, followed by Shimakonde, most spoken on the plateau, with 21.8%, and in third place, Kimwani, most spoken on the coast, with 6.1%. The main economic base of the population of Cabo Delgado is subsistence agriculture, mainly practiced in traditional ways. The province has an area of about 5.6 million hectares of arable land, of which about 1 million hectares (ha) are in use by approximately 527,324 producer families and 87,338 ha owned by the business sector, producing around 2.8 million tons of food crops (in 1,055,076 ha) and 93,547 tons of cash crops (grown in 126,020 ha). Fishing is widely practiced along the coast and constitutes a source of diet and income for the population.

2.2 Data Collection and Survey Design

To achieve the research goals, we developed a set of procedures supported by quantitative assessment as a mechanism for collecting information. We developed a combined questionnaire (closed and open) corresponding to 31 items, containing about 19 questions, divided into two main parts. The first part was limited to the characterisation of the respondents, composed of the following items: (a) date, (b) province, (c) dialect, (d) gender, (e) religion, (f) head of household, (g) age, (h) can read or write, (i) education level, (j) total number of households (men, women, those who are under or over 18), and (k) have always lived here. While the second part was related to socio-economic aspects, survival, use and services of ecosystems, composed of the following questions: (a) what is the main activity, (b) what is the alternative activity, (c) what type of housing do you have, (d) which is the material used for housing construction, (e) which is the main source of food, (f) if they have a source of water, (g) currently there is more availability of water, (h) own an agricultural area, (i) what method or technique for opening of the agricultural area uses, (j) if they use fire to clear the agricultural area, (k) it has produced more or less, (l) what are the main ecosystem services it benefits from, (m) if they recognise other benefits, values, importance of ecosystem services, (n) if yes, say which, (o) to mention ecosystem services that have been decreasing in the last 5 years, (p) to believe that ecosystem services in this region are finite, (q) ecosystem services from this region is being explored in a sustainable or unsustainable way, (r) would you like to have more information on the correct exploitation and sustainable use of ecosystem services, (s) would you be receptive to support any local or regional plan to ensure the sustainable exploitation of ecosystem services, (t) would this support require some remuneration or condition to this plan.

Our focus was to apply questionnaires to men and women in equal proportion. However, women were more timid to participate while men were more willing to answer questions. We carried out simple random sampling for a sample of N = 1000 participants from the province of Cabo Delgado, northern Mozambique. This task ran between May and August 2019.

We always favoured an application of home questionnaires and, where it was not possible, we survey people in their own places (for instance at the farm, road, market, work). The time recorded for each respondent was 24 min. To guarantee independence and reduce interference in responses, six university students were trained to apply the questionnaires. All students could speak in addition to Portuguese at least two main local languages (Emakhuwa, Shimakonde, and Kimwani). Due to the level of literacy and limited knowledge of communities on technical and scientific terms such as natural resources, ecosystem services, sustainable and unsustainable, beneficial, values and importance of ecosystem services, management, and conservation, we translated all terms into a plain language, always highlighting and demonstrating with examples. A pilot test was applied to a sample of 5% (N = 1000), corresponding to 50 participants. The students applied the test and we aimed to understand (a) posture, objectivity and ethics of the students, (b) receptivity of the respondents or respondents, (c) coherence, clarity, complexity of the questions and time required for each respondent (Van Teijlingen and Hundley 2002). Before applying the questionnaires, we used to have meetings with the heads of the districts, villages, or towns to explain the purpose of the research and obtain authorization to carry out the tasks. In addition, for all participants, before answering the questionnaires, we clearly explained them the objectives of the work and we obtained the respective verbal consent.

Regarding preference, value, and importance of ecosystem services our approach was exploratory (allowing people to say how much they knew about local services). We did not previously identify a limited number of services, values, or importance so that subsequently the communities could question or recognise them as addressed by the authors He et al. (2018). These authors predefined or limited the number of ecosystem services to 15 and based on these services (previously defined) the populations responded or recognised their importance/preference. Also, the authors Zhang et al. (2016) embarked on the same approach and identified about 29 ecosystem services (divided into three groups) and questioned the communities about their perception. However, our approach was the opposite as we clearly intended to verify whether the populations understood or had "traditional or local" knowledge related to ecosystem services without first exposing them or inducing them to respond or identify a certain previously established service. Then, all responses were compiled and categorised according to the logical framework of Millennium Ecosystem Services (2005) and Intergovernmental Science-Policy Platform on Biodiversity and Ecosystem Services (IPBES) (2019a, b) (Table 1).

2.3 Data Analysis

The total population of the province of Cabo Delgado is around 2,333,278 inhabitants (51.4% men and 48.6% women) (INE 2017). The survey sample was about 1000 inhabitants, three times the number needed (approximately 385 inhabitants, according to the sample calculation performed through raosoft (http://www.raosoft.com/samplesize.html)). We adopted a 5% margin of error, a 95% confidence level, a 50% response distribution rate. Approximately 63% of the questions required closed answers (example: yes or no; would or would not support; do or do not have; reduced or increased; infinite or finite, sustainable or unsustainable). However, to facilitate the analysis, all responses were coded as 1 (affirmative) and 0 (negative). For some questions, such as level of education, the codification was as follows (0 = no education or illiterate, 1 = primary education, 2 = high school, 3 = university). Nevertheless, questions with answers: maybe, no, yes, the following codifications were assigned 0, 1, 2, respectively. Thus, we used a more descriptive assessment to characterize the demographic, social, economic, and perception/importance issues about respondents' ecosystem services. After collecting the data, the results were translated (from Portuguese to English),

Table 1 Ecosystem services and human wellbeing framework

Ecosystem services		Constituents of well being	
Supporting	Nutrient cycling	Security	Personal safety
	Soil formation		Secure resources access
	Primary production		Security for disaster
	Habitat provision	Basic material for good life	Adequate livelihoods
Provisioning	Food production		Sufficient nutrition for food
	Freshwater		Shelter
	Wood	Health	Access to goods
	Timber		Strength
	Water		Feeling well
	Fuel		Access to clear air and water
Regulating	Climate regulation	Good social relations	Social cohesion
	Flood regulation		Mutual respect
	Disease regulation		Ability to help other
	Water purification		
Cultural	Aesthetic		
	Spiritual		
	Educational		
	Recreational		

Source Adapted from MEA (2005), IPBES (2019a, b)

transcribed and subsequently stored on an Excel sheet to then be exported and analysed using Statistical Package for the Social Sciences (SPSS version 25) Software. We applied Pearson's chi-square test (χ^2) to assess the association of variables on preference and knowledge of ecosystem services, importance, as well as protection and sustainability of local ecosystem services.

3 Results

3.1 Characterisation of Respondents

Table 2 shows the results of the characterisation of the survey respondents. There was linguistic diversity among populations. We have identified about six (6) languages and they are present in the province. More than 93% of the respondents speak Emakhuwa, considering it the most dominant local language in the region.

However, Shimakonde and Kimwani are the other two most spoken languages with 40% and 24% respectively. Although Portuguese is the country's official language, only 30% of the population speak Portuguese and 70% share the different local languages. The proportion of women in the study was relatively significant, estimated at 31%, while men were around 69%. In addition, our analysis shows that respondents are mostly from the Islamic religion, with 62%, while the Christian religion comprises about 38% of the respondents. Nevertheless, we have found that among the respondents, young people under 18 years old (with about 4%) assume the responsibilities of heads of families. However, most of the time men assume the responsibility of the head of the family (with around 88%). Although in a relatively smaller proportion (around 12%), women are a fundamental and core part of the leadership of families in the province. In general, young people aged 18–30 years old assume the heads of families most of the time (with about 52.5%).

We also found that the level of education is relatively low. About 71% of the respondents have primary education and 17.1% have never studied. However, about 10.6% of the respondents attended high school and only 1.3% have higher education. In the universe of the respondents, about 68.4% are native (local) and 31.6% migrated from other provinces in the north of the country such as Nampula and Niassa.

3.2 *Socioeconomic and Well-Being Characterisation of the Communities*

Regarding the respondents' socioeconomic conditions and well-being, Table 3 and Fig. 2 describe and present the results. However, we have found that more than 61% of the population has agriculture as the main activity. A minority, represented by 3.6%, have no job (they carry out domestic activities), while the other 3.60% are students. Another relatively significant part of 26% of the population develops qualified jobs. We understand that in addition to the greater number of the population practicing agriculture as a source of income and subsistence, only about 35% of the population has alternative sources of subsistence, which means that more than 64% depends exclusively on a single source for subsistence. Mostly, 97.6% of the population lives in very precarious housing. At the same time, 98% build their homes using local material and products from ecosystem services (bamboo, wood, fibre, grass, sticks). Although it is not piped or potable water (usually consisting of traditional wells and small rivers around the village), the majority of the population (98%) has access to water. Distances for obtaining water vary from 0 m to more than 1000 m. However, about 69% of the population obtains water over a distance of between 0 and 500 m. In addition, despite being accessible, 96.5% of the population understands that water is in short supply. All respondents are engaged in the agricultural practice or activity or at least each individual has an area for agricultural practice. Thus, about 53% of the respondents have an area equal to 1

Table 2 Plotting characterization of research respondents (N = 1000)

Respondents profile	N (Number)	Percentage (%)
Dialect (Local language)		
Emakhuwa	933	93.30
Echuwabo	1	0.10
Kimwani	24	2.40
Shimakonde	40	4.00
Cinyungwe	1	0.10
Kiswahili	1	0.10
Portuguese language		
Portuguese	300	30.00
Local Language	700	70.00
Gender		
Female	312	31.20
Male	688	68.80
Religion		
Christian	380	38.00
Islam	620	62.00
Head of household		
Female	120	12.00
Male	880	88.00
Age		
≤ 18 years old	44	4.40
19–30 years old	481	48.10
31–60 years old	454	45.40
61–86 years old	21	2.10
Education level		
No education	171	17.1
Primary school	710	71.0
Secondary school	106	10.6
University degree	13	1.3
Have you always lived here		
Yes	684	68.4
No	316	31.6

Source Elaborated by the authors

hectare, and more than 46% have an agricultural area between 2 and 10 ha. A very small proportion, about 0.10%, has an area of more than 10 ha.

Agricultural activity is carried out in a traditional, itinerant and subsistence way. Burning has been the means for cleaning up cultivation areas. However, 76.20% use fire to clean the areas, while 23.80% adopt other practices (weeding, felling). The productivity or production of agricultural land has been quite low. 75.70% understands that productivity is low, while around 24.30% has high productivity. Products from agriculture and purchase consisting of cereal grains (73.48%), fresh

Table 3 Plotting socioeconomic and well-being characterization of respondents (N = 1000)

Respondents profile	N (Number)	Percentage (%)
Core business/activity		
Agriculture/Farming	619	61.90
Technicians	172	17.20
Qualified Workers	88	8.80
Students	85	8.50
Non employed	36	3.60
Alternative business/activity		
No	647	64.70
Agriculture/Farming	276	27.60
Labourer	24	2.40
Tradesman	51	5.10
Students	2	0.20
Housing types		
Precarious	976	97.6
Conventional	24	2.40
Types of building materials		
Local and forest product	981	98.10
Cement block/zinc coating	19	1.90
Water sources		
Yes	998	99.8
No	2	0.20
Distance covered to access water (m)		
1–50	46	4.60
51–100	81	8.10
101–500	567	56.70
501–1000	210	21.10
>1000	96	9.60
Water availability		
Less	965	96.5
More	35	3.5
Agricultural land		
Yes	1000	100.00
Hectare (ha)		
= 1	535	53.50
2–5	461	46.10
6–10	30	0.30
>10	10	0.10
Fire (for land clearing)		
Yes	762	76.20
No	238	23.80
Productivity		

(continued)

Table 3 (continued)

Respondents profile	N (Number)	Percentage (%)
Less	757	75.70
More	243	24.30
Disease		
Malaria	540	54.00
Cholera	155	15.50
HIV	4	0.40
Bilharzia	22	2.20
Measles	61	6.10
STD	20	2.00
Migraine	15	1.50
Tuberculosis	138	13.80
Diarrhoea	45	4.50

Source Elaborated by the authors

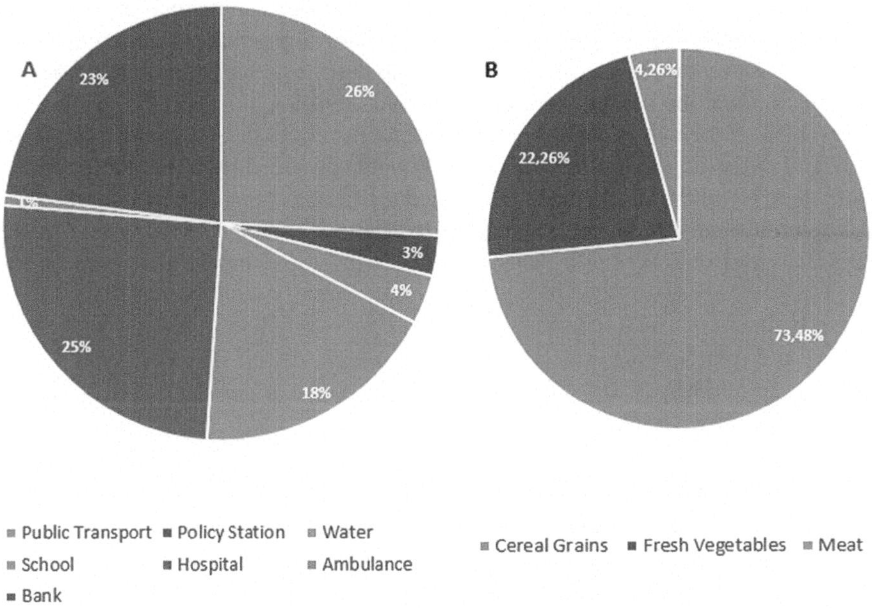

Fig. 2 Plotting socioeconomic and well-being characterization of respondents (A = critical public services and assets, B = main food sources) (N = 1000). *Source* Elaborated by the authors

vegetables (22.26%), and meat (4.26%) insure the population's survival or food source (Fig. 2).

There are about nine (9) most common types of diseases in the region. However, malaria, with 54% of the respondents, is considered the most frequent disease, followed by cholera and bilharzia with 15.50% and 13.80% respectively. Most of the basic social and public services that guarantee the well-being of the population are in very limited numbers and, in some cases, absent. About 26% understand that there is no public transport, 25% (hospitals), 23% (banks), 18% (schools), 4% (water), 3% (police centre) and only 1% (ambulance) (Fig. 2).

3.3 Ecosystem Services Benefits and Sustainability, Perceptions, and Values

Although a complex set of environmental interactions (ecosystem services) supports the survival of the human population and the maintenance of local biodiversity, our results show that provisioning services are the respondents' preference and the most beneficial to communities (Fig. 3). More than 20% of the respondents understand that wild animals (for food and commerce) are the biggest beneficiaries they get from regional ecosystems. However, about 13% of the respondents say that the benefits of ecosystem services come from wood (for commercial purposes and housing construction), as well as firewood for charcoal production (for commercial purposes). Nevertheless, about 10% of the respondents prefer bamboo (for housing construction and the manufacture of household utensils).

Over 9% of the respondents mentioned water resources (to safeguard agriculture, and daily domestic services) as one of the benefits provided by ecosystem services. More than 7% of the respondents mentioned medicinal plants (for treating diseases and spiritual issues), as well as precious stones (for commercial purposes) as the most preferred resources and benefits provided by the ecosystem. Generally, the

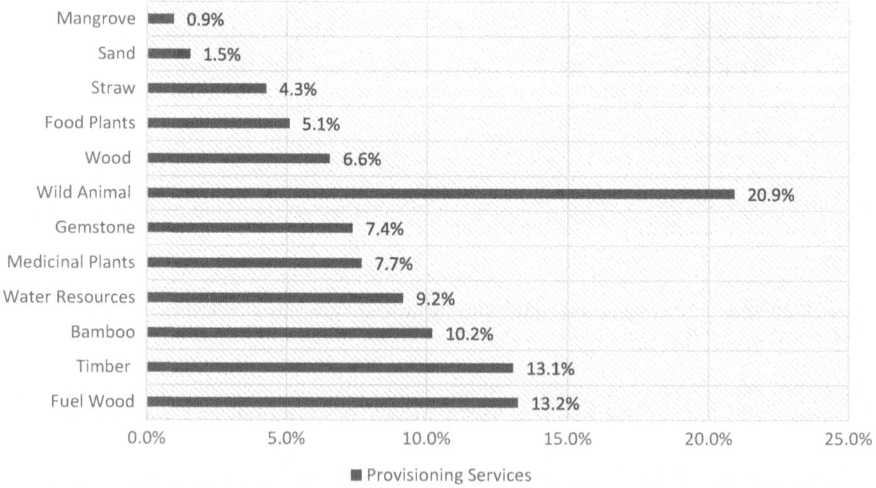

Fig. 3 Plotting the main ecosystem services that most benefit and used by community day-by-day (N = 1000). *Source* Elaborated by the authors

sticks used for the construction of protective fences (corrals, farms, and housing), construction of housing, domestic fuel is highlighted by more than 6% of the population as most preferred and beneficial. About 5% of respondents believe that wild plants (for food) are the most beneficial to local communities. In this regard, more than 4% of the respondents consider Straw (for housing coverage) to be of greater benefit and importance. Finally, more than 1.5 and 0.9% of the respondents consider sand (for construction) as well as mangrove forests respectively to have the greatest benefits.

Figure 4 reveals the respondents' perception of other values and importance that local ecosystem services offer or provide to communities. The results illustrate that about 94% of the respondents revealed that they did not know about other importance or values that local ecosystem services could provide. However, only about 6% (corresponding to 62 respondents) reported having more information or knowing other values of ecosystem services. Thus, the group only recognised provisioning and cultural services. However, about 33% of the population said that tourism and water storage in the basement is one of the results of services provided by ecosystems. In addition, about 23% highlighted that creating conditions for the survival or housing of animals is a value or importance of the role of local ecosystems. 8% of the respondents considered that the existence of deposits, that is the storage of precious stones underground at the local level, is one of the values or importance attributed to ecosystem services. Finally, 3% of the population stressed that the existence of oil and gas deposits are the result of the role and actions of ecosystems.

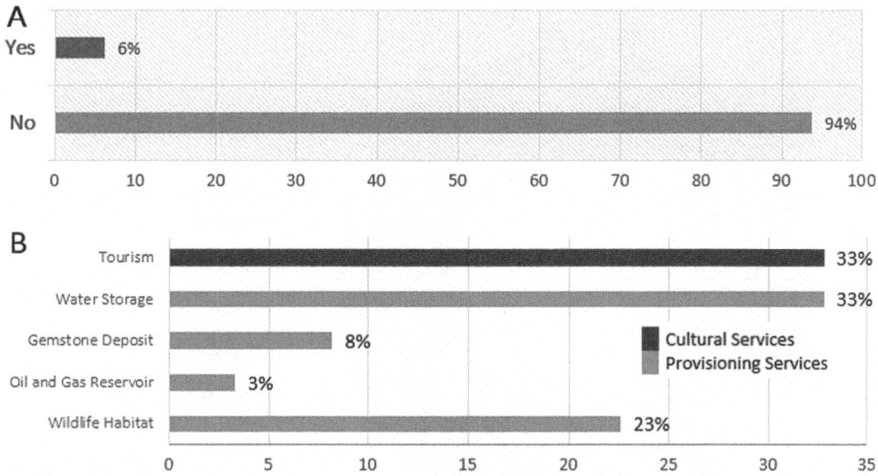

Fig. 4 Plotting the ecosystem services values by communities. **a** Shows the respondents results who recognize other value and importance of ecosystem services (Yes = affirmative, No = negative for N = 1000). **b** Reveal other values or importance recognized by respondents for N = 62). *Source* Elaborated by the authors

Figure 5 shows the respondents' perception regarding the reduction of ecosystem services in the last five years. The results show that more than 28% of the respondents stated that there is a reduction in the availability of water in the region. While more than 21% agree that wild animals have declined (probably evidenced by the limited number of catches and sightings). Nevertheless, more than 18% believe that wood has reduced, while more than 16% said that soil productivity has reduced significantly. A part of the respondents stated that forest fires with 9.63% of responses, precious stones (3.21%), frequency of rains (1.30%), and pests (1.30%) have reduced considerably. However, more than 63% of the respondents recognised that ecosystem services are limited, while 36.9% understand that services are infinite (Fig. 6a). Regarding the sustainable use of ecosystem services (Fig. 6a), more than 65% of respondents do not know if they are being exploited in a sustainable or unsustainable way. Nevertheless, 20% of the population realised that ecosystem services are being exploited in an unsustainable way, and 13.5% said they are being exploited in a sustainable way.

Table 4 presents the respondents' results on their receptivity to support any plan/ project to improve the planning and protection of local ecosystem services. The results show that 89.5% of the respondents would be receptive to support any plan that could improve and raise the sustainability of local ecosystem services, while 10.5% would not support it. However, if they demanded conditions to support these plans, about 58.3% do not know whether they would require conditions for their involvement. In addition, 27.1% responded negatively, while 14.6% responded positively.

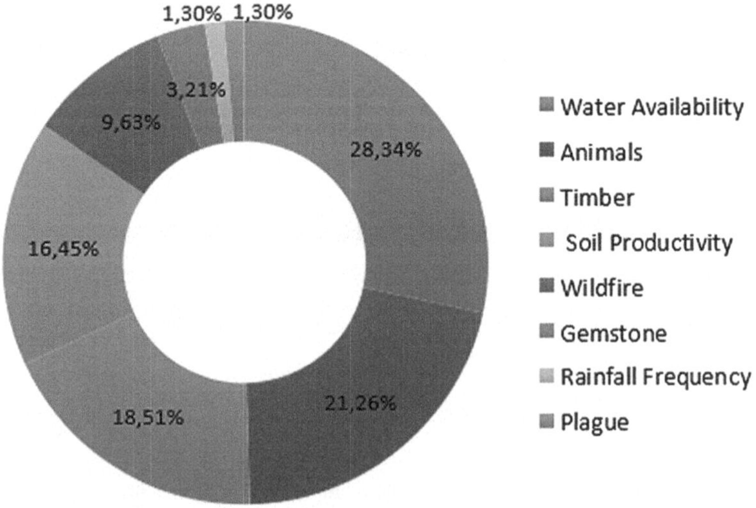

Fig. 5 Plotting respondent's responses on ecosystem services decreases in the last 5 years (n = 1000). *Source* Elaborated by the authors

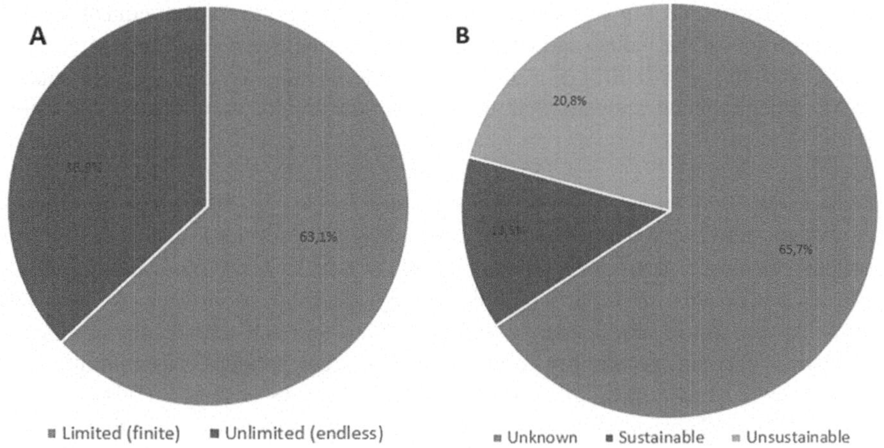

Fig. 6 Plotting respondents' perception on ecosystem services (A = Ecosystem services are limited or not; B = Ecosystem services exploitation for N = 1000). *Source* Elaborated by the authors

Table 4 Plotting respondents' receptivity to support the ecosystem services sustainability and conditions for engagement

		Desired conditions to support?			Total
		Unknown	No	Yes	
Receptivity to support ecosystem services sustainability	No	70	35	0	105 (10.5%)
	Yes	513	236	146	895 (89.5%)
Total		583 (58.3%)	271 (27.1%)	146 (14.6%)	1000 (100%)

Source Elaborated by the authors

3.4 *Ecosystem Services Perception: Pearson's Chi-Squared Test Result*

In this section, we assess whether the recognition of the values and importance of ecosystem services was associated with the level of education and if the respondents interested in supporting any local plan for the protection and sustainability of ecosystem services were related to: (a) reduction of ecosystem services (food and commerce), (b) reduced productivity and (c) reduced water availability. However, we describe our hypotheses as follows: H0: There is no association; H1: There is an association between categorical variables. In this context, Table 5 (supplementary material) presents the test value and the associated significance for the first case (education and recognition of values and importance of ecosystem values). This first analysis allows us to conclude that the chi-square test of independence showed that there is a statistically significant association between the level of education and the recognition of the values and importance of ecosystem services, since $p \leq 0.05$ [X2 (2) = 9.904; p < 0.007]. For Table 6 (supplementary material) the chi-square test of independence showed that there is a statistically significant association between receptivity to support ecosystem services sustainability and water available nowadays, since [X2 (1) = 4.255; p < 0.039]. Whereas Table 7 (supplementary material), the chi-square test of independence reveals a statistically significant association between receptivity to support ecosystem services sustainability and services decrease: gemstones, since [X2 (1) = 8.022; p < 0.005]. In the other analysis, with results in Table 8 (supplementary material), it allows us to state that the chi-square test of independence indicated that there is a statistically significant association between receptivity to support ecosystem services sustainability and services decrease: animals, since $p \leq 0.05$ [X2 (1) = 58.111; p < 0.001].

The result of Table 9 (supplementary material) allowed us to verify that there is a statistically significant association between receptivity to support ecosystem services sustainability and services decrease: water availability, since [X2 (1) = 61.455; p < 0.001]. At the same time, we have found a statistically significant association between receptivity to support ecosystem services sustainability and services decrease: Wood, since [X2 (1) = 43.200; p < 0.001] (Table 10, supplementary material). Finally, our analysis of the association of Person, for Table 11 (supplementary material) proved to be positive or statistically significant between the variables categories receptivity to support ecosystem services sustainability and services decrease: agriculture productivity, since [X2 (1) = 57.263; p < 0.001].

4 Discussion

We frame our discussion in the context of rural communities' livelihoods, protection, sustainability, and decision-making regarding local ecosystem services for the benefit of rural communities, the environment, decision-makers, and biodiversity. However, it is imperative to note and understand that the presentation, description, and analysis of Sects. 3.1 and 3.2 proves the importance of these variables in increasing the understanding and deepening of complex social dynamics and their advantages for the sustainability of ecosystem services. In this context, we understand that there is a relationship of direct dependence on respondents for local ecosystem services. This dependence verified at all social levels (e.g.: connection to spiritual and medicinal issues, food and construction of improvements, Table 3, Figs. 2, 3, 4) must be improved scientifically to raise evidence and mechanisms capable of triggering promising strategies for the sustainability of ecosystem services based on local communities. However, the approach was to understand how our results can be framed in the top-down and bottom-up decision-making process in order to guarantee and to integrate local communities as key actors and main vehicle in the decision-making process, planning, management and protection of community-based ecosystem services as ecosystem services are present in abundance at community or local level. With this, we understand that regardless of the approach we will adopt, local stakeholders should play an important role in the entire decision-making process. We found that over 93% of the respondents speak the local Emakhuwa language. This can be an indicator for planning local inclusive and comprehensive programs and actions. The planning and management of community-based ecosystem services should focus on integration and participation without exception for all local actors, especially those who exercise and perform family leadership roles, from youth, women, and men (Table 2). We understand that the stakeholders achieve results are more quickly when they clearly understand what is intended and when they see benefits from their involvement (Verburg et al. 2016). We can see the success of these actions (results) in the form and level of communication between decision makers and rural communities. Communication mechanisms based on the specific characteristics of communities can simplify steps in the decision-making process. Despite the insignificant and limited research that address this issue, there is scientific evidence concluding that various actions or interventions for the protection and sustainability of ecosystem services have failed due to the exclusion and ignorance of local assumptions (integration of key local groups, failure to overcome language barriers, communication and specific characteristics of local communities) (Martinez-Harms et al. 2015; Kabisch 2015; Thompson et al. 2016; Verburg et al. 2016; Keenan et al. 2019; Johnson et al. 2018; Morea 2019).

Our assessment concludes that about 17.1% of the respondents have no education (illiterate). Despite being relatively smaller, compared to respondents with schooling, it is essential to ensure that decisions strengthen local communities without access to education. They often consider this group a minority, unfortunate and excluded from many local participation and decision-making processes (Sangha et al. 2019). However, they play an important role at the local level. A relatively significant part is made up of community, traditional, local, spiritual, and religious leaders considered as advisers, decision makers, and influencers of local social, spiritual, and economic life. Empowering and strengthening education actions to communities can guarantee greater and better participation, engagement, consensus, cooperation, sharing of results and benefits (Cárcamo et al. 2014).

In Mozambique, most of the decision makers and rural communities' conflicts in the process of exploiting ecosystem services is the absence or limited understanding, if not ignorance on the part of decision makers, that local communities with their local or traditional knowledge, it has protected and played a key role for years in safeguarding and protecting local ecosystem services. Thus, disseminating knowledge to this group would ensure better understanding and knowledge about values, the importance of ecosystem services, consequently guaranteeing broad participation, transparency of the actors in sustainability and protection of ecosystem services. However, local decision makers choose the opposite as a strategy to manipulate communities in the area of acquiring areas to exploit ecosystem services, signing agreements, contracts and sharing local benefits with advantages for decision makers and losses for communities.

Around 89.5% (Table 3) of the respondent's practise agriculture, and 61.90% have agriculture as their main source of income. In addition, about 98% of the respondents build their homes and improvements using local material from forests, products from ecosystem services (bamboo, wood, fibre, grass, sticks) (Table 3). However, the percentages of the populations that directly exploit the services are quite significant and it may jeopardise the maintenance and continuity of ecosystem services, which can then condition the life and survival of the populations and biodiversity if considered that the population has been increasing considerably at an annual rate equal to 3% (INE 2017). Moreover, guaranteeing protection and sustainability based on communities and benefiting biodiversity should ensure that decisions directed at agricultural activities, human settlements, uses and exploitation of local services are integrated into a regional context of land management good sustainable practices to help improve the sustainability of ecosystem services. This path can guide agricultural actions to more sustainable levels such as conservation agriculture, in which it can significantly contribute to improving productivity and protecting important areas for biodiversity conservation and the production of local ecosystem services (wood, fibre, firewood, grass and others). At the same time, decisions must be based on the fundamental social and economic needs of rural communities (such as access and sharing of water resources, access to formal markets, health centres, transport systems to ensure the flow of products from communities, access banks (Table 3)).

Finally, intervene in alternative sources of subsistence for rural communities, promoting microcredit programs and financing community programs that favour other alternatives unlike agriculture and excessive collection of ecosystem services for their subsistence.

They reported about nine (9) diseases (Table 3). Diseases such as Malaria, Cholera and Tuberculosis (TB) were most cited. Diseases must deserve special attention and perfectly framework them in the decision-making process for the protection and sustainability of community-based ecosystem services. In-depth knowledge of local diseases, ways of dissemination, prevention and treatment when integrated into the management and sustainability system of local ecosystem services can reduce risks to the health of communities, flora and fauna, and ensure the continuity of the productivity of local ecosystem services. The communities less exposed to diseases can strengthen and participate effectively in the decision-making process as well as in the management of ecosystem services. Neglecting to framework issues related to health and diseases in the context of protection and sustainability of ecosystem services can have unprecedented consequences for the economy, communities, and biodiversity.

We found that provisioning services are the most preferential and indispensable in the daily lives of local communities (Figs. 3 and 4). We speculate that limitations in preferences or use of ecosystem services may be associated with the traditional way of life in rural communities, especially in Africa, in which the survival and well-being of communities is simply limited to the supply of basic daily needs (food and water) and protection (housing) (conditioned by widespread poverty and economic constraints). We also consider that the reasons bellow influence considerably in the non-exploitation of other ecosystem services and values:

(a) Absence or limited information, or knowledge of the importance and usefulness of local ecosystem services.
(b) Absence of local actions aimed at promoting ecosystem services as sources of survival and well-being for local communities.

Nevertheless, we understand that the limited level of recognition, attribution of importance, use or values of local ecosystem services by communities (Figs. 3 and 4) is framed in the methodological context of the research. Our approach made it possible to extract information from communities (blindly) without exposing or displaying them to a predefined list of local services. This procedure was the opposite of the one developed by the authors Zhang et al. (2016) and He et al. (2018). In their studies, the authors showed respondents a predefined list containing ecosystem services so that their respective values and importance were recognised and attributed. However, based on this procedure, they obtained from respondents all levels of ecosystem services (supporting, provisioning, regulating, and cultural) according to their needs.

Despite extracting critical and important information, we understand that the procedure applied by Zhang et al. (2016) and He et al. (2018) can be advantageous. In this context, so far, our procedure is relatively disadvantageous as it confronts a rural target audience with low literacy and education to, by themselves, recognise the values, importance, and preference of the services of the relatively complex ecosystems to be addressed. Nevertheless, it is imperative to test the methodology or procedure used by Zhang et al. (2016), and He et al. (2018) in our context to validate the premises.

However, we have found that limitations in the recognition of the values, usefulness and importance of ecosystem services by rural communities can trigger short and medium term (considering the rate of local population growth) quite significant impacts on ecosystems, allowing or facilitating the community promote inadequate, disorderly, or even destructive local actions to exploit ecosystem services as a pretext of contributing nothing, or being meaningless to their survival and well-being.

The balance found by the way respondents use provisioning services (Fig. 3) can be justified by the net balance of the gains and benefits that communities obtain in their day-to-day lives as well as satisfaction and full supply of their daily needs through the exploited services. However, we consider it necessary, urgent, and important that the context of decision-making frame the perceptions or limitations of the communities mainly in the planning process with the aim of reducing asymmetries and guaranteeing greater appreciation, enjoyment of ecosystem services by rural communities.

A small number of respondents (around 6%) recognised other values and importance of ecosystem services. The main services were provisioning and cultural services (Fig. 4). Nevertheless, we agree that the attributes/characteristics for the perception of other values and importance of ecosystem services by communities are associated with:

(a) Respondents with a higher or advanced level of education or education (secondary or university).
(b) Experience or knowledge of the local situation, and access to privileged information on local economic development.

To support these premises, note that the province of Cabo Delgado is one of the richest in the country in terms of natural resources. It has one of the largest graphite reserves in the world, estimated at 200 million tons of graphite, as well as one of the largest reserves of precious stones, mainly ruby, estimated production at $760 million per year. However, in 2012, it was considered the region with the largest reserves of natural gas in the world, estimated at 200 billion cubic feet. It has the second largest biodiversity conservation area in Mozambique, known as the Quirimbas National Park, with an estimated area of 7,500 km^2.

Finally, it presents two of the main and important rivers in Mozambique, the Messalo and Montepuez rivers. In this context, respondents are likely to have made a direct link between existing resources (most common and reported) and their value or importance. Thus, we understand that the respondents when affirming the importance or value "tourism and habitat for animals" were relating the actions or activities carried out within the Quirimbas National Park. However, in relation to the importance or value of "water deposits", they were relating the existence of the Montepuez and Messalo rivers. In addition, the importance or value "deposits of precious stones" were related to the graphite and ruby deposits identified in the region. As well as "gas reserves" related to the large gas reserves found in the region.

However, our brief defence revealed to us the existence of a spectrum and diversifying ways to perceive and understand the values or importance of local ecosystem services. Thus, we argue that in the decision-making process it is critical to rethink and discuss how these groups can integrate and play a role in the dissemination, knowledge promotion or regional information on the values and importance of ecosystem services to local communities (as agents of change) with the aim of raising community awareness. However, researchers have found different perceptions in relation to the decline in ecosystem services (Fig. 5) in the last five years. They mostly mentioned reduction of water availability for agricultural practices, reduction of biodiversity (as a source of food and commercial activity), reduction of forest resources, and reduction of soil availability with about 28%, 21%, 18% and 16% respectively. However, there is limited evidence or local scientific studies that prove or corroborate the perceptions of communities.

Nevertheless, an assessment developed by Mucova et al. (2018) found that the Quirimbas National Park (Cabo Delgado) lost about 301,761.7 ha, corresponding to 41.67% of the total QNP coverage land in the last 38 years. Which means around 10.42% of the vegetated lands every 10 years corresponding to 74,431.1 (ha) (Mucova et al. 2018). The main causes are associated with intensive agriculture, human settlements, population growth, illegal exploitation of forest resources and miners (Mucova et al. 2018). In this context, we can establish some relationship between the communities' perceptions and the scientific evidence found and consider that the reduction of ecosystem services is framed in the local socioeconomic context and survival of rural communities, whose exploitation is intensive, unruly and unsustainable. Thus, we found that differences in perceptions about reductions in ecosystem services could provide an important understanding of the role of different community groups in the exploitation of ecosystem services as well as the way they react, visualize or explain the decreases in the local ecosystem services. Deepening this theme can support the design of specific activities that converge to different community needs and consequently promote greater social inclusion and local groups.

Decisions and paths for the planning, exploitation and sustainability of ecosystem services based on rural communities must focus on their fundamental needs based on their beliefs as well as on their commitments for engagement, protection, and safeguarding. Figure 6 shows how local policies for protecting ecosystem services should be analysed and discussed. About 65% do not know whether local services are being operated in a sustainable or unsustainable way. Still, more than 35% understand that the resources are infinite. Both groups are critical in decision-making as they can play an important role in the success or failure of any sustainability plan and action for ecosystem services. However, reducing the number of partners or communities with limited information on ecosystem services to minimums or absolute zero should be the way forward. Our evaluations made it possible to understand that the higher the level of education of local communities the better the comprehension and understanding of the communities on the importance, values of ecosystem services. The Person test revealed a positive correspondence between the recognition of the values and importance of ecosystem services and level of education. At the same time, the tests showed that there is a relationship between receptivity to support ecosystem services sustainability and reduction of local ecosystem services. The analysis shows that the community's social characteristics significantly influenced the form and mode proposed for the protection of local ecosystem services, hence the sustainability of ecosystem services will be achieved when integrated and considering all social variables that may contribute to success or failure of local plans.

5 Conclusion

As this paper has demonstrated, among four ecosystem services (supporting, provisioning, regulating, and cultural) the sampled populations only recognised two services (provisioning and cultural) as being directly relevant to them, revealing relatively low levels of knowledge, value, and importance of their local services.

Since provisioning services seem to be the most prevalent, local communities should further exploit them, for instance:

(a) animal hunting (food and commerce),
(b) firewood (for charcoal production),
(c) wood and bamboo (for construction services),
(d) water (for consumption and agriculture),
(e) medicinal plants (for healing diseases and spiritual issues) and
(f) prospecting for precious stones, an activity to be exercised with care, due to its potential environmental impacts.

This paper has some limitations. The first one is the fact that the recognition and the attribution of values and importance of local ecosystem services are framed in the context of population literacy and in the methodological process applied in the research. In addition, the size of the sample and its geographical location cannot be representative of trends across the country.

Despite these limitations the study has led to useful insights. For instance, the reduction of water availability for agricultural practice, reduction of biodiversity (as a source of food and commercial activity), reduction of forest resources, and reduction of soil productivity are recognised and considered the main threats to the reduction of ecosystem services in the last five years. Also, it identified the fact that the local socioeconomic and survival context, whose exploitation of ecosystems is intensive, unruly, and unsustainable, frames and associates with the causes for reductions in services.

Nevertheless, a worrying trend is that most of the population does not seem to understand whether ecosystem services are being exploited in a sustainable or unsustainable way. Our evaluations have concluded that the higher the level of education of local communities the better their comprehension and understanding of local services. The Person test revealed a positive correspondence between the recognition of values, the importance of ecosystem services and level of education. It also allowed us to verify the existence of a relationship between receptivity to support ecosystem services sustainability and the reduction of local ecosystem services. However, decision-making for the protection and sustainability of ecosystem services based on rural communities should focus on the economic and social fabric of populations (agricultural activities, beliefs, human settlements, uses of ecosystem services) and frame it in the local context and regional good sustainable land management practices to help improve the sustainability of ecosystem services.

The analyses allow us to verify that the community's social characteristics significantly influenced the form and way in which it is intends to approach plan and protect local ecosystem services. There is a chance that the sustainability of local ecosystem services will be positively pursued when all social variables are integrated and considered likely to contribute to the success of local plans.

Acknowledgements First and corresponding author would like to deeply thank WWF Russell E. Train Education for Nature Program (EFN) for support and fellowship granted.

Thanks are due for the financial support to CESAM (UID/AMB/50017/2019), to FCT/MCTES through national funds, and the co-funding by the FEDER, within the PT2020 Partnership Agreement and Compete 2020.

Declaration of Competing Interest

The authors declare that they have no known competing financial interests or personal relationships that could have appeared to influence the work reported in this paper.

Appendix

Supplementary Material: Pearson's chi-squared test tables.

See Tables 5, 6, 7, 8, 9, 10, and 11.

Table 5 Education * recognition of values and importance of ecosystem values crosstabulation

			Recognition of values and importance of ecosystem services		Total
			No	Yes	
Education	No education	Count	152	19	171
		Expected count	160,4	10,6	171,0
		% within education	88.9%	11.1%	100.0%
		% within recognition of values and importance of ecosystem services	16.2%	30.6%	17.1%
	Primary Education	Count	676	34	710
		Expected count	666,0	44,0	710,0
		% within education	95.2%	4.8%	100.0%
		% within recognition of values and importance of ecosystem services	72.1%	54.8%	71.0%
	Secondary Education and University Degree	Count	110	9	119
		Expected count	111,6	7,4	119,0
		% within education	92.4%	7.6%	100.0%
		% within recognition of values and importance of ecosystem services	11.7%	14.5%	11.9%
Total		Count	938	62	1000
		Expected count	938,0	62,0	1000,0
		% within education	93.8%	6.2%	100.0%
		% within recognition of values and importance of ecosystem services	100.0%	100.0%	100.0%

Chi-square tests

	Value	df	Asymp. Sig. (2-sided)
Pearson chi-square	9,904[a]	2	0.007
Likelihood ratio	8,804	2	0.012
Linear-by-linear association	2,745	1	0.098
Number of valid cases	1000		

Source Elaborated by the authors

[a]0 cells (0.0%) have expected count less than 5. The minimum expected count is 7,38

Table 6 Receptivity to support ecosystem services sustainability * water available nowadays crosstabulation

			Water available nowadays		Total
			No	Yes	
Receptivity to support ecosystem services sustainability	No	Count	105	0	105
		Expected count	101,3	3,7	105,0
		% within receptivity to support ecosystem services sustainability	100.0%	0.0%	100.0%
		% within water available nowadays	10.9%	0.0%	10.5%
	Yes	Count	860	35	895
		Expected count	863,7	31,3	895,0
		% within receptivity to support ecosystem services sustainability	96.1%	3.9%	100.0%
		% within water available nowadays	89.1%	100.0%	89.5%
Total		Count	965	35	1000
		Expected count	965,0	35,0	1000,0
		% within receptivity to support ecosystem services sustainability	96.5%	3.5%	100.0%
		% within water available nowadays	100.0%	100.0%	100.0%

Chi-square tests

	Value	df	Asymp. Sig. (2-sided)	Asymp. Sig. (2-sided)	Asymp. Sig. (1-sided)
Pearson chi-square	4,255[a]	1	0.039		
Continuity Correction[b]	3,176	1	9.075		
Likelihood ratio	7,913	1	0.005		
Fisher's exact test				0.043	0.019
Linear-by-linear association	4,251	1	0.039		
Number of valid cases	1000				

Source Elaborated by the authors

[a]1 cells (25.0%) have expected count less than 5. The minimum expected count is 3,68

[b]Computed only for a 2 × 2 table

Table 7 Receptivity to support ecosystem services sustainability * services decrease: gemstones crosstabulation

			Services decrease: Gemstones		Total
			No	Yes	
Receptivity to support ecosystem services sustainability	No	Count	105	0	105
		Expected count	98,3	6,7	105,0
		% within receptivity to support ecosystem services sustainability	100.0%	0.0%	100.0%
		% within resources decrease: Gemstones	11.2%	0.0%	10.5%
	Yes	Count	831	64	895
		Expected count	837,7	57,3	895,0
		% within receptivity to support ecosystem services sustainability	92.8%	7.2%	100.0%
		% within resources decrease: Gemstones	88.8%	100.0%	89.5%
Total		Count	936	64	1000
		Expected count	936,0	64,0	1000,0
		% within receptivity to support ecosystem services sustainability	93.6%	6.4%	100.0%
		% within resources decrease: Gemstones	100.0%	100.0%	100..0%

Chi-square tests

	Value	df	Asymp. Sig. (2-sided)	Asymp. Sig. (2-sided)	Asymp. Sig. (1-sided)
Pearson chi-square	8,022[a]	1	0.005		
Continuity correction[b]	6,872	1	0.009		
Likelihood ratio	14,703	1	0.000		
Fisher's exact test				0.001	0.001
Linear-by-linear association	8,014	1	0.005		
Number of valid cases	1000				

Source Elaborated by the authors

[a]0 cells (00.0%) have expected count less than 5. The minimum expected count is 6,72

[b]Computed only for a 2 × 2 table

Table 8 Receptivity to support ecosystem services sustainability * services decrease: animals crosstabulation

			Services decrease: Animals		Total
			No	Yes	
Receptivity to support ecosystem services sustainability	No	Count	97	8	105
		Expected count	60,5	44,5	105,0
		% within receptivity to support ecosystem services sustainability	92.4%	7.6%	100.0%
		% within resources decrease: animals	16.8%	1.9%	10.5%
	Yes	Count	479	416	895
		Expected count	515,5	379,5	895,0
		% within receptivity to support ecosystem services sustainability	53.5%	46.5%	100.0%
		% within resources decrease: animals	83.2%	98.1%	89.5%
Total		Count	576	424	1000
		Expected count	576,0	424,0	1000,0
		% within receptivity to support ecosystem services sustainability	57.6%	42.4%	100.0%
		% within resources decrease: animals	100.0%	100.0%	100.0%

Chi-square tests

	Value	df	Asymp. Sig. (2-sided)	Asymp. Sig. (2-sided)	Asymp. Sig. (1-sided)
Pearson chi-square	58,111[a]	1	0.000		
Continuity correction[b]	56,531	1	0.000		
Likelihood ratio	70,239	1	0.000		
Fisher's exact test				0.000	0.000
Linear-by-linear association	58,053	1	0.000		
Number of valid cases	1000				

Source Elaborated by the authors
[a]0 cells (00.0%) have expected count less than 5. The minimum expected count is 44,52
[b]Computed only for a 2 × 2 table

Table 9 Receptivity to support ecosystem services sustainability * services decrease: water availability crosstabulation

			Services decrease: Water availability		Total
			No	Yes	
Receptivity to support ecosystem services sustainability	No	Count	8	97	105
		Expected count	45,7	59,3	105,0
		% within receptivity to support ecosystem services sustainability	7.6%	92.4%	100.0%
		% within services decrease: water availability	1.8%	17.2%	10.5%
	Yes	Count	427	468	895
		Expected count	389,3	505,7	895,0
		% within receptivity to support ecosystem services sustainability	47.7%	52.3%	100.0%
		% within services decrease: water availability	98.2%	82.8%	89.5%
Total		Count	435	565	1000
		Expected count	435,0	565,0	1000,0
		% within receptivity to support ecosystem services sustainability	43.5%	56.5%	100.0%
		% within services decrease: water availability	100.0%	100.0%	100.0%

Chi-square tests

	Value	df	Asymp. Sig. (2-sided)	Asymp. Sig. (2-sided)	Asymp. Sig. (1-sided)
Pearson chi-square	61,455[a]	1	0.000		
Continuity correction[b]	59,835	1	0.000		
Likelihood ratio	73,925	1	0.000		
Fisher's exact test				0.000	0.000
Linear-by-linear association	61,393	1	0.000		
Number of valid cases	1000				

Source Elaborated by the authors

[a]0 cells (00.0%) have expected count less than 5. The minimum expected count is 45,68

[b]Computed only for a 2 × 2 table

Table 10 Receptivity to support ecosystem services sustainability * services decrease: Wood cross tabulation

			Services decrease: Wood		Total
			No	Yes	
Receptivity to support ecosystem services sustainability	No	Count	97	8	105
		Expected count	66,3	38,7	105,0
		% within receptivity to support ecosystem services sustainability	92.4%	7.6%	100.0%
		% within services decrease: Wood	15.4%	2.2%	10.5%
	Yes	Count	534	361	895
		Expected count	564,7	330,3	895,0
		% within receptivity to support ecosystem services sustainability	59.7%	40.3%	100.0%
		% within services decrease: Wood	84.6%	97.8%	89.5%
Total		Count	631	369	1000
		Expected count	631,0	369,0	1000,0
		% within receptivity to support ecosystem services sustainability	63.1%	36.9%	100.0%
		% within services decrease: Wood	100.0%	100.0%	100.0%

Chi-square tests

	Value	df	Asymp. Sig. (2-sided)	Asymp. Sig. (2-sided)	Asymp. Sig. (1-sided)
Pearson chi-square	43,200[a]	1	0.000		
Continuity correction[b]	41,806	1	0.000		
Likelihood ratio	53,194	1	0.000		
Fisher's exact test				0.000	0.000
Linear-by-linear association	43,157	1	0.000		
Number of valid cases	1000				

Source Elaborated by the authors
[a]0 cells (00.0%) have expected count less than 5. The minimum expected count is 38,75
[b]Computed only for a 2 × 2 table

Table 11 Receptivity to support ecosystem services sustainability * services decrease: Agriculture productivity crosstabulation

			Services decrease: Agriculture productivity		Total
			No	Yes	
Receptivity to support ecosystem services sustainability	No	Count	105	0	105
		Expected Count	70,6	34,4	105,0
		% within receptivity to support ecosystem services sustainability	100.0%	0.0%	100.0%
		% within services decrease: agriculture productivity	15.6%	0.0%	10.5%
	Yes	Count	567	328	895
		Expected count	601,4	293,6	895,0
		% within receptivity to support ecosystem services sustainability	63.4%	36.6%	100.0%
		% within services decrease: agriculture productivity	84.4%	100.0%	89.5%
Total		Count	672	328	1000
		Expected count	672,0	328,0	1000,0
		% within receptivity to support ecosystem services sustainability	67.2%	32.8%	100..0%
		% within services decrease: agriculture productivity	100.0%	100.0%	100.0%

Chi-square tests

	Value	df	Asymp. Sig. (2-sided)	Asymp. Sig. (2-sided)	Asymp. Sig. (1-sided)
Pearson chi-square	57,263[a]	1	0.000		
Continuity correction[b]	55,612	1	0.000		
Likelihood ratio	89,376	1	0.000		
Fisher's exact test				0.000	0.000
Linear-by-linear association	57,205	1	0.000		
Number of valid cases	1000				

Source Elaborated by the authors
[a]0 cells (00.0%) have expected count less than 5. The minimum expected count is 34,44
[b]Computed only for a 2 × 2 table

References

Barbier EB, Hacker SD, Koch EW, Stier AC, Silliman BR (2011) Estuarine and coastal ecosystems and their services. In: Treatise on estuarine and coastal science, pp 109–127. https://doi.org/10.1016/b978-0-12-374711-2.01206-7

Blasiak R, Yagi N, Kurokura H, Ichikawa K, Wakita K, Mori A (2015) Marine ecosystem services: perceptions of indispensability and pathways to engaging citizens in their sustainable use. Mar Policy 61:155–163. https://doi.org/10.1016/j.marpol.2015.08.005

Blythe J, Armitage D, Alonso G, Campbell D, Esteves Dias AC, Epstein G, Marschke M, Nayak P (2020) Frontiers in coastal well-being and ecosystem services research: a systematic review. Ocean Coast Manage:105028. https://doi.org/10.1016/j.ocecoaman.2019.105028

Boerema A, Meire P (2017) Management for estuarine ecosystem services: a review. Ecol Eng 98:172–182. https://doi.org/10.1016/j.ecoleng.2016.10.051

Cabana D, Ryfield F, Crowe T, Brannigan J (2020) Evaluating and communicating cultural ecosystem services. Ecosyst Serv 42:101085. https://doi.org/10.1016/j.ecoser.2020.101085

Campbell E, Marks R, Conn C (2020) Spatial modeling of the biophysical and economic values of ecosystem services in Maryland, USA. Ecosyst Serv 43: 101093. https://doi.org/10.1016/j.ecoser.2020.101093

Cárcamo PF, Garay-Flühmann R, Gaymer CF (2014) Collaboration and knowledge networks in coastal resources management: how critical stakeholders interact for multiple-use marine protected area implementation. Ocean Coast Manag 91:5–16. https://doi.org/10.1016/j.ocecoaman.2014.01.007

Chakraborty S, Gasparatos A (2019) Community values and traditional knowledge for coastal ecosystem services management in the "satoumi" seascape of Himeshima island, Japan. Ecosyst Serv 37: 100940. https://doi.org/10.1016/j.ecoser.2019.100940

Depellegrin D, Menegon S, Gusatu L, Roy S, Misiunė I (2020) Assessing marine ecosystem services richness and exposure to anthropogenic threats in small sea areas: a case study for the Lithuanian sea space. Ecol Ind 108: 105730. https://doi.org/10.1016/j.ecolind.2019.105730

Ellis EC, Pascual U, Mertz O (2019) Ecosystem services and nature's contribution to people: negotiating diverse values and trade-offs in land systems. Curr Opin Environ Sustain 38:86–94. https://doi.org/10.1016/j.cosust.2019.05.001

Filho WL, Barbir J, Nagy GJ, Sima M, Kalbus A, Paletta A, Nikolova M, Bonoli A (2020) Reviewing the role of ecosystems services in the sustainability of the urban environment: a multi-country analysis. J Cleaner Prod:121338. https://doi.org/10.1016/j.jclepro.2020.121338

Grima N, Singh SJ (2020) The self-(in)sufficiency of the Caribbean: ecosystem services potential Index (ESPI) as a measure for sustainability. Ecosyst Serv 42: 101087. https://doi.org/10.1016/j.ecoser.2020.101087

Hardaker A, Pagella T, Rayment M (2020) Integrated assessment, valuation and mapping of ecosystem services and dis-services from upland land use in Wales. Ecosyst Serv 43: 101098. https://doi.org/10.1016/j.ecoser.2020.101098

Hasan SS, Zhen L, Miah MG, Ahamed T, Samie A (2020) Impact of land use change on ecosystem services: a review. Environ Dev:100527. https://doi.org/10.1016/j.envdev.2020.100527

Havinga I, Bogaart PW, Hein L, Tuia D (2020) Defining and spatially modelling cultural ecosystem services using crowdsourced data. Ecosyst Serv 43: 101091. https://doi.org/10.1016/j.ecoser.2020.101091

He S, Gallagher L, Su Y, Wang L, Cheng H (2018) Identification and assessment of ecosystem services for protected area planning: a case in rural communities of Wuyishan national park pilot. Ecosyst Serv 31:169–180. https://doi.org/10.1016/j.ecoser.2018.04.001

Hugé J, Rochette AJ, de Béthune S, Parra Paitan CC, Vanderhaegen K, Vandervelden T, Janssens I, Janssens de Bisthoven L (2020) Ecosystem services assessment tools for African Biosphere Reserves: a review and user-informed classification. Ecosys Serv 42: 101079. https://doi.org/10.1016/j.ecoser.2020.101079

Instituto Nacional de Estatística—INE (2017) Recenseamento Geral da População e Habitação-Divulgação de Resultados Preliminares do IV RGPH 2017. Maputo, Mozambique. http://www.ine.gov.mz/operacoes-estatisticas/censos/censo-2007/censo-2017/divulgacao-de-resultados-preliminares-do-iv-rgph-2017.pdf/view

IPBES (2018) The IPBES regional assessment report on biodiversity and ecosystem services for Africa. In: Archer E, Dziba L, Mulongoy KJ, Maoela MA, Walters M (eds) Secretariat of the intergovernmental science-policy platform on biodiversity and ecosystem services. Bonn, Germany, pp 492

IPBES (2019a) Summary for policymakers of the global assessment report on biodiversity and ecosystem services of the Intergovernmental Science-Policy Platform on Biodiversity and Ecosystem Services. In: Díaz S, Settele J, Brondízio ES, Ngo HT, Guèze M, Agard J, Arneth A, Balvanera P, Brauman KA, Butchart SHM, Chan KMA, Garibaldi LA, Ichii K, Liu J, Subramanian SM, Midgley GF, Miloslavich P, Molnár Z, Obura D, Pfaff A, Polasky S, Purvis A, Razzaque J, Reyers B, Roy Chowdhury R, Shin YJ, Visseren-Hamakers IJ, Willis KJ, Zayas CN (eds) IPBES secretariat, Bonn, Germany, pp 56. https://doi.org/10.5281/zenodo.3553579

IPBES (2019b) Global assessment report on biodiversity and ecosystem services of the Intergovernmental Science-Policy Platform on Biodiversity and Ecosystem Services. In: Brondizio ES, Settele J, Díaz S, Ngo HT (eds) IPBES secretariat, Bonn, Germany

Johnson MK, Lien AM, Sherman NR, López-Hoffman L (2018) Barriers to PES programs in Indigenous communities: a lesson in land tenure insecurity from the Hopi Indian reservation. Ecosyst Serv 32:62–69. https://doi.org/10.1016/j.ecoser.2018.05.009

Kabisch N (2015) Ecosystem service implementation and governance challenges in urban green space planning—the case of Berlin, Germany. Land Use Policy 42:557–567. https://doi.org/10.1016/j.landusepol.2014.09.005

Keenan RJ, Pozza G, Fitzsimons JA (2019) Ecosystem services in environmental policy: barriers and opportunities for increased adoption. Ecosyst Serv 38: 100943. https://doi.org/10.1016/j.ecoser.2019.100943

Li J, Bai Y, Alatalo JM (2020) Impacts of rural tourism-driven land use change on ecosystems services provision in Erhai Lake Basin, China. Ecosyst Serv 42: 101081. https://doi.org/10.1016/j.ecoser.2020.101081

Liu L, Wang Z, Wang Y, Zhang Y, Shen J, Qin D, Li S (2019) Trade-off analyses of multiple mountain ecosystem services along elevation, vegetation cover and precipitation gradients: a case study in the Taihang Mountains. Ecol Ind 103:94–104. https://doi.org/10.1016/j.ecolind.2019.03.034

Martin CL, Momtaz S, Gaston T, Moltschaniwskyj NA (2020) Estuarine cultural ecosystem services valued by local people in New South Wales, Australia, and attributes important for continued supply. Ocean Coast Manag 190: 105160. https://doi.org/10.1016/j.ocecoaman.2020.105160

Martinez-Harms MJ, Bryan BA, Balvanera P, Law EA, Rhodes JR, Possingham HP, Wilson KA (2015) Making decisions for managing ecosystem services. Biol Cons 184:229–238. https://doi.org/10.1016/j.biocon.2015.01.024

Millennium Ecosystem Assessment (MEA) (2005) Ecosystems and human well-being: synthesis. Island Press, Washington, DC

Morea JP (2019) A framework for improving the management of protected areas from a social perspective: the case of Bahía de San Antonio Protected Natural Area, Argentina. Land Use Policy 87: 104044. https://doi.org/10.1016/j.landusepol.2019.104044

Mucova SAR, Filho WL, Azeiteiro UM, Pereira MJ (2018) Assessment of land use and land cover changes from 1979 to 2017 and biodiversity and land management approach Quirimbas National Park, Northern Mozambique, Africa. Glob Ecol Conserv:16. https://doi.org/10.1016/j.gecco.2018.e00447

Müller F, Bicking S, Ahrendt K, Kinh Bac D, Blindow I, Fürst C, Schubert H, Zeleny J (2020) Assessing ecosystem service potentials to evaluate terrestrial, coastal, and marine ecosystem types in Northern Germany—an expert-based matrix approach. Ecol Ind 112: 106116. https://doi.org/10.1016/j.ecolind.2020.106116

Nahuelhual L, Vergara X, Bozzeda F, Campos G, Subida MD, Outeiro L, Villasante S, Fernández M (2020) Exploring gaps in mapping marine ecosystem services: a benchmark analysis. Ocean Coast Manag 192: 105193. https://doi.org/10.1016/j.ocecoaman.2020.105193

Obeng EA, Oduro KA, Obiri BD, Abukari H, Guuroh RT, Djagbletey GD, Appiah-Korang J, Appiah M (2019) Impact of illegal mining activities on forest ecosystem services: local communities' attitudes and willingness to participate in restoration activities in Ghana. Heliyon 5(10): e02617. https://doi.org/10.1016/j.heliyon.2019.e02617

Pogue SJ, Kröbel R, Janzen HH, Beauchemin KA, Legesse G, de Souza DM, Byrne J, McAllister TA (2017) Beef production and ecosystem services in Canada's prairie provinces: a review. Agric Syst 166:152–172. https://doi.org/10.1016/j.agsy.2018.06.011

Quevedo JM, Uchiyama Y, Kohsaka R (2020) Perceptions of the seagrass ecosystems for the local communities of Eastern Samar, Philippines: preliminary results and prospects of blue carbon services. Ocean Coast Manag 191: 105181. https://doi.org/10.1016/j.ocecoaman.2020.105181

Reynolds C, Venter N, Cowie BW, Marlin D, Mayonde S, Tocco C, Byrne MJ (2020) Mapping the socio-ecological impacts of invasive plants in South Africa: are poorer households with high ecosystem service use most at risk? Ecosyst Serv 42: 101075. https://doi.org/10.1016/j.ecoser.2020.101075

Riegels N, Lynggaard-Jensen A, Jensen JK, Gerner NV, Anzaldua G, Mark O, Butts M, Birk S (2020) Making the ecosystem services approach operational: a case study application to the Aarhus River, Denmark. Sci Environ:135836. https://doi.org/10.1016/j.scitotenv.2019.135836

Sangha KK, Russell-Smith J, Costanza R (2019) Mainstreaming indigenous and local communities' connections with nature for policy decision-making. Glob Ecol Conserv: e00668. https://doi.org/10.1016/j.gecco.2019.e00668

Schirpke U, Leitinger G, Tasser E, Rüdisser J, Fontana V, Tappeiner U (2020) Functional spatial units are fundamental for modelling ecosystem services in mountain regions. Appl Geogr 118: 102200. https://doi.org/10.1016/j.apgeog.2020.102200

Shi Y, Shi D, Zhou L, Fang R (2020) Identification of ecosystem services supply and demand areas and simulation of ecosystem service flows in Shanghai. Ecol Ind 115: 106418. https://doi.org/10.1016/j.ecolind.2020.106418

TEEB (2010) The Economics of Ecosystems and Biodiversity Ecological and Economic Foundations (TEEB). In: Pushpam Kumar (ed) The economics of ecosystems and biodiversity ecological and economic foundations. Earthscan, London and Washington, DC

Thompson BS, Primavera JH, Friess DA (2016) Governance and implementation challenges for mangrove forest Payments for Ecosystem Services (PES): empirical evidence from the Philippines. Ecosyst Serv 23:146–155. https://doi.org/10.1016/j.ecoser.2016.12.007

Tusznio J, Pietrzyk-Kaszyńska A, Rechciński M, Olszańska A, Grodzińska-Jurczak M (2020) Application of the ecosystem services concept at the local level—challenges, opportunities, and limitations. Ecosyst Serv 42: 101077. https://doi.org/10.1016/j.ecoser.2020.101077

Van Teijlingen E, Hundley V (2002) The importance of pilot studies. Nursing Standard (Royal College of Nursing (Great Britain): 1987). 16:33–6. https://doi.org/10.7748/ns2002.06.16.40.33.c3214

Verburg R, Selnes T, Verweij P (2016) Governing ecosystem services: national and local lessons from policy appraisal and implementation. Ecosyst Serv 18:186–197. https://doi.org/10.1016/j.ecoser.2016.03.006

Xu C, Jiang W, Huang Q, Wang Y (2020) Ecosystem services response to rural-urban transitions in coastal and island cities: a comparison between Shenzhen and Hong Kong, China. J Cleaner Prod:121033. https://doi.org/10.1016/j.jclepro.2020.121033

Xu X, Chen M, Yang G, Jiang B, Zhang J (2020b) Wetland ecosystem services research: a critical review. Glob Ecol Conserv. https://doi.org/10.1016/j.gecco.2020.e01027

Yang Q, Liu G, Hao Y, Zhang L, Giannetti BF, Wang J, Casazza M (2019) Donor-side evaluation of coastal and marine ecosystem services. Water Res 166: 115028. https://doi.org/10.1016/j.watres.2019.115028

Zhang W, Kato E, Bhandary P, Nkonya E, Ibrahim HI, Agbonlahor M, Ibrahim HY, Cox C (2016) Awareness and perceptions of ecosystem services in relation to land use types: evidence from rural communities in Nigeria. Ecosyst Serv 22:150–160. https://doi.org/10.1016/j.ecoser.2016.10.011

Zhou L, Guan D, Huang X, Yuan X, Zhang M (2020) Evaluation of the cultural ecosystem services of wetland park. Ecol Ind 114: 106286. https://doi.org/10.1016/j.ecolind.2020.106286

Addressing Rural Community's Risk Perceptions, Vulnerability, and Adaptation Strategies to Climate Change in Mozambique, Africa

Serafino Afonso Rui Mucova, Ulisses Miranda Azeiteiro, Antonio García Vinuesa, Walter Leal Filho, and Mário Jorge Verde Pereira

1 Introduction

Climate change has reached quite challenging social, political, economic, and environmental dimensions of the century. These challenges and concerns have been increasing as progressive trends in increasing concentrations of greenhouse gases have been confirmed, which consequently have significantly altered patterns of precipitation, temperature, and extreme global events (Hardwick et al. 2010; Allen et al. 2018). For instance, concentrations of nitrogen dioxide (N_2O) in the atmosphere have increased by more than 20% from 270 parts per billion (ppb) in 1750 to 331 ppb in 2018, one of the highest rates in the past 50 years (Prinn et al. 2018; Tian et al. 2020). While carbon dioxide (CO_2) reached 409.8 ± 0.1 parts per million (ppm), at 2.5 ± 0.1 ppm increase from 2018; and CH4 reached 1866.6 ± 0.9 ppb in 2019, a 9.2 ± 0.9 ppb increase from 2018 (Dunn et al. 2020). These dramatic increases in greenhouse gases forced record increases in global warming, where the average global surface temperature reached 0.87 °C above pre-industrial values in the decade of 2006–2015 (Hoegh-Guldberg et al. 2018).

S. A. R. Mucova (✉)
Faculty of Natural Sciences, Lúrio University, Pemba 958, Mozambique

S. A. R. Mucova · U. M. Azeiteiro · M. J. V. Pereira
Department of Biology & CESAM (Center for Environmental and Marine Studies), University of Aveiro, Aveiro, Portugal

A. G. Vinuesa
Facultade de Ciencias da Educación, Universidade de Santiago de Compostela, Santiago de Compostela, Spain

W. L. Filho
Hamburg University of Applied Sciences, Hamburg, Germany

W. L. Filho
School of Science and the Environment, Manchester Metropolitan University, Manchester, England

W. Leal Filho et al. (eds.), *Sustainability in Natural Resources Management and Land Planning*, World Sustainability Series,
https://doi.org/10.1007/978-3-030-76624-5_17

Consequently, the world has become conditioned to those factors and it has been experiencing relatively insecure and fragile standards of living and survival. Extreme events become frequent and more violent, changing the geography of the main survival assets, and threatening human lives, especially in unprepared countries or regions with limited actions to cope (Yaduvanshi et al. 2020).

Africa is repeatedly cited as one of the regions in the world less prepared to deal with the impacts of climate change and extreme cyclical events, and it is also the region that will suffer the most due to the global dynamics caused by the increase in greenhouse gases (Masipa 2017; Filho et al. 2020; Tegegne et al. 2020). The dimensions of the shocks are expected to vary from region to region (Ayanlade et al. 2018; Baarsch et al. 2020; Dube and Nhamo 2020), but also to socio-economic groups, with rural communities being mostly affected (Nkomwa et al. 2014; Evariste et al. 2018; Adzawla et al. 2019; Assan et al. 2020).

Rural communities represent about 56% of the continent's population (United Nations 2019). This group is particularly vulnerable and most affected by extreme events due to its geographical location, income activities and biophysical assets dependent on environmental factors, general poverty, and reduced capacity for adaptation (Artur and Hilhorst 2012, 2014; Dendir and Simane 2019; Zakaria et al. 2020; Atuoye et al. 2020).

These communities experience challenging survival implications, as extreme climatic events, alter the agricultural production cycle (Adhikari et al. 2015; Alaanuloluwa et al. 2020), limit the ability to acquire water resources for agriculture (Gebrechorkos et al. 2019; Ahmed 2020; Olabanji et al. 2020), impose marked habitability risks (Hoegh-Guldberg et al. 2018), reduce areas of important biophysical assets for survival (Kry et al. 2020; Ogada et al. 2020), change the migratory patterns of communities (Anderson and Silva 2020), increase poverty (Zhou et al. 2017; Azzarri and Signorelli 2020), raise social and gender differences (Antwi-Agyei et al. 2020; Tsige et al. 2020), alter living standards (Hoegh-Guldberg et al. 2018), contribute to increases in food insecurity (Braitstein et al. 2017; Nyiwul 2020), and they undermine the productivity of rural ecosystem services ais (Hoegh-Guldberg et al. 2018).

The combined effects of the consequences and impacts caused by climate change, as well as extreme climatic events, considerably increase the vulnerability of rural communities due to their high exposure, sensitivity, and reduced capacities to cope with environmental shocks (Dumenu and Obeng 2016; Wichern et al. 2019; Pasquini et al. 2020).

Mozambique is one of the Africa's most vulnerable countries to climate change and extreme weather events (characterised by tropical cyclones, floods, and droughts). It is the 31st most vulnerable and the 32nd least ready to adapt to climate change (WMO 2019). Since 1911, Mozambique has recorded more than 75 (seventy-five) extreme climatic events, the most frequent and most violent have been observed in the last 20 years (INGC 2009; INGC 2019).

On average, every 2 (two) years, a tropical storm or tropical cyclone hits Mozambique. Also, every 3 (three) or 4 (four) years, severe droughts are cyclically frequent in the country. Floods of different magnitudes occur every year in parts of the country and they are exacerbated by the fact that Mozambique is downstream from 9 of the 15 main river basins in southern Africa (WMO 2019).

The country's climatic and environmental scenario imposes life and survival challenges on rural communities that mostly have the poor housing, infrastructure, as well as livelihood activities and local assets dependent on environmental conditions for survival, such as rain-fed farming, livestock, and fishing (INGC 2009; Baez et al. 2019). Studies show that about 18.6 million people, more than 66% of the population, live in rural areas (INE 2017). Yet, a considerable group of the rural communities lives in coastal areas (in many places this consists of lowlands with sandy beaches, estuaries, and mangroves).

The impacts of environmental shocks on these rural and coastal communities have brought unprecedented damage and destruction. For instance, the severe tropical cyclones Idai and Keneth experienced in 2019, as well as the floods and inundations recorded in 2000. Both were responsible for the deaths of more than 1300 people, more than 4,000,000 affected and displaced, about 2,146,135 hectares of crop areas destroyed, more than $ 1.5 billion in economic damage (Christie and Hanlon 2002; INGC 2013; UNICEF 2019; OCHA 2019).

Experiencing any of the environmental shocks such as droughts, floods and cyclones in Mozambique leads to a drop of up to 25–30% in food consumption per capita and 0.4 fewer meals per day per person in rural communities. Therefore, poverty increased by 12 and 17.5% points after the 2005 drought and the 2008 Cyclone Jokwe (Baez et al. 2019).

These adversities and chaotic scenarios for rural communities demand urgent local responses to cope. They must be based on impacts, ways of life, experiences, and community organisation, as well as their current and future needs. In this context, it is urgent to understand how this population group deals with local environmental changes or shocks and what are the local coping capacities. The systematisation of this knowledge allows defining the profile and characterising the local knowledge system (IKS), which is fundamental for defining community intervention strategies (Breidlid 2009; Nkomwa et al. 2014; Syafwina 2014; Smith et al. 2017).

Mismatches and failures in implementing policies to adapt to environmental shocks based on rural communities could compromise the sustainability of ecosystems and natural resources in rural and coastal areas, limiting the reach of the objectives of the 2030 agenda consisting of:

Goal 2: To end hunger, achieve food security and improved nutrition, and to promote sustainable agriculture.

2. a) To increase investment, including through enhanced international cooperation, in rural infrastructure, agricultural research and extension services, technology development and plant, and livestock gene banks in order to enhance agricultural productive capacity in developing countries, in particular the least developed countries.

Goal 11: To make cities and human settlements including, safe, resilient, and sustainable.

11. a) To support positive economic, social, and environmental links between urban, peri-urban, and rural areas by strengthening national and regional development planning.

A better and deeper understanding of local knowledge can better inform planners and decision makers about sustainability at the community level (Parker 1995). Considering that the risk reduction, as well as the poverty reduction of rural communities, is strongly linked to the natural resources and available ecosystems services, accompanied by land planning, capable of reducing vulnerability of resources and local communities (Pandit et al. 2018).

The growing focus on reducing vulnerability, as well as increasing the adaptive capacity of local communities, focuses on land planning and natural resource management issues based on reality, experiences, and local knowledge (Van Lier 1998; Namatama 2020). For instance, planning for resilient build hospitals, schools, infrastructure and adapted to extreme events; regions for the resilient agriculture and conservation practices; regions for mangrove reforestation (to combat coastal erosion, as well as to contain sea levels rising); water supply and exclusive regions for community settlements, is partly supported by the strong needs of communities, as well as lessons with environmental chocks experienced by local communities (Jaarsma 1997; Namatama 2020).

In this setting, land planning and natural resources management, not aligned to the needs, knowledge, and experiences of local communities, is susceptible to failure, or even aggravating local social and environmental risks.

It is in this perspective that we set out to investigate this topic in order to identify and describe the local knowledge system (perception, vulnerability, risks, and adaptation) to climate change and extreme events in rural communities in northern Mozambique. We advance in this pathway as a contribution to improve and to expand the understanding of the theme and support in the formulation of actions for adaptation, land planning and reduction of environmental shocks to rural communities in the country.

2 Methods

2.1 Study Area

The study was carried out in the northern region of Mozambique, specifically in the province of Cabo Delgado. Cabo Delgado is a subdivision of Mozambique. It is located in the extreme northeast of the country and is part of the three provinces that make up the northern region of the country (Fig. 1). The province has an area of 82,625 km^2 and, in 2017, it had a population of 2,333,278 inhabitants. It is divided into 17 districts and it has, since 2013, five municipalities (INE 2017).

Cabo Delgado province has immense natural and mineral resources for global exploration and production. In 2014, the world's largest deposit was discovered in

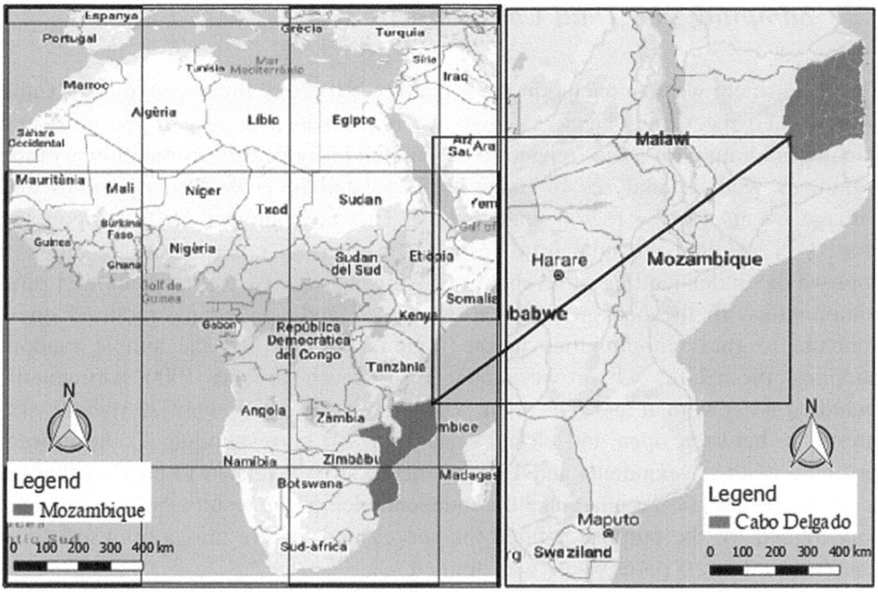

Fig. 1 Geographic location of the study area, Cabo Delgado, northern Mozambique. *Source* Elaborated by the authors

this province, which should contain more than 115.9 million tons of graphite (Observador 2014). In 2013, approximately 65 trillion cubic feet of recoverable natural gas were discovered there (Anadarko Petroleum Corporation 2018). In addition to mineral resources, the province is considered the third province with the largest protected and biodiversity conservation area in the country, the Quirimbas National Park. This area (Quirimbas National Park) was created in 2002, with a land and marine extension of about 7500 km^2 (ANAC 2012).

Cabo Delgado is the second province with low indicators of housing quality, both in urban and rural areas. The province is still considered the third province with the highest malnutrition rates in the country. Although poverty has reduced significantly since 2002, this province is the country's second poorest province (Ministério de Planificação e Desenvolvimento 2010; INE 2017).

Due to its geographical location, it is one of the most exposed to extreme climatic events. In 2019, one of the most severe cyclones hit the province, Cyclone Keneth (INGC 2019).

Due to the presence of large concentrations of mineral and natural resources, since October 2017, it has been suffering from terrorist attacks that affected and displaced more than 2,500,000 and killed more than 1500 people (European Parliament 2020).

2.2 Sampling and Data Collection

This assessment was conducted in the second half of 2019 in the province of Cabo Delgado. To meet our objectives, first, we developed a review of technical and scientific literature on issues related to the vulnerability of rural communities in the region, as well as analyses of manuscripts related to ecosystem dynamics and climatic events for the region under study. This task allowed us to support the process of validating results from respondents in rural communities in order to improve the understanding of levels of exposure, sensitivity, and adaptation of rural communities. In the background, we developed and applied a structured questionnaire to rural communities in the region. We adopted the simple random sampling procedure, where we managed to reach around 1000 participants, including 41% women and 69% men. The survey presented about 26 (twenty-six) questions, between open and closed ones, nine (9) corresponding to the characterisation of the respondents and 17 (seventeen) strictly related to climate change, and extreme events. To guarantee the independence of the results, field technicians duly trained for the purpose applied the survey and, before this, a pilot test for a sample equal to 50 participants was applied.

The perception of vulnerability to climate change in rural communities was assessed considering the following indicators: (a) exposure, (b) sensitivity, and (c) adaptation capacity. In this context, the main historical changes in the climatic variables (precipitation), biophysical assets (forest cover change), and occurrences of extreme climatic events were considered exposure indicators. While losses of assets and physical infrastructure (areas of crops, housing, and incomes) related to environmental climate shocks were considered indicators of sensitivity. Finally, the adaptation of rural communities was assessed using the following criteria: (a) human development (training and advanced training, information, and awareness of climate change); (b) social assets (leadership and high roles in local and regional organisations, access to finance); (c) physical assets and infrastructure (quality of housing, transportation, and communication).

2.3 Data Analysis

Quantitative data from the collection of the scientific and technical literature review was analysed according to the descriptive procedure. The main results were organised and summarised in tables and graphs according to their categories (e.g.: forest cover change; rainfall historical changes; records of extreme events). Subsequently, they were compared to results or responses from rural communities. This procedure supported in the process of validating and determining the exposure, sensitivity, and adaptation of rural communities.

Regarding qualitative information (survey), we applied descriptive statistics to questions related to the characterisation of respondents, as well as climate change and extreme events. We used Microsoft Excel's mathematical tools, SPPS (version 25) to calculate the averages, frequencies, percentages, standard deviations, as well as applying Pearson's chi-square test ($\chi2$) to determine the association of variables.

The rainfall data obtained from Climate Change Knowledge Portal (CCKP) (https://climateknowledgeportal.worldbank.org/country-profiles) were analysed using the Ms-Excel in the form of rainfall curves. These data, together with forest cover change provided by the global forest watch (https://www.globalforestwatch.org/dashboards/country/MOZ), and community survey was used to explain exposure. And results from the community survey were used to explain the sensitivity and adaptive capacity.

3 Results

3.1 Characterisation and Profile of Rural Communities (Respondents)

The analysis of Table 1 shows that the rural community of Cabo Delgado can be dominated mainly by the Emakhuwa language community speaking (a). Almost 99.3% said they speak that language fluently, while the region is not very representative in the Cinyungue and Kiswahili languages, both with 0.1% of the population.

Table 1 Description of the profile and characterisation of respondents from Cabo Delgado. *Source* Elaborated by the authors

(a) Dialect (local language)	(%)	*(b) Language*	(%)	*(c) Religion*	(%)
Emakhuwa	99.3	Portuguese	30.0	Christian	38.0
Echuwabo	1.0	Local Language	70.0	Islam	62.0
Kimwani	2.4	*(d) Gender*		*(e) Have you always lived here*	
Shimakonde	4.0	Female	31.2	Yes	64.4
Cinyungwe	0.1	Male	68.8	No	31.6
Kiswahili	0.1	*(f) Age*		*(g) Education level*	
(h) Head of household		$\leq$ 18 years old	4.4	No education	17.1
Female	12.0	19–30 years old	48.1	Primary school	71.0
Male	88.0	31–60 years old	45.4	Secondary school	10.6
		61–86 years old	2.1	University degree	1.3

In general, about 70% of the rural community speak Portuguese (the official language of Mozambique) as well as the local language. Nevertheless, a relatively significant part revealed to speak only the local language, and without skills to communicate in the official language (b), consisting of 30% of the population.

Two religions were found to be quite dominant in the province. About 62% of the population profess the Islamic religion and 38.0% the Catholic religion (c). We found that the participation of women was relatively lower compared to men. So, about 31% were women and more than 68% men (d).

Our research confirmed that 64% of the study population was born and resident in the province of Cabo Delgado, however, more than 31% came from other regions or provinces of Mozambique (e).

We found that the leadership or head of the family was mostly attributed to men. About 12% of women have the responsibility of leading their families, while about 88% are the responsibility of men (h).

The population's participation in the study was disproportionate between the age groups. The most affected group was aged 19–30 years, with about 48%, and 31–60 years with more than 45% participation (f).

Our assessment shows that the community's illiteracy rate may be relatively low, as around 71% of the population has completed primary education and about 17% have completed secondary education. A much lower part, composed of 1% of the population, completed higher education (g).

3.2 *Analysis of Exposure*

3.2.1 Occurrence of Extreme Climate Events

The records of the occurrence of extreme climate events in northern Mozambique have been dated from April 1952, when it first reached the province of Niassa, a neighbourhood with the province of Cabo Delgado (our case study). However, the occurrences for Cabo Delgado date back to December 9, 1959, according to the Table 2. The province has recorded about 7 extreme climate events since 1959, reaching the province between the months of November and April. The results demonstrate that extreme climate events changed their behaviour since 1988. In 1999, the province experienced, for the first time, a severe tropical storm, and then it was hardly hit by a very intense tropical cyclone in 2014. In 2019, the province experienced again a second very intense tropical cyclone, known as Cyclone Kenneth. The results suggest that, in the last 20 years, cyclones have become more frequent and violent. However, between 2002 and 2006, the province of Cabo Delgado was partially influenced by two tropical storms (Atang and Anita, respectively) that passed along the Mozambique Channel, as well as the border region between Mozambique and Tanzania.

Table 2 Record of occurrence of extreme climatic events in Cabo Delgado, Mozambique (1959–2019). *Source* Elaborated by the authors

I.D	Event category	Year	Name	Peak intensity	Source
1	Tropical depression	December 1959	Unknown	Unknown	[1]
2	Tropical storm	November 1969	Corrine	85 km/h	[2]
3	Moderate tropical storm	December 1980	Bettina	85 km/h	[3]
4	Moderate tropical storm	November 1988	Unknown	Unknown	[4]
5	Severe tropical storm	December 1999	Astride	95 km/h	[5]
6	Very intense tropical cyclone	March 2014	Hellen	230 km/h	[6]
7	Very intense tropical cyclone	April 2019	Kenneth	2015 km/h	[7]

3.2.2 Forest Cover Change

The results of the forest cover change in Cabo Delgado province suggest that since 2001—2019 the province has lost about 470 kha of forest cover. These figures are equivalent to a reduction of about 10% in forest cover since 2000.

However, the years with the highest records of forest cover losses were 2017 (42.6 kha), 2008 (40.2 kha), 2013 (34.3 kha), 2009 (31.1 kha), and 2014 (25.1 kha). While the years with the lowest loss of forest cover are: 2007 (8.1 kha), 2005 (8.9 kha) and 2001 (9.4 kha) (Figs. 2, 3, 4).

3.2.3 Rainfall Variability

Our results converge that, in the last 55 years (1961–2016), the province of Cabo Delgado registered a decrease in precipitation in seven months of the year (April, May, June, July, August, November, and December). This meant a variation between 5 and 23.36 mm per month, corresponding to 12–19% reduction in precipitation.

3.2.4 Adaptive Capacity

Our results in Table 1 show that around 71.1% of the population has at least primary education and the illiteracy rate is around 17.1%, corresponding to people without access to education.

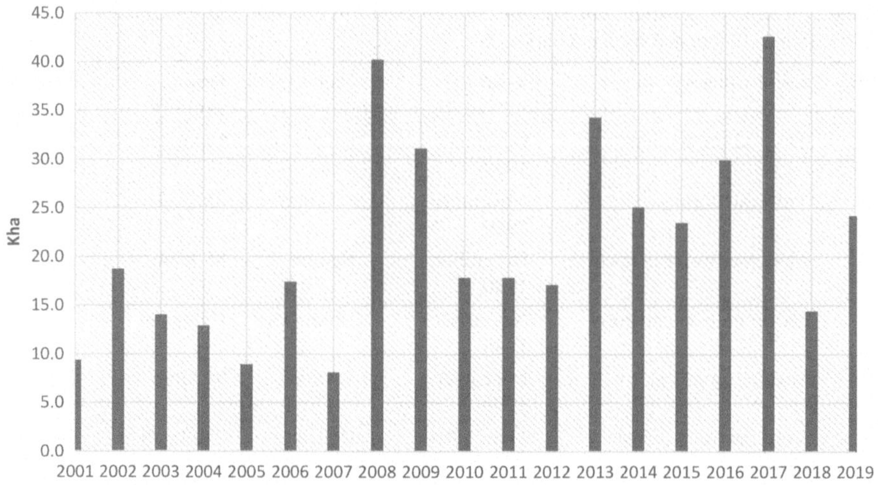

Fig. 2 Plotting forest cover loss in Cabo Delgado, Mozambique (2001–2019). *Source* Compiled by the authors, data from the global forest watch https://www.globalforestwatch.org/dashboards/country/MOZ

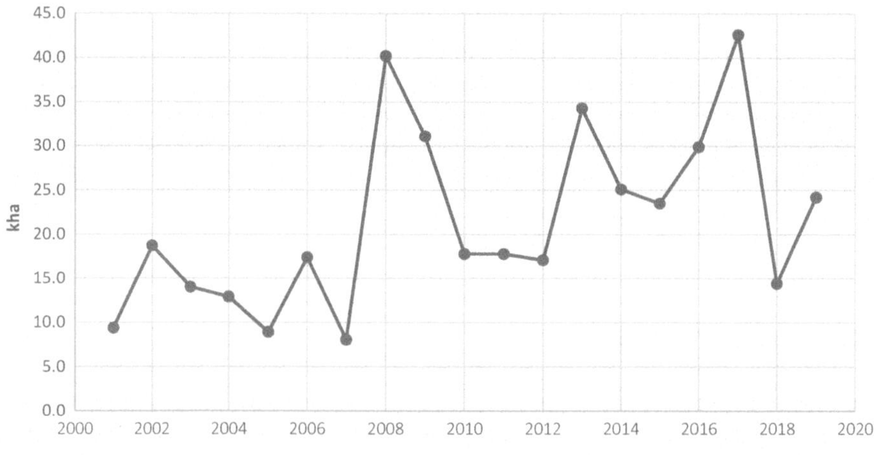

Fig. 3 Plotting detailed analysis between 2 year of forest cover loss in Cabo Delgado, Mozambique (2001–2019). *Source* Compiled by the authors, data from the global forest watch https://www.globalforestwatch.org/dashboards/country/MOZ

Nevertheless, we found that most respondents have a main activity, income, and assets that are mostly dependent on local climate factors. Agriculture, with around 61.90% (Fig. 5), has been the main source of livelihood for local communities.

We found that about 97.6% of the population lives in housing that is generally of a precarious type and a relatively insignificant number have conventional housing.

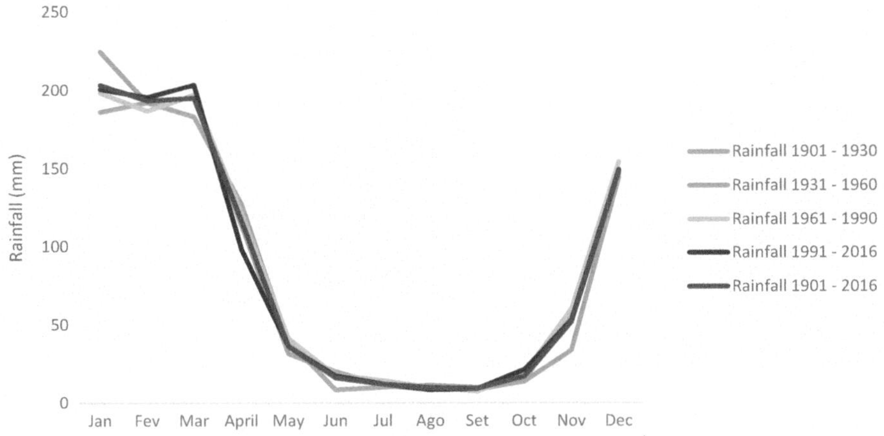

Fig. 4 Plotting the average monthly rainfall of Cabo Delgado for 1901–2016. *Source* Compiled by the authors, data from https://climateknowledgeportal.worldbank.org/country/mozambique/climate-data-historical

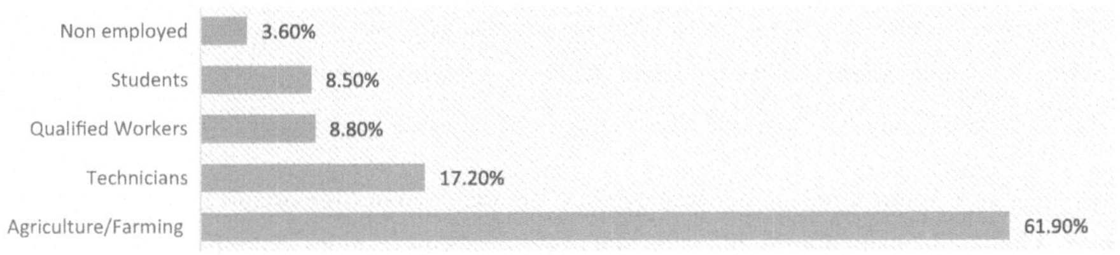

Fig. 5 Plotting the main activity and source of family incomes. *Source* Elaborated by the authors

The types and forms of construction vary in two ways, about 98% use local materials from the forests for construction and 1.9% choose to use cement block with zinc coating (Fig. 6).

Figure 7 shows the percentage distribution of the main needs or priority infrastructure of local communities. About 26% consider that there is a lack of public transport, while almost 25% believe that it is important to increase health services. Nevertheless, nearly 18% say that schools are limited, while 23% recognise that there is a deficit in banking institutions.

The level of awareness or even knowledge of climate change is quite limited. About 70% of rural communities do not have access to information or basic knowledge about climate change (Fig. 8). However, when analysed between educational levels, we found that the largest group with access to information has primary education. The number of respondents with higher or secondary education revealed an absence of information or knowledge. Nevertheless, the group of respondents without access to education proved to have basic knowledge on the subject.

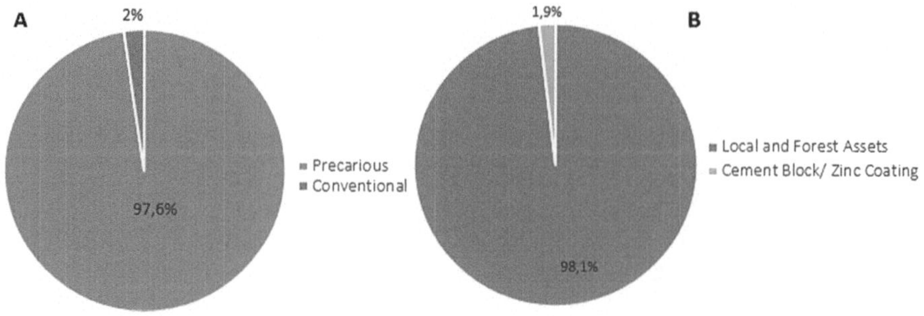

Fig. 6 Plotting the types of housing and shape, building material. **A** Type of housing. **B** Types of building materials. *Source* Elaborated by the authors

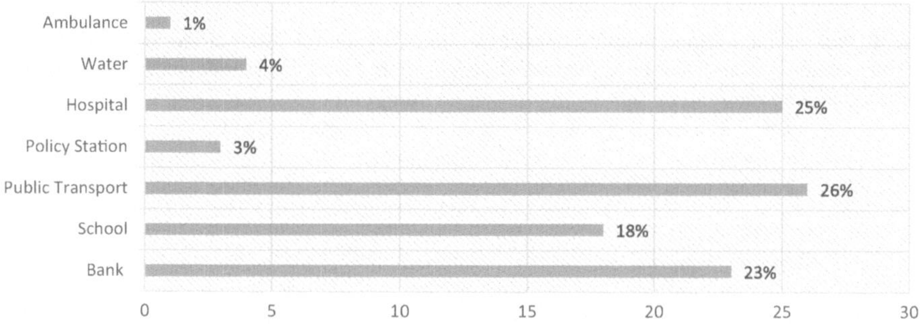

Fig. 7 Plotting the main priority infrastructure of the rural communities. *Source* Elaborated by the authors

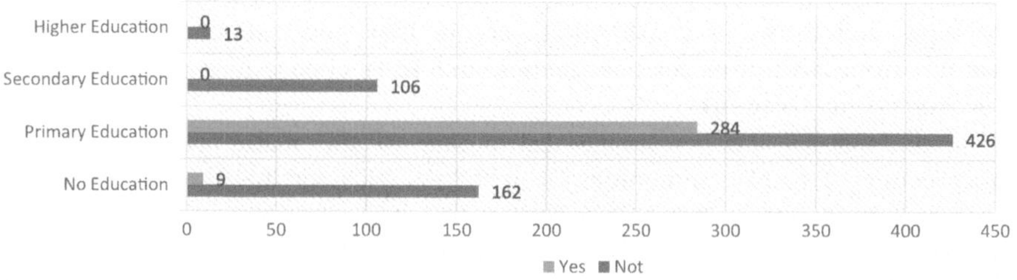

Fig. 8 Plotting the climate change awareness and knowledge by rural communities (numbers). *Source* Elaborated by the authors

3.2.5 Perception of Climate Change and Climate Extremes Events

The perception of climate change, as well as the extreme climate events observed on a daily basis by communities in the province of Cabo Delgado, varies significantly. About 100% of the community said there was a reduction in rainfall.

Nevertheless, about 100% understand that there is increasing or frequency of floods in the region.

Yet, 30.9% affirm that the intensity of the heat has increased, while 69.10% understand that it has reduced.

Regarding cyclones, more than 76% say that cyclones are increasing in intensity, while more than 23% say the opposite. For about 77% of the population, the winds in the region have become increasingly strong and 23% perceive that they are less intense (Fig. 9).

We found that most respondents, around 84.60%, recognise that changes or climate variations in the region are uncommon. However, about 15.4% say that the variations are normal or common (Fig. 10B).

When asked why local environmental variations were uncommon, about 61.4% stated that it does not rain regularly. Another 22.2% indicated that the rainy season in the region has lagged considerably. In addition, 4.8% said that the rivers are drying up, while 3.6% say that the months from January to March have been raining torrentially, in an abnormal way and, because of these rains, there is destruction of houses, roads, and farming areas (Fig. 10A).

Our assessment allowed us to understand how rural communities interpret such environmental changes. We found that around 65.9% do not know whether environmental changes are natural or man-made. Nevertheless, 25.6% of the respondents considered that man caused the verified variations. Finally, 8.5% of rural communities attributed it to natural factors (Fig. 11).

We found that the community understands that climate change and extreme events can have some influence on their daily lives. In this context, about 16.8% said that climate change could contribute to the increase in disease. Nevertheless, 93.4% agree that climate change can contribute to the increase in poverty. Over 95% understand that climate change can increase people's hunger (Fig. 12).

Our analysis also confirmed that about 90% of communities understand that local changes can contribute to the availability of water for consumption, as well as agricultural practices. Finally, 57.7% find that agricultural production can be significantly affected due to environmental factors (Fig. 12).

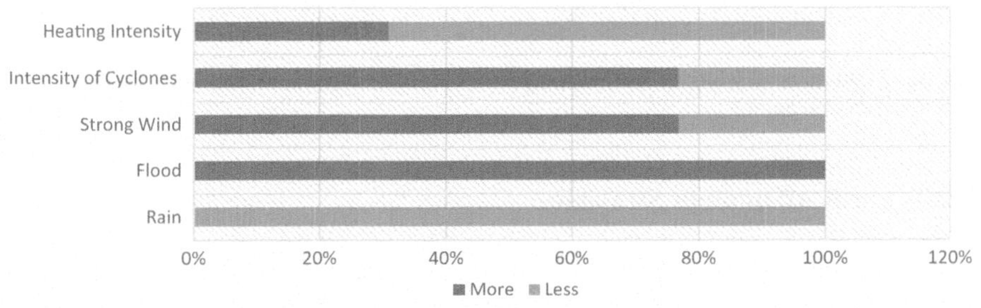

Fig. 9 Plotting the perception of the rural communities related to climate issues. *Source* Elaborated by the authors

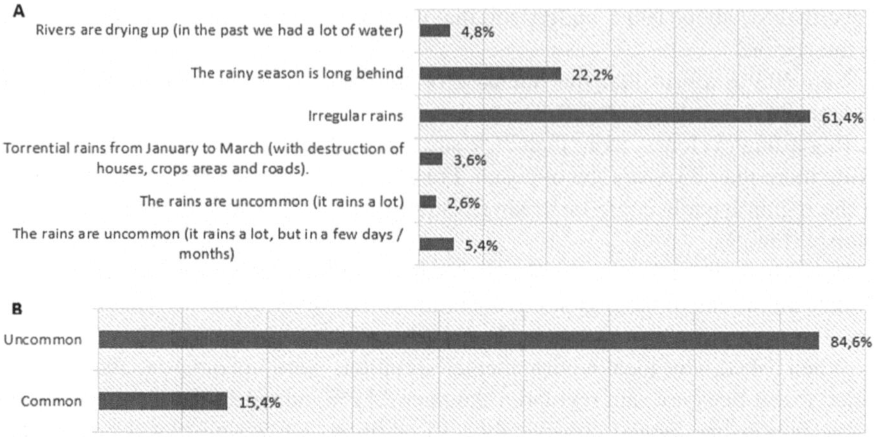

Fig. 10 Plotting the perception of the local communities based on environmental indicators. *Source* Elaborated by the authors

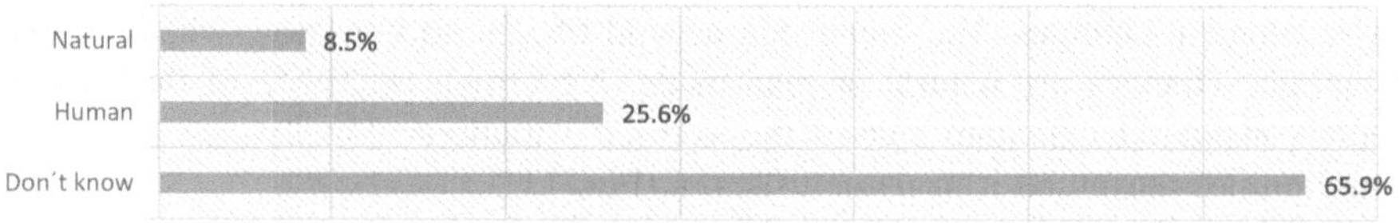

Fig. 11 Perception of the respondents regarding to the source of local environmental changes. *Source* Elaborated by the authors

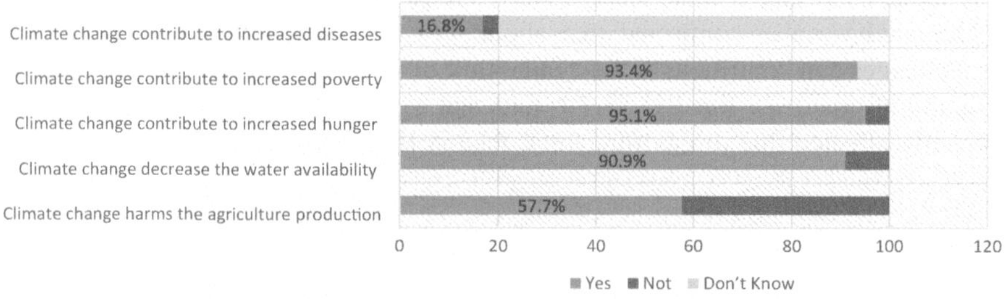

Fig. 12 Plotting perception of the respondents about the impacts of climate change. *Source* Elaborated by the authors

3.2.6 Local Adaptation Strategies

In this section, we consider local strategies for adapting to climate change and extreme climatic events, and all defined and implemented actions that are currently taking place in Cabo Delgado to deal with the environmental shocks that have actually been recognised by rural communities.

Nevertheless, we understand that the implementation of any action with a view to tackling environmental shocks is the result of an understanding or knowledge of the current situation, based on evidence, but introduced in order to change the scenario with the perspective of providing a well-being balanced and adjusted to the local reality.

In this setting, when asked whether people should be concerned about climate change, about 88.80% said "yes" and the remaining 11.20% considered it not relevant (Fig. 13).

More than 45% of respondents agree that preparing to deal with climate change is a waste of time. However, about 54.50% believe that preparation for coping is essential.

Regarding measures to tackle water scarcity or drought, about 98.4% have no alternative measures to deal with water scarcity. In addition, more than 95% of the community has no plan to implement in a scenario of progressive reduction in productivity and agricultural production (Fig. 13).

Figure 14 explores the understanding of local communities in an adverse context of local environmental changes (climate change and extreme climate events). In this context, we question the following "What if local environmental or climate conditions prevail or increase, what would you do?". About 481 respondents (50.7% M and 42.3% F) stated that they did not know exactly how to proceed in a similar scenario. However, 309 (43.3% F and 25.3% M) converged that they would stay in the same place. Almost 166 (19.2% M and 10.9% F) would migrate to other places to live. Finally, 44 (4.8% M and 3.5% F) claims that they would find other areas for agricultural practice.

Another analysis based on school education allowed us to verify that the absence of alternatives in the face of a possible adverse scenario was related to the level of education. Figure 15 shows that the alternative of the answer "Don't know" was observed in all social groups, however, it was significantly/considerably frequent in the group of people with primary education level (51.1%), secondary (38.5%), and university (48.1%) when compared to the group of people with no education (27.5%).

However, the response of the alternative "I would stay here, because I have nowhere else to live" was significantly observed in the group of people with no education (42.1%). Finally, the response of the alternative "Migrate looking for other safe places" was mostly observed in the group of people with secondary education (38.5%).

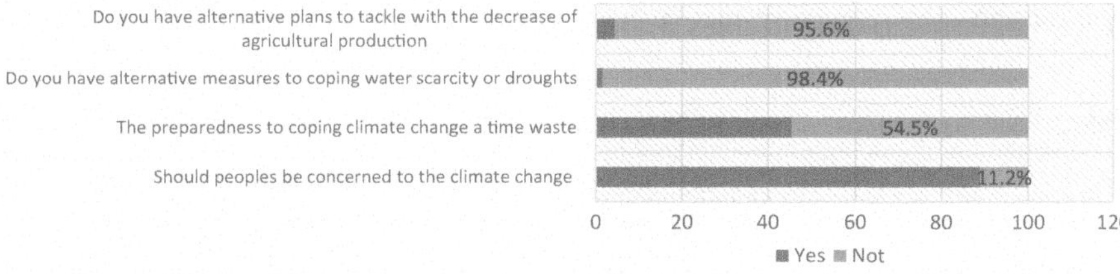

Fig. 13 Plotting the perception of the preparedness to climate change and climate extreme events. *Source* Elaborated by the authors

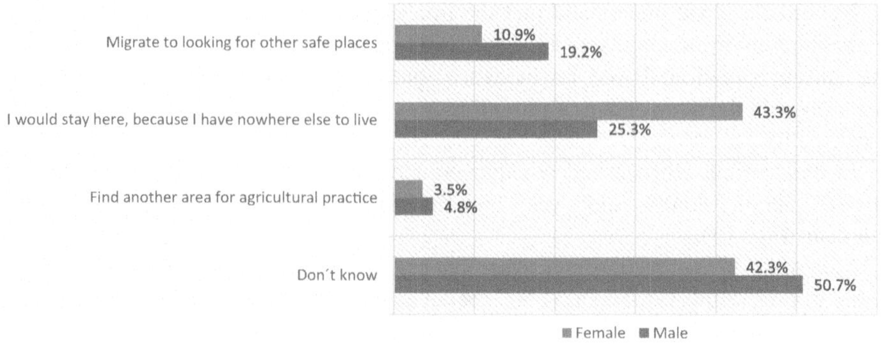

Fig. 14 Plotting the perception of the communities in a possible adverse scenario of increased variability and climate change/extreme events. *Source* Elaborated by the authors

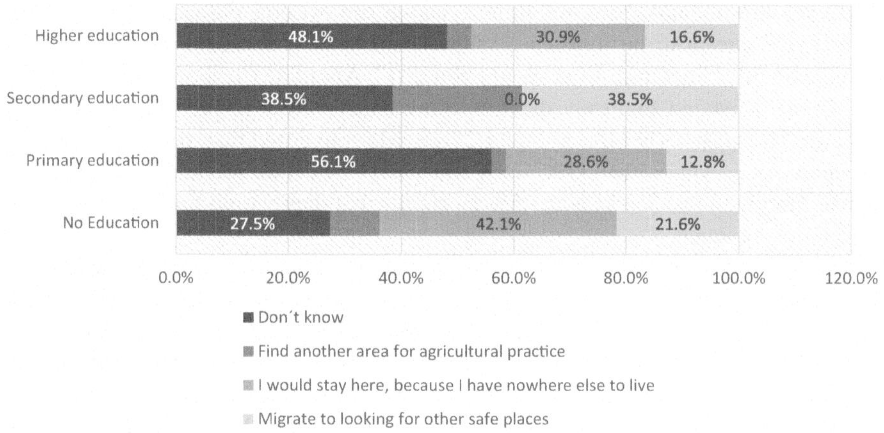

Fig. 15 Plotting the perception of the communities in a possible adverse scenario of increased variability and climate change/extreme events. *Source* Elaborated by the authors

Notwithstanding, it allows us to state that the chi-square test of independence indicated that there is a statistically significant association between educational level and coping responses in a possible adverse scenario, since $p \leq 0, 05$ [X2 (9) = 94.470; $p < 0.001$] (Tables 3, 4).

Table 3 Chi–Square tests. *Source* Elaborated by the authors

	Value	df	Asymp. Sig. (2-sided)
Pearson Chi–Square	96.470[a]	9	0.000
Likelihood ratio	93.601	9	0.000
Linear-by-linear association	0.404	1	0.525
Number of valid cases	1000		

[a]4 cells (25.0%) have expected count less than 5. The minimum expected count is 0.57

Table 4 Educational level* A possible adverse scenario of increased variability and climate change/extreme events crosstabulation. *Source* Elaborated by the authors

			A possible adverse scenario of increased variability and climate change/extreme events				Total
			Don't know	Find another area for agricultural practice	I would stay here because I have nowhere else to live	Migrate to looking for other safe places	
Educational level	No education	Count	47	15	72	37	171
		Expected count	82.3	7.5	52.8	28.4	171.0
		% within educational level	27.5	8.8	42.1	21.6	100.0
		% within a possible adverse scenario of increased variability and climate change/extreme events	9.8	34.1	23.3	22.3	17.1
		% del total	4.7	1.5	7.2	3.7	17.1
	Primary education	Count	398	18	203	91	710
		Expected count	341.5	31.2	219.4	117.9	710.0
		% within educational level	56.1	2.5	28.6	12.8	100.0
		% within a possible adverse scenario of increased variability and climate change/extreme events	82.7	40.9	65.7	54.8	71.0
		% Total	39.8	1.8	20.3	9.1	71.0
	Secondary education	Count	31	8	34	33	106
		Expected count	51.0	4.7	32.8	17.6	106.0
		% within educational level	29.2	7.5	32.1	31.1	100.0
		% within a possible adverse scenario of increased variability and climate change/extreme events	6.4	18.2	11.0	19.9	10.6
		% Total	3.1	0.8	3.4	3.3	10.6
	Higher education	Count	5	3	0	5	13
		Expected count	6.3	0.6	4.0	2.2	13.0
		% within educational level	38.5	23.1	0.0	38.5	100.0
		% within a possible adverse scenario of increased variability and climate change/extreme events	1.0	6.8	0.0	3.0	1.3
		% Total	0.5	0.3	0.0	0.5	1.3

(continued)

Table 4 (continued)

		A possible adverse scenario of increased variability and climate change/extreme events				Total
		Don't know	Find another area for agricultural practice	I would stay here because I have nowhere else to live	Migrate to looking for other safe places	
Total	Count	481	44	309	166	1000
	Expected Count	481.0	44.0	309.0	166.0	1000.0
	% within educational level	48.1	4.4	30.9	16.6	100.0
	% within a possible adverse scenario of increased variability and climate change/ extreme events	100.0	100.0	100.0	100.0	100.0
	% Total	48.1	4.4	30.9	16.6	100.0

4 Discussion

4.1 *Exposure, Sensitivity, Risks and Perception of Rural Communities on Climate Change and Climate Extreme Events*

According to Intergovernmental Panel on Climate Change (IPCC), vulnerability is "The degree to which a system is susceptible to, or unable to cope with, adverse effects of climate change, including climate variability and extremes. Vulnerability is a function of the character, magnitude, and rate of climate variation to which a system is exposed, its sensitivity, and its adaptive capacity." (IPCC 2014).

Notwithstanding, exposure is defined as "the nature and degree to which a system is exposed to significant climatic variations" (Folland et al. 2001; IPCC 2014). While sensitivity is "the degree to which a system is affected, either adversely or beneficially, by climate-related stimuli" (Folland et al. 2001; IPCC 2014). In addition, adaptive capacity is "the ability of a system to adjust to climate change (including climate variability and extremes), to take advantage of opportunities or to cope with the consequences" (Folland et al. 2001; IPCC 2014).

In this context, our assessment shows that Cabo Delgado province, as well as rural communities, are highly vulnerable to the impacts of climate change and extreme climate events (Table 2; Figs. 2, 4, 5, 6, 7, 8, 9, and 13) which is exacerbated by the geographical location of the province and limited coping measures (INGC 2009; Artur et al. 2018).

Cabo Delgado has been hit by about seven extreme climate events and the last two (Cyclone Tropical Hellen 2014 and Keneth 2019, Table 2) have severely threatened the lives and survival of rural communities. Both were responsible for more than 45 deaths, affecting more than 27,000 rural people, destroying hundreds of homes, as well as important public infrastructure (United Nations Office for the Coordination of Humanitarian Affairs—OCHA 2014, 2019).

However, changes in the behavioural patterns of extreme climate events in Cabo Delgado are perfectly matched to recent trends in records and occurrences of extreme climate events in Southern Africa (Mavume et al. 2009; Moyo and Nangombe 2015; Shi et al. 2020; Cattiaux et al. 2020). Research warns that extreme climate events will become more violent and more frequent in the coming decades (Handmer et al. 2012). Thus, regions such as Cabo Delgado, and with a similar population characteristic (Table 1, Figs. 6 and 7), should experience dramatic and unprecedented clashes.

We understand that the loss of forest cover in Cabo Delgado poses a threat in reducing vulnerability in the context of extreme climate events. Trends suggest that, from 2001 to 2019 (Fig. 2), the province has lost about 47,000 hectares of forest cover. Studies show that forests play a critical role in the context of extreme climatic events. These provide goods and services to rural communities during extreme climatic events (Robledo et al. 2012). Yet, they act as source of cash for coping with weather-related crop failure (Fisher et al. 2010) and as defensive barriers to storms and strong winds.

Nevertheless, despite the lack of local studies that demonstrate the potential of forests to mitigate the impacts of extreme climatic events, our results (Table 1, Figs. 5, 6 and 7) suggest that the increase in the reduction of forest cover will significantly increase the exposure and sensitivity of communities to extreme climatic events.

We found no evidence of diversification of income sources to compensate for the limited production and productivity of agricultural soils conditioned by environmental factors. About 61.90% of the respondents (Fig. 5) have agriculture as their main income activity. In Mozambique, about 75% of the population depends on agriculture for their survival (Mosca 2017). For its productivity, agriculture depends considerably on climate factors such as precipitation. However, Cabo Delgado has recorded a decrease in precipitation of 5–23.36 mm per month for the past 55 years (Fig. 4). These variations can increase significantly if global efforts to reduce greenhouse gas emissions are not fully met (IPCC 2019). In this context, we understand that, in a context of limited diversification of income activities of the communities and in a scenario of increased reduction of precipitation in the province, it can condition the communities to cope with extreme climatic events in addition to increasing food insecurity and foster poverty.

We find that there is a direct correspondence between the responses of the communities in relation to the reduction of the frequency of rains, records of occurrences of floods, intensities of cyclones and strong winds (Fig. 9) with the empirical data on the records of cyclones, and reduction of precipitation (Table 2, Fig. 4). This convergence evidence demonstrates that traditional or indigenous knowledge, when systematised, can play a fundamental role in understanding spatial and temporal occurrence of extreme climatic events, climate variability, as well as supporting the definition of actions to reduce risks, vulnerability, and measures for coping.

Nevertheless, our research found that communities clearly understand that extreme climatic events, as well as climate change, can contribute to increased disease, hunger, reduced availability of water resources, and agricultural production (Fig. 12). We argue that this knowledge is not associated with a type of scientific knowledge as such; however, it may be the result of daily experiences and observations of various local environmental standards.

The destruction of housing, crops, public infrastructure, and reduced availability of water resources were identified by the communities because of local environmental changes (Fig. 10). The loss or reduction of biophysical assets in Cabo Delgado suggests a high degree of sensitivity to which the community is exposed. Thus, we argue that in the absence of real actions affecting communities, this level of sensitivity can contribute considerably to increasing the vulnerability of communities, economic infrastructure, services of local ecosystems, and threatening the lives and survival of several families.

Despite the existence of important local knowledge to support local decision-making, we found that there are deficits or lack of common or basic knowledge related to environmental issues, mainly about climate change/extreme climatic events (Fig. 11). However, we are unable to find direct factors that contribute to this. Nevertheless, we observed that the low level of education, associated with limited learning conditions, as well as limited access to information can be considered as conditioning factors.

4.2 Adaptation Strategies and Implications for Adaptation Planning

Adaptation strategies are non-existent and the adaptive capacity of communities to cope with the impacts of extreme climatic events is considerably low (Figs. 5, 6, 7, 13, and 14). This suggests the existence of a high level of local vulnerability, which implies the introduction of integrated medium- and long-term responses, addressing issues such as: infrastructure, education, agriculture, gender, human capital, governance, biodiversity, diversification of sources income and ecosystem services. Thus, the current scenario in the province of Cabo Delgado allows us to discuss and reflect on the following aspects:

(a) An adaptation strategy focusing on the profile of traditional knowledge of communities on extreme climatic events/climate change, as well as their specific characteristics and current and future needs for survival patterns, is urgent.
(b) It is necessary to plan mitigation actions designed according to specific regional characteristics (districts). The distribution of rainfall in the province is not uniform across all districts. The reduction in forest cover rates varies from

district to district. Although tropic cyclones have an impact at the provincial scale, there are references that generally indicate the districts and regions frequently affected, routes used by extreme events and points of dissipation. This information must be considered in the planning process, with the aim of identifying and knowing critical areas that may justify long-term action interventions.

(c) The agricultural sector must be rethought in a strategic way. It will be necessary to define regions with favourable micro ecological characteristics to promote concentrated or block farming activities, sustainable (conservation agriculture) and large scale. However, it will be necessary to empower rural communities and allow them to be key players in the decision-making process.

(d) The planning must foresee the introduction of the early warning system of extreme climatic events/climate changes, appropriate to the characteristics of the communities.

(e) Adaptation strategies should focus on training, education in the short, medium and long term of the communities so that essential skills and tools are assigned, capable of supporting the sustainable decision-making process in an adverse context, and or of preparation.

(f) Financing community activities and small associations can help diversify alternative sources of income generation. However, a set of infrastructures or services such as micro banks, communication, adequate access routes must be available to support the initiatives.

5 Conclusion

Our research provided useful information to increase local social understanding, with the aim of providing tools for local planning and natural resources management for risk and vulnerability reduction to climate change/extreme events.

We note that there is a lack of tools and capacities to increase the adaptive capacity of communities, as well as strategies for adapting to extreme events. The geographic location of communities, and not accompanied by strict land planning based on the local implications and consequences of extreme events, may increase the local vulnerability. Specifically, we found that the illiteracy rate of the community can be relatively low, as around 71% has completed primary education, and about 17% have completed secondary education. A much lower number composed of 1% has been completed higher education.

Cabo Delgado province is particularly vulnerable to extreme weather events and climate change. Since 1959, it has recorded about seven extreme climatic events, the last two being more violent. It lost about 470000 ha, of forest cover from 2001 to 2019. Precipitation decreased, with a variable rate of 5 and 23.36 mm per month, in the last 55 years.

We find that perceptions of climate change, vary substantially among gender and education. The study revealed that communities have noticed the changes as declining rainfall, increasing droughts, storms, cyclones, and rising temperature. The gradual loss of income, properties, decline of agricultural production, and increased hunger were identified as indicators of sensitivity to climate change/ extreme events. Yet adaptive capacities are low, more than 90% of respondents do not use any adaptation strategy to cope climate change/extreme events. We conclude that communities are very exposed and vulnerable to the possible shocks of climate change/extreme events. So, income diversification, strong education system and awareness on climate change mater, and advanced agricultural policy should be considered as priorities for adaptation strategies.

This paper has some limitations. The first one is the fact that the perception of risk and vulnerability of rural communities are framed in the context of climate change and extreme events to direct local actions to increase adaptive capacities and improve land planning to reducing risks and vulnerabilities. In addition, the size of the sample and its geographical location cannot be representative of trends across the country.

6 Supplementary Information

The information provided here is related to Table 2 of the results, associated with the "source" sub-column.

1. http://www.atms.unca.edu/ibtracs/ibtracs_current/index.php?name=v04r00-1959335S12065.
2. https://en.wikipedia.org/wiki/1969%E2%80%9370_South-West_Indian_Ocean_cyclone_season.
3. https://en.wikipedia.org/wiki/1980%E2%80%9381_South-West_Indian_Ocean_cyclone_season.
4. https://en.wikipedia.org/wiki/1987%E2%80%9388_South-West_Indian_Ocean_cyclone_season.
5. Kenneth R. Knapp; Michael C. Kruk; David H. Levinson; Howard J. Diamond; Charles J. Neumann (2010). 2000 Astride (1999357S08076). The International Best Track Archive for Climate Stewardship (IBTrACS): Unifying tropical cyclone best track data (Report). Bulletin of the American Meteorological Society.
6. https://en.wikipedia.org/wiki/2013%E2%80%9314_South-West_Indian_Ocean_cyclone_season
7. https://en.wikipedia.org/wiki/2018%E2%80%9319_South-West_Indian_Ocean_cyclone_season.

References

Adhikari U, Nejadhashemi AP, Woznicki SA (2015) Climate change and eastern Africa: a review of impact on major crops. Food Energy Secur 4(2):110–132. https://doi.org/10.1002/fes3.61

Adzawla W, Kudadze S, Mohammed AR, Ibrahim II (2019) Climate perceptions, farmers' willingness-to-insure farms and resilience to climate change in Northern region, Ghana. Environ Dev 100466. https://doi.org/10.1016/j.envdev.2019.100466

Ahmed SM (2020) Impacts of drought, food security policy and climate change on performance of irrigation schemes in Sub-saharan Africa: the case of Sudan. Agric Water Manag 232:106064. https://doi.org/10.1016/j.agwat.2020.106064

Alaanuloluwa Ikhuoso O, Adegbeye MJ, Elghandour MMY, Mellado M, Al-Dobaib SN, Salem AZM (2020) Climate change and agriculture: the competition for limited resources amidst crop farmers-livestock herding conflict in Nigeria—a review. J Clean Prod 123104. https://doi.org/10.1016/j.jclepro.2020.123104

Allen MR, Dube OP, Solecki W, Aragón-Durand F, Cramer W, Humphreys S, Kainuma M, Kala J, Mahowald N, Mulugetta Y, Perez R, Wairiu M, Zickfeld K (2018) Framing and context. In: Masson-Delmotte V, Zhai P, Pörtner H-O, Roberts D, Skea J, Shukla PR, Pirani A, Moufouma-Okia W, Péan C, Pidcock R, Connors S, Matthews JBR, Chen Y, Zhou X, Gomis MI, Lonnoy E, Maycock T, Tignor M, Waterfield T (eds) Global Warming of 1.5 °C. An IPCC Special Report on the impacts of global warming of 1.5 °C above pre-industrial levels and related global greenhouse gas emission pathways, in the context of strengthening the global response to the threat of climate change, sustainable development, and efforts to eradicate poverty (In Press)

Anadarko Petroleum Corporation (2018) Mozambique liquefied natural gas project. Retrieved at https://www.mzlng.total.com/en/project/resource

Anderson KJ, Silva JA (2020) Weather-related influences on rural-to-urban migration: a spectrum of attribution in Beira, Mozambique. Global Environ Change 65:102193. https://doi.org/10.1016/j.gloenvcha.2020.102193

Antwi-Agyei P, Amanor K, Hogarh JN, Dougill AJ (2020) Predictors of access to and willingness to pay for climate information services in north-eastern Ghana: a gendered perspective. Environ Dev 100580. https://doi.org/10.1016/j.envdev.2020.100580

Artur L, Hilhorst D (2012) Everyday realities of climate change adaptation in Mozambique. Glob Environ Change 22(2):529–536. https://doi.org/10.1016/j.gloenvcha.2011.11.013

Artur L, Hilhorst D (2014) Floods, resettlement and land access and use in the lower Zambezi, Mozambique. Land Use Policy 36:361–368. https://doi.org/10.1016/j.landusepol.2013.08.017

Artur L, António C, Bata O, Afonso F, Muai G (2018) Embedding adaptation into development planning and decision making process at the municipal levels in Mozambique. In: Alves F, Leal Filho W, Azeiteiro U (eds) Theory and practice of climate adaptation. Climate change management. Springer, Cham. https://doi.org/10.1007/978-3-319-72874-2_9

Assan E, Suvedi M, Schmitt Olabisi L, Bansah KJ (2020) Climate change perceptions and challenges to adaptation among smallholder farmers in semi-arid Ghana: a gender analysis. J Arid Environ 182:104247. https://doi.org/10.1016/j.jaridenv.2020.104247

Atuoye KN, Luginaah I, Hambati H, Campbell G (2020) Who are the losers? Gendered-migration, climate change, and the impact of large scale land acquisitions on food security in coastal Tanzania, Land Use Policy. ISSN 105154:0264–8377. https://doi.org/10.1016/j.landusepol.2020.105154

Ayanlade A, Radeny M, Morton JF, Muchaba T (2018) Rainfall variability and drought characteristics in two agro-climatic zones: an assessment of climate change challenges in Africa. Sci Total Environ 630:728–737. https://doi.org/10.1016/j.scitotenv.2018.02.196

Azzarri C, Signorelli S (2020) Climate and poverty in Africa South of the Sahara. World Dev 125:104691. https://doi.org/10.1016/j.worlddev.2019.104691

Baarsch F, Granadillos JR, Hare W, Knaus M, Krapp M, Schaeffer M, Lotze-Campen H (2020) The impact of climate change on incomes and convergence in Africa. World Dev 126:104699. https://doi.org/10.1016/j.worlddev.2019.104699

Baez JE, Caruso G, Niu C (2019) Extreme weather and poverty risk: evidence from multiple shocks in Mozambique. Econ Disasters Clim Change. https://doi.org/10.1007/s41885-019-00049-9

Braitstein P, Lama T, Keino S, Gladanac B, Yego F, Cole D, Cortinois A, Fox A (2017) Increasing food security and nutrition resilience in response to climate change in east Africa: findings from a multisectoral symposium. Lancet Glob Health 5:S23. https://doi.org/10.1016/s2214-109x(17)30130-4

Breidlid A (2009) Culture, indigenous knowledge systems and sustainable development: a critical view of education in an African context. Int J Educ Dev 29(2):140–148. https://doi.org/10.1016/j.ijedudev.2008.09.009

Cattiaux J, Chauvin F, Bousquet O, Malardel S, Tsai C-L (2020) Projected changes in the Southern Indian Ocean Cyclone activity assessed from high-resolution experiments and CMIP5 models. J Clim 33:12. https://doi.org/10.1175/JCLI-D-19-0591.1. https://ui.adsabs.harvard.edu/abs/2020JCli...33.4975C

Christie F, Hanlon J (2002) Mozambique & the great flood of 2000. Can J Afr Stud/Revue Canadienne des Études Africaines 36. https://doi.org/10.2307/4107342

Dendir Z, Simane B (2019) Livelihood vulnerability to climate variability and change in different agroecological zones of Gurage Administrative Zone, Ethiopia. Prog Disaster Sci 3:100035. https://doi.org/10.1016/j.pdisas.2019.100035

Dube K, Nhamo G (2020) Evidence and impact of climate change on South African national parks. Potential implications for tourism in the Kruger National Park. Environ Dev. https://doi.org/10.1016/j.envdev.2019.100485

Dumenu WK, Obeng EA (2016) Climate change and rural communities in Ghana: social vulnerability, impacts, adaptations and policy implications. Environ Sci Policy 55:208–217. https://doi.org/10.1016/j.envsci.2015.10.010

Dunn RJH, Stanitski DM, Gobron N, Willett KM (eds) (2020) Global climate [in "State of the Climate in 2019"]. Bull Am Meteor 101(8):S9–S127. https://doi.org/10.1175/BAMSD-20-0104.1

European Parliament (2020) European parliament resolution on the humanitarian situation in Mozambique (2020/2784(RSP)), 6. Retreived at: https://www.europarl.europa.eu/doceo/document/B-9-2020-0303_EN.html

Evariste FF, Denis Jean S, Victor K, Claudia M (2018) Assessing climate change vulnerability and local adaptation strategies in adjacent communities of the Kribi-Campo coastal ecosystems, South Cameroon. Urban Clim 24:1037–1051. https://doi.org/10.1016/j.uclim.2017.12.007

Filho WL, Taddese H, Balehegn M, Nzengya D, Debela NA, Abayineh E, Mworozi S, Osei DY, Nagy GJ, Yannick N, Kimu S, Balogun AL, Alemu EA, Li C, Idsaph HS, Wolf F (2020) Introducing experiences from African pastoralist communities to cope with climate change risks, hazards and extremes: fostering poverty reduction. Int J Disaster Risk Reduct https://doi.org/10.1016/j.ijdrr.2020.101738

Fisher M, Chaudhury M, McCusker B (2010) Do forests help rural households adapt to climate variability? Evidence from Southern Malawi. World Dev 38(9):1241–1250. https://doi.org/10.1016/j.worlddev.2010.03.005

Folland CK, Karl TR, Christy JR, Clarke RA, Gruza GV, Jouzel J, Mann ME, Oerlemans J, Salinger MJ, Wang S-W (2001) Observed climate variability and change. In: Houghton JT, Ding Y, Griggs DJ, Noguer M, van der Linden PJ, Dai X, Maskell K, Johnson CA (eds) Climate change 2001: the scientific basis. Contribution of Working Group I to the Third Assessment Report of the Intergovernmental Panel on Climate Change. Cambridge University Press, Cambridge, United Kingdom and New York, NY, USA, p 881

Gebrechorkos SH, Bernhofer C, Hülsmann S (2019) Impacts of projected change in climate on water balance in basins of East Africa. Sci Total Environ. https://doi.org/10.1016/j.scitotenv.2019.05.053

Handmer J, Honda Y, Kundzewicz ZW, Arnell N, Benito G, Hatfield GJ, Mohamed IF, Peduzzi P, Wu S, Sherstyukov B, Takahashi K, Yan Z (2012) Changes in impacts of climate extremes: human systems and ecosystems. In: Field CB, Barros V, Stocker TF, Qin D, Dokken DJ, Ebi KL, Mastrandrea MD, Mach KJ, Plattner G-K, Allen SK, Tignor M, Midgley PM (eds) Managing the risks of extreme events and disasters to advance climate change adaptation. A special report of working groups I and II of the intergovernmental panel on climate change (IPCC). Cambridge University Press, Cambridge, UK, and New York, NY, USA, pp 231–290

Hardwick JR, Westra S, Sharma A (2010) Observed relationships between extreme sub-daily precipitation, surface temperature, and relative humidity. Geophys Res Lett 37(22):n/a–n/a. https://doi.org/10.1029/2010gl04508

Hoegh-Guldberg O, Jacob D, Taylor M, Bindi M, Brown S, Camilloni I, Diedhiou A, Djalante R, Ebi KL, Engelbrecht F, Guiot J, Hijioka Y, Mehrotra S, Payne A, Seneviratne SI, Thomas A, Warren R, Zhou G (2018) Impacts of 1.5 ºC global warming on natural and human systems. In: Masson-Delmotte V, Zhai P, Pörtner H-O, Roberts D, Skea J, Shukla PR, Pirani A, Moufouma-Okia W, Péan C, Pidcock R, Connors S, Matthews JBR, Chen Y, Zhou X, Gomis MI, Lonnoy E, Maycock T, Tignor M, Waterfield T (eds) Global warming of 1.5 °C. An IPCC special report on the impacts of global warming of 1.5 °C above pre-industrial levels and related global greenhouse gas emission pathways, in the context of strengthening the global response to the threat of climate change, sustainable development, and efforts to eradicate poverty (In Press)

INGC (2009) INGC climate change report: study on the impact of climate change on disaster risk in Mozambique. In: Asante K, Brito R, Brundrit G, Epstein P, Fernandes A, Marques MR, Mavume A, Metzger M, Patt A, Queface A, Sanchez del Valle R, Tadross M, Brito R (eds) Instituto Nacional de Gestão de Calamidades, Mozambique

INGC (2013) Asante K, Brito R, Brundrit G, Epstein P, Fernandes A, Marques MR, Mavume A, Metzger M, Patt A, Queface A, Sanchez del Valle R, Tadross M, Brito R (eds) Main report: INGC climate change report: study on the impact of climate change on disaster risk in Mozambique. Maputo, Mozambique

INGC (2019) Avaliação preliminar de ocorrência do ciclone IDAI e Kanetth. Relatório de Governo de Moçambique (Relatório da Situação Nacional 1). Maputo, Moçambique, p 17. https://www.afro.who.int/sites/default/files/201905/NationalSitRep1Mozambique10May2019_PORT.pdf

Instituto Nacional de Estatística (INE) (2017). IV Recenseamento Geral da População e Habitação de Mocambique. http://www.ine.gov.mz/

IPCC (2014) Core Writing Team, Pachauri RK, Meyer LA (eds) Climate change 2014: synthesis report. Contribution of working groups I, II and III to the fifth assessment report of the intergovernmental panel on climate change. IPCC, Geneva, Switzerland, p 151. Retrieved at https://www.ipcc.ch/report/ar5/syr/

IPCC (2019) Pörtner H-O, Roberts DC, Masson-Delmotte V, Zhai P, Tignor M, Poloczanska E, Mintenbeck K, Alegría A, Nicolai M, Okem A, Petzold J, Rama B, Weyer NM (eds) IPCC special report on the ocean and cryosphere in a changing climate (In press)

Jaarsma CF (1997) Approaches for the planning of rural road networks according to sustainable land use planning. Landsc Urban Plan 39(1):47–54. https://doi.org/10.1016/s0169-2046(97)00067-4

Kry S, Sasaki N, Datta A, Abe I, Ken S, Tsusaka TW (2020) Assessment of the changing levels of livelihood assets in the Kampong Phluk community with implications for community-based ecotourism. Tour Manag Perspect 34:100664

Masipa TS (2017) The impact of climate change on food security in South Africa: current realities and challenges ahead. Jàmbá: J Disaster Risk Stud 9(1). https://doi.org/10.4102/jamba.v9i1.41

Mavume A, Rydberg L, Rouault M, Lutjeharms J (2009) Climatology and landfall of tropical cyclones in the South-West indian ocean. West Indian Ocean J Mar Sci 8(1).https://doi.org/10.4314/wiojms.v8i1.56672

Ministério de Planificação e Desenvolvimento (2010) Pobreza e Bem Estar em Moçambique: Terceira Avaliação Nacional. Maputo, Moçambique, p 158. https://www.open.ac.uk/technology/mozambique/sites/www.open.ac.uk.technology.mozambique/files/files/IOF%20-

Mosca J (2017) Agricultura familiar em Moçambique: ideologias e políticas. Observatório do Meio Rural (OMR). Maputo, Moçambique, p 38. https://revista.fct.unesp.br/index.php/nera/article/viewFile/5296/4057

Moyo EN, Nangombe SS (2015) Southern Africa's 2012–13 violent storms: role of climate change. Procedia IUTAM 17:69–78. https://doi.org/10.1016/j.piutam.2015.06.011

Namatama N (2020) An assessment of stakeholders' participation in land use planning process of Luapula Province planning authority. Land Use Policy 97:104735. https://doi.org/10.1016/j.landusepol.2020.104735

Nkomwa EC, Joshua MK, Ngongondo C, Monjerezi M, Chipungu F (2014) Assessing indigenous knowledge systems and climate change adaptation strategies in agriculture: a case study of Chagaka Village, Chikhwawa, Southern Malawi. Phys Chem Earth Parts A/B/C 67–69:164–172. https://doi.org/10.1016/j.pce.2013.10.002

Nyiwul L (2020) Climate change adaptation and inequality in africa: case of water, energy and food insecurity. J Clean Prod 123393. https://doi.org/10.1016/j.jclepro.2020.123393

Observador (2014) Descoberto maior depósito mundial de grafite em Moçambique. Retrieved at https://observador.pt/2014/10/24/descoberto-maior-deposito-mundial-de-grafite-em-mocambique/

Ogada MJ, Rao EJO, Radeny M, Recha JW, Solomon D (2020) Climate-smart agriculture, household income and asset accumulation among smallholder farmers in the Nyando basin of Kenya. World Dev Perspect 100203

Olabanji MF, Ndarana T, Davis N, Archer E (2020) Climate change impact on water availability in the olifants catchment (South Africa) with potential adaptation strategies. Phys Chem Earth Parts A/B/C 102939. https://doi.org/10.1016/j.pce.2020.102939

Pandit R, Parrota J, Anker Y, Coudel E, Diaz Morejón CF, Harris J, Karlen DL, Kertész Á, Mariño De Posada JL, Ntshotsho Simelane P, Tamin NM, Vieira DLM (2018) Chapter 6: responses to halt land degradation and to restore degraded land. In: Montanarella L, Scholes R, Brainich A (eds) IPBES: the IPBES assessment report on land degradation and restoration. Secretariat of the intergovernmental science-policy platform on biodiversity and ecosystem services, Bonn, Germany, pp 435–528

Parker P (1995) From sustainable development objectives to indicators of progress: options for New Zealand communities. NZ Geogr 51(2):50–57

Pasquini L, van Aardenne L, Godsmark CN, Lee J, Jack C (2020) Emerging climate change-related public health challenges in Africa: a case study of the heat-health vulnerability of informal settlement residents in Dar es Salaam, Tanzania. Sci Total Environ 141355. https://doi.org/10.1016/j.scitotenv.2020.141355

Prinn RG, Weiss RF, Arduini J, Arnold T, DeWitt HL, Fraser PJ, Ganesan AL, Gasore J, Harth CM, Hermansen O, Kim J, Krummel PB, Li S, Loh ZM, Lunder CR, Maione M, Manning AJ, Miller BR, Mitrevski B, Mühle J, O'Doherty S, Park S, Reimann S, Rigby M, Saito T, Salameh PK, Schmidt R, Simmonds PG, Steele LP, Vollmer MK, Wang RH, Yao B, Yokouchi Y, Young D, Zhou L (2018) History of chemically and radiatively important atmospheric gases from the Advanced Global Atmospheric Gases Experiment (AGAGE). Earth Syst Sci Data 10:985–1018. https://doi.org/10.5194/essd-10-985-2018

Robledo C, Clot N, Hammill A, Riché B (2012) The role of forest ecosystems in community-based coping strategies to climate hazards: three examples from rural areas in Africa. Forest Policy Econ 24:20–28. https://doi.org/10.1016/j.forpol.2011.04.006

Shi X, Chen J, Gu L, Xu C-Y, Chen H, Zhang L (2020) Impacts and socioeconomic exposures of global extreme precipitation events in 1.5 and 2.0 °C warmer climates. Sci Total Environ 142665. https://doi.org/10.1016/j.scitotenv.2020.142665

Smith BM, Chakrabarti P, Chatterjee A, Chatterjee S, Dey UK, Dicks LV, Majhi KR, Basu P (2017) Collating and validating indigenous and local knowledge to apply multiple knowledge systems to an environmental challenge: a case-study of pollinators in India. Biol Cons 211:20–28. https://doi.org/10.1016/j.biocon.2017.04.032

Syafwina (2014) Recognizing indigenous knowledge for disaster management: smong, early warning system from Simeulue Island, Aceh. Procedia Environ Sci 20:573–582. https://doi.org/10.1016/j.proenv.2014.03.070

Tegegne G, Melesse AM, Alamirew T (2020) Projected changes in extreme precipitation indices from CORDEX simulations over Ethiopia, East Africa. Atmos Res 105156. https://doi.org/10.1016/j.atmosres.2020.105156

Tian H, Xu R, Canadell JG, Thompson RL, Winiwarter W, Suntharalingam P, Yang J, Janssens-Maenhout G (2020) A comprehensive quantification of global nitrous oxide sources and sinks. Nature 586(7828):248–256. https://doi.org/10.1038/s41586-020-2780-0

Tsige M, Synnevåg G, Aune JB (2020) Gendered constraints for adopting climate-smart agriculture amongst smallholder Ethiopian women farmers. Sci Afr 7:e00250. https://doi.org/10.1016/j.sciaf.2019.e00250

UNICEF (2019). Cyclone Idai and Kenneth cause devastation and suffering in Mozambique. Retrieved at https://www.unicef.org/mozambique/en/cyclone-idai-and-kenneth

United Nations—Department of Economics and Social Affairs (2019) World population prospects. https://population.un.org/wpp/

United Nations Office for the Coordination of Humanitarian Affairs (2014). Tropical cyclone Hellen—Mar 2014. https://reliefweb.int/disaster/tc-2014-000043-mdg

Van Lier HN (1998) The role of land use planning in sustainable rural systems. Landsc Urban Plan 41(2):83–91. https://doi.org/10.1016/s0169-2046(97)00061-3

Wichern J, Descheemaeker K, Giller KE, Ebanyat P, Taulya G, van Wijk MT (2019) Vulnerability and adaptation options to climate change for rural livelihoods—a country-wide analysis for Uganda. Agric Syst 176:102663. https://doi.org/10.1016/j.agsy.2019.102663

WMO (2019) Reducing vulnerability to extreme hydro-meteorological hazards in Mozambique after Cyclone IDAI. Mission report following tropical cyclone IDAI (29 April–7 May 2019)

Yaduvanshi A, Nkemelang T, Bendapudi R, New M (2020) Temperature and rainfall extremes change under current, 1.5 °C, 2.0 °C and 3.0 °C global warming across Indian climate zones. Weather Clim Extremes 100291. https://doi.org/10.1016/j.wace.2020.10029

Zakaria A, Azumah SB, Appiah-Twumasi M, Dagunga G (2020) Adoption of climate-smart agricultural practices among farm households in Trans-Volta Togoland, Ghana: the role of farmer participation in training programmes. Technol Soc 101338. https://doi.org/10.1016/j.techsoc.2020.101338

Zhou X, Chen J, Li Z, Wang G, Zhang F (2017) Impact assessment of climate change on poverty reduction: a global perspective. Phys Chem Earth Parts A/B/C 101:214–223. https://doi.org/10.1016/j.pce.2017.06.011

Adaptive Management as a Vehicle to Achieve Sustainability of Boreal Forests: A Historical Review from Fennoscandia to Minnesota

Jan Kunnas and Bruno Borsari

1 Introduction

Boreal forests form a distinctive biome of the northern hemisphere (50° to 65° N latitude), extending from Scandinavia, through Russia, across Siberia, to Alaska, through Canada (Fig. 1), and cover over 8% of earth's terrestrial surface (FAO 2020). One major challenge for the management of northern forests is the long growing time necessary for trees, to reach harvestable size in this cold environment. More challenges are caused by climate change, as boreal forests are particularly vulnerable to the long-term effects of climate change, and its more immediate disturbances like drought, insect infestations and fire (Fischlin et al. 2007).

The purpose of this study consisted in reviewing the use and approaches to land management in two selected regions of the boreal forest biome: Finland and Sweden, Europe and Minnesota, USA. This review presented a historical documents analysis of forest lands use over the past 200 years, with the intent of understanding differences and/or similarities in management approaches and policy making, that regulated the use of forests in the selected study regions. It aimed at documenting what has changed in managing boreal forests and what emphasis if any, has been given to sustainability. Conserving the productivity of these distinctive ecosystems without affecting too drastically, their ecological integrity and regenerative capabilities justified the need for this study.

J. Kunnas (✉)
School of Resource Wisdom, University of Jyväskylä, Jyväskylä, Finland

B. Borsari
Department of Biology, Winona State University, Winona, MN 55987, USA
e-mail: bborsari@winona.edu

W. Leal Filho et al. (eds.), *Sustainability in Natural Resources Management and Land Planning*, World Sustainability Series,
https://doi.org/10.1007/978-3-030-76624-5_18

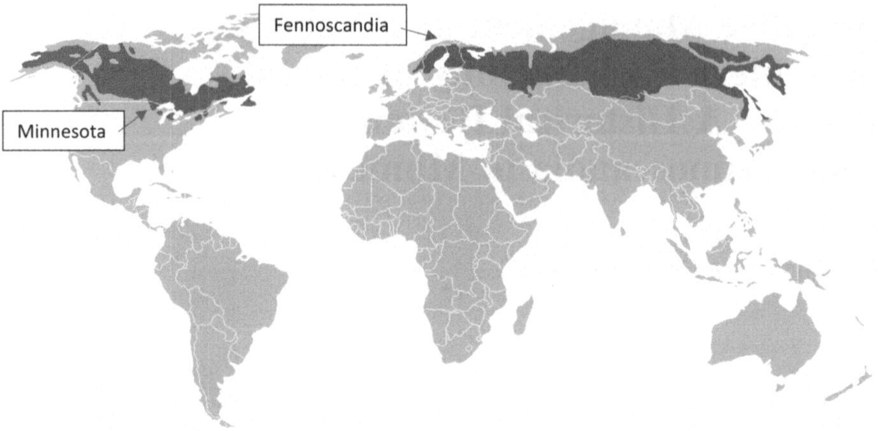

Fig. 1 Distribution of boreal forests and study regions (Modified after: Vzb83 (CC BY-SA 3.0)). Available at: https://upload.wikimedia.org/wikipedia/commons/3/31/Taiga.png

2 Methodology

Staley (2007) suggested looking at the present as a harbinger of something that may happen in the future, when studying historical events. Kunnas et al. (2019) and Kunnas and Myllyntaus (2020) applied Staley's framework on recorded patterns of past forest management and uses in Finland and North Sweden.

These works inspired the methodological approach adopted in this study, whose nature was nonexperimental. Through this research endeavor we examined data to understand the past without looking solely, at developing a chronological list of facts and dates. Rather, we studied the dynamics of history in managing boreal forests, and by reviewing historical evidence, we tried to explain anthropogenic activities and patterns to develop new hypotheses and paradigms, supportive of a sustainable forest management.

3 Results

3.1 Forest Management in Sweden and Finland in the Nineteenth Century

Concerns about massive deforestation, which characterised the policy of the Swedish Government for centuries, remained in Finland, after its annexation to the Russian empire as an autonomous grand duchy, in 1809. Its sawmilling industry became suppressed by tight regulations that supported mercantilist doctrines, wanting to save the forests in favour of future energy needs of the mining industry

(Hanho 1915; Kunnas et al. 2019). A new law in 1851 implemented more drastic restrictions limiting sawmilling operating time to a certain season, each year. It heralded the idea that logging was priority activity for state-owned forests. Unauthorised slash-and-burn cultivation on crown lands became prohibited and, unlike earlier prohibitions, this law was enforced. In Northern Sweden, crown lands in the 1820s were considered of marginal value and thus, suitable for an establishment of homesteads to promote agriculture and population growth. However, a firm establishment of the wood industry brought with it enormous values to crown lands, that had been considered "useless" just a couple of decades earlier (Kuisma 2006; Aarnio 1999; Schager 1925).

Sawmilling became free from restrictive regulations in Finland after the mid-nineteenth century, as liberal ideas gained a foothold, also to strengthen state finances in the aftermath of the Crimean war. Better knowledge about Finland's forest resources worked in favour of sawmills, too. The head of the New Forest Agency, C.W. Gylden, estimated in 1853 that Finnish forests' net annual, timber production was 30 million m^3, whereas consumption only 16 million m^3. Permissions were granted for building several steam sawmills across the country between 1859–1861, with Finland's first, operating in March 1860. Sweden had established already its first steam sawmill in Norrland in 1849 (Kuisma 2006; Kuusterä 1989; Meinander 1945; Heckscher 1968).

An increasing economic value of forests triggered also criticism about the generous allocations of forest lands to peasants in Finland and Sweden. For example, the Governor of Swedish Norrbotten County, P. H. Widmark, argued in 1860 that people establishing new homesteads in forested lands had reduced their farming to a minimum while focusing instead, on timber sale without concerns for insuring a regeneration of forest stands (Carlgren et al. 1925). At the same time, a contiguous series of crop failures in the 1860s speeded up emigration to America, which reached its peak in the 1880s, when one every twelfth Swede emigrated. It was hoped that settlement and industrialization of Norrland would stop this trend. To support agriculture further under these difficult circumstances, the allocations of forest lands for new settlements became even larger (Betänkande 1931).

Soon, the need for wood as raw material increased, as sawmilling diversified in other wood-processing industries, such as mechanical and chemical pulping, spool and plywood production. Sweden's first groundwood mill for mechanical pulping was built in 1857 at Trollhättan Falls (Heckscher 1968), whereas Finland's established in 1859 in the municipality of Vyborg and the second in Tampere in 1866. This expansion driven by high demands for timber resulted in severe cases of deforestation. Growing demands for forest products and increasing trees harvesting, worried the Swedish sawmill industry of exhausting soon, timber availability. Pressure from various sawmill owners to protect the industry against competition from pit props sales, pulp and paper industry lead to regulation enactment in 1874, with measures against logging the young forest in Norrbotten county (dimension law), which banned harvesting and shipment of timber, that did not have required dimensions. In 1882, these provisions applied also in Västerbotten (Holmström 1988).

At this time, individual farmers in Finland and Sweden held large tracts of forest and to secure timber sawmills and pulp mill companies leased logging rights from them. Initially, the leases could be for 50 years in Sweden, but by 1905 these had been reduced to five years. Also, companies began to buy whole farms to secure more timber, as it was not allowed the purchase of forests alone. By the beginning of the twentieth century companies owned over a third of all individual homestead lands in Dalarna and Norrland, Sweden. Thus, in 35 years, forest companies in central Norrland acquired nearly 90% of their present-day forest holdings. These purchases started an intense and prolonged debate about the companies' right to continued forest land acquisition, generating concerns about the future of peasants' land ownership. The bids received from forestry companies were considered great deals and only later, farmers realized they had been cheated for having sold their forest plot, or logging rights at low prices. In 1906 a law was introduced that banned companies from buying forests in the four northernmost counties of Sweden (Holmström 1988; Rentzhog 1991). Similar provisions were adopted in Finland, from 1885 to 1915, when land acquisitions by logging companies became restricted (Karjalainen 2000).

Due to fast deforestation, sawmilling profits fell as the harvested size of the pine trees diminished, forcing mills to acquire timber further away from their facilities. Around the turn of century logs with a diameter of 15 cm, or less had become acceptable for harvest, whereas a few decades earlier a timber of 25 cm would have been rejected. The economic salvation of the forest industry occurred with the sulphite cellulose process (discovered in 1867) and its first use in Sweden in 1874, at Bergvik sulphite mill in Gävleborg County. As the sulphite process used spruce trees, it caused a re-evaluation of what a forest meant in the catchment areas for the factories (Petterson 2015).

3.2 Twentieth and Twenty First Century: Changing Values and Climate

In 1874, the Swedish government enacted provisions compelling all private owners to ensure forests' regeneration and the 1903 Forestry act established as a law, which emerged from those early regulations. However, only private owned lands had to comply, and not those in Norrbotten nor Västerbotten, where the "dimensions law" specifying minimum diameters of trees continued to allow these to be felled. As young growing forests were cut to increasing demand for firewood during World War I, the 1903 law was revised in 1923, requiring in addition to re-growth after harvesting also the protection of young forests. After 1925 this law was enacted in Norrbotten and Västerbotten as well, when the dimensions law was repealed (Nylund 2009; Bernes and Lundgren 2009).

On December 6, 1917 Finland declared its independence from Russia and independence brought change in Finnish forest policy, especially regarding state-owned forests. Finnish director of the organization of state-owned forest (Metsähallitus), A.K. Cajander stated, that as an independent country, Finland needed more income, and this should be obtained from its forests. He believed that forest revenues were expandable by increasing logging, improving their management, and increasing the drainage of peatlands, thus promoting forests' growth. This new view was presented to forest officials in 1919, suggesting that diameter standards for trees to be cut were to be abandoned. Experience had shown that this practice was not effective in reforestation efforts and in the long run it would have lost its economic viability. Instead, restorative logging practices were to be employed, including sequential cutting, through the seed-tree cutting method. Cajander's forest management theory recognised forest sites and types according to vegetation structure and composition and it is still used today in forestry planning and management (Parpola 2014; Laitakari 1960).

World War II brought significant changes in the harvesting methods of Finland′s state-owned forests. To obtain wood as quickly as possible, felling occurred alongside transport routes, and parcels were clear-cut to provide maximum amount of wood per unit area. Serious drawbacks were noted in this this plan, but Metsähallitus stated that it was prepared to carry out logging in manners and quantities that did not conform to rational forest management (Parpola 2014). The practice of clear cutting, introduced under exceptional circumstances, became the norm in 1948 with the so-called "selective cutting declaration". Eventually, forest owners who wanted to manage their forests differently were prosecuted, so that their properties could have been forcibly managed in accordance with current provisions (Lähde 2015).

As the Finnish forest industry resumed its post-world war activities in the mid-1950s, the demand for wood exceeded annual growth rate of the trees. Consequently, the total standing stock shrunk from 1540 million m^3 in the early 1950s to 1490 million m^3 by 1970. This gave impetus to several timber production programmes that succeeded, as indicated by the latest forest inventory, showing a standing stock of 2470 million m^3 in 2017, 1.6 times that of 1970. Simultaneously, annual growth increased from 55.2 million m^3 in the early 1950s to 107 million m^3 in 2017 (Tomppo and Henttonen 1996; Luonnonvarakeskus 2018). A major contribution to this growth in timber volumes was the drainage of forested peatlands, accompanied by fertilizers use. However, the effect of clear-cut followed by tree planting has been heavily disputed. Some compared this practice to the previous selective diameter-limit cutting, others instead argued that the initial slow growth of saplings outweighed the benefits of clear-cuts. Simultaneously, the effect of ending forest grazing has gained less notion (Henttonen et al. 2020; Huikari 1998; Kuusela 1999; Lähde et al. 1999). Regardless of what component yielded most timber volume, the measures undertaken to achieve this had major effects on the structure, dynamics and biodiversity of forests. According to the fifth national endangerment estimate, for 733 species (representing 27.5% of the endangered species), it was found that the primary cause of loss was attributable to changes in forest habitats, caused by logging. A decline of decaying wood in old-growth forests and large trees constituted primary reasons of loss, for more than 50% of these species,

Fig. 2 Decaying wood in an old-growth, boreal forest in Pyhä-Häkki, Finland (Photo Jan Kunnas)

whereas for more than 25% of the species it was forest regeneration and management (Hyvärinen et al. 2019) (Fig. 2).

Also in Northern Sweden decades of uncontrolled logging caused timber dimensions to shrink, forcing a relocation of sawmilling back south, where it remains today. By the end of the 1920s, Norrland's share of timber production decreased to 42% and further, to 27% by the 1950s (Petterson 2015). Although the epicentre of the forest industry moved southwards, the pressure towards northern forests did not diminish. The skewed age structure of forests with prevailing populations of young trees in lieu of mature forests, forced timber acquisitions from mountain near forests, the last remaining frontier. This is a thousand kilometres long belt of forested mountains from Dalarna to Lapland. It had been protected from logging since the early 1950s, as forest regrowth after harvesting was uncertain, due to extreme climatic conditions. However, in 1982 logging restrictions in this region were removed, causing an acrimonious debate that followed the implementation of logging plans and road building on public and private lands (Byström 1986).

The growing environmental awareness manifested in opposition to logging of forested mountain lands materialized in a new policy that was enacted by the Swedish Parliament in 1993. It was characterized by two equally important goals; the production of valuable timber yields, while conserving biodiversity (Regeringen 1993). Similar stipulations about biodiversity protection were introduced to Finnish forest legislation (Finnish Parliament 1996). As a measure of balance between these two goals, Sweden achieved protection of 3.2% of its forest lands as national parks and nature reserves, 0.2% as habitats and conservation agreements, and 5% as voluntary allocations required by forest certification schemes. Consequently, 91.6% of all productive forest land is used for timber production. In Finland, protected forest areas cover 5.7% of all forest lands, leaving 94.3% for forestry. The success of these provisions was substantiated by assessing forests' species diversity and richness, through periodic surveys. In Sweden for example, it was discovered that of 1.787 known forests' species, about half (861), were designated as endangered, making this information vital in managing forests sustainably (Luke 2019; Swedish Forest Industries Federation 2011).

In the 1950s birch, aspen and other broadleaved tree species had not much use beyond firewood. They were regarded as weeds and treated with phenoxyacetic acids (herbicides similar to Agent Orange, used in the Vietnam war by the US troops) until these were banned in the 1980s. Birch was also re-assessed commercially, as new technology allowed the making of pulp from it, and the plywood industry expanded. Later, aspen and other broadleaved trees were revaluated, as important components of forests' biodiversity (Enander 2007). The non-native lodgepole pine (*Pinus contorta*) introduced to Sweden in the 1920s from North America and used in large scale replanting projects since the early 1970s was also re-evaluated. Approximately, 40.000 hectares were planted with it in 1984, but thereafter, its use dwindled as large areas where damaged by the fungus *Gremeneniella abietina.* The infection spread also to plantations of domestic pine (*Pinus sylvestris*), inducing experts to argue that the real problem was clear-cutting and tree planting in climatic conditions that favoured the fungus. The scale of this infection became devastating to the reputation of *Pinus contorta* being planted and marketed as a durable alternative for growing timber species in severe climatic conditions. When its planting ended in the early 1990s, lodgepole pine-covered about 4% of the productive forest land in Sweden (Hagner 2005; Widenfalk 2015).

3.3 Management of Boreal Forests in Minnesota

The northern forests of Minnesota are part of the southern edge of the boreal forest biome (Pleticha et al. 2019). The trees community is dominated by conifers such as black spruce, white spruce, tamarack, jack pine, and balsam fir, all well adapted to the region's cold winters and shallow soils (Fig. 3). Hardwood trees like birch, aspen, maple and basswood are often interspersed with the conifer species, according to soil fertility conditions, which in Spodosols are typically very limited and nitrogen deficient (Goldblum and Rigg 2010).

Minnesota forests began to be harvested in the mid-nineteenth century when European colonists looking for new sources of high-quality timber moved west, from Maine, through Wisconsin. By 1849 logging was well established in the new, territory and quickly spread inland every winter, near pristine stands of pine (*Pinus strobus* L.). Winter was logging season and the icy roads facilitated the transport of logs to the nearest river, where these were unloaded into the riverbanks, waiting for the spring thaw, to be driven to sawmills, downriver. This transportation method

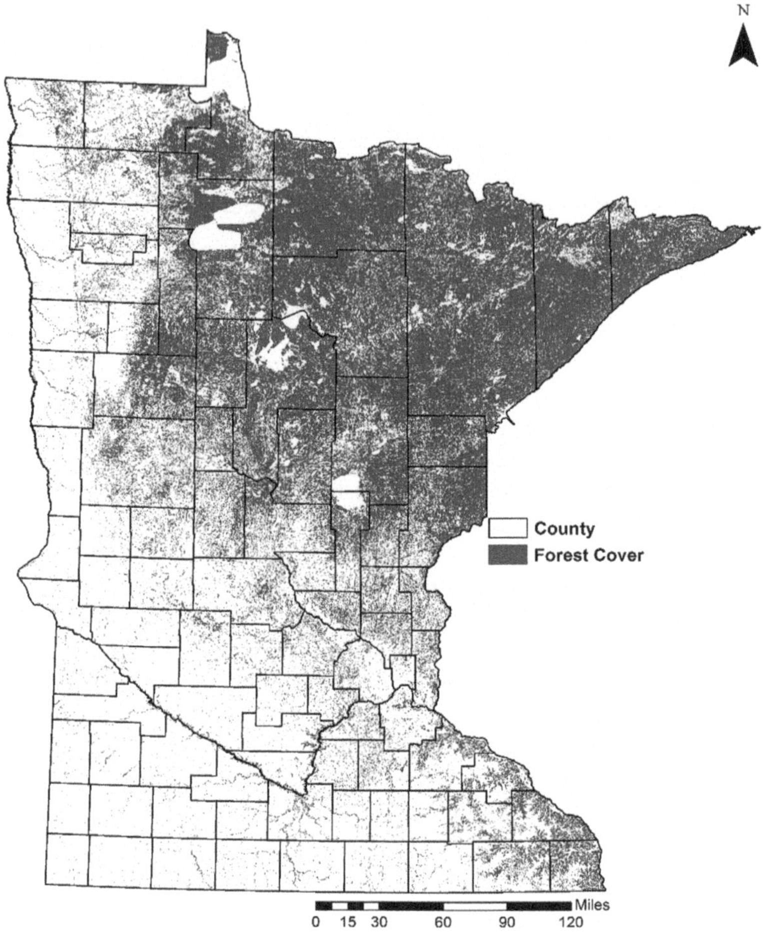

Fig. 3 Forests (7.1 million ha.) concentrate in the northeastern part of Minnesota (Hillard et al. 2018, 145)

prevailed until the 1890s when railroads were built and reached inland territories (MNHS 2020).

Through the 1800s, timber production and sawmilling were inextricably intertwined. Sawmills were built next to rivers, which were both the "highways" to deliver timber from the woods to the mills and their source of transportation power. The first commercial sawmill in Minnesota opened in 1839 at Marine on the St. Croix river, to which followed one in Stillwater, Minneapolis and Winona (Havighurst 2005). Steam power started in sawmilling during the 1870s, replacing water as energy source and allowing sawmills to move elsewhere, besides river towns. Further expansions of the logging industry occurred in Minnesota by 1880 with the growth of railroad networks and improvements of steam engine technology yet, the peak of lumber production was achieved in the early years of the twentieth century, when lumberjacks could harvest 610 million board meter per year. At this rate, however, the prime pine stands became exhausted by the 1920s, causing the demise of the sawmilling industry in 1929. With the crisis of the timber industry, companies relocated to the Pacific Northwest or the Southern states. Lumber companies that remained in Minnesota shifted production from logs to pulp, paper and various building materials. The last log drive in Minnesota occurred on the Little Fork River in 1937 (MNHS 2020).

In the mid-1960s, more potent machines like skidders and crane loaders were employed to clear cut the new, mixed boreal forests of aspen, spruce, birch, which remains the preferred method for these types of harvests. Chen and Popadiouk (2002) identified these forests as boreal mixed woods (BMWs), suggesting that a good understanding of their ecology is very important for developing sustainable management of this and similar landscapes. Also, a new timber harvester (Cut-to-length) has been used in Minnesota woods, since the early 1990s, reducing the number of machines at logging sites, while minimizing risks of damaging the forest floor.

Data from a study about disturbance frequency and patch structure from pre-European settlement to present in the forests of Minnesota suggested that management practices have a greater influence than natural processes in generating landscape patterns and this information can be used by land managers to restore spatial pattern variability in managed forest landscapes (White and Host 2008). Friedman and Reich (2005) conceded that a significant change (85%) has occurred in the relative abundance and dominance of tree species in northern Minnesota forests, in more than 100 years. Therefore, present-day logging industry is not itinerant anymore and it had to adapt to environmental changes, demanding that professionals employed in this industry possess multiple skills, spanning from ecological land management, to marketing and an ability to operate and maintain complex multipurpose machinery.

A longitudinal study (1840–2005) of Canadian forests showed that twentieth century forestry practices which included fire, generated a forest landscape where younger forest habitats began to dominate this environment. It concluded that human-caused fires expected to increase wildfire activity in the boreal forests of eastern North America and these, in conjunction with continued forest management, could jeopardize recovery and resilience of boreal forests (Boucher et al. 2014). Also, climate change is affecting the forests of North America with clear visible impacts, especially in Northern Minnesota, where average temperatures have at most increased by over 3 °C degrees from the period 1901–1960 to 1991–2012.[1] A similar situation characterizes the Nordic countries, like Finland, where the average annual temperature has risen by about 2 °C from the 1880s to the early 2010s (Mikkonen et al. 2015).

[1]https://www.mprnews.org/story/2015/02/02/climate-change-primer (accessed 3.12.2020).

Frelich and Reich (2010) predicted that current environmental changes will determine major shifts to large swaths of forests along its southern boundary, in the next 50–100 years. More specifically, future climate conditions will cause higher mortality among mature trees, because of prolonged droughts, unpredictable fires, windstorms, and diseases by insect pests and other pathogens. Increasing populations of herbivores and invasions by exotic earthworm species will hamper tree seedlings' growth and determine a rapid change to more northeastward latitudes of the prairie–forest border.

4 Discussion

The valuation and use of forests changed several times in the last two hundred years, especially when considering their rotation period (Fig. 4). For example, in the mid-nineteenth century, only two decades before the unprecedented growth of the sawmill industry, forests were envisioned as an energy source (charcoal) for the mining industry. In more recent times, sharp changes have occurred in the attitude and economic valuation of forests, to include the cultivation of broadleaved and even, non-native tree species. Although it may be difficult to anticipate future human needs and forests' use, it remains imperative to adapt forests' management to anthropogenic induced disturbances, to conserve their regenerative capabilities.

Climate change adds uncertainties regarding the future of forests, and it is likely to exacerbate conflicts about different views of future forests' use, as well. Over the course of the twenty-first century, boreal forests are expected to experience the highest increase in temperatures among all forest biomes and continuous extraction of resources will likely impose more pressures on boreal forests' health (Frelich and

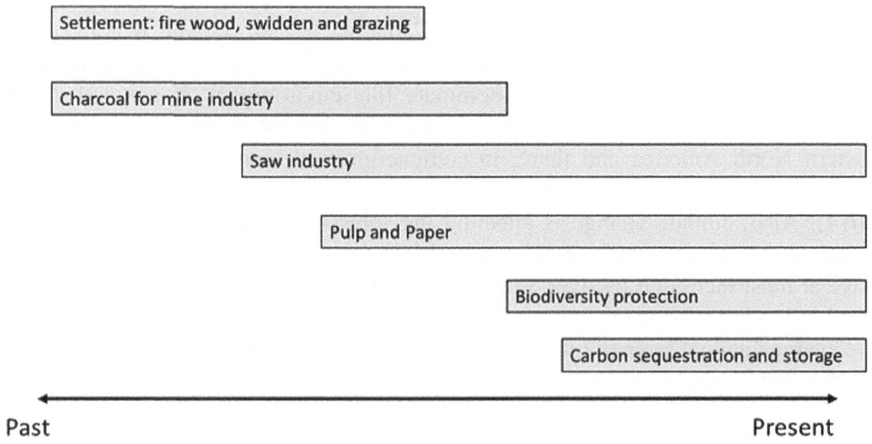

Fig. 4 Shifting emphases in forest uses, from past to present times (Adapted from: Kunnas and Myllyntaus 2020)

Reich 2010). Climate change will lead to greater variability in temperature and precipitation, which will result in more unpredictable seasonal shifts and increased risks of outbreaks of insects' infestations, zoonotic diseases and influxes of invasive species. Increased flooding due to extreme weather patterns may lead to increased leaching of nutrients from harvested timber into aquifers and damaging water quality (Gauthier et al. 2015).

Therefore, management that is adaptive to change will become a keystone tool in maintaining forests' capabilities to recover from natural, or anthropogenic disturbances, without jeopardizing biological productivity. One valuable approach to achieve this, is the "climate-wise classification of forests" used by Metsähallitus, the caretaking organization of state-owned forests in Finland. This method classifies every forest as either carbon sink or carbon storage, based on an inventory of soil, land use and ecological data. This classification serves to design distinctive management practices for each forest class. Methods used for carbon sequestration include fertilization, regeneration with selectively bred seeds and seedlings, regeneration of underproductive forests and afforestation. Methods used for carbon storage are improving forest density, prolonging the rotation period between wood harvests and restricting forestry operations to facilitate other forms of forest use (Mäntyranta 2018). In practice, this classification provides little change, as in general, compartments, where forest use was already restricted, are treated as carbon storages, while compartments designated for normal forest use are treated as carbon sinks.

By increasing rotation length and by decreasing thinning intensity it would be possible to enhance forest carbon stocks by a factor of 1.5–2 without diminishing wood yields when compared to current practices. From the climate and carbon balance points of view, the dilemma of long rotations in forestry is the cyclic nature of the carbon from the atmosphere to trees biomass. To maximize carbon uptake tree stands must be cut regularly, making scientists suggest an application of a mixed strategy, where a large forest area is devoted to carbon sequestration, and commercial forests are harvested in rotation, to guarantee satisfactory timber productions (Pingoud et al. 2018).

Continuous cover forestry provides a cost-efficient alternative for carbon sequestration, as in conventional even-aged rotation forest management, a stand is a source of carbon after the clear-cut. For example, Peura et al. (2017) showed that a 'continuous cover forestry' approach favours timber net value, carbon sequestration, bilberry production, scenic beauty and abundance of large trees. In addition to this, the proportion of saw logs compared with pulpwood is higher (Pukkala 2014). This provides increased possibilities to replace energy-intensive materials like steel and cement in buildings, with timber that in addition provides a long-lasting carbon sink. This approach provides advantages regarding uncertainty issues about forests' future. It has the potential to maintain habitat connectivity, providing corridors for the northward migration of species due to more frequent warming temperatures occurring at northern latitudes (Pukkala et al. 2012). Continuous cover management also reduces wind damage (Pukkala et al. 2016). In windy areas, it may though be reasonable to replace spruce with deciduous trees, as they are less sensitive to wind damage. This danger materialized when hurricane Gudrun hit southern Sweden in January 2005, damaging 75 million m^3 of spruce-dominated forest (Felton et al. 2016).

All three management approaches (prolonged rotation periods, continuous cover forestry and increased share of broadleaved trees), are consistent with biodiversity conservation goals (Felton et al. 2016). In the end, the method of felling itself is not as important, as conserving certain, structural features of the forest, such as dead wood, large trees and a mixture of deciduous and coniferous tree species (Koivula and Vanha-Majamaa 2020).

These strategies of forest conservation and management are clear demonstrations of adaptive management to uncertainties and change, brought about by a future unknown. In order to cope with uncertainty and changing needs and values, we propose employing a distinctive adaptive management model for present and future uses of boreal forests (Fig. 5).

This model is cyclical and articulated in four distinctive steps, leading to making decisions that emphasize on multiple data sources. Externally located, between each step category are four variables that at different levels of magnitude and frequency, will influence the whole process, from one step to the next. Monitoring or observation phase (Step 1) produces quantitative and qualitative data that allow an evaluation (Step 2) of forests' conditions and health. Disturbance, which is the variable between steps 1 and 2 will affect the data collected during a specific timeframe, due to a variation of its scale, at the beginning of every, new cycle. Resilience is the variable between assessment (Step 2) and planning (Step 3) and with its effect, it will mitigate change caused by disturbance.

Knowledge of forests' natural history or ecological memory provides stakeholders with an additional set of data that will enhance the accuracy of step 2 (Assessment) and the efficacy of step 3 (Planning), prior to enacting the action plan. This approach to managing boreal forests was inspired by the idea that an ecological memory is pivotal to understand ecosystems' responses to disturbance. This is supported by information legacies (how species adapt to disturbance) and

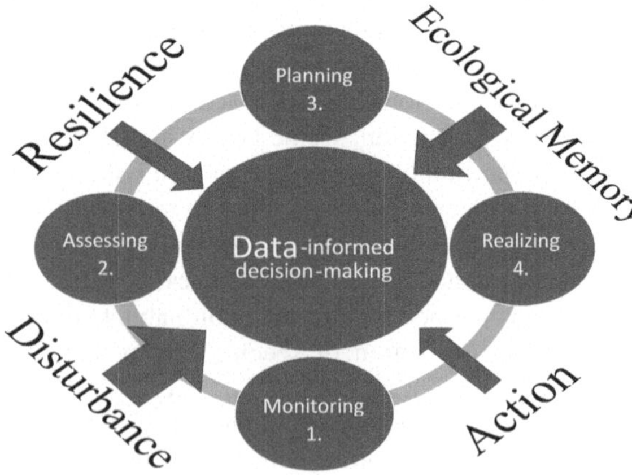

Fig. 5 Boreal forests, adaptive management model

material legacies (abiotic/biotic structures generated by one disturbance at the time), making ecosystems' resilience subordinate to disturbances that are conservative of these legacies (Johnstone et al. 2016).

The knowledge acquired from applying the proposed model will add data and expand the records base about the ecological memory of forests, which will drive further planning and actions in forests conservation and management. Species conservation demands adaptive approaches in the management of natural resources, as the one here proposed. The model suggests a valid framework for enhancing forests management with stewardship and conservation as keystone priorities.

5 Conclusions

This review was not free from limitations because the study areas represented a minuscule section of the boreal forest biome. Another constraint that affected this work consisted in having to summarize selectively, a plethora of historical data from the three regions, over two hundred years.

Nonetheless, one clear conclusion emerged from our study and this is the inability to predict with accuracy the future of boreal forests even through time spans as short as those needed by trees to reach harvest maturity. Unpredictable weather patterns caused by a changing climate add further uncertainties to prediction-making. Therefore, aiming at preserving the ecological integrity and productivity of boreal forests remains the holistic goal for achieving their sustainable management, worldwide. Our historical analysis revealed a flux of methodologies and policies that especially during the last century strove at avoiding the possibility of exhausting forests' resources, if their extraction carried on, unrestrained. The work of Leopold (1949) with its 'land ethic' set the pace for the restoration movement, which driven by ecological knowledge added values that remain pivotal in preparing resource managers. Achieving a balance between forests use and conservation demands that stakeholders employ a broad spectrum of management practices, supported by the evidence obtained from a multitude of quantitative and qualitative data. This study advocated for an inclusion of data derived also from environmental history, to complement assessments of forests, leading to design effective action plans that become truly, restorative and regenerative. As recommended by our model, preservation of the ecological memory of boreal forests stands as an imperative condition for achieving this higher-level management for these ecosystems, their associated resources and services, in pursuit of sustainability. Hopefully, the model here proposed will facilitate also communication and better understanding across the cultural boundaries of science, economy, policy and forestry, that characterize the community of professionals engaged in the present and future management of forests and natural resources.

Acknowledgements Thanks to Scott Hillard and Val Cervenka at the Minnesota Department of Natural Resources for providing an updated map of the Minnesota forests.

References

Aarnio, J. (1999). Kaskiviljelystä metsätöihin. Joensuu: Joensuun yliopiston maantieteen laitos

Bernes C, Lundgren LJ (2009) Bruk och missbruk av naturens resurser. Naturvårdsverket, Stockholm

Betänkande med förslag till lag om vård av vissa skogar inom Västerbottens och Norrbottens läns Lappmarker (1931). Jordbruksdepartementet, Stockholm

Boucher Y, Grondin P, Auger I (2014) Land use history (1840–2005) and physiography as determinants of southern boreal forests. Landscape Ecol 29:437–450

Byström M (1986) Vad har hänt. Furan 2:26–29

Carlgren M, Eckermann H, Hellström O, Högbom AG, Kempe F, Kinnman G, Nyström K (1925) Skogsbruk och skogsindustrier i Norra Sverige. Almqvist & Wicksells Boktryckeri, Uppsala

Chen HYH, Popadiouk RV (2002) Dynamics of North American boreal mixedwoods. Environ Rev 10:137–166

Enander K-G (2007) Skogsbruk på samhällets villkor. Sveriges lantbruksuniversitet, Umeå

FAO (2020) Global forest resources assessment 2020—key findings. Rome. https://doi.org/10.4060/ca8753en

Felton A, Gustafsson L, Roberge JM, Ranius T, Hjältén J, Rudolphi J, Lindbladh M, Weslien J, Rist L, Brunet J, Felton AM (2016) How climate change adaptation and mitigation strategies can threaten or enhance the biodiversity of production forests: insights from Sweden. Biol Cons 194:11–20

Finnish Parliament (1996) Metsälaki (Forest Act) 12.12.1996/1093. http://finlex.fi/fi/laki/ajantasa/1996/19961093

Fischlin A, Midgley GF, Price JT, Leemans R, Gopal B, Turley C, Rounsevell MDA, Dube OP, Tarazona J, Velichko AA (2007) Ecosystems, their properties, goods, and services. In: Parry ML, Canziani OF, Palutikof JP, van der Linden PJ, Hanson CE (eds) Climate change 2007: impacts, adaptation and vulnerability. Contribution of working group II to the fourth assessment report of the intergovernmental panel on climate change. Cambridge University Press, Cambridge, pp 211–272

Frelich LE, Reich PB (2010) Will environmental changes reinforce the impact of global warming on the prairie–forest border of central North America? Front Ecol Environ 8:371–378

Friedman SK, Reich PB (2005) Regional legacies of logging: departure from presettlement forest conditions in Northern Minnesota. Ecol Appl 15(2):726–744

Gauthier S, Bernier P, Kuuluvainen T, Shvidenko AZ, Schepaschenko DG (2015) Boreal forest health and global change. Science 349(6250):819–822. https://doi.org/10.1126/science.aaa9092

Goldblum D, Rigg LS (2010) The deciduous forest—boreal forest ecotone. Geogr Compass 4 (7):701–717

Hagner S (2005) Skog i förändring. Kungl. Skogs- och Lantbruksakademien, Stockholm

Hanho JT (1915) Tutkimuksia Suomen metsätalouden historiasta 19. Vuosisadalla, I Sahateollisuudesta ja sahantuotteiden viennistä 1840- ja 1850-luvulla. Helsinki

Havighurst W (2005) Voices on the river. The story of the Mississippi Waterways. Castle Books, Edison, NJ

Heckscher E (1968) An economic history of Sweden, third printing. Oxford University Press, London

Henttonen HM, Nöjd P, Suvanto S, Heikkinen J, Mäkinen H (2020) Size-class structure of the forests of Finland during 1921–2013. Eur J Forest Res 139:279–293

Hillard S, Bergstrand K, Burns S, Deckard D (2018) Minnesota's forest resources. Grand rapids: department of natural resources; division of forestry. https://files.dnr.state.mn.us/forestry/um/forest-resources-report-2018.pdf. Accessed 3.12.2020

Holmström P (1988) Bruksmakt och maktbruk. Robertsfors AB 1897–1968. Umeå: Umeå Studies in Economic History 9

Huikari O (1998) Arktisten metsien kasvun ihme. Terra Cognita, Helsinki

Hyvärinen E, Juslén A, Kemppainen E, Uddström A, Liukko U (2019) The 2019 red list of finnish species. Ministry of the Environment & Finnish Environment Institute, Helsinki
Johnstone JF, Allen CD, Franklin JF, Frelich LE, Harvey BJ, Higuera PE, Mack MC, Meentemeyer RK, Metz MR, Perry GLW, Schoennage T, Turner MG (2016) Changing disturbance regimes, ecological memory, and forest resilience. Front Ecol Environ 14(7):369–378
Karjalainen T (2000) Puutavarayhtiöiden maanhankinta ja -omistus Pohjois-Suomessa vuosina 1885–1939. Department of History, University of Oulu, Oulu
Koivula M, Vanha-Majamaa I (2020) Experimental evidence on biodiversity impacts of variable retention forestry, prescribed burning, and deadwood manipulation in Fennoscandia. Ecol Process 9(11)
Kuisma M (2006) Metsäteollisuuden maa. SKS, Helsinki
Kunnas J, Myllyntaus T (2020) Lessons from the past? Finnish forest utilization from the mid-18th century to the present. Environ Hist. https://doi.org/10.3197/096734020X15900760737121 (Fast Track)
Kunnas J, Keskitalo CH, Pettersson M, Stjernström O (2019) The institutionalization of forestry as a primary land use in Sweden. In: Keskitalo C (ed) The politics of arctic resources, change and continuity in the "Old North" of Northern Europe. Routledge, London, pp 62–77
Kuusela K (1999) Metsän leiviskät, Metsänsuunnittelu ja saavutukset 1947–1996. Atena, Jyväskylä
Kuusterä A (1989) Valtion sijoitustoiminta pääomamarkkinoiden murroksessa 1859–1913. SHS, Helsinki
Lähde E (2015) Suomalainen metsäsota. Into, Helsinki
Lähde E, Laiho O, Norokorpi Y (1999) Diversity-oriented silviculture in the Boreal Zone of Europe. For Ecol Manage 118(1–3):223
Laitakari E (1960) Metsähallinnon vuosisataistaival 1859–1959. Suomen metsätieteellinen seura, Helsinki
Leopold A (1949) A sand county almanac and sketches here and there. Oxford University Press, New York
Luke (2019) Statistics database. http://bit.ly/2WdpOO0. Accessed 30 Aug 2019
Luonnonvarakeskus (2018) Valtakunnan metsien 12. inventointi (VMI12): Puuvarat kasvavat edelleen, Tiedote 9.10.2018, Liite 1. https://www.luke.fi/wp-content/uploads/2018/10/Tiedote-vmi-2018-liite_1.pdf. Accessed 2 Sept 2019
Meinander N (1945) Vesisahan tarina. Otava, Helsinki
Mikkonen S, Laine M, Mäkelä HM, Gregow H, Tuomenvirta H, Lahtinen M, Laaksonen A (2015) Trends in the average temperature in Finland, 1847–2013. Stoch Environ Res Risk Assess 29:1521–1529
MNHS (2020) Minnesota historical society, forestry history center. Available at: https://www.mnhs.org/foresthistory/learn/conservation
Mäntyranta H (2018) Metsähallitus categorised state-owned commercial forests to combat climate change—the world's first climate-wise forest classification. https://forest.fi/article/metsahallitus-categorised-state-owned-commercial-forests-to-combat-climate-change-the-worlds-first-climate-wise-forest-classification/. Accessed 27 Aug 2020
Nylund J-E (2009) Forestry legislation in Sweden. The Swedish University of Agricultural Sciences, Uppsala
Parpola A (2014) Uinuvat metsävaramme käytön piiriin. Ph.D. diss, University of Helsinki
Petterson R (2015) Sågad skog för välstånd, Den Svenska sågverksindustrins historia 1850–2010. Kungl. Skogs- och Lantbruksakademien, Stockholm
Peura M, Burgas Riera D, Eyvindson K, Repo A, Mönkkönen M (2017) Continuous cover forestry is a cost-efficient tool to increase multifunctionality of boreal production forests in Fennoscandia. Biol Cons 217:104–112
Pingoud K, Ekholm T, Sievänen R, Huuskonen S, Hynynen J (2018) Trade-offs between forest carbon stocks and harvests in a steady state—a multi-criteria analysis. J Environ Manage 210:96–103

Pleticha K, Kerber AK, Hoff M (2019) Standing tall. Forestry. State of Minnesota, Department of Natural Resources. FOR-642, St. Paul, MN. Available at: https://www.dnr.state.mn.us/publications/forestry/index.html

Regeringen (1993) Regeringens proposition 1992/93:226 om en ny skogspolitik. https://data.riksdagen.se/fil/A0AE3402-7DB4-4E92-8B13-A24F1FD077EF

Pukkala T (2014) Does biofuel harvesting and continuous cover management increase carbon sequestration? Forest Policy Econ 43:41–50

Pukkala T, Sulkava R, Jaakkola L, Lähde E (2012) Relationships between economic profitability and habitat quality of Siberian jay in uneven-aged Norway spruce forest. For Ecol Manage 276:224–230

Pukkala T, Laiho O, Lähde E (2016) Continuous cover management reduces wind damage. For Ecol Manage 372:120–127

Rentzhog T (1991) Norrlandsfrågan, Från 90-tal till 90-tal. In: Medelius H, Rentzhog S (eds) 90-tal: Visioner och vägval. Nordiska Museet, Stockholm, pp 169–182

Schager N (1925) Det svenska skogsbrukets förutsättningar och historia. Socialiseringsnämnden, Stockholm

Staley DJ (2007) History and future: using historical thinking to imagine the future. Lexington Books, Plymouth

Swedish Forest Industries Federation (2011). Living forests: 2010–2011. http://skyddadskog.se/wp-content/uploads/Living_Forests_201109131.pdf. Accessed 31.1.2018

Tomppo E, Henttonen H (1996) Suomen metsävarat 1989–1994 ja niiden muutokset vuodesta 1951 lähtien. Metsäntutkimuslaitos, Helsinki

White MA, Host GE (2008) Forest disturbance frequency and patch structure from pre-European settlement to present in the Mixed Forest Province of Minnesota, USA. Can J Forest Resour 38:2212–2226

Widenfalk O (2015) Contortatall i Sverige—En kunskapssammanställning och riskbedömning. Svenska FSC, Stockholm

Environmental Issues and Urban Expansion—A Study of the Atibaia River Basin in São Paulo, Brazil

João Luiz de Moraes Hoefel, Peter Martins Ribeiro, José Ricardo Filho Ramos, Maria Carolina Las-Casas e Novaes, Maikon Pinheiro, Walan Clésios Basso Simões, Debra Barbarick, and Deborah Nickel

1 Introduction

Traditionally, urbanization processes have significantly aggravated several socioeconomic and environmental problems in cities. It was with the beginning of the industrial revolution, as well as the transportation and agricultural revolutions that followed that urbanization surpassed the local scale and began to accelerate (Ugeda 2014). After the 1950s, it became evident on a worldwide scale that the most striking aspect of the urbanization process was the speed at which it was occurring in developing countries, regardless of the industrialization process.

This has resulted in urban land expansion (Schewenius et al. 2014; Mohammed et al. 2014; Suhartini and Jones 2019). All over the world, for a growing part of the global population, cities provide the day-to-day living environment. Asia, Africa, and Latin America have all had an unprecedented rate of people moving into their cities that resulted in an expressive urban expansion (Schewenius et al. 2014; Mohammed et al. 2014; Suhartini and Jones 2019).

Infrastructure development has emerged as a popular strategy for attracting private capital. It is also a factor in urbanization and, within this context, large-scale urban development, or mega projects. Mega projects, including highway construction, have been described as some of the most visible and ubiquitous urban revitalization strategies. Such strategies are initiated by city elites in search of

In Memorian. This author José Ricardo Filho Ramos died in February, 15th 2021

J. L. de Moraes Hoefel (✉) · P. M. Ribeiro · J. R. F. Ramos · M. C. L. e. Novaes · M. Pinheiro · W. C. B. Simões
Center for Sustainability and Cultural Studies, NESC/CEPE/UNIFAAT), UNIFAAT University Center, Estrada Municipal Juca Sanches 1050, Atibaia, São Paulo, CEP 12954-070, Brazil

D. Barbarick · D. Nickel
Sierra Water Protectors, 345 Linden Ave, Auburn, CA 95603, USA

W. Leal Filho et al. (eds.), *Sustainability in Natural Resources Management and Land Planning*, World Sustainability Series,
https://doi.org/10.1007/978-3-030-76624-5_19

economic growth and market competitiveness (Adama 2018; Seixas and Hoefel 2019; Vernalha et al. 2019). Adama (2018) highlights that mega cities in the global South face an increasing crisis in the establishment of basic infrastructure.

It is worth mentioning that the relationship between urbanization and sustainability has long been a focus of study. The interaction between urbanization and sustainability has been embodied in an idea of ecological sustainability. This ecological sustainability is intricately linked to ubiquitous urban sprawl and social sustainability that are correlated with continuous urban population growth. From an ecological perspective, the past several decades have seen widespread urban expansion in developing countries (ex. China, Brazil and India). This expansion has put enormous strain on the supporting ecological systems (Zeng et al. 2016) and according to Shen et al. (2016) and Wey (2018) urgent action is needed to develop sustainable urbanization practices in order to address these challenges.

Goal 11 from the 17 Sustainable Development Goals (SDGs), "*Make cities and human settlements inclusive, safe, resilient and sustainable*" targets relevant urban issues such as housing, transportation, water, air quality, development planning. Also targeted is integration between national and regional planning and international agreements so that cities may grow as living, resilient, and sustainable cities (United Nations 2015).

This chapter will look at environmental impact and sustainability in order to analyse current environmental changes and their effects on the transformation processes occurring in the Cantareira System Environmental Protected Area (EPA Cantareira) and the upper portion of the Atibaia River Basin located in São Paulo, Brazil. The chapter will also indicate environmentally appropriate measures for this region.

2 Urbanization and Socio-Environmental Issues

Depending on the perspective, urbanization and cities will be either key components in the transition to sustainability, or major threats to sustainability. The differing views are partly a result of the wide range of urban conditions and the uneven urbanization processes around the world. Urban areas can be sites of innovation and production of knowledge and wealth. They can provide widespread access to employment, education, sanitation, and modern energy. On the other hand, urban areas can also have high levels of pollution, social exclusion, and environmental degradation. This can in turn cause unintended consequences outside of the urban boundaries. All of these outcomes can occur simultaneously through the same urbanization process (Seto et al. 2017).

In Brazil, the urbanization process was the main factor that brought about the rural exodus. This occurred not only because of the job opportunities available in cities created by industrialization and commerce, but also because of better access to health care, education, and an escape from the precarious working and living conditions in the countryside and ultimately, the possibility for a better life (Ugeda 2014).

The rural exodus originally caused by the harsh living conditions in the countryside, resulted in population concentrations in cities. This caused a tendency for these largest urban concentrations to be in underdeveloped countries. It is precisely in these countries that the urban labour market and the existing labour force are unbalanced and access to basic consumer goods is restricted by low wages. What follows is a lack of public services, sound public policy making, and planning processes. Population concentrations that congregate and multiply on the outskirts of large cities, characteristically have precarious housing, transportation and sanitation (Alves et al. 2010).

A lack of adequate planning and effective management strategies, and the inability of government to accommodate the high influx of hopeful people coming from rural areas for what they perceive to be a better life, has resulted in serious pressure on both the supporting social economic infrastructure and the environment. For example, urbanization has been identified as the cause of a great number of environmental problems which include but are not limited to air, water, land and noise pollution, deforestation, local climate alteration, flooding and traffic congestion (Hooper and Hubbart 2014; Ohwo and Abotutu 2015; Ndiaye et al. 2016; Krueger et al. 2019; Ray and Shaw 2019).

From this perspective Ugeda (2014), states:

> In Brazil the intense rural exodus and the lack of jobs in the secondary and tertiary sectors, brought consequences such as the expansion of slums, the growth of an informal economy and in many cases an increase in the contingent of the poor population. Migration to cities which was initially seen as a possibility of improving quality of life now exposes a significant portion of the population to poverty and more precarious conditions than those experienced in the countryside (Ugeda 2014, p. 105).

It has also been noted that an association between urbanization and environmental degradation is not uncommon. It is especially true that when the growth of the urban population intensifies, and resources are scarce, there is an increase in pollution and a decrease in the quality of life. The dynamics of population distribution in the main urban concentrations of many countries, point to a new scenario where the fall of population growth rates is confronted by new ways of occupying space. This changes the intra-urban dynamics and impacts the environment bringing new challenges for sustainability related to urban expansion (Ojima 2006; Mohanty 2019).

As an example of Brazilian Urbanization, this chapter presents a case study carried out in the Cantareira System Environmental Protected Area (EPA Cantareira) and the upper portion of the Atibaia River Basin located in the northern portion of the São Paulo Metropolitan Region. The results of this study have highlighted the changes in the use of space and point out the challenges for sustainable urbanization in recent urban concentrations. Evidence confirms the theoretical propositions of a new stage in the development of modern society and the challenges for environmental issues in an urban context.

Studies show that disorderly urban growth represents a reality common to most Brazilian municipalities. This fact is beyond the control of many institutions that plan the use of urban land and has generated new challenges to local governance. These challenges include the mitigation of the many social, environmental and economic problems that have emerged in relationships triggered by the absence of structural interventions and concrete action based on public policy (Alves et al. 2010).

In this context, this chapter identifies how these phenomena are associated and particularly how urbanization processes generate new situations of vulnerability and social environmental impact. The chapter takes into special consideration water resources, both natural systems and the structural systems necessary for urban growth and emphasizes the growing concern that different areas of society have for these issues. It also demonstrates that there is a need for understanding the dynamics of environmental transformation within the context of urbanization processes in cities and to find solutions to the problems that arise (Alves et al. 2008).

In recent years, the urbanization process has significantly expanded and increased social needs. These social needs are a sum of population growth, housing demands, political and economic development, and scientific-technological expansion. According to Bispo and Levino (2011), the advance of global urban degradation, especially in Brazil, is the result of artificial changes and/or disturbances to the environment caused by human activities. The idea of progress is often limited to meeting needs without taking into consideration the social environmental problems created by urbanization, or the requirements that need to be met in order to create a sustainable city. The definition of urban space according to Bispo and Levino (2011) is the consequence of the relationship that human beings establish between a constructed space and nature—especially when population concentrations appear, and human activities diversify.

It is easily observed that population and economic growth, together with growing demand for housing, ends up generating other kinds of problems in the urban landscape. The formation of a city takes place due to the inherent need of human beings to associate, interrelate and organize for the common good (Bispo and Levino 2011). In this context, cities encompass the manifestation of society and concentrate the population as well as the instruments of production, capital, and other needs (Vestena and Schmidt 2009).

Since 1950's mankind has known and lived with the risk of deterioration because of its activities and different manifestations but it was only in the 1980's that society began to be concerned with the environmental crisis. It was then that society began looking for solutions that were suitable for protecting natural systems and for helping systems that were already weakened.

The effort to maintain environmental sustainability requires an integrated and interdisciplinary approach. Both are indispensable in understanding environmental problems. Social environmental impacts started to have an important social significance only in the last decades. This occurred mainly after the intense growth of the means of production was coupled with unrestrained consumption of natural resources.

Given these facts, it is clear that urban spaces are extraordinarily complex, and that well-defined planning is crucial. Creating a structure that allows direct connection between human beings and the environment, and that also takes into consideration economic, social, territorial, ecological, and administrative factors, is essential. For Bispo and Levino (2011) the objective of this multi-sectored view is to conserve both environmental resources and provide quality of life. Following this same reasoning, Grostein (2001) states that the advancement, scale, and speed of urbanization are not problems in themselves, but the lack of effective planning that makes them problematic.

Analysing improvements and monitoring public authority related to urban infrastructure shows that public policy making is aimed at demographic growth and a denser population concentration in urban settings. This urban growth has a direct relationship to environmental accidents and is mainly a result of disorderly land occupation and impermeable soil. Also contributing to social environmental impact is land occupation in areas of risk.

To prevent these problems from occurring and to try to organize in an orderly fashion, it would be necessary to carry out environmental impact studies that support rational planning and management for the use of urban land. According to Freitas (2017) there is an unfair process in the production of urban space that restricts legalized urban occupation to a minority that has the means to pay the high price of this commodity. This situation encourages and brings about serious damage to environmental and urban sustainability.

According to Carvalho and Idelfonso (2010), urban management involves combining the planning processes to be adopted and implemented, with the capacity to monitor and reorient the procedures, in order to be able to achieve previously defined objectives. Also, any urban intervention project initiated by the government should aim to bring about sustainable urban development and a better quality of life.

3 Study Area

The disorderly land occupation analysed in the Cantareira System Environmental Protected Area (EPA Cantareira) and the upper portion of the Atibaia River Basin, has exposed the population to harm from soil compaction. Soil compaction prevents water infiltration, increasing the problems of urban soil erosion. This combination of elements leads to an increase in the frequency of flooding. Lack of sewage treatment and inadequate disposal of solid waste are two other factors related to basic infrastructure that have caused social environmental impacts. Both have contributed to the contamination of rivers and damaged water quality and are topics, in this research, that are directly related to population and urban growth.

Large population growth has caused a significant increase of urbanization in Brazil. Analysis made by the Brazilian Institute of Geography and Statistics—IBGE

(Brasil 2019)—shows in the 2010 population census there was a population of 190,755,799 and in 2019 the population had grown to 210,147,125.

Rapid population growth, the disorderly land occupation that follows, and the absence of planning are cause for concern. These issues negatively impact the physical, biotic and atrophic environments. The frequent occupation of floodplains and riverbanks that are Permanent Preservation Areas (PPA) has caused adverse problems for both humans and nature.

Law 12.651/2012 of the new Brazilian Forest Code in Art. 3 provides a concept of a Permanent Preservation Area (Brasil 2012). A Permanent Preservation Area is defined as a protected area covered or not by native vegetation; with the environmental function of preserving water resources, the landscape, geological stability and biodiversity; to facilitate the gene flow of fauna and flora; to protect the soil; and to ensure the well-being of human populations.

It is understood that the purpose of Permanent Preservation Areas is to preserve, among other things, water resources, biodiversity, and soil. Thus Article 4 of Law 12.651/2012 provides for strips of watershed along the margins of perennial and intermittent natural water courses (Brasil 2012):

> I – The watershed areas of any perennial or intermittent natural watercourse excluding the ephemeral ones from the edge of the regular water bed are required to be a minimum width of: a) 30 (thirty) meters for water courses less than 10 (ten) meters wide; b) 50 (fifty) meters for water courses that are 10 (ten) to 50 (fifty) meters wide; c) 100 (one hundred) meters for water courses that are 50 (fifty) to 200 (two hundred) meters wide; d) 200 (two hundred) meters for watercourses that are 200 (two hundred) to 600 (six hundred) meters wide; e) 500 (five hundred) meters for water courses that are wider than 600 (six hundred) meters.

It is noted that there are laws in force that protect PPA areas located along riverbanks. However, for legislation to be more effective, both urban and sustainability planning processes are necessary. Such planning could bring many benefits to the population, including improved water quality.

This expressive population growth and urbanization is a national concern. There is also concern at the municipal level because urbanization brings about diverse kinds of environmental impact, including impact on water resources.

An example presented by Campos and Carneiro (2013, p. 23) about the municipality of Atibaia, which is in the study area of this chapter, points out that over the years there was a significant population increase in Atibaia. This occurred mainly from the 1950s to the present day. The population of the municipality is 126,603 inhabitants and 115,229 are concentrated in the urban area and 11,374 in the rural area with a demographic density of 264.61 inhabitants per km^2 and an urbanization rate of 91%.

The municipality of Atibaia is in the Atibaia River Basin that was formed by the confluence of the Cachoeira and Atibainha rivers in the city of Bom Jesus dos Perdões. This is also a subject of analysis in this chapter. The Atibaia River basin is of great importance to the State of São Paulo because it is one of the watersheds that supplies water to the Cantareira Water System. This reservoir system is responsible for about half of the water used by the municipality of São Paulo, the largest urban centre in Latin America. The system also supplies almost all the water used by the

Metropolitan Region of Campinas (Irrigart 2007). Because of these factors, there is a great concern for water resources and their management in the municipalities described in this study and of special concern are the serious problems associated with the silting up of water courses and the quality of water for human consumption (Silva 2001).

Campos and Carneiro (2013) also emphasize that the problem of flooding is recurrent and increasingly harms society. Flooding and rivers go hand-in-hand, but this process ends up causing great and far reaching social impact when it is associated with unregulated land occupation and illegal, or even legal, appropriation of areas subject to flooding.

In 2009 and 2010, Atibaia was affected by major floods and, according to data cited by Campos and Carneiro (2013), 1639 people were directly and indirectly affected. In the 2009 floods 13 neighbourhoods were affected and in 2010 56 neighbourhoods were affected by floods.

It is clear that the lack of urban planning affects both water resources and the general population. Lack of urban planning also causes other serious environmental problems. To minimize these problems, it is important to create and implement plans and projects that bring together both current policies and sustainable social environmental planning of municipalities to benefit the population.

4 Methods

To achieve the objectives proposed in this research to characterize current environmental changes and their effects on the transformation processes of the Cantareira EPA and in the Atibaia River Basin, specifically in three municipalities, Nazaré Paulista, Bom Jesus dos Perdões and Atibaia (Fig. 1), analyses of land cover fragments, obtained through remote sensing mapping using satellite images and IBGE data, were used to compare data provided by the municipalities involved and to establish a relationship between the growth of urban occupation and social-environmental degradation.

According to Florenzano (2011), remote sensing is a technology that uses artificial satellites to capture energy from the Sun reflected by the Earth's surface. These signals are transformed into data that provides information about the earth's surface.

First a point of built-up areas in the State of São Paulo was produced on the DATAGEO—Sistema Ambiental Paulista website produced by the São Paulo State Environmental Department/Environmental Planning Coordination (SMA/CPLA) based on the classification of 2005 IRS satellite images (Indian Remote Sensing Satellite). This point was generated through remote sensing developed by CPLA in 2009 from the classification of monochromatic scenes from the IRS satellite merged with bands 5, 4 and 3 of the LANDSAT 7 satellites in 2005 (Datageo 2019). To obtain a comparison with the 2005 built-up area point, an image taken from the catalogue of the National Institute for Space Research (INPE) Orbital/Point = 76/219,

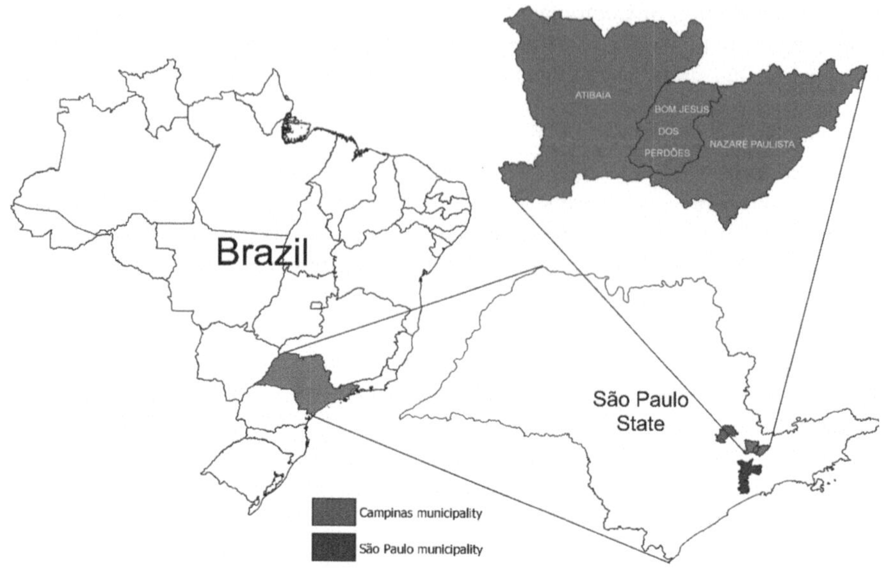

Fig. 1 Study area location in São Paulo, Brazil. *Source* Elaborated by the authors

"UTM" projection dated 01/21/2019 which covers the areas of the municipalities studied was used.

To manipulate the images and obtain data the software QGIS—A Free and Open Source Geographic Information System was used in conjunction with the SCP Plugin (Semi-Automatic Classification *Plugin*) installed in QGIS. First the RGB 4, 3, 2 channels (Red, Green and Blue channels) with 30 m resolution were interpolated into the original 2019 image and the PAN channel 8 (Panchromatic channel 8) which obtains the near infrared range with 15 m resolution was interpolated to improve the image resolution and consequently improving the mapping accuracy.

Afterwards, a mask was created on the image using the limits of the municipalities obtained on the Datageo website. The title Municipal Boundary/SP (Datageo 2019) was used to separate the study region from the rest of the image. Next, a list of classes with types of soil vegetation cover was created. Associated with these classes, polygons were drawn on the image for the software to have a spectral sampling of the types of cover for mapping, and to use the supervised classification technique. Classes are defined by the analyst, and he must identify pixels (samples) belonging to previously created classes and leave the task of identifying the rest of the pixels belonging to classes to an algorithm (Florenzano 2011). The types of classes determined were vegetation, bare rock, bare soil, water, and urbanized coverage.

After the SCP Plugin, and in order to perform the supervised classification of the image, the points corresponding to the urbanized coverage were isolated and converted into polygons. Using satellite images with higher resolution visualized through Web Map Services connection (WMS), the polygon perimeters of the urbanized patches were manually corrected and built-up. Areas not mapped by the SPC algorithm were added. With the polygons corresponding to the built-up area stain in the 2019 image corrected, it was possible to compare the 2019 image with the built-up area stain polygons of 2005. It was also possible to calculate the dimension of occupation by these areas in relation to the territorial area of the municipalities. This data could also be contrasted with data obtained from city halls.

In 2019, data regarding land subdivisions, their regulation status, and their infrastructure was collected from city halls. Using QGIS, this data facilitated location and mapping work of the perimeters of regulated subdivisions. In the case of Bom Jesus dos Perdões, this was not possible because the municipality did not pass on the data. It was possible, however, to measure (in square meters) the total area regulated up to 2019 in both municipalities. This measurement was compared with the expansion of the urban area in recent years—thus obtaining the urban growth of the area in relation to both regulated and unregulated area, and the distribution within both municipality territories and Conservation Areas that exist in the study area.

5 Results and Discussion

According to Figs. 2 and 3, between 2005 and 2019, the total area of urban occupation in thc municipalities analysed corresponded to a growth of 48.92% in Bom Jesus dos Perdões, 74.37% in Atibaia and 90.49% in Nazaré Paulista. The rapid urbanization process in this region was influenced by the expansion of the Metropolitan Regions of São Paulo and Campinas, and the construction of the Dom Pedro I and Fernão Dias Highways, two major highways of great regional influence. In regard to sustainability, this expansion has caused a number of impacts and various kinds of problems in the region.

Analysis of demographic census taken between 2010 and 2019 showed that the population growth in Atibaia had an estimated increase of 16,158 inhabitants, Bom Jesus dos Perdões an estimated increase of 5740 new inhabitants and Nazaré Paulista an estimated increase of 2110 inhabitants (Brazil/IBGE 2019).

Secondary data provided by City Halls indicated that regulated subdivisions in Atibaia in the last 14 years correspond to 47.86% of the total area, and that the urbanized area expanded 39.82 km^2 compared to the occupied area in 2005. Therefore, it is possible to conclude that in Atibaia real estate speculation is the most intense among the municipalities analysed. The municipality of Nazaré Paulista covers only 11.52% of the regulated area of the total area currently occupied. Bom Jesus dos Perdões did not present data about regulated subdivisions, existing illegal settlements, and possible investments in urban, social, and environmental planning. It is difficult to identify unregulated subdivisions because they are not registered in official institutions, particularly in the Municipalities analysed.

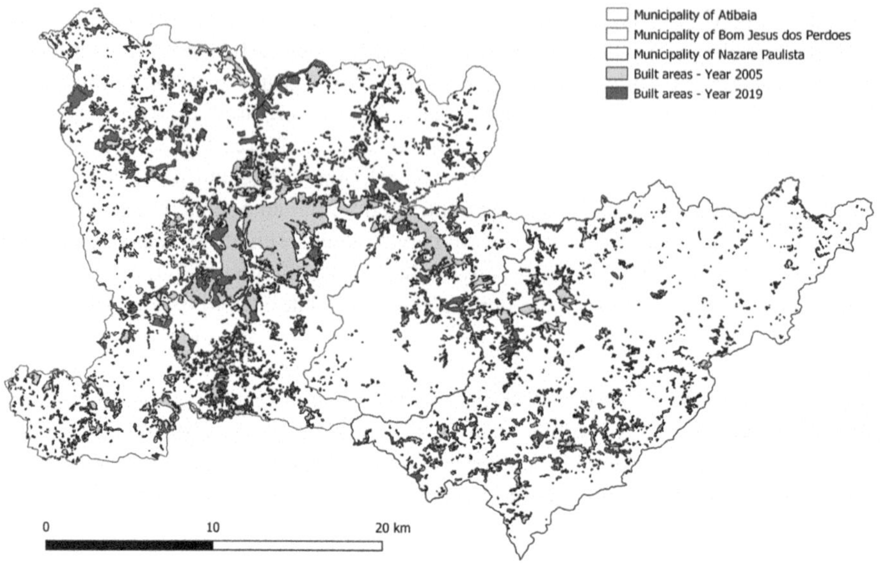

Fig. 2 Urban expansion in the years 2005–2019. *Source* Elaborated by the authors

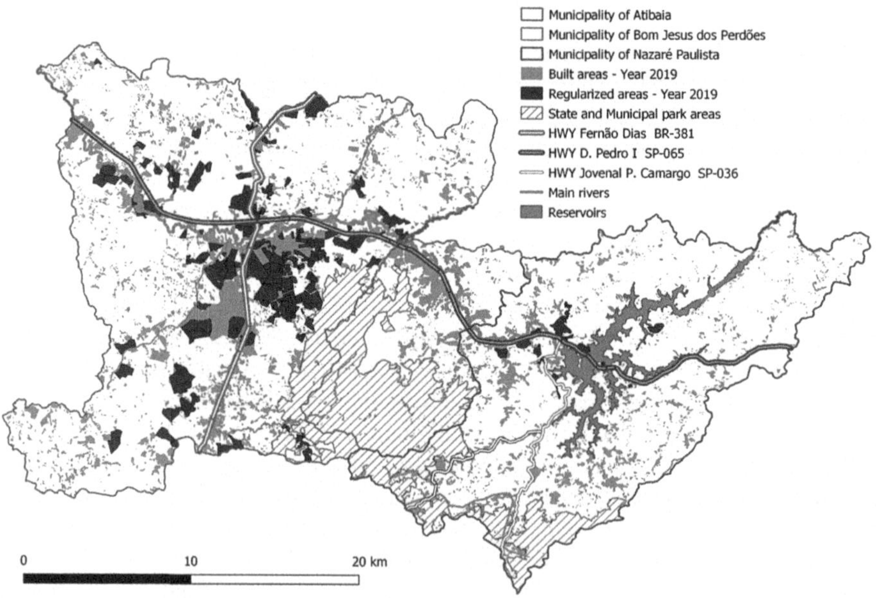

Fig. 3 Location of urbanized areas in relation to regional highways. *Source* Elaborated by the authors

The dynamics of real estate speculation in Atibaia have begun to change the city's features. Spaces previously occupied by residences have begun to verticalize as a way of dealing with the saturation of urban spaces. This verticalization has mostly taken place at the north and south ends of the city. It was also observed that there was a greater degree of urbanization along the margins of the main highway that runs through the city (Fernão Dias Highway). Changes to the highway environment include, among other things, changes to air quality, environmental vulnerability, and an increase in industrialization processes (Hoefel et al. 2015).

Urbanization processes in Atibaia, that do not incorporate proposals that take sustainability into consideration, have led to the creation of a fragmented and segregated city. This is especially true in areas where there are horizontal closed condominiums, areas with precarious, improvised housing, neighbourhoods lacking infrastructure, and more recently housing estates located in the outskirts of the city. According to Mohanty (2020a) and as previously mentioned, informal urbanization causes complex environmental, social, economic, and spatial challenges, is unsustainable in the long run, and is a persistent challenge to achieving sustainable urban development, environmental quality and the building of a healthy city. Similar situations are described by Maia (2010) and Suhartini and Jones (2019) when they analysed informal urbanization and settlements, and by Dalla Libera and Melo (2020) when they highlighted the principal urban planning problems in small cities in Brazil, including the lack of sustainable guidelines for city growth.

According to Belizaire (2020) there is no single model for a sustainable city, but rather there are choices of solutions designed to support long-term ecological balance. However, the concept of sustainable cities takes into consideration the pace of urbanization, rate of energy consumption, and the use of natural resources. It also emphasizes the mechanisms of proper governance, planning, and infrastructure to ensure the highest quality of life for citizens.

In Nazaré Paulista, the southern region of the municipality was the one that grew the most. This was especially true along the axis that connects the municipality with the city of Guarulhos. Close to the Atibainha Reservoir, which is part of the Cantareira Water System, there has also been significant urbanization in areas—even though new construction is illegal. According to Souza (2007), the biggest problems caused by a growing occupation in areas close to a drinking water reservoir, are problems caused by overflow, especially of sewage and the subsequent increase of pollution. One factor leading to the demographic density of Nazaré Paulista has been its growth dynamics. The official consolidated urban area is relatively small, and the greatest density has occurred in a dispersed and irregular manner in rural areas. This has been facilitated by the construction of paved roads into areas where there is a lack of necessary infrastructure. In the city of Bom Jesus dos Perdões, the greatest density occurred close to the city centre and is characterized by local commerce and the implementation of industrial areas, which have stimulated migration to the municipality.

Another point that was considered is that the study region contains the following important Conservation Areas: (CA's) Piracicaba-Juqueri-Mirim Area II Environmental Protected Area (EPA Piracicaba); the Cantareira System

Environmental Protected Area (EPA Cantareira); Itapetinga State Park; Itaberaba State Park; and Pedra Grande Natural Monument. The creation of EPA's Piracicaba and Cantareira is linked to the conservation of regional water resources that supply the Cantareira Water Supply System. The other Conservation Areas were created to protect regional water resources and because of the extreme relevance of protecting biodiversity in areas of Atlantic Forest and the Brazilian Cerrado. These Conservation Areas contain extensive forested areas with large rocky outcrops (Gaspareto and Furlan 2014).

This research shows that safeguarding good water and land governance are extremely important in the study area. In addition, as mentioned by Ortega-Montoya and Tejeda-Gonzalez (2020), there is a need to integrate water resource policies with policies relevant to soil usage and to minimize vulnerabilities and social and environmental impacts.

As shown in Figs. 2 and 3, the greatest urban growth in protected areas is in Bom Jesus dos Perdões (208.8%). This is followed by Atibaia (135.94%) and Nazaré Paulista (102.13%). In Nazaré Paulista and Atibaia the occupation has doubled, and in Bom Jesus dos Perdões it has practically tripled. The situation in Nazaré Paulista is still quite serious because 5.57% of the municipality area is designated to the Cantareira System Water Reservoir, and 9.81% is reserved for the Itaberaba State Park. As previously mentioned, this occupation is relevant to the sustainability of regional resources.

According to Villaschi (2003) and Adama (2018), the consequences of disorderly urban growth processes such as the lack of basic sanitation, an absence of essential services, occupation of inadequate areas, the destruction of resources of ecological value, environmental pollution and precarious housing all lead to poor living conditions. With the great increase in population, the demand for water consumption, and the release of urban sewage into rivers without immediate measures taken, the prospects for the future of water resources are grave. In addition to these factors, when public water services fail, citizens adapt by buying water from private suppliers or storing water in tanks on rooftops (Krueger et al. 2019).

Due to the high degree of human impact on the upper portion of the Atibaia River Basin and in the Cantareira System EPA especially in areas designated as PPAs, the urbanization process is leading to environmental degradation, high water consumption, and the discharge of domestic effluents, that is resulting in the deterioration of water bodies (Demanboro et al. 2013). Once again, it is important to emphasize that the Atibaia River water basin is a water source for the São Paulo and Campinas Metropolitan areas and is of regional significance. Proposals for creating sustainable urbanization are vitally necessary (Zeng et al. 2016; Shen et al. 2016; Wey 2018).

According to Mohanty (2020a, b), sustainable urban planning is a challenge that requires the development of strategies and practices in cities to improve the quality of both human life and the environment. When considering the characteristics of this study area, the diverse points presented in the analysis should be understood as a means of stimulating sustainable urbanization planning.

6 Final Considerations

Given the magnitude of current urban transitions and the global reach of production–consumption systems, there is a need to generate robust scientific knowledge of cross-scale interactions in order to guide urban development in a more sustainable direction (Seto et al. 2017).

This study indicates that significant unplanned urban expansion is occurring. Lack of planning has resulted in the appearance of peripheries and urban voids; has facilitated real estate speculation; and has generated an increase in the displacement of the population between housing, work, and leisure. Occupation patterns in the study area also revealed diverse kinds of environmental impact generated by urban growth. Such patterns are characterized by unsustainable residential and recreation centers around the reservoir and in protected areas.

Quality of life in cities is declining in this process of transformation, expansion, and social spatial segregation. There is a need to incorporate the concepts of urban sustainability associated with technically structured development models in this study area. These models and concepts should be based on environmental conservation and the search for better use of already occupied areas with the aim of avoiding uncontrolled expansion and pollution of the regional attributes, as well as the incorporation of sustainable actions and proposals.

For the purposes of urban sustainability, it remains necessary to consider other issues not addressed in this chapter. Issues related to economic, social, cultural, demographic, political and institutional aspects, require further research to complement this study.

References

Adama O (2018) Urban imaginaries: funding mega infrastructure projects in Lagos, Nigeria. GeoJournal 83:257–274. https://doi.org/10.1007/s10708-016-9761-8

Alves CD, Alves H, Pereira MN (2008) Análise dos Processos de Expansão Urbana e das Situações de Vulnerabilidade Socioambiental em Escala Intra-urbana. IV Encontro Nacional da ANPPAS, 4 a 6 de junho de 2008, Brasília/DF/Brasil

Alves HPF, Alves CD, Pereira MN, Monteiro AMV (2010) Dinâmicas de urbanização na hiperperiferia da metrópole de São Paulo: análise dos processos de expansão urbana e das situações de vulnerabilidade socioambiental em escala intraurbana. R Bras Est Pop 27(1):141–159

Belizaire M (2020) Sustainable urban governance: power of partnerships, planning process, public partnerships, and performance parameters. In: Leal W, Azul A, Brandli L, Özuyar P, Wall T (eds) Sustainable cities and communities. Encyclopedia of the UN Sustainable Development Goals. Springer, Cham. https://doi.org/10.1007/978-3-319-95717-3_63

Bispo TC, Levino NDA (2011) Impactos Ambientais decorrentes do uso e ocupação desordenada do solo: um estudo da região da periferia de Maceió/AL. XXXI Encontro Nacional de Engenharia de Produção: Inovação Tecnológica e Propriedade Intelectual: Desafios da Engenharia de Produção na Consolidação do Brasil no Cenário Econômico Mundial, 1–13, out/2011. Available at: http://www.abepro.org.br/biblioteca/enegep2011_tn_sto_143_901_18402.pdf. Accessed 12 Nov 2019

Brasil (2012) Lei nº 12.651, de 25 de maio de 2012. Dispõe sobre a proteção da vegetação nativa. Available at: http://www.planalto.gov.br/ccivil_03/_Ato2011-2014/2012/Lei/L12651.htm. Accessed 20 Oct 2019
Brasil. Instituto Brasileiro de Geografia e Estatística-IBGE (2019) IBGE Cidades. Brasília, IBGE. Available at: http://www.ibge.gov.br/cidadesat/topwindow.htm?1. Accessed 5 Oct 2019
Campos RD, Carneiro CDR (2013) Geologia da região de Atibaia e possíveis causas das inundações de 2009 e 2010. Terrae 10(1–2):21–35
Carvalho RGD, Idelfonso ICS (2010) Problemas de uso e ocupação do solo e degradação ambiental no núcleo urbano de Tibau/RN, Mossoró/RN. Available at: http://www.geomorfologia.ufv.br/simposio/simposio/trabalhos/trabalhos_completos/eixo11/077.pdf. Accessed 22 Oct 2019
Dalla Libera LT, Melo EFRQ (2020) Planning small cities toward being inclusive, safe, resilient, and sustainable: the case of a city in Rio Grande Do Sul, Brazil. In: Leal Filho W, Marisa Azul A, Brandli L, Gökçin Özuyar P, Wall T (eds) Sustainable cities and communities. Encyclopedia of the UN Sustainable Development Goals. Springer, Cham. https://doi.org/10.1007/978-3-319-95717-3_9
Datageo Sistema Ambiental Paulista (2019) Infraestrutura de dados espaciais ambientais do Estado de São Paulo. Available at: http://datageo.ambiente.sp.gov.br. Accessed 07 Feb 2019
Demanboro AC, Laurentis GL, Bettine SC (2013) Cenários ambientais na bacia do rio Atibaia. Engenharia Sanitária e Ambiental 18(1):27–37
Florenzano TG (2011) Iniciação em sensoriamento remoto. Oficina de Textos, São Paulo
Freitas CFS (2017) Regulações territoriais e expansão urbana informal: é possível preservar e incluir? Paranoá Cadernos de Arquitetura e Urbanismo 19:1–15. https://doi.org/10.18830/issn.1679-0944.n19.2017.02
Gaspareto TC, Furlan SA (2014) Pressão urbana e Unidades de Conservação na região metropolitana de São Paulo: análise ecológica da paisagem. Available at: http://www.fau.ufal.br/evento/pluris2016/files/Tema%201%20%20Ambiente%20e%20Energia/Paper717.pdf. Accessed 02 Dec 2019
Grostein MD (2001) Metrópole e Expansão Urbana: a persistência de processos "insustentáveis". São Paulo em Perspectiva 15(1):13–19. Available at: http://www.scielo.br/scielo.php?script=sci_arttext&pid=S0102-88392001000100003. Accessed 12 Nov 2019
Hoefel João Luiz M, Seixas SRC, Oliveira KES, Rocha J, Lima FB (2015) Urbanização e mudanças no uso do solo no eixo do corredor de exportação—Rodovias D. Pedro I/Tamoios-SP. Anais ENANPUR 16(1):1–19
Hooper LW, Hubbart JA (2014) Water management in Kenya: toward an ethic of sustainability. J Sci Res Rep 3(9):1144–1152
Irrigart. Engenharia e Consultoria em Recursos Hídricos (2007). Bacias Hidrográficas dos rios Piracicaba, Capivari e Jundiaí: situação dos recursos hídricos 2004/2006, relatório síntese. Piracicaba, FEHIDRO/PCJ/CBJ
Krueger EH, Borchardt D, Jawitz JW, Klammler H, Yang S, Zischg J, Rao PSC (2019) Resilience dynamics of urban water supply security and potential of tipping points. Earth's Future 7:1–25. https://doi.org/10.1029/2019EF001306
Maia DS (2010) A periferização e a fragmentação da cidade: loteamentos fechados, conjuntos habitacionais populares e loteamentos irregulares na cidade de Campina Grande-PB, Brasil. Scripta Nova. Revista Electrónica de Geografía y Ciencias Sociales, 14(331):80
Mohammed KS, Samat N, Elhadary YAE (2014) Bottom up approach: urbanization in the perception of the local communities of Balik Pulau, Penang Island, Malaysia. Br J Appl Sci Technol 4(32):4533–4549
Mohanty M (2019) Globalization and Cities. In: Leal W, Azul A, Brandli L, Özuyar P, Wall T (eds) Sustainable cities and communities. Encyclopedia of the UN sustainable development goals. Springer, Cham, pp 1–11. https://doi.org/10.1007/978-3-319-71061-7_52-1
Mohanty M (2020a) Challenges of Informal Urbanization. In: Leal W, Azul A, Brandli L, Özuyar P, Wall T (eds) Sustainable cities and communities. Encyclopedia of the UN sustainable development goals. Springer, Cham. https://doi.org/10.1007/978-3-319-95717-3

Mohanty M (2020b) Sustainable urban planning and making sustainable cities. In: Leal W, Azul A, Brandli L, Özuyar P, Wall T (eds) Sustainable cities and communities. Encyclopedia of the UN sustainable development goals. Springer, Cham. https://doi.org/10.1007/978-3-319-95717-3

Ndiaye ML, Ndiaye HAD, Traore VB, Tahirou DA, Chedikh BA (2016) Impacts of rainfall variability and urban sprawl on the environment and the population well-being at Ouakam Commune, Dakar, Senegal. Arch Curr Res Int 6(2):1–12

Ohwo O, Abotutu A (2015) Environmental impact of urbanization in Nigeria. Br J Appl Sci Technol 9(3):212–221

Ojima, R. (2006). A produção e o consumo do espaço nas aglomerações urbanas brasileiras: desafios para uma urbanização sustentável. XV Encontro Nacional de Estudos Populacionais (ABEP), Caxambú/MG, 18–22 de Setembro de 2006, pp 1–19

Ortega-Montoya Y, Tejeda-Gonzalez JC (2020) Sustainable, fair, and democratic water management. In: Leal W, Azul A, Brandli L, Özuyar P, Wall T (eds) Sustainable cities and communities. Encyclopedia of the UN sustainable development goals. Springer, Cham. https://doi.org/10.1007/978-3-319-95717-3

Ray B, Shaw R (2019) Water insecurity in Asian cities. In: Ray B, Shaw R (eds) Urban drought, disaster risk reduction. Springer Nature Singapore Pte Ltd., Singapore, pp 17–32. https://doi.org/10.1007/978-981-10-8947-3_2

Schewenius M, Tengö M, Nagendra H, Katti M, Elmqvi T (2014) What is the scope of nature in the city? Curr Conserv 8(1):4–5

Seixas SRC, Hoefel JLM (2019) Megaprojects—socioeconomic and environmental dynamics in D. Pedro I-Tamoios Road Axis, São Paulo, Brazil. Adv Res 18(6):1–15

Seto KC, Golden JS, Alberti M, Turner BL II (2017). Sustainability in an urbanizing planet. PNAS 114(34):8935–893. https://doi.org/10.1073/pnas.1606037114

Shen L, Shuai C, Jiao L, Tan Y, Song X (2016) A global perspective on the sustainable performance of urbanization. Sustainability 8:783. https://doi.org/10.3390/su8080783

Silva WS (2001) Identificação de unidades ambientais no município de Atibaia—SP. Dissertação (Mestrado em Geografia)—Departamento de Geografia, Universidade de São Paulo, São Paulo, Brasil, p 158

Souza JFV (2007) Responsabilidade ambiental e a ocupação desordenada do solo: um olhar sobre os reservatórios da Região Bragantina. Sequência: Estudos Jurídicos e Políticos 28(54):47–73

Suhartini N, Jones P (2019) Urbanization and urban governance in developing countries. In: Suhartini N, Jones P (eds) Urban governance and informal settlements. Springer Nature, The Urban Book Series, Cham, pp 13–40. https://doi.org/10.1007/978-3-030-06094-7_2

Ugeda JC Jr (2014) Planejamento da Paisagem e Planejamento Urbano: Reflexões sobre a Urbanização Brasileira. Revista Mato-Grossense De Geografia (cuiabá) 17(1):101–116

United Nations (2015) The millennium development goals report 2015. United Nations, New York

Vernalha EBR, Seixas SRC, Hoefel JLM (2019) Vehicular fleet expansion and accidents variation numbers: a contribution to the analysis of the D. Pedro I-Tamoios exporter road axis. Curr J Appl Sci Technol 37(3):1–11

Vestena LR, Schmidt LP (2009) Algumas reflexões sobre a urbanização e os problemas socioambientais no centro- socioambientais no centro-sul Paranaense. Acta Scientiarum. Hum Soc Sci 37(1):67–73. Available at: https://www.redalyc.org/pdf/3073/307325328009.pdf. Accessed: 20 Oct 2019

Villaschi J (2003) Novas modalidades de urbanização e sua inserção regional: o condomínio Alphaville Lagoa dos Ingleses, Nova Lima-MG. In: Castriota LB (Org). Urbanização brasileira: redescobertas. Belo Horizonte: C/Arte, pp 159–174

Wey W-M (2018) A commentary on sustainably built environments and urban growth management. Sustainability 10:3898. https://doi.org/10.3390/su10113898

Zeng C, Deng X, Dong J, Hu P (2016) Urbanization and sustainability: comparison of the processes in "BIC" countries. Sustainability 8:400. https://doi.org/10.3390/su8040400

Prioritizing the Implementation of Walkable Green Spaces: An Analysis Under Climate Change in a Global South Megacity

Diego Rafael Galvão Cesar Braga and Robson da Silva Moreno

1 Introduction

The problems arising from the profound and continuous anthropic changes, carried out by the modern conception of "city", have been the source of several environmental issues. These problems range from heat islands, changes in the hydrological cycle and the air quality, and other elements that affect human health (Pitman et al. 2015; Reid et al. 2017; Tzoulas et al. 2007). Climate change enhances the impacts of weather events like droughts, heatwaves and rainfall in urban regions (Demuzere et al. 2014; Gill et al. 2007). Amplified by climate change, these events compromise the health and even the physical integrity of urban populations.

Urban planning has been increasingly concerned with the effects that heatwaves have on human health. Climate change is increasing the frequency of high-intensity heatwaves in urban agglomerations (Milan and Creutzig 2015). Heatwaves threaten the health of vulnerable people exposed to their effects (Geirinhas et al. 2017; Mitchell et al. 2018). The level of vulnerability varies according to the population's susceptibility and exposure conditions. It involves factors such as personal physiological, social and economic conditions, and the configuration of the nearby urban space (Milan and Creutzig 2015). The elderly population is especially vulnerable to variations in the environmental temperature due to the development of physical and mental problems brought about by old age and their specific social context (Casas et al. 2016; Geib 2012; Hajat et al. 2014).

Urban green spaces are relevant for mitigating the effects of heatwaves on health. At the local level, they reduce the strength of winds and floods and promote thermal comfort by shading and cooling surfaces (Pitman et al. 2015; Reid et al. 2017). They also benefit citizens' health and quality of life in direct and indirect ways, regardless of the socio-economic condition (Pitman et al. 2015; Takano et al. 2002).

D. R. G. C. Braga (✉) · R. da Silva Moreno
Engineering, Modeling, and Applied Social Sciences Center, Federal University of ABC, São Bernardo do Campo, Brazil

W. Leal Filho et al. (eds.), *Sustainability in Natural Resources Management and Land Planning*, World Sustainability Series,
https://doi.org/10.1007/978-3-030-76624-5_20

For those reasons, the expansion and equitable distribution of green spaces is a valuable strategy for climate adaptation aimed at promoting public health in urban areas (Gill et al. 2007; Tzoulas et al. 2007). Cities can do that by increasing squares, parks and viable urban arborization areas, also called Walkable Green Spaces (WGS). These green spaces can promote public health conditions and mitigate the effects of climate change and extreme weather events, especially for the elderly population (Takano et al. 2002; Tamosiunas et al. 2014). Ageing with quality of life and climate change are both issues demanding specific planning and public policies (Geirinhas et al. 2017; Guo et al. 2018).

Promoting changes in the urban fabric for the implementation of green spaces is complex. Planners need to take full advantage of the beneficial effects of these spaces, strategically implementing them in critical locations in the city. (Gill et al. 2007). In the Global South, the "greening" processes of urban areas, common in the Global North, rarely occur (Haase et al. 2014). Aiming at promoting and facilitating urban planning as a way to improve the adaptation of the most vulnerable elderly citizens to climate change, we developed a low-cost analysis method using geo-processing tools and social-environmental open data and indicators. This method helps to identify intra-urban regions to prioritize for creating or building walkable green spaces with a focus on local adaptation of the most vulnerable elderly population to the effects of heatwaves. In the later sections, we discuss and associate the spatial distribution of WGS with its accessibility, urban planning, and the results obtained for the Brazilian megacity of São Paulo.

2 Methodology

2.1 Study Area

The Global South has been experiencing increasing numbers in urbanization since the 1950 decade. The cities expansion has suppressed important vegetation areas in places of rich biodiversity (Nagendra et al. 2018). The Global South is notable for its considerable percentage of residents that, to have access to the city, live in precarious settlements with no formal urban planning or management processes (Adegun 2017; Miraftab 2009). Latin America is the second most urbanized region in the world. 4 megacities stand out: São Paulo, Rio de Janeiro, Buenos Aires and Mexico City. São Paulo is the largest Brazilian city, established on a plateau and close to the coast, in a place once occupied by Atlantic Rainforest vegetation. Its population growth began to take place at the beginning of the twentieth century when the city became a national centre of industries and services.

Current studies show the need to improve the response to climatic variations in different sectors of local administration linked to health and urban planning. Geirinhas et al. (2017) demonstrate that São Paulo is a city with a great tendency of raising the frequency and duration of heatwaves in the country. Diniz et al. (2020)

reinforce this statement by showing that the mortality of people above 65 years old will increase in the region during heatwaves. The city also presents several places of high vulnerability to thermal stress for the elderly population (Lapola et al. 2019), which call for more studies on how urban planning can influence the improvement of the living conditions for these people throughout extreme weather events.

This chapter follows the line established by planners and academics who point out the expansion of vegetation cover in urban areas as an adaptation measure. This expansion must incorporate the design and planning processes in which green and open spaces, composing an articulated system and surpassing the restricted vision of the twentieth century of residual or island treatment of green spaces (Lyle 1998). We believe that progress in urban planning is brought by aggregating qualitative and quantitative information at the city and local levels.

2.2 *Prioritizing Places for Building Walkable Green Spaces*

To identify the priority urban areas for WGS implementation policies, we generated a spatialized Priority Index (*PI*) based on the elderly population's climate vulnerability and risk of public health disasters during heatwaves. We calculated the PI by quantifying social-environmental and spatial data using the geoprocessing tool ESRI ArcGIS Desktop 10.5 (ESRI 2016).

We consider "climate vulnerability" as in Lapola et al. (2019, p. 479): "a combination of exposure, sensitivity and adaptive capacity". The population's vulnerability is based on spatial, socioeconomic, institutional and environmental characteristics. Individual characteristics such as income, health condition, age and level of education, among others, are used to measure adaptive capacity and sensitivity to extreme weather events. Exposure involves presence in a spatial location where there is a potential danger factor to a person. It is possible to have exposure in a location and not have the vulnerability, but to have vulnerability there must be exposure. The risk of a disaster occurring is an aggregation of danger and vulnerability. Therefore, the vulnerability of the elements (elderly people) exposed to danger (heatwaves) favours the risk of disasters (damage to health) (Cardona et al. 2012; Lapola et al. 2019).

To simplify the model, we consider the presence of elderly people an exposure factor to danger. We adopted the concept of danger as being the occurrence of a "physical event" that may harm vulnerable elements (Cardona et al. 2012). This physical event corresponds to the high surface temperature, and the harmful effect is the deterioration of the health condition of elderly people caused by these high temperatures.

In this study, census tracts were used as spatial units (i), as they can provide data on a neighbourhood scale (PNUD et al. 2013). For sensitivity and adaptive capacity, the Human Development Index (HDI). The HDI doesn't explicitly distinguish sensitivity and adaptive capacity but allows for the quantification of this information as an aggregation of both (Lapola et al. 2019). The environmental data correspond to the WGS and thermal data quantified for each spatial unit. We acknowledge that street trees, squares and parks present differents health and social benefits directly and indirectly (Takano et al. 2002; Tamosiunas et al. 2014), adopting different weights for those variables. Derkzen et al. (2017) point out that the urban population has a higher perception of thermal comfort in parks and squares than in tree-lined streets. The index was developed specifically for urban areas, disregarding the areas with native vegetation and conservation units, as they are not considered WGS according to the concept given in Takano et al. (2002). Therefore, we do not take into account data that are not exclusively about the urban area.

We defined the model to calculate the PI as in Eq. 1 below:

$$PI = T_i * \left[\frac{S_i * E_i}{1 + \left(\frac{2}{3} * P_i\right) * \left(\frac{1}{3} * A_i\right)} \right] \quad (1)$$

The model calculates an index that seeks to relate each spatial unit in the urban area of a municipality, generating proportional rather than absolute values. In Eq. 1, socioeconomic data (*S*) correspond to sensitivity and adaptability. (*S*) are the normalized values (ranging from 0 to 1) of the inverted HDI of each spatial unit (*i*). These data are specific to the elderly population.

The temperature (*T*) corresponds to the normalized values of the surface temperature data. The data on WGS are: (*P*) the normalized value of the ratio of the total area of squares and parks to the spatial unit area, and (*A*) the normalized value of the proportion between trees and the spatial unit's urban road network length. Exposure (*E*) is a binary value (1 or 0) for the presence (1) or absence (0) of the elderly population in the spatial unit.

We consider that the inhabitants of places with a higher HDI to have higher adaptive capacities and lower vulnerability and exposure to heatwaves. Temperature is a factor of external influence due to a physical event (Milan and Creutzig 2015), so even in regions where there are low (high) temperature peaks, the index values may be higher (lower) due to the socioeconomic indicators and WGS values.

2.3 Socioeconomic Data

The socioeconomic data (*S* in Eq. 1) are the HDI values on a census tract spatial scale, from the official Brazilian demographic census of 2010 (PNUD et al. 2013). The city has 1594 census tracts varying in size, with an average area of 7.73 km^2 (standard deviation of 58.4 km^2) (Lapola et al. 2019). The HDI does not include variables linked to adaptive capacities, such as community cohesion and residential infrastructure (Harlan et al. 2006). However, the HDI data relate to the health quality of the elderly and their ability to cope with heatwaves. Life expectancy also derives from health conditions. The elderly usually have health problems brought about by age, such as breathing or cardiac issues. These problems worsen during large temperature variations (Hajat et al. 2014). Other factors like the social isolation that tends to occur with ageing, especially in communities with less interactivity, also influence the state of physical and mental health (Geib 2012; Harlan et al. 2006).

According to Geib (2012), socioeconomic inequalities are indirect determinants for health. Education and per capita income are related to factors such as poverty and social vulnerability, which are strong definers of health condition, life expectancy, and vulnerability to extreme weather events (Geib 2012; Lapola et al. 2019). Education is responsible for promoting knowledge about health and the environment, improving a person's understanding and ability to adapt to climate change (Landin and Giatti 2016), and income allows the purchase of material goods that can increase resilience and decrease the exposure (McMichael 2003). We assume that the HDI is not an index that completely covers the possible variables for measuring the population's adaptive capacity, mainly because it provides average data for several individuals in a given sector. Although not perfect, the HDI works well when used to assess the risk to climate-related events (Lapola et al. 2019), in addition to being an open and easy to acquire information.

2.4 Environmental Data

Environmental data are both the maximum surface temperature and the quantification of WGS. The maximum surface temperature of the urban area (T in Eq. 1) was calculated based on mosaics of daytime images from Landsat 8 TIR sensor, at a resolution of 100 by 100 m resampled to 30 by 30 m. Landsat 8 imagery is also open and free to obtain. We generated 12 mosaics for the hottest months of the year in the city of São Paulo, from December to February, for the years 2013 to 2019. The surface temperature was calculated using the mosaics, according to the method described by Jimenez-Munoz et al. (2014). The calculation resulted in a raster image containing the maximum surface temperature (T) presented over the seven years.

Although the use of near-surface temperature data is more accurate to attest the air temperature, in the context of the index presented here, we opted for surface temperature data because of its fine-scale. The surface temperature and the air temperature have a high linear correlation, which supports the use of the surface temperature data as an approximation to the air temperature, allowing its use for this index under the proposed conditions (Kawashima et al. 2000; Lapola et al. 2019).

The WGS data correspond to the spatial area of parks and squares (P in Eq. 1) and the presence of street trees (A in Eq. 1). These data were obtained from the Geosampa portal, which belongs to the São Paulo city Prefecture (http://geosampa.prefeitura.sp.gov.br/). We calculated (P) as the ratio of the total area of squares and parks to the spatial unit area. For street trees (A), we calculate the total number of trees in every 100 m of public urban streets in each spatial unit proportionally to the streets' length. We considered exclusively the trees found on sidewalks, flower beds, roundabouts and square sidewalks. Street trees are of great importance to mitigate heatwave and heat island effects (Rosenzweig et al. 2006). Although squares and parks may vary in terms of vegetation cover, we assumed for simplicity that they all have the same. Due to the lack of open data on tree size and treetop cover, we opted to use only data based on the presence and non-presence of trees.

3 Results and Discussion

In Brazil, inequalities are linked to bad income distribution which restricts the access to urban services and infrastructure to a portion of the population. This bad income distribution, allied to the lack of effective fiscal policies and the high rates of informality and unemployment, creates a scenario in which health is degraded throughout life, affecting the vulnerability and longevity of the elderly population (Geib 2012). The city of São Paulo is a place with a very diverse social structure, a wide variation in the HDI range and a multifaceted organization of the public space.

The method produced both statistical and spatial results that allow perceiving the disparities among very close spatial units. It is also quite evident that the central region has most of the best environmental quality and higher quantity of WGS. This central region also houses most of the high-income population of São Paulo. One possible explanation is the history of the urbanization process that took place in the city.

At the beginning of the twentieth century, through hygienist policies, public authorities promoted afforestation and gardening of urban streets and squares in the central region and its surroundings. The central region has come to be occupied by São Paulo elite. The workers and more impoverished people occupied the peripheries, in a model of extensive occupation of the territory. This model was strongly influenced by the private sector, mainly by the City Company and the real state market, with the municipal government acquiescence. Another factor that still influences the urban landscape today was the adoption of the Garden City model. A model of urban pattern with a focus on green landscaping, presence of squares, and intense street afforestation made exclusively in high-standard land plots (Di Giuseppe 1998; Leme 2003; Osello 1983). The peripheral urban regions have had a low degree of planning, with a focus on low-quality housing for renting. This pattern later culminated in the self-construction models (Leme 2003). There was no concern for the natural environment by the public authorities and the private sector in those peripheral regions, which is one of the reasons for the smaller presence of WGS (Fig. 1).

Table 1 shows that the reason for prioritization can differ for each region: high temperatures, socioeconomic conditions, lack of WGS presence or a combination of all factors. It is relevant to highlight the difference in the values in each census tract, especially the proportion between total urbanized area and WGS, for the high

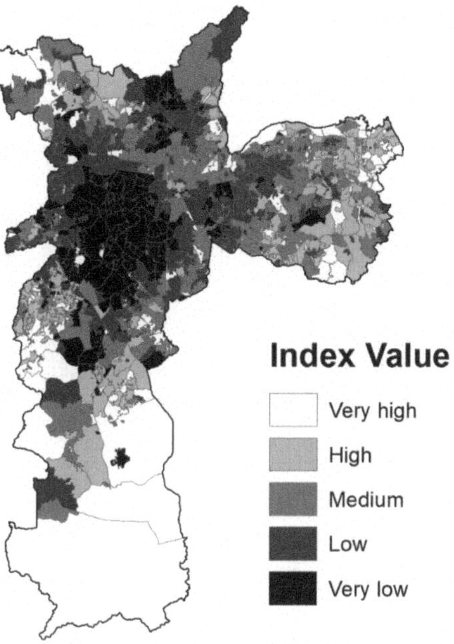

Fig. 1 Priority Index (PI) calculated for each census tract. Higher values mean a higher necessity of WGS to improve the elderly population's health condition. *Source* Elaborated by the authors

positioned spatial units in the index ranking. Also notable, the census tract of lowest vulnerability does not need the presence of WGS to mitigate the effect of heat-waves, indicating that the vulnerability is also strongly associated with socioeconomic factors and surface temperature control.

The index denotes that despite the large presence of natural vegetation or forestry in some regions, WGS is neglected, especially in the extreme south and north regions of the city.

The spatial distribution of WGS is also associated with temperature control. Figure 2 shows the maximum surface temperature registered by satellite imagery from 2013 to 2019. The central regions are those with the highest vegetation coverage and lowest temperatures in the whole urban area, denoting a relationship between these two factors.

Figure 3 denotes the lack of public policies and planning to correct the distribution of WGS in the city, maintaining its historical concentration in high HDI regions instead of promoting WGS implementation in the others. The 10% richest have a WGS area approximately equal to that of the 50% poorest (Fig. 3a). The distribution and the number of squares and parks, and the presence of street trees is related to the HDI value of the census tract. Street trees follow a similar spatial distribution as the one for the parks and squares (Fig. 3b). Highlights to Fig. 3c and Fig. 3d, which show that even though the highest HDI range corresponds to the smallest absolute number of squares, it still links to the largest area of all.

Table 1 Urban, social and environmental characteristics for the five higher and lower positions in the Priority Index ranking

Census tract	Total urbanized area (m^1)	Total square and park area w	Street tree count (mean)	HDI	Life expectancy (years)	Household per capita income (R$)	Maximum surface temperature (°C)	Index position
Jaguaré: Conjunto Cingapura do	370,681.08	0.00	0.090	0.635	70.14	448.78	39.81	1°
Jaguaré Paraisópolis	484,806.14	0.00	0.561	0.710	72.82	518.74	44.00	2°
Lageado: Centro Educacional Unificado Jambeiro (CEU)	345,486.41	0.00	1.997	0.638	69.58	415.46	39.56	3°
Perus: Escola Municipal de Ensino Fundamental Fernanda Gracioso/Jardim Russo	1,177,895.2	0.00	3.200	0.639	71.45	469.78	39.19	4°
Real Parque: Escola Municipal de Edução Infantil Pero Neto	190,177.58	603.50	2.307	0.683	72.50	510.82	40.45	5°
Berrini/Vila Funchal: Estação Berrini	706,261.50	35,138.65	5.978	0.965	81.82	5778.61	36.68	1590°
Vila Madalena: Estasao Vila Madalena	576,874.93	20,273.26	9.702	0.965	81.82	5773.61	36.52	1591°
Jardim Paulistano: Delegacia de Polícia Participativa/14° Delegacia de Polícia	188,314.90	6.14	7.076	0.965	81.82	5773.61	36.43	1592°
Vila Madalena: Estação Santuárto Nossa Senhora de Fátima/Sumaré	545,797.85	17,178.22	7.060	0.965	81.82	5773.61	36.18	1593°
Parque Savoy: Condomínio Residencial City Park	1212.09	0.00	0.000	0.914	81.38	4631.37	30.52	1594°

Compiled by the authors

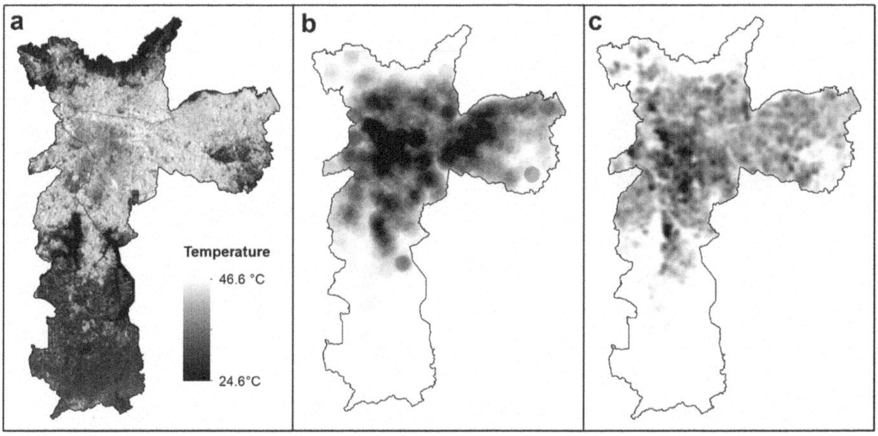

Fig. 2 **a** shows the maximum surface temperature from 2013 to 2019 in the city of São Paulo; **b** shows the concentration of urban parks; **c** the concentration of street trees. *Source* Elaborated by the authors

3.1 Elderly Health and Heatwaves in the City of São Paulo

Diniz et al. (2020) demonstrate that the mortality of elderly residents is higher and tends to increase in the city of São Paulo during heatwaves. Other studies also prove the increased risk of death of the exposed and vulnerable elderly in urban areas during heatwaves, including for the city of São Paulo (Astrom et al. 2011; Diniz et al. 2020; Gasparini et al. 2017). Zhao et al. (2019) show that the southeastern region of Brazil registers the highest number of elderly people hospitalization related to heart and breathing problem throughout heatwaves. Although vulnerability widely varies according to sex, educational level and other factors (Bell et al. 2008), the elderly residents are the most vulnerable, even when they have the same degree of exposure as the rest of the population. Gouveia et al. (2003) and Geib (2012) argue that the health condition of the elderly is determined by their socioeconomic characteristics, being an aggravating factor for the mortality of this population stratum. The results allowed to identify the places in which elderly people are more susceptible to the impact of heatwaves. It is also visible that planning processes do not see regions far from the city centre and surrounded by natural vegetation as places lacking urban green infrastructure. This perception allows for the implementation of WGS to neighbourhoods that are usually neglected by conventional urban planning.

Building walkable green spaces in high-temperature regions with a lower PI value can still contribute to the reduction of temperature and thermal comfort in the city, especially in places where there is little to no WGS (Demuzere et al. 2014). Either by creating or expanding accessible parks and squares or by increasing urban afforestation, WGS can improve the physical and mental health, quality of life, and

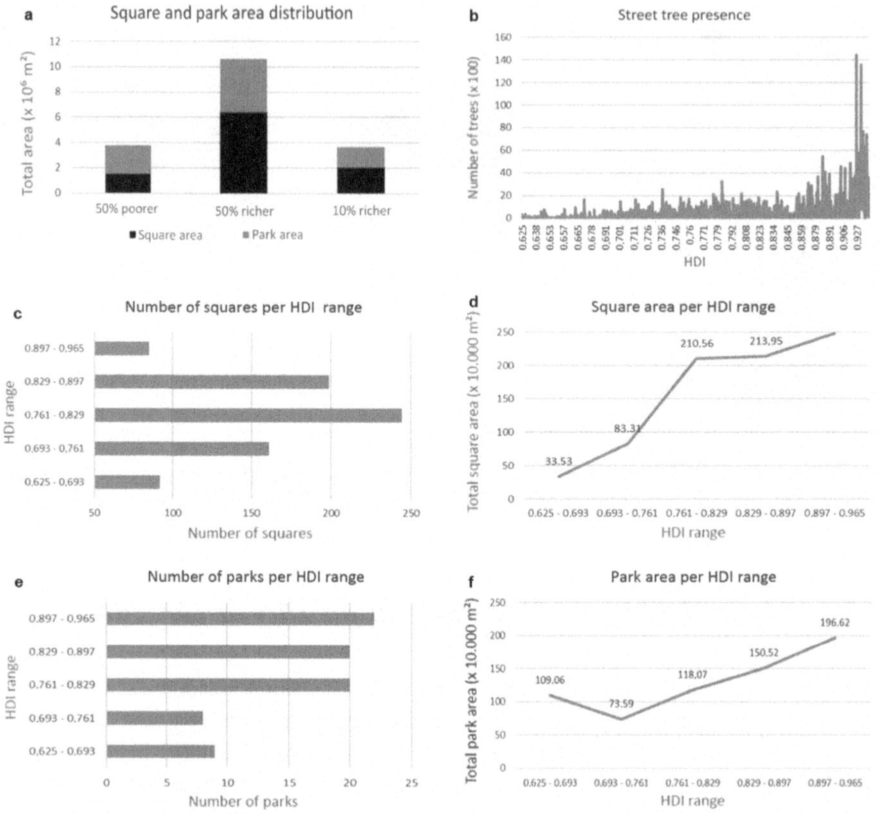

Fig. 3 Spatial data of WGS, associating its distribution and the population's income and HDI ranges. In **a** there is a clear distinction in the distribution of WGS between the 50% poorer and the 50% richer; **b** associates the presence of street tree cover and the value of HDI; **c** and **e** show the absolute distribution of squares and parks, respectively, by HDI range; and **d** and **f** compare the total area of squares and parks, respectively, to the HDI range. *Source* Elaborated by the authors

longevity, bringing benefits to the elderly (Geib 2012; Milan and Creutzig 2015; Takano et al. 2002). That is particularly true for poor people and those living in the peripheral areas of a city. Because they have a lower capacity to adapt to heatwaves and, therefore, higher vulnerability, they depend on the commitment and actions of the public power to raise their adaptation capacity.

Urban planning aimed at an ecologically balanced and socially just city is considered a significant ally to cope with climate change (Milan and Creutzig 2015). In São Paulo, the mitigation of high temperatures during heatwaves and the microclimate stabilization provided by the WGS can bring many social-environmental benefits to public health (Lee et al. 2015).

The results show that the causes of vulnerability differ for each census tract. The response pattern to public policies aimed at increasing the climate adaptation capacity of the elderly population, in a megacity like São Paulo, also differs. Public authorities should consider local socio-cultural and economic characteristics individually before developing urban adaptation policies. For this reason, despite their recognized socio-environmental benefits, authors like Ahern et al. (2012) emphasize the need to carry out experimental projects, which take in local characteristics for the urban planning and policy processes.

3.2 Contributions to Urban Planning

Lyle (1998) states that dealing with urban green spaces refers to finding balance point amidst green spaces, housing expansion, and environmental impacts. An emblematic example that reinforces the evidence of São Paulo's social-environmental disparities, as pointed out by Geib (2012), is the case of the Paraisópolis slum, which represents a situation seen in several other parts of the city. Paraisópolis is the second-largest slum of São Paulo, being located in the bottom area of a valley and surrounded by a high-income neighbourhood, called Morumbi (Travassos et al. 2017). Studies published by São Paulo's municipal government (São Paulo 2008; Silva et al. 2011) attest to the high environmental quality of this surrounding neighbourhood, mainly due to its excellent vegetation cover. However, this situation is quite different when we observe the precarious settlement of Paraisópolis, with a high population density, little afforestation, high rates of respiratory diseases, among other negative indicators (Silva et al. 2011).

Identifying and acting upon extreme cases like Paraisópolis is one of the main challenges planners and academics have to face. These situations call for the development of green and inclusive technologies which allow improving the social-environmental quality of precarious urban settlements, endowed with different urban parameters from those of the formal city (Adegun 2017). These challenges reflect in public policies, development of urban green infrastructures and other planning tools which support an ecological transition. They impose that a planning strategy needs to involve the community, the government and universities in the implementation of adaptation processes focused on the local social-environmental characteristics to be truly effective (Ahern et al. 2012).

4 Conclusion

We have developed this model on a scholarly basis, without in loco evaluation of the quantitative and qualitative data used and with socioeconomic indicators from the year 2010. Despite that, it still is a robust and versatile model, since a person can feed it with data from any origin and period. It can also be quickly modified to detect regions lacking green spaces and street trees, or urban sectors with low adaptive capacity to extreme weather events. The model usage can strengthen the actions taken by public administrations, planners, academics and local populations on finding areas that need to improve the adaptive capacity of the elderly through the implementation of walkable green spaces. Its proposal to urban planning is to facilitate projects seeking to raise the living standards of the local population and the effectiveness of actions aimed at mitigating and adapting the city to the effects of climate change.

Urban planning processes have changed over the past few decades, adapting to the growing weight of the climate change issue. Planning upon urban formations presupposes dealing with situations where various factors interact. Other elements must also be considered, such as understanding the cause of the social-environmental disparities within a territory. These disparities are increasing in a world in which the prevailing economic system is experiencing successive crises, in a scenario of high concentration of income, and the precarization of the living conditions for large proportions of the urban population.

Thus, this chapter proposed a method for calculating an index to help planners and related professionals perceiving which urban sectors lack WGS, focusing on improving the capacities of the elderly population to better cope with climate change and heatwaves. As well demonstrated, the application of this method in a city where social-environmental inequalities occur exposes the contrasting situations that emerge among neighbourhoods. This tool could be useful to the various cities in the Global South facing similar circumstances.

The priority index presented here, although simple, is a tool working towards strengthening social and environmental equity. Because it is simple to understand, people with no academic expertise can use it. Extreme weather events call for innovative public policy policies and a more participatory model of governance. The index also allows for greater precision in the implementation of projects and programs aimed at urban ecological balance. Seeking for environmental justice is an action taken at the municipal level that makes an enormous contribution in combating climate change, mitigating the effects of extreme weather events and improving the thermal comfort and the quality of life for the older citizens.

References

Adegun OB (2017) Green infrastructure in relation to informal urban settlements. J Archit Urban 41(1):22–23

Ahern J, Pelllegrino P, Becker N (2012) Infraestrutura verde: desempenho, estética, custos e método. In: Costa L, Machado D (org) Conectividade e Resiliência: estratégias de projeto para a metrópole. Rio Book's, Rio de Janeiro

Åström DO, Forsberg B, Rocklov J (2011) Heat wave impact on morbidity and mortality in the elderly population: A review of recent studies. Maturitas 69:99–105

Bell ML, O'Neill MS, Ranjist N, Borja-Aburto VH, Cifuentes LA, Gouveia NC (2008) Vulnerability to heat-related mortality in Latin America: a case-crossover study in Sao Paulo, Brazil, Santiago, Chile and Mexico City, Mexico. Int J Epidemiol 37:796–804

Cardona OD, van Aalst MK, Birkmann J, Fordham M, McGregor G et al (2012) Determinants of risk: exposure and vulnerability. In: Field CB, Barros V, Stocker TF, Qin D, Dokken DJ,

Ebi K-L et al (eds) Managing the risks of extreme events and disasters to advance climate change adaptation—a special report of Working Groups I and II of the Intergovernmental Panel on Climate Change (IPCC). Cambridge University Press, Cambridge
Casas ALF, Santos GMD, Chiocheti NB, de Andrade M (2016) Effects of temperature variation on the human cardiovascular system: a systematic review. In: Leal Filho W, Azeiteiro UM, Alves F (eds) Climate change and health: improving resilience and reducing risks. Springer International Publishing, Cham
Demuzere M, Orru K, Heidrich O, Olazabal E, Geneletti D, Orru H et al (2014) Mitigating and adapting to climate change: multi-functional and multi-scale assessment of green urban infrastructure. J Environ Manage 146:107–115
Derkzen ML, van Teeffelen AJA, Venburd PH (2017) Green infrastructure for urban climate adaptation: how do residents' views on climate impacts and green infrastructure shape adaptation preferences? Landsc Urban Plan 157:106–130
Di Giuseppe DT (1998) A Crise do Planejamento Urbano—Uma experiência alternativa em São Paulo (O caso dos Núcleos Regionais de Planejamento). FGV/EAESP, São Paulo
Diniz FR, Gonçalves FLT, Sheridan S (2020) Heat wave and elderly mortality: historical analysis and future projection for metropolitan region of São Paulo, Brazil. Atmosphere 11(9):933
ESRI (2016) ArcGIS desktop: release 10.5. Environmental Systems Research Institute, Redlands, CA
Gasparini A, Guo Y, Sera F, Vicedo-Cabrera AM, Huber V, Tong S et al (2017) Lancet Planet Health 1:e360–e367
Geib LTC (2012) Determinantes sociais da saúde do idoso. Cien Saude Colet 17(1):123–133
Geirinhas JL, Trigo RM, Libonati R, Coelho CAS, Palmeira AC (2017) Climatic and synoptic characterization of heat waves in Brazil. Int J Climatol 38(4):1760–1776
Gill SE, Handley JF, Ennos AR, Pauleit S (2007) Adapting cities for climate change: the role of the green infrastructure. Built Environ 33(1):115–133
Gouveia N, Hajat S, Armstrong B (2003) Socioeconomic differentials in the temperatura-mortality relationship in São Paulo, Brazil. Int J Epidemiol 32:390–397
Guo Y, Gasparrini A, Li S, Sera F, Vicedo-Cabrera AM, et al (2018) Quantifying excess deaths related to heatwaves under climate change scenarios: a multicountry time series modelling study. PLoS Med 15(7):e1002629
Haase D, Frantzeskaki N, Elmqvist T (2014) Ecosystem services in urban landscapes. Pract Appl Gov Implic 43:407–412
Hajat S, Vardoulakis S, Heaviside C, Eggen B (2014) Climate change effects on human health: projections of temperature-related mortality for the UK during the 2020s, 2050s and 2080s. J Epidemiol Community Health 68:641–648
Harlan SL, Brazel AJ, Prashad L, Stefanov WL, Larsen L (2006) Neighborhood microclimates and vulnerability to heat stress. Soc Sci Med 63:2847–2863
Jimenez-Munoz JC, Sobrino J, Skokovic D, Mattar C, Rosselló JC (2014) Land surface temperature retrieval methods from Landsat-8 thermal infrared sensor data. IEEE Geosci Remote Sens Lett 11(10):1840–1843
Kawashima S, Ishida T, Minomura M, Miwa T (2000) Relations between surface temperature and air temperature on a local scale during winter nights. J Appl Meteorol Climatol 39(9):1570–1579
Landin R, Giatti LL (2016) The health sector in an adaptive dialectic strategy: the case of the São Paulo's municipal policy on climate change. In: Leal Filho W, Azeiteiro UM, Alves F (eds) Climate change and health: improving resilience and reducing risks. Springer International Publishing, Cham
Lapola DM, Braga DR, Di Giulio GM, Torres RR, Vasconcellos MP (2019) Heat stress vulnerability and risk at the (super) local scale in six Brazilian capitals. Clim Change 154:477–492
Lee ACK, Jordan HC, Horsley J (2015) Value of urban green spaces in promoting healthy living and wellbeing: prospects for planning. Risk Manag Healthc Policy 8:131–137

Leme MCS (2003) O impacto da globalização em São Paulo e a precarização das condições de vida. Revista EURE 29(87):23–36

Lyle JT (1998) Design for human ecosystems. Landscape, land use, and natural resources. Van Nostrand Reinhold Company, New York

McMichael AJ (2003) Global climate change and health: an old story writ large. In: McMichael AJ et al (eds) Climate change and human health: risks and responses. World Health Organization, Geneva

Milan BF, Creutzig F (2015) Reducing urban heat wave risk in the 21st century. Curr Opin Environ Sustain 14:221–231

Miraftab F (2009) Insurgent planning: situating radical planning in the Global South. Plan Theory Spec Issue Strangely Fam 8(1):32–50

Mitchell D, Heaviside C, Schaller N, Allen M, Ebi KL, Fischer et al (2018) Nat Clim Change 8:551–553

Nagendra H, Bai X, Brondizio ES, Lwasa S (2018) The urban south and the predicament of global sustainability. Nat Sustain 1(7):341–349

Osello MA (1983) Planejamento urbano em São Paulo (1899–1961): introdução ao estudo dos planos e realizações. EAESP/FGV, São Paulo

Pitman SD, Daniels CB, Ely ME (2015) Green infrastructure as life support: urban nature and climate change. Trans R Soc S Aust 139(1):97–112

PNUD. Programa das Nações Unidas para o Desenvolvimento, IPEA. Instituto de Pesquisa Econômica Aplicada, and FJP. Fundação João Pinheiro. (2013) Atlas do Desenvolvimento Humano no Brasil de 2013. http://atlasbrasil.org.br/2013/pt/radar-idhm/

Reid CE, Clougherty JE, Shmool JLC, Kubzansky LD (2017) Is all urban green space the same? a comparison of the health benefits of trees and grass in New York City. Int J Environ Res Public Health 14(11)

Rosenzweig C, Solecki W, Parshall L, Gaffin S, Lynn B, Goldberd R et al (2006) Mitigating New York City's heat island with urban forestry, living roofs, and light surfaces. New York City regional heat island initiative final report. New York State Energy Research and Development Authority, New York

São Paulo. Secretaria Municipal do Verde e do Meio Ambiente e Centros de Estudos da Metrópole (2008) Indicadores ambientais e gestão urbana: desafios para a construção da sustentabilidade na cidade de São Paulo.

Silva E, Pesquero C, Ribeiro H, Assunção J (2011) Qualidade do ar na favela Paraisópolis, SP, e possíveis implicações à saúde. Revista Do Departamento De Geografia 18:60–66

Takano T, Nakamura K, Watanabe M (2002) Urban residential environments and senior citizens' longevity in megacity areas: the importance of walkable green spaces. J Epidemiol Community Health 56:913–918

Tamosiunas A, Grazuleviciene R, Luksiene D, Dedele A, Reklaitiene R et al (2014) Accessibility and use of urban green spaces, and cardiovascular health: findings from a Kaunas cohort study. Environ Health 13(20)

Travassos L, Penteado CLC, Fortunato I (2017) Urbanização desigual: rios, mídia e modernização ecológica. Espacio Abierto Cuaderno Venezolano de Sociología 26(2)

Tzoulas K, Korpela K, Venn S et al (2007) Promoting ecosystem and human health in urban areas using green infrastructure: a literature review. Landsc Urban Plan 81:167–178

Zhao Q, Li S, Coelho MSZS, Saldiva PHN, Xu R et al (2019) Ambient heat and hospitalisation for COPD in Brazil: a nationwide case-crossover study. Thorax 74(11)

The Impacts of the COVID-19 Pandemic on Biodiversity Conservation in Mozambique

Serafino Afonso Rui Mucova, Ulisses Miranda Azeiteiro, Walter Leal Filho, and Mário Jorge Verde Pereira

1 Introduction

Biological diversity or biodiversity plays a critical role in the survival of human populations and in the provision of essential services for society. It is estimated that about 2 billion people rely on wood fuel to meet their primary energy needs, and approximately 4 billion people rely primarily on natural medicines for their health care. Moreover, nearly 70% of drugs used for cancer are natural (IPBES 2019). Thus, our dependence on biodiversity and ecosystem services has increased exponentially due to economic and social factors such as, population expansion and economic growth (Cimon-Morin et al. 2013; Van der Biest et al. 2019). This trend of growth and dependence on biological diversity immediately suggests the implementation of adequate responses to offset social needs and the continuation of services provided by biodiversity.

The implementation of responses, mainly in the context of sustainable development, has progressed (IPBES 2019), but not necessarily enough to reach the Aichi Biodiversity Targets for 2020. Nevertheless, in a crucial moment, in which it was necessary insightful and advanced interventions to respond to global challenges, they were largely compromised due to an unknown disease caused by a virus (World Health Organization 2020a). This virus and disease were first reported

S. A. R. Mucova (✉)
Faculty of Natural Sciences, Lúrio University, P.O. Box 958, Pemba, Mozambique

S. A. R. Mucova · U. M. Azeiteiro · M. J. V. Pereira
Department of Biology & CESAM (Center for Environmental and Marine Studies), University of Aveiro, 3810-193 Aveiro, Portugal

W. L. Filho
Hamburg University of Applied Sciences, Hamburg, Germany

W. L. Filho
School of Science and the Environment, Manchester Metropolitan University, Manchester, England

W. Leal Filho et al. (eds.), *Sustainability in Natural Resources Management and Land Planning*, World Sustainability Series,
https://doi.org/10.1007/978-3-030-76624-5_21

on December 31, 2019, in the Chinese city of Wuhan, Hubai province (World Health Organization 2020a).

Some phylogenetic evidence supports that the unknown disease started between 6 October 2019 and 11 December 2019. A probable period that it has passed from an intermediate organism to humans (Van Dorp et al. 2020).

Notwithstanding, by January 3, 2020, about 44 people had been infected with pneumonia of unknown etiology (World Health Organization 2020a). In this setting, until 12 January 2020, suspicions indicated for infection and contagious disease through exposure to a seafood wet market in Wuhan (Zhu et al. 2020; Chan et al. 2020).

Nevertheless, cases were quickly reported from January 13, Thailand (2 cases); January 15, Japan (1 case) and January 20, Republic of Korea (1 case). These were the first imported cases of pneumonia infection of unknown etiology in the world (World Health Organization 2020a). Along the Africa, Egypt was known as the first country with a case of imported infection, reported on February 14, 2020 (World Health Organization 2020c). Mozambique, the health authorities confirmed the first imported infection on March 22, 2020 (MISAU 2020).

When infected countries were preparing to halt the spread of transmission, the first six cases of deaths caused by the virus were reported, on 21 January 2020, in the city of Wuhan (World Health Organization 2020b). Therefore, on March 11, 2020, the World Health Organization declared the disease a global pandemic (World Health Organization 2020e). Thus, it was the beginning of a catastrophic and unprecedented period of the twenty-first century of world society, as well as global efforts leading to the biodiversity conservation.

Two months later, after the first confirmations recorded in Wuhan, the World Health Organization, on February 11, 2020, named the disease as COVID-19; it stands for Corona Virus Disease 2019. The virus that causes the disease COVID-19 has been identified as Severe Acute Respiratory Syndrome Coronavirus 2 (SARS-CoV-2) (World Health Organization 2020d).

Historically, seven types of the coronavirus family have been identified worldwide. However, SARS-CoV-2 is the seventh virus that is easily transmitted to humans (Andersen et al. 2020; Yang and Wang 2020). The two other severe diseases-causing coronaviruses are known as SARS-CoV (Severe Acute Respiratory Syndrome Coronavirus), identified in southern China in 2003, and MERS CoV (Middle East Respiratory Syndrome Coronavirus), identified in Saudi Arabia in 2012 (WHO 2019; Andersen et al. 2020; Yang and Wang 2020).

Due to its lethality rate and rapid transmission among humans, it has rushed some theories associated with the origin of SARS-CoV-2. Some international personalities and media have speculated and increased rumours that the virus has been manipulated laboratory. However, research carried out by Kristian G. Andersen and his colleagues quickly dispelled doubts and dismantled the theory (Andersen et al. 2020). Nevertheless, gaps and doubts remain to be clarified, such as: What is the animal source of SARS-CoV-2? Who is the intermediary? How does the virus move from animals to humans?

Ongoing research, although not yet conclusive, but sustained by strong scientific evidence, points to Pangolin as the intermediate animal of the SARS-CoV-2 virus (Zhang et al. 2020; Han 2020; Lam et al. 2020). Even so, some scientific sources deny this view, claiming that Pangolins seems to be victims infected by COV (Choo et al. 2020).

Despite of, the United Nations Office on Drugs and Crime (UNODC), in its recent report, framed the current evidence (transmission of COVID-19 via Pangolin) in the context of international trafficking of the species from Africa (UNODC 2020). The scenario shows that about 25 tons of African Pangolin scales were seized in a mega operation carried out last year, representing almost 500,000 killed Pangolins. Yet, between 2014 and 2018, approximately 370,000 Pangolins were seized worldwide (UNODC 2020).

In Mozambique, analyses show that, in the last 16 years, there has been a considerable increase in cases of Pangolin trafficking, with 2019 being the worst year of Pangolin trafficking, where from February to July the authorities seized about 48.9 tonnes of Pangolin (ANAC 2020).

The scientists warn that Illegal and international wildlife trade are now one of the fourth lucrative illegal businesses, worth an estimated $ 5–23 billion per year (May 2017; Wildlife trafficking Alliance 2021). According to Scheffers et al. (2019), almost 24% of terrestrial bird, mammal, amphibian, and squamate reptile species, are traded globally. And it is likely that future international trade will affect up to 4064 species at risk of extinction (Scheffers et al. 2019).

At the same time, as scientists seek to answer important questions to control the pandemic, the number of COVID-19 cases is increasing dramatically. Today (07 March 2021), roughly 116,639,024 were confirmed positive, and approximately 2,590,159 of deaths were registered worldwide, distributed in 188 countries and 5 continents (Johns Hopkins University 2021). Thus, Mozambique participates with 62,131 positive cases, and 686 registered deaths (MISAU 2021; Johns Hopkins University, 2021). The projections suggest that the country will register almost 20,000,000 cases (corresponding to 69.3% of the entire population), if appropriate measures are not implemented (worst case scenario) and the pandemic peak was reached in January 2021.

The global impacts of this pandemic have been suggested, analysed, and discussed, but they are not yet conclusive. The United Nations Educational, Scientific and Cultural Organization analysed that about 90% of students worldwide were impacted (United Nations Educational, Scientific and Cultural Organization 2020). Around 2.7 billion workers are being affected (International Labour Organization 2020), representing more than 81% of the world's workforce. The United Nations reported that the pandemic would have substantial implications for the implementation of the 2030 Sustainable Development Agenda (United Nations 2020a). Yet, United Nations warns that the COVID-19 caused a drop-off of about $ 50 billion of global exports in February (United Nations 2020b). Also, nearly 130 million people could face acute food insecurity by the end of 2020 (United Nations World Food Program 2020).

The pandemic has led to the cancellation of COP26 UN Climate Change Conference set to take place in Glasgow in November 2020; and Bonn Climate Change Conference (SB 52) scheduled to take place in Bonn in October 2020. The scientists claim that these cancellations have consequences on the planned efforts to address climate change and biodiversity conservation challenges (Corlett et al. 2020).

The recent report by the African Union indicated that negative and harmful impacts are expected along the African economy, where the African continent is estimated to have lost more than USD 65 billion due to falling oil prices on the world market (African Union 2020).

The United Nations University report suggests that the number of populations living in poverty is expected to increase between 420 and 580 million if the worst scenario of a 20% contraction in population income is reached (Sumner et al. 2020). Furthermore, it is expected that global GDP will decline by 4.5% in 2020 (OECD 2020).

Notwithstanding, the global impacts of the COVID-19 on biodiversity conservation are poorly known and available data are insufficient to assess the implications worldwide. A central and critical question was previously asked by the authors Corlett et al. (2020) "How is the pandemic affecting biodiversity now?". Answering that question satisfactorily requires adding a set of information and in-depth knowledge of the situation in each region, which it is currently not widely available.

A well-positioned paper indicated some immediate implications such as, cancelling strategic meetings, postponing conferences, and training, reducing funding, limiting access to research infrastructures, postponing, and cancelling field missions (Corlett et al. 2020).

An report prepared by IUCN for Africa concluded that during a pandemic, there were considerable reductions and suspensions of protected area operations, visitor facilities closed to the public, workplaces shut, and important supply chains disrupted, with negative implications for biodiversity conservation (Waithaka 2020).

Further on, Lindsey et al. (2020), demonstrated that the pandemic will create a storm of reduced international funding for Africa, as well as actions restrictions by international agencies. Calculating that the impacts of COVID-19 on biodiversity conservation will be severe and negative (Lindsey et al. 2020).

Despite of some praiseworthy efforts to draw the current picture of the stage of biodiversity conservation in Africa, more and urgent assessments are needed. Thus, we agree that the urgency is to increase as much scientific evidence as possible about the practical implications of the pandemic in different latitudes (Bates et al. 2020; Rutz et al. 2020). Then, to systematise it with the aim of providing a global understanding of the situation with the goals of supporting in the definition of effective and sustainable actions to cope this pandemic and other zoonotic diseases (Geijzendorffer et al. 2017; Bates et al. 2020; Corlett et al. 2020).

The main goal of this research was to assess the practical impacts of the COVID-19 on biodiversity conservation actions in Mozambique with the purpose of increasing the evidence, knowledge and understanding of COVID-19, and informing its local implications, as well as the actions needed to cope. Yet to understand the role of ongoing armed conflicts in the protected area of Mozambique, as a factor contributing to the biodiversity pressure.

2 Material and Methods

2.1 National Context

Mozambique is a country on the eastern coast of southern Africa, and it is limited to the north by Tanzania; to the northwest by Malawi and Zambia; to the west of Zimbabwe, South Africa, and Swaziland; to the south-by-South Africa; and to the east by the Indian Ocean (Fig. 1). The country has an estimated population of 28,861,863 inhabitants, of which about 66% live and work in rural areas. Its demographic density is 33.5 inhabitants per Km^2 (National Statistics Institute—INE 2017). The population growth rate is 2.8% per year, while the gross birth rate (per thousand) is 37.9 inhabitants.

Nevertheless, has an estimated area of 801,590 km^2 and a coastal extension of 2500 kms (MITADER 2015). The GDP growth rate is 3.7%, while the GDP per capita is USD 466.18 (INE 2017). The biodiversity conservation, as well as the management of conservation areas, is regulated by the National Administration of Conservation Areas (ANAC), a public institution, under the supervision of the Ministry of Land and Environment (MITA, former MITADER—Ministry of Land, Environment and Rural Development). ANAC, founded in 2011, has the mandate to oversee and ensure the biodiversity conservation and the sustainable development of ecotourism.

Biodiversity conservation areas occupy about 25% of the national territory, including 7 National Parks, 7 National Reserves and 70 game hunting areas (ANAC 2015; MITADER 2015). Available data suggest that Mozambique has more than 6,500 species of wild plants, of which more than 300 species of plants are on the IUCN red list and 22% are endemic (ANAC 2015; MITADER 2015). The country has a terrestrial fauna represented by more than 726 species of birds, 171 species of reptiles, 85 of amphibians and 3075 species of insects. Furthermore, has 7 (seven) ecological regions of global importance such as: the coast influenced by Agulhas Current System, Ecological Region of the East Coast of Africa, Lakes of the Rift Valley, Forests of the Southern Rift Mountains, Mangroves of Southern Africa, Eastern and Southern Forests of Miombo and Wetlands of the Zambezi Delta (MITADER 2015).

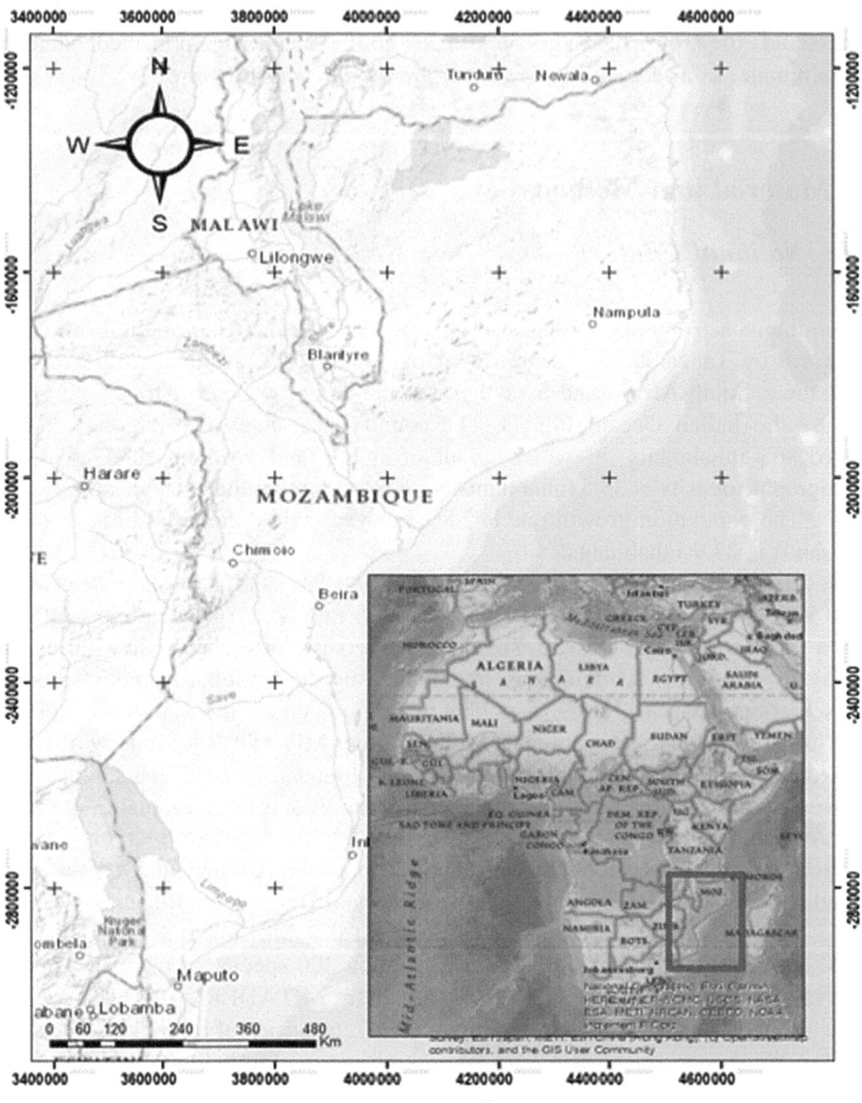

Fig. 1 Geographic location of Mozambique, Africa. *Source* Elaborated by the authors

2.2 Methods and Analyses

For the purpose of this study, one online survey was created, and a transverse survey of respondents was employed. The survey had two (2) main components. The first component consists of four (4) questions (consisting of socio-demographic issues), and the second consists of twelve (12) questions (related to biodiversity conservation action question). The survey was prepared in Portuguese and the results were translated into English (https://forms.gle/S2TDp99FQzcqNzJd7).

This survey was employed exclusively to institutions that are directly involved in biodiversity conservation actions in Mozambique (e.g. government, civil society, international organisations, non-governmental organizations, academies, research centre, foundations, and United Nations agency). Almost 31 institutions, approximately 90% of the existing institutions were assessed. The head representatives of the institutions were the main target, and, in the impossibility, the survey reached senior specialists, officers, and researchers with very solid knowledge of the institution.

The database of the institutions was obtained through the government of Mozambique, specifically at the Ministry of Land and Environment (https://www.mitader.gov.mz/) and National Administration of Conservation Areas (http://www.anac.gov.mz). Furthermore, the obtained database was incomplete, with a very limited number of institutions, and without providing important information such as, name of the national representative, contacts, among others. Thus, to fill this gap, an information search was developed through the institutions' web pages, technical documents, scientific paper, social networks such as Facebook, LinkedIn, YouTube, as well as national and international electronic newspapers. After this, at least two persons from each institution were contacted and sent the survey to answer. We noted that at least one of them answered our survey.

Notwithstanding, to complement our empirical assessment and in order to achieve a closer perception of Mozambique reality, we decided to carefully analyse Mozambique's preparedness and efforts to reduce the main future risks of transmission of zoonotic diseases in the context of biodiversity conservation. Doing so, we have examined two main documents that guide national and international efforts dedicated to the biodiversity conservation prepared by the government of Mozambique, namely:

1. National Strategy and Action Plan of Biological Diversity of Mozambique (2015–2035: http://www.biofund.org.mz/wp-content/uploads/2017/03/NBSP-Biodiversty-Strategy-and-Plan-Mocambique-2015-2035-Ingl.pdf).
2. Strategic Plan for the National Administration of Conservation Areas (2015–2024: http://www.anac.gov.mz/wp-content/uploads/2017/07/Plano-Estrategico-da-ANAC-2015-2024-1.pdf).

Our strategy was to identify, through the documents, clear evidence of proactive thinking to increase knowledge about the role of diseases, actions to monitor and control diseases and their threats as a fundamental condition to protect public health, people, economy, and national biodiversity.

So, we decided to transcribe directly, putting the paragraphs in question in quotes, cited and including the respective pages of any specific actions that demonstrate gaps or show a proactive thinking to identify, monitor, reduce the spread of zoonotic or emerging diseases in biodiversity conservation arena.

To conclude, we reflected on the role of ongoing armed conflicts (during the COVID-19 pandemic) in northern and central Mozambique in mitigating the impacts of COVID-19 in biodiversity conservation actions, as well as in the spread of future zoonotic diseases in Mozambique.

3 Results

3.1 Characterisation of Respondents, State of Emergency and Actions Taken During the COVID-19 Pandemic

The survey aggregated 55 participants from 31 institutions nationwide. Approximately 23% came from higher education institutions and 12% shared by the United Nations Agency, Foundation, International Cooperation Agency, and Research Centre (Fig. 2). Almost 19% came from national parks, reserves, and 13% from NGOs. Furthermore, about 10% came from civil society and 7% from government or decision makers (Fig. 2). We found that, of the 55 participants, about 74.5% were Men and 25.5% were Women.

Our research shows that approximately 94.0% of the institutions had their actions and projects considerably affected since the enactment of the global

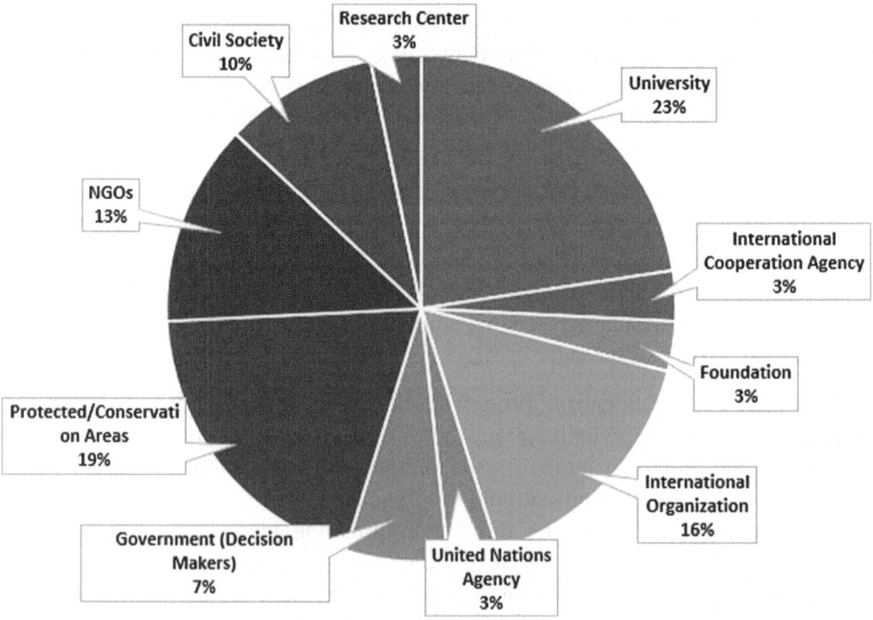

Fig. 2 Representation and proportion of institutions participating in the research. *Source* Elaborated by the authors

lockdown and the State of Emergency enacted by the government of Mozambique (Fig. 3a). At the same time, we found that nearly 90.9% had an immediate shock with at least a minimum duration of one (1) month, registering quite significant disruptions in terms of the management system and a sudden slowdown in actions in the field (Fig. 3b). Nevertheless, from that universe, about 30.9% had an impact lasting more than two (2) months (Fig. 3b), whose impacts have been felt to date.

Despite the fact that most of the institutions are drastically affected by the impacts of COVID-19, none of them has completely interrupted their actions. The results show that 18.2% worked full time and without limitation, 40.0% worked part time at home (telework) and finally about 41.8% worked part time and with restrictions in the office (telework) (Fig. 3c).

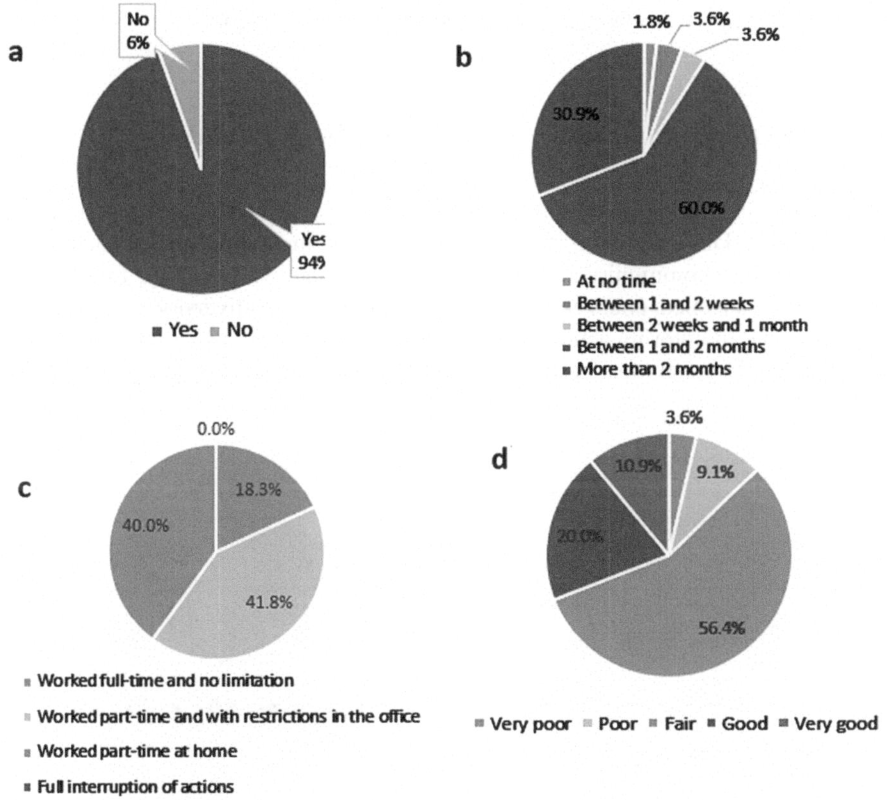

Fig. 3 It shows the different responses from respondents based on the following questions: **a** Do you consider that the State of Emergency enacted by the Mozambican government affected the normal development of your actions? **b** How long have you been affected by the State of Emergency and global lockdown? **c** During the State of Emergency and general lockdown, did you work? **d** Considering the challenges of working outside the office, how do you assess your available conditions to carry out your work tasks at home. *Source* Elaborated by the authors

Regarding the infrastructures (software, database, storage devices, office, communication networks and data sharing) for the development of biodiversity conservation actions in teleworking (Fig. 3d), about 56.4% said they had reasonable conditions, 20.0% were in good condition, 10.9% were very good and the remaining 11.7% responded between poor (9.1%) and very poor (3.6%) (Fig. 3d).

We found that communication matter (voice and data) or the quality of internet for the development actions in teleworking is not good enough to increase biodiversity conservation efforts. We note that roughly 49.1% recognised the communication challenges in the country as high and they understand that is minimally reasonable (Fig. 4a). Nevertheless, about 23.7% understand that communication is between good and very good, and 27.3% found that the quality of communication is between poor and very poor (Fig. 4a).

Regarding digital skills to develop biodiversity conservation actions remotely, most institutions are prepared enough to implement and adopt advanced technologies to support biodiversity conservation efforts (Fig. 4b).

3.2 *Practical Implications of the COVID-19 Pandemic*

The pandemic poses enormous challenges for biodiversity conservation in Mozambique. During and to date, around 10% of institutions have completely interrupted their community actions throughout the national territory. In addition, about 11% cancelled courses and specific training to biodiversity conservation. Patrolling and monitoring actions were completely suspended (8%), and about 32% reported that their activities intrinsically linked to the field were interrupted (Fig. 5).

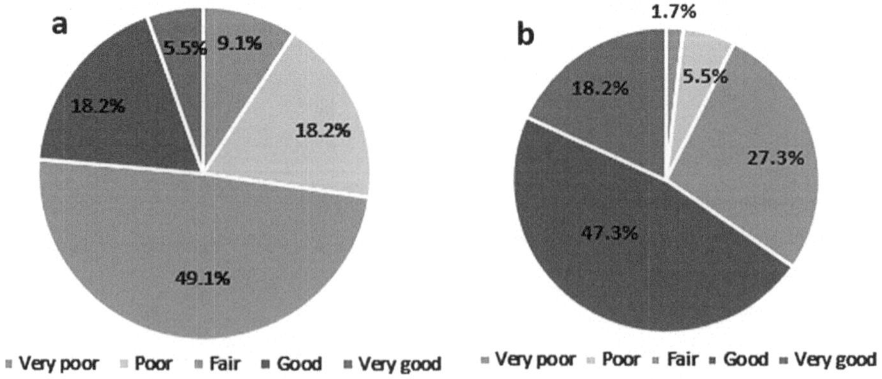

Fig. 4 It comprises the respondents' information based on the answers: **a** how do you assess the quality of the communication (internet or data) for the fulfilment of teleworking? **b** How do you assess the digital skills of the company's employees to perform teleworking tasks? *Source* Elaborated by the authors

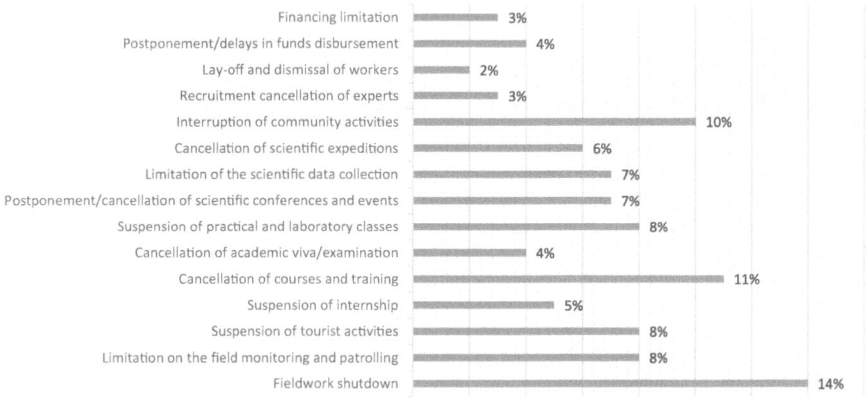

Fig. 5 Plotting the main practical implications registered/occurred after the global lockdown and State of Emergency in the country. *Source* Elaborated by the authors

Approximately, 7% complained that funding was being reduced by donors and the limited financial resources available would reach institutions late. Scientific tasks related to the expeditions (with about 6% of the institutions) and data collection in the field (with about 7%) were significantly affected.

Furthermore, teaching and research institutions have cancelled their laboratory practice programs, on-site examinations and internship programs within institutions, and for conservation areas. Yet nearly 3% interrupted the process of hiring specialists, and 2% dismissed technicians and field workers (Fig. 5).

Regarding of the implications shown in Fig. 6, about 72.7% (Fig. 6a) understand that the challenges imposed by COVID-19 are negatively critical to the progress of biodiversity conservation efforts in the country. Approximately, 21.8% stated that the same challenges are much more critical (Fig. 6a). A relatively insignificant fraction, comprising around 5.5% (Fig. 6a), indicated that the implications are not critical for the conservation projects and actions of its institutions.

When asked if the implications reported in Fig. 6 would compromise the institution's goals, 62.0% (Fig. 6b) reported that "yes" and affected their plans and goals by up to 50%. Yet, about 20.0% understand that, in the current context, their efforts may be affected by more than 50% and it may even reach 100% (Fig. 6b). Finally, around 18.0% stated that the current implications caused by COVID-19 are expected to interfere negatively in around 25% of its actions.

When analysed the mitigation strategies adopted by the institutions, about 45% (Fig. 6c) understand that the preventive measures adapted would help to contain the negative impacts of COVID-19 in their programs. Nevertheless, about 11.0% stated that the mitigation measures adopted are not sufficient and efficient to halt the negative impacts. In addition, 36.3% considered that they were not sure whether the adapted internal plans resulted in positive actions to control the negative impacts of the COVID-19 (Fig. 6c). Figure 6d shows that almost 95.0% concluded that the negative impact of COVID-19, associated with restrictive measures implemented

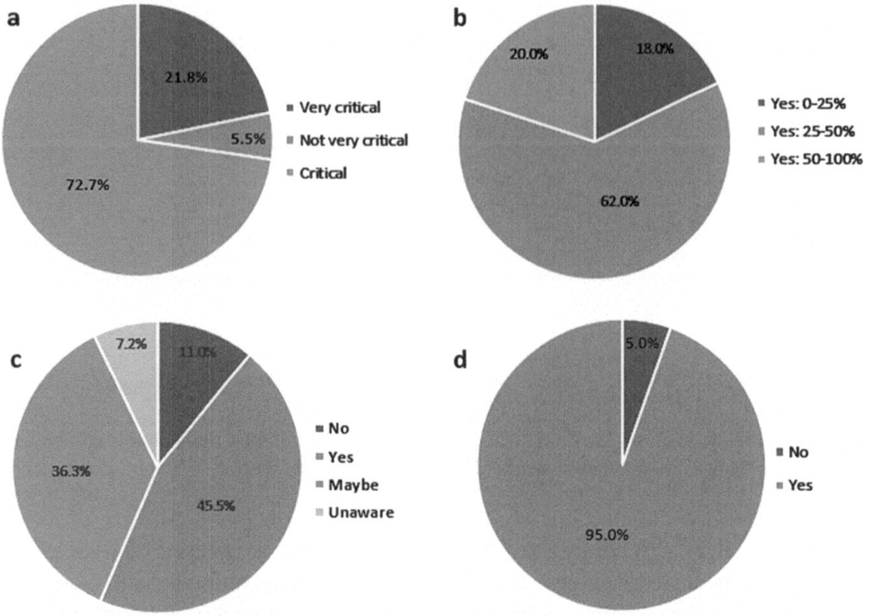

Fig. 6 The graphs present answers from the following questions: **a** How do you assess the importance of those implications in biodiversity conservation actions? **b** Do you understand that the COVID- 19 will compromise the goals, or the achievement of the results outlined by your institution for the current year? **c** Do you consider the measures implemented by your institution sufficient to mitigate the impacts of COVID-19 on biodiversity conservation actions? **d** Do you consider that the COVID -19 pandemic, associated with the State of Emergency in the country, had negative economic implications for the institution? *Source* Elaborated by the authors

by governments, had a very negative economic impact and mainly for biodiversity conservation efforts.

3.3 Capacities and Strategies to Mitigate the Impacts of Zoonotic and Emerging Diseases

Analysis performed through the Strategic Plan of the National Administration of Conservation Areas of Mozambique and the Strategy and Action Plan for the Conservation of Biological Diversity in Mozambique allowed us to state:

- We could not find clear evidence of capabilities and proactive thinking to increase knowledge and identification of diseases (zoonotic and emerging), impacts, as well as strategies for controlling and mitigating. Yet, we could not see strategic traits that supported the design of policies and plans to respond to the emergence of public health crises, natural disasters, as well as environmental crises.

Notwithstanding, the Strategic Plan of the National Administration of Conservation Areas proved to be quite fundamental to support actions and efforts to protect biodiversity in the future, however it presented enough elements to support our statement, such as:

- "There is still no internal capacity to develop an applied research, although this is a fundamental instrument to define species, populations and habitat's management measures" (ANAC 2015, 83 p).
- "There is lack of technical staff and equipment in conservation areas, taking into account the size and the complexity of the work that is to be done" (ANAC 2015, 83 p).
- "This situation affects ANAC capacity to: (i) perform CA management duties with due diligence; (ii) design, implement and monitor strategic plans and management plans; (iii) apply the conservation legislation; and (iv) collect information on basic and financial management, to guide the decision-making process" (ANAC 2015, 85 p).
- "The link between State institutions working in monitoring is weak. That has a reflection on the impunity of environmental crimes and on the increase of transnational poaching levels" (ANAC 2015, 86 p).
- "ANAC is currently consolidating itself and it still has a weak administration and management capacity for the CAs due to the lack of qualified personnel, proper equipment and facilities" (ANAC 2015, 87 p).
- "There are still missing regulations, norms and procedures for an effective management of the CAs" (ANAC 2015, 87 p).
- "Limited financial resources for biodiversity conservation" (ANAC 2015, 89 p).
- "Lack of mechanisms to compensate for biodiversity conservation and protection" (ANAC 2015, 89 p).
- "Limited modern technologies such as biotechnology, information technology, GPS systems that can positively affect ANAC's actions" (ANAC 2015, 89 p).

This means that many of the efforts conceived in ANAC's strategic plan until 2024 are mostly aimed at bridging the gaps and flaws identified above, improving, and increasing the institution's capacities to better serve in the biodiversity conservation arena.

Nevertheless, the Action Plan for the Conservation of Biological Diversity in Mozambique (2015–2035) proved to be quite incipient to support technical and scientific actions with the aim of supporting the identification, reduction, monitoring, and mitigation of possible impacts of zoonotic or emerging diseases.

4 Discussion

4.1 Implications of the COVID 19 Pandemic on Biodiversity Conservation

Our research found significant reductions and suspensions of actions, which has resulted in negative impacts, for the time being, in the short term in conservation efforts (Figs. 3, 5, 6). The implications found was due to the implementation of measures such as, social isolation, physical distancing, movement restrictions to give rise to actions to control and cope with the COVID-19 pandemic.

The sudden shocks observed in the institutions were largely due to the lack of knowledge of the disease, fear of contagion, the absence of mitigation and control plans for zoonotic diseases related to the SARS-COV-2 virus, and timidity motivated by drastic increases in case of infections and deaths in South Africa, China, the European continent, as well as the American.

The measures adopted by the Government of Mozambique, despite being implemented later, were fundamental to ensure the protection of the health and life of all workers who are at the forefront of safeguarding natural resources and biodiversity conservation. As a positive implication, it allowed to register reduced deaths of workers due to COVID-19.

Nevertheless, we found that the measures imposed have neutralised the local action, forcing a reduction in interventions in the field and in the community environment. Our study concludes that 94% were affected by the pandemic (Fig. 3a). These figures reveal the degree of impact caused by the COVID-19, but also the level of organisation and preparation of institutions to face this calamity. Our results are aligned with current global trends, which prove that the pandemic has hit financial institutions and protected areas in Africa hard, delaying in some way the achievement of the goal of global and national agenda (Waithaka 2020; Lindsey et al. 2020).

However, the regional and global dimension of the impacts of the COVID-19 on key sectors of biodiversity conservation suggests that the preparation, mitigation responses today and for coping in the post-COVID-19 phase should be rethought and implemented in the same scale. Some actions will include: (a) identifying threats from countries; (b) understanding the flow of threats inside and outside countries; (c) verifying the level of mitigation of threats by countries; (d) assessing the country's level of preparedness in the face of threats, (e) designing and implementing joint integrated programs to cope; and (f) developing innovative techniques to detect sudden changes of the threats and their containment mechanism (Neethirajan 2017; Astill et al. 2018).

Although more than 87.3% agreed that they had acceptable conditions for delivering actions in the teleworking, and 92.8% had sufficient digital skills, almost 72.8% converged that the poor quality of communication did not show encouraging signs for reaching its goals (Fig. 3d, 4a, 4b). In this context, we note that the pandemic revealed weaknesses and deficits for the conservation sector that must be addressed as an integrated response to the COVID-19. Yet, it is urgent to invest heavily in adequate work infrastructures, as well as to enhance the quality of communication significantly so that biodiversity conservation actions do not suffer drastic ruptures in a possible scenario of disaster and environmental crises.

Approximately 10% completely interrupted their regular community interventions at the interface and within the conservation areas (Fig. 5). These institutions collaborate and assist families living in buffer zones. So, the absence of community action could result in negative implications for biodiversity management efforts, as well as for the survival of rural communities located on the border of those areas. It is estimated that about 50,000 people are living around the country's conservation areas (World Bank 2018) and are dependent on those resources, as well as on existing jobs there for survival, increased family income, and food security.

In Mozambique, tourism activity lost more than 95% of revenues, and affected almost 12 thousand workers (Ministry of Tourism and Culture 2020). With the drastic fall in nature-based tourism (which led to the dismissal of more unskilled and local workers), a reduction in activities to exploit natural resources (which resulted in reduced rates and incentives for communities) (Waithaka 2020), interruption in the assistance of sustainable actions of the communities, it is hoped that the situation may threaten the survival of populations, increase poverty, and force communities to find new ways for their survival.

The new survival paradigms may include the presence and direct extraction of resources within conservation areas, thus increasing pressure on ecosystem services, challenging, and reducing actions for the biodiversity conservation. A similar conclusion was presented by Waithaka (2020) and warns that the local population will need to be better supported and economically trained to mitigate the impacts of COVID-19 on conservation (Waithaka 2020).

We found evidence that the training of future biologists, ecologists and conservationists will undergo considerable changes. The Fig. 5 shows that 11% largely cancelled their on-site training programs. In addition, 7% have discontinued their scientific expedition and data collection activities in the field. The cancellation of conferences, courses, advanced training, practical classes, professional internships, expeditions, and research work can contribute to increasing deficits in essential tools and knowledge for many students. However, these gaps may have implications for the systematization of young scientists' technical and practical knowledge. And we recognise that the scenario caused by COVID-19 can lead to dropouts, discouragement (in the courses), and an absence of passion for biodiversity conservation (Corlett et al. 2020).

Another approach focuses on the following point: Less trained students whose needs were aimed at increasing practical experiences and direct contacts with the reality in the country's conservation sector may lose confidence and opportunities to complement their training and improve their technical and scientific skills. Thus, when assessing the situation in an integrated context, and in the medium term, we understand that fewer senior technicians may be available to assist in the multiple actions of biodiversity conservation.

Biodiversity conservation is effective when multiple efforts work in an integrated manner. In the current scenario, some of these efforts have been temporarily halted. The findings showed that almost 22% stopped their patrolling and monitoring actions (Fig. 5) and 3% postponed the hiring of specialists to support the country's conservation efforts. In addition, 2% dismissed part of their technicians and field workers (Fig. 5). In this setting, there is strong evidence showing that the achievement of the goals of nature conservation for 2020 will go backwards, and we justify: The limitation in patrolling areas and the reduction of technicians in the service of biodiversity conservation in a sector with limited human resources to cover huge conservation areas (ANAC 2015) may condition the limitation of actions to certain regions. For conservation areas with enormous challenges such as poaching, illegal logging, invasion of communities, and human-animal conflicts, coupled with the absence of effective responses, can result in increased pressure, degradation, and vulnerability of biodiversity.

The Fig. 6 shows a very worrying picture for the biodiversity conservation. About 94.5% considered the impacts critical and more than 82.0% converged that their actions could be critically affected between 50% and 100%. In the current situation, the results suggest that the achievement of many national and international goals has been and will be compromised. However, the commitment to safeguard biodiversity must be on the country's agenda.

4.2 Cooperation and Integrated Planning on Biodiversity Conservation Action

The spread of the COVID-19 pandemic must be framed in a context of international trade and trafficking of Wildlife. Where border weaknesses (control and monitoring); limited local, regional, and global commitments; widespread corruption, especially in poor and developing countries, are proving to be one of the main weaknesses, threats, and barriers to prevent trafficking and massive destruction of biological diversity.

However, more than ever, the consequences and implications of COVID-19, which unfortunately claimed lives and weakened economic and social systems, must be clearly understood in response to inadequate and limited efforts to fully preserve nature and protect biodiversity.

Today, the world is preparing for a new phase (post-COVID-19 recovery). This phase will bound to fail if the biodiversity conservation and nature protection is not integrated into an economic, social and development context of countries. Thus, countries need to be prepared to make more dramatic and unprecedented decisions in favour of protecting biological diversity. Yet, national, and international agenda, and commitments should focus strongly and largely on cooperation, which should include: (a) strengthening borders, (b) banning the market and illegal trade in wild species, (c) raising the ability to track and detect genetic and biological resources at the borders, (d) sophistication of resources and methods for identifying and locating traffickers of prohibited species, as well as penalising and criminalising offenders with exemplary penalties.

A complementary approach will involve integrating health and disease issues in the context of biodiversity conservation and nature protection: (a) Increasing scientific evidence and knowledge within conservation areas, (b) training and development of human capacities for monitoring and controlling, and (c) enhancing communication and response capacities.

4.2.1 Perception for Mozambique

Strategically, the country urgently needs to review its strategy and action plan for the biodiversity conservation (2015–2035) to allow the framing of more proactive and integrated thoughts and actions to halt the cross-border trafficking and illegal trade of animals and forestry resources, which is considered one of the determining factors for the spread of diseases to humans and *vice-versa*.

In addition, we consider that effectively responding to the challenges for mitigation or control of zoonotic diseases, COVID-19 and other emerging diseases is considerably challenging in a short time scale the medium term. However, it will be necessary to overcome barriers that severely threaten biodiversity conservation actions in the country, such as armed and military conflicts.

Armed and Military Conflicts (What Role Do They Assume?)

The wars had implications and severe damages for the biodiversity conservation (Daskin and Pringle 2018). Wildlife was drastically decimated to feed the troops, especially medium and large mammals. In those periods, animal trophies were indiscriminately collected for illicit national and international trade (Dias 1971; Hatton et al. 2001). The predatory slaughter of animals was responsible for the fall in the greater number of herbivores and carnivores within and outside the conservation areas (Daskin and Pringle 2018; Stalmans et al. 2019). Thus, 26 years after the end of the civil war, Mozambique has faced yet another armed and military conflict. Since 2017 until then, this conflict has killed more than 1500 people and displaced more than 2,500,000 people (UNHCR 2020). The conflicts take place in a province in the north of Mozambique, and it is extremely important for the biodiversity conservation in the country and in Africa. The province of Cabo Delgado includes three important Ecoregions on the planet (Olson et al. 2001). In 2018, the first Mozambican biosphere reserve, called the Quirimbas Biosphere Reserve, was established by the government of Mozambique and UNESCO. There is also located the Quirimbas National Park, the third largest conservation area in Mozambique, with a land and marine extension of about 7500 km^2.

Notwithstanding, a second event of armed conflict was unleashed in 2013. According to data released in October 2020, they indicate that an additional

125,000 people were affected, and dozens of people were considered dead (CIP 2020). This conflict is confined to two main provinces in central Mozambique, namely: Sofala and Manica Province. The Sofala Province is home to two conservation areas, Gorongosa National Park and Marromeu National Reserve. While in the Province of Manica, the Chimanimani National Park is located.

Facing the dramatic scenario of conflicts that continues until then, one of our biggest concerns are focused on understanding (questions not yet answered):

1. How is biodiversity today in the Quirimbas National Park, Chimanimani National Park, as well as in the Marromeu National Reserve?
2. How are the combined factors (armed conflicts, COVID-19) currently affecting those conservation areas?

We note that the conflicts in the province of Cabo Delgado forced the total closure of biodiversity management, and conservation operations in the Quirimbas National Park. As a result, an immediate set of actions was triggered, such as:

1. Deactivation of control and inspection sites.
2. Immediate withdrawal of rangers who were in the front line.
3. Deactivation of biodiversity monitoring infrastructure.
4. Cancellation of community actions.

Consequently, the territory, as well as biodiversity, has been seen fully unprotected. Given this scenario there are strong possibilities that the Quirimbas National Park are being critically pressured by:

1. Presence of poachers looking for animal and forest species of commercial value for trafficking.
2. Presence of illegal miners looking for precious mineral resources for smuggling.
3. The presence of communities looking for safe regions or places to hide and resources (animals and vegetables) for the supply of immediate meals.

All the internal dynamics in the Quirimbas National Park conditioned by the conflicts can lead to a forced change in the geography of local occurrence/presence of animals that, in the next phase, can, with relative ease, come into direct contact with pets and people.

In addition, the translocation of specimens for the trafficking, slaughter and illegal hunting can play a fundamental role in the transmission of diseases to humans.

Thus, we consider that eliminating armed conflicts or moving them away from the conservation areas can represent a won battle for the biodiversity conservation and to control future diseases in Mozambique.

4.3 Building Strong Partnerships and Foster Institutional Networking: National Approach is the Only Pathways

The COVID-19 showed that challenges can be overcome relatively easily when actions are combined and perfectly coordinated between stakeholders. The world saw such solidarity that had no borders. Unilateral efforts have proved to be quite weak and insufficient to cope the pandemic in many social and economic sectors.

The biodiversity conservation can be considered effective when integrated in a multidisciplinary and interdisciplinary context, but also when supported by technologies capable of offering an understanding very close to reality in order to strengthen more coherent and assertive decisions.

The existence of strategically consolidated and robust partnerships could support to mitigate the implications of COVID-19 in the biodiversity conservation actions in the country. The results of our research revealed that patrols and enforcement in conservation areas were limited or stopped; the collection of scientific data and field missions have been postponed; and the monitoring of biodiversity was conditioned (Fig. 5). These actions are extremely important for the maintenance of biodiversity, as well as ecosystem services that support the survival of the country's communities and economy.

However, we understand that establishing strong partnerships and cooperation to build a national multidisciplinary centre for sea and land observation, highly equipped with remote sensing technologies and unmanned devices, would play a crucial role in increasing actions for biodiversity conservation during the pandemic. Through this centre and using available technologies, it would be possible remotely and with less human intervention for national scientists and institutions to request and collect very relevant information from their intervention sites with multiple advantages such as, sharing information in real time. One could guarantee the patrolling of the conservation areas of the most critical places, as well as continue to monitor biodiversity without significant interruptions in the planning. In addition, it would enable scientists and institutions to obtain data containing a variety of parameters (chemical, biological, geological, and environmental) even if in a preliminary way.

These advances would allow strategically continuing to make important decisions for the control, monitoring of protected areas, and strengthening of realer actions for the biodiversity conservation in the country.

However, we argue that the picture presented cannot simply be framed in the current context of the COVID-19 pandemic, but in other complex parallel contexts such as, those of armed conflicts and the occurrence of extreme weather events.

Every 4 years, extreme weather events (of high magnitudes) affect Mozambique. In March and April 2019, the country was hit by one of the worst and most drastic extreme weather events (in the Southern Africa region) caused by Cyclone Kenneth and Idai (Mucova et al. 2021).

In practice, extreme climatic events (floods, inundations—caused by tropical cyclones), in addition to destroying and harming immense areas and ecosystems important for the survival and maintenance of biodiversity, can largely condition the accessibility to conservation areas (the roads connecting conservation areas are weak and highly precarious). The occurrence of these natural phenomena can significantly contribute to the interruption of monitoring, patrolling, missions, and scientific activities, significantly altering institutional agenda with a high negative impact on the reduction of actions to protect local or country biodiversity.

In both contexts, whether in the current scenario of the COVID-19 pandemic and armed conflicts, as well as the occurrence of extreme weather events, it is crucial to rethink future and long-term solutions in order to mitigate impacts on biodiversity conservation.

5 Conclusion

The facts and evidence allow us to conclude that the progressive reduction of habitats and ecosystems, associated with the international trade and trafficking in prohibited wild species and at risk of extinction may be in the context of COVID-19.

Combined analysis of the country's biodiversity conservation actors allowed us to understand that a total of 94.0% institutions was considerably affected. A total of 10.0% interrupted their interventions with rural communities that are within and at the interface of biodiversity conservation areas.

Nevertheless, despite almost the majority of field works being completely interrupted, about 18.0% worked without interruption from the offices, 40.0% developed their actions under a combined regime (office and telework), and more than 48.0% restricted all its interventions to an exclusive teleworking regime.

The actions developed under teleworking proved to be inefficient to mitigate the impacts of COVID-19 and to continue to support biodiversity conservation efforts in Mozambique, due to the poor quality of communication in the country, as well as the absence of support infrastructures outside of the offices.

Funding for about 7.0% of institutions will be significantly shortened, and available budget to support future actions will arrive later. Approximately 3.0% planned to recruit senior specialists, however, this activity was cancelled due to the impacts of the pandemic. In addition, 2.0% reduced their field workforce.

The impacts of COVID-19 on the biodiversity conservation are critical, about 80.0% consider that their actions could be critically affected between 50 and 100%. At the same time, about 95.0% are and will be financially hard hit.

Armed conflicts, mainly in northern Mozambique (Biosphere Reserve and Quirimbas National Park) may have immediate and future implications for the spread of zoonotic diseases, and challenging negative impacts to monitor, control and mitigate the spread of COVID-19 in this conservation area.

Almost 54.5 do not fully understand whether the measures introduced or existing would help to mitigate the impacts of COVID-19 on biodiversity conservation.

There is a lack of technical, scientific, financial and intersectoral coordination capacities to monitor and mitigate the COVID-19 pandemic in the biodiversity conservation arena, as well as to avoid, reduce the risks of transmission and spread of other zoonotic diseases. In this setting, we understand that the actions and commitments established by the institutions for the year 2020 will be mostly interrupted, and the country's national and international goals will backwards.

One of our contributions is that Mozambique will urgently need to embed nature protection and biodiversity conservation at the center of political agenda and integrate it unconditionally into the country's development context.

This action will allow to discuss and analyse the priorities for biodiversity conservation in the same proportion as for safety, health, agriculture, among others. Yet, it will allow the allocation of financial resources to the dimension of the challenges and needs. Finally, it will allow to function as an important mechanism to bring the parties together and negotiate positions to avoid instability and armed conflicts within and close to the country's conservation areas.

According to the current scenario accused by the COVID-19 pandemic, and the signs that nature has given, we warn that the absence, limited and inadequate actions to reduce armed conflicts, illegal trafficking in wild species, and to significantly increase efforts to biodiversity conservation, may imply relatively occurrences of environmental crises and shocks, with negative impacts on the economy and peoples.

Acknowledgements First and corresponding author would like to deeply thank WWF Russell E. Train Education for Nature Program (EFN) for support and fellowship granted.

Declaration of Competing Interest The authors declare that they have no known competing financial interests or personal relationships that could have appeared to influence the work reported in this paper.

References

Administração Nacional das Áreas de Conservação de Moçambique—ANAC (2015). Plano Estratégico da Administração Nacional das Áreas de Conservação 2015–2024, Maputo, p 172. http://www.anac.gov.mz/wp-content/uploads/2017/07/Plano-Estrategico-da-ANAC-2015-2024-1.pdf

Administração Nacional das Áreas de Conservação de Moçambique—ANAC (2020). In: http://www.biofund.org.mz/semana-de-comemoracao-do-dia-mundial-do-pangolim-mocambique-focado-na-proteccao-do-mamifero/

African Union (2020) Impact of the Coronavirus (COVID-19) on the African economy. https://www.tralac.org/documents/resources/covid-19/3218-impact-of-the-coronavirus-covid-19-on-the-african-economy-african-union-report-april-2020/file.html

Andersen KG, Rambaut A, Lipkin WI, Holmes EC, Garry RF (2020) The proximal origin of SARS-CoV-2. Nat Med 26(2020):450–455. https://doi.org/10.1038/s41591-020-0820-9

Astill J, Dara RA, Fraser EDG, Sharif S (2018) Detecting and predicting emerging disease in poultry with the implementation of new technologies and big data: a focus on avian influenza virus. Front Vet Sci 5:263. https://doi.org/10.3389/fvets.2018.00263

Bates AE, Primack RB, Moraga P, Duarte CM (2020) COVID-19 pandemic and associated lockdown as a "Global Human Confinement Experiment" to investigate biodiversity conservation. Biol Cons 248. https://doi.org/10.1016/j.biocon.2020.108665

Centro de Integridade Pública—CIP (2020) Número de deslocados internos em Moçambique cresceu em cerca de 2 700 % em dois anos, Maputo, p 30. https://cipmoz.org/wp-content/uploads/2020/10/Nu%CC%81mero-de-deslocados-em-Moc%CC%A7ambique-1-5.pdf

Chan JF, Yuan S, Kok KH, To KK, Chu H, Yang J, Xing F, Liu J, Yip CC, Poon RW, Tsoi HW, Lo SK, Chan KH, Poon VK, Chan WM, Ip JD, Cai JP, Cheng VC, Chen H, Hui CK, Yuen KY (2020) A familial cluster of pneumonia associated with the 2019 novel coronavirus indicating person-to-person transmission: a study of a family cluster. Lancet 395:514–523. https://doi.org/10.1016/S0140-6736(20)30154-9

Choo SW, Zhou J, Tian X, Zhang S, Qiang S, O'Brien SJ, Tan KY, Platto S, Koepfli KP, Antunes A, Sitam FT (2020) Are pangolins scapegoats of the COVID-19 outbreak-CoV transmission and pathology evidence? Conserv Lett 13(6):1–12. https://doi.org/10.1111/conl.12754

Cimon-Morin J, Darveau M, Poulin M (2013) Fostering synergies between ecosystem services and biodiversity in conservation planning: a review. Biol Cons 166:144–154. https://doi.org/10.1016/j.biocon.2013.06.023

Corlett RT, Primack RB, Devictor V, Maas B, Goswami VR, Bates AE, Koh LP, Regan TJ, Loyola R, Pakeman RJ, Cumming GS, Pidgeon A, Johns D, Roth R (2020) Impacts of the coronavirus pandemic on biodiversity conservation. Biolog Cons 246. https://doi.org/10.1016/j.biocon.2020.108571

Daskin JH, Pringle RM (2018) Warfare and wildlife declines in Africa's protected areas. Nature 553:328–332. https://doi.org/10.1038/nature25194

Dias JATS (1971) Terão justificação os abates indiscriminados da caça como medida de luta contra a mosca tsé-tsé? Uma análise do caso de Moçambique. Revista De Ciências Veterinárias 4:87–100

Geijzendorffer IR, van Teeffelen AJ, Allison H, Braun D, Horgan K, Iturrate-Garcia M, Santos MJ, Pellissier L, Prieur-Richard A-H, Sakai S, Quatrini S, Zuppinger-Dingley D (2017) How can global conventions for biodiversity and ecosystem services guide local conservation actions? Curr Opin Environ Sustain 29:145–150. https://doi.org/10.1016/j.cosust.2017.12.011

Han GZ (2020) Pangolins harbor SARS-CoV-2-related Coronaviruses. Trends Microbiol 28 (7):515–517. https://doi.org/10.1016/j.tim.2020.04.001

Hatton J, Couto M, Oglethorpe J, Shambaugh J (2001) Biodiversity and war: a case study of Mozambique. Biodiversity Support Program, Washington, D.C., p 84

Instituto Nacional de Estatística (INE) (2017). IV Recenseamento Geral da População e Habitação de Mocambique, Maputo. http://www.ine.gov.mz/

International Labour Organization—ILO (2020) COVID-19: pandemic in the world of work. https://www.ilo.org/global/abouttheilo/newsroom/news/WCMS_739961/lang–en/index.htm

IPBES (2019): Summary for policymakers of the global assessment report on biodiversity and ecosystem services of the Intergovernmental Science-Policy Platform on Biodiversity and Ecosystem Services. In: Díaz S, Settele J, Brondízio ES, Ngo HT, Guèze M, Agard J, Arneth A, Balvanera P, Brauman KA, Butchart SHM, Chan KMA, Garibaldi LA, Ichii K, Liu J, Subramanian SM, Midgley GF, Miloslavich P, Molnár Z, Obura D, Pfaff A, Polasky S, Purvis A, Razzaque J, Reyers B, Roy Chowdhury R, Shin YJ, Visseren-Hamakers IJ, Willis KJ, Zayas CN (eds) IPBES secretariat, Bonn, Germany, p 56

Johns Hopkins University (2021). COVID-19 dashboard by the center for systems science and engineering (CSSE). https://coronavirus.jhu.edu/map.html

Lam TT, Jia N, Zhang YW, Shum MH, Zhu H (2020) Identifying SARS-CoV-2 related coronaviruses in Malayan pangolins. Nature 583. https://doi.org/10.1038/s41586-020-2169-0

Lindsey P, Allan J, Brehony P, Dickman A, Robson A, Begg C, Bhammar H, Blanken L, Breuer T, Fitzgerald K, Flyman M, Tyrrell P (2020) Conserving Africa's wildlife and wildlands through the COVID-19 crisis and beyond. Nat Ecol Evol 4(2020):1300–1310. https://doi.org/10.1038/s41559-020-1275-6

May C (2017) Transnational crime and the developing world channing may, Global Financial Integrity, p 166. https://respect.international/wp-content/uploads/2017/03/Transnational-Crime-and-the-Developing-World-Global-Financial-Integrity-2017

Ministério de Saúde de Moçambique—MISAU (2020). Informação sobre coronavirus—COVID-19. http://www.misau.gov.mz/index.php/informacao-sobre-coronavirus-covid-19

Ministry of Tourism and Culture (2020). Impacto da COVID-19 no turismo em Moçambique. https://www.tvm.co.mz/index.php?option=com_k2&view=item&id=7096:impacto-da-covid-19-no-turismo-sector-perdeu-mais-de-95-das-receitas-em-6-meses&Itemid=278

MITADER (2015) National strategy and action plan of biological diversity of Mozambique (2015–2035), Maputo, p 116. http://www.biofund.org.mz/wp-content/uploads/2017/03/NBSP-Biodiversty-Strategy-and-Plan-Mocambique-2015-2035-Ingl.pdf

Mucova SAR, Azeiteiro UM, Filho WL, Lopes CL, Dias JM, Pereira MJ (2021) Approaching Sea-Level Rise (SLR) change: strengthening local responses to sea-level rise and coping with climate change in Northern Mozambique. J Marine Sci Eng 9(2):205. https://doi.org/10.3390/jmse9020205

Neethirajan S (2017) Recent advances in wearable sensors for animal health management. Sens Bio-Sens Res 12:15–29. https://doi.org/10.1016/j.sbsr.2016.11.004

OECD (2020) OECD economic outlook, interim report September 2020. OECD Publ. https://doi.org/10.1787/34ffc900-en

Olson DM, Dinerstein E, Wikramanayake ED, Burgess ND, Powell GVN, Underwood EC, D'Amico JA, Itoua I, Strand HE, Morrison JC, Loucks CJ, Allnutt TF, Ricketts TH, Kura Y, Lamoreux JF, Wettengel WW, Hedao P, Kassem KR (2001) Terrestrial ecoregions of the world: a new map of life on earth. Bioscience 51(11):933–938. https://doi.org/10.1641/0006-3568(2001)051

Rutz C, Loretto MC, Bates AE (2020) COVID-19 lockdown allows researchers to quantify the effects of human activity on wildlife. Nat Ecol Evol 4:1156–1159. https://doi.org/10.1038/s41559-020-1237-z

Scheffers BR, Oliveira BF, Lamb I, Edwards DP (2019) Global wildlife trade across the tree of life. Science 366(6461):71–76. https://doi.org/10.1126/science.aav5327

Stalmans ME, Massad TJ, Peel MJS, Tarnita CE, Pringle RM (2019) War-induced collapse and asymmetric recovery of large-mammal populations in Gorongosa National Park, Mozambique. PLoS ONE 14(3):e0212864. https://doi.org/10.1371/journal.pone.0212864

Sumner A, Hoy C, Ortiz-Juarez E (2020) Estimates of the impact of COVID-19 on global poverty. WIDER Working Paper 2020/43. https://doi.org/10.35188/UNU-WIDER/2020/800-9

UNHCR—UN Refugee Agency (2020) Fresh violence in northern Mozambique forces thousands to flee. https://www.unhcr.org/news/briefing/2020/2/5e3d2d8f4/fresh-violence-northern-mozambique-forces-thousands-flee.html

United Nations (2020a) Shared responsibility, global solidarity: responding to the socio-economic impacts of COVID-19. https://reliefweb.int/sites/reliefweb.int/files/resources/sg_report_socio-economic_impact_of_covid19.pdf

United Nations (2020b) Coronavirus COVID-19 wipes $50 billion off global exports in February alone, as IMF pledges support for vulnerable nations. https://news.un.org/en/story/2020/03/1058601

United Nations Educational, Scientific and Cultural Organization (2020). Education: from disruption to recovery. https://en.unesco.org/covid19/educationresponse

United Nations University (2020). Estimates of the impact of COVID-19 on global poverty. WIDER Working Paper 2020/43. https://doi.org/10.35188/UNU-WIDER/2020/800-9.

United Nations World Food Program (2020) Covid-19 will double number of people facing food crises unless swift action is taken, April 21. https://www.wfp.org/news/covid-19-will-double-number-people-facing-food-crises-unless-swift-action-taken

UNODC (2020) Wildlife crime: pangolin scales. United nations office on drugs and crime.https://www.unodc.org/documents/wwcr/2020/Wildlife_crime_Pangolin_UNODC

Van der Biest K, Meire P, Schellekens T, D'hondt B, Bonte D, Vanagt T, Ysebaert T (2019) Aligning biodiversity conservation and ecosystem services in spatial planning: focus on ecosystem processes. Sci Total Environ 712. https://doi.org/10.1016/j.scitotenv.2019.136350

Van Dorp L, Acman M, Richard D, Shaw LP, Ford CE, Ormond L, Pang J, Owen GJ, Balloux F (2020) Emergence of genomic diversity and recurrent mutations in SARS-CoV-2. Infect Genet Evol 83. https://doi.org/10.1016/j.meegid.2020.104351

Waithaka J (2020) The impact of covid-19 pandemic on Africa's protected areas operations and programmes. The International Union for Conservation of Nature, p 44. https://www.iucn.org/sites/dev/files/content/documents/2020/report_on_the_impact_of_covid_19_doc_july_10.pdf

Wildlife Trafficking Alliance (2021) Illegal Wildlife Trade. Retrieved at https://wildlifetraffickingalliance.org/

World Bank (2018) Banco Mundial Aprova USD $45 Milhões para Fortalecer Áreas de Conservação de Moçambique e Aumentar Resiliência Rural. Washington, 20 de Setembro de 2018. https://www.worldbank.org/pt/news/press-release/2018/09/20/world-bank-approves-45-million-to-strengthen-mozambiques-conservation-areas-and-increase-rural-resilience

World Health Organization (2019) Middle East respiratory syndrome coronavirus (MERS-CoV). https://www.who.int/news-room/fact-sheets/detail/middle-east-respiratory-syndrome-coronavirus-(mers-cov)

World Health Organization (2020a) Novel Coronavirus (2019-nCoV) situation report—1. Data as reported by: 20 January 2020. https://www.who.int/docs/default-source/coronaviruse/situation-reports/20200121-sitrep-1-2019-ncov.pdf?sfvrsn=20a99c10_4

World Health Organization (2020b) Novel Coronavirus (2019-nCoV) situation report—2. Data as reported by: 22 January 2020. https://www.who.int/docs/default-source/coronaviruse/situation-reports/20200122-sitrep-2-2019-ncov.pdf?sfvrsn=4d5bcbca_2

World Health Organization (2020c) Novel Coronavirus (2019-nCoV) situation report—26. Data as reported by 15 February 2020. https://www.who.int/docs/default-source/coronaviruse/situation-reports/20200215-sitrep-26-covid-19.pdf?sfvrsn=a4cc6787_2

World Health Organization (2020d) Novel Coronavirus (2019-nCoV) situation report—22. Data as reported by 11 February 2020. https://www.who.int/docs/default-source/coronaviruse/situation-reports/20200211-sitrep-22-ncov.pdf?sfvrsn=fb6d49b1_2

World Health Organization (2020e) Novel Coronavirus (2019-nCoV) situation report—51. Data as reported by 11 March 2020. https://www.who.int/docs/default-source/coronaviruse/situation-reports/20200311-sitrep-51-covid-19.pdf?sfvrsn=1ba62e57_10

Yang P, Wang X (2020) COVID-19: a new challenge for human beings. Cell Mol Immunol 17:555–557. https://doi.org/10.1038/s41423-020-0407-x

Zhang T, Wu Q, Zhang Z (2020) Probable pangolin origin of SARS-CoV-2 associated with the COVID-19 outbreak. Curr Biol 33(7):1346–1351. https://doi.org/10.1016/j.cub.2020.03.022

Zhu N, Zhang D, Wang W, Li X, Yang B, Song J, Zhao X, Huang B, Shi W, Lu R, Niu P, Zhan F, Ma X, Wang D, Xu W, Wu G, Gao GF, Tan W (2020) A novel coronavirus from patients with pneumonia in China, 2019. N Engl J Med 382:727–733. https://doi.org/10.1056/NEJMoa2001017

Elevation Based Sea-Level Rise Impact Assessment for Sustainable Coastal Natural Resources Management and Coastal Adaptation Planning

A. Saleem Khan, A. Ramachandran, V. Selvam, K. Palanivelu, S. Punitha, P. Thirumurugan, S. Jayakumar, and P. Malini

1 Introduction

Coastal ecosystems are drastically impacted by the changing climate. Increase in extreme events, sea-level rise (SLR), warming of the sea surface temperatures and ocean acidification will have profound impacts on coastal and marine ecosystems (WII 2015). Accelerating SLR compounded by growing frequency and magnitude of sea-born hazards like storm surges and hurricanes threaten the low elevation coastal zone (LECZ) including small islands and deltas (UNU 2009). Inundation to SLR makes the low-elevated coastal regions as more vulnerable to the impacts of climate change (Khan et al. 2012a). Adaptation is considered as one of the significant measures that reduce the coastal vulnerability to actual or expected climate-change effects such as SLR (Khan et al. 2012b). However, sustainable management of coastal natural resources and land planning warrants urgent attention in response to the impacts of rising sea-levels by planning suitable coastal adaptation strategies.

The vulnerability of the low-lying coastal regions to inundation due to SLR and flooding events are determined mainly by elevation (Gesch 2012). The impending

A. Saleem Khan (✉)
Centre for Climate Change and Sustainability Studies, School of Habitat Studies, Tata Institute of Social Sciences (TISS), Deonar, Mumbai, Maharashtra 400088, India

A. Saleem Khan · A. Ramachandran · K. Palanivelu · P. Thirumurugan · P. Malini
Centre for Climate Change and Disaster Management, Anna University, Sardar Patel Road, Chennai, Tamil Nadu 600025, India

V. Selvam · S. Punitha
Coastal System Research, M.S. Swaminathan Research Foundation, Institutional Area, Taramani, Chennai, Tamil Nadu 600113, India

S. Jayakumar
Department of Ecology and Environmental Sciences, School of Life Sciences, Pondichery University, Puducherry Union Territory 605014, India

W. Leal Filho et al. (eds.), *Sustainability in Natural Resources Management and Land Planning*, World Sustainability Series,
https://doi.org/10.1007/978-3-030-76624-5_22

effects of the predicted SLR, especially the area of land that could be inundated have been quantified widely by using coastal elevation data (Gesch 2009). The quality and efficacy of SLR impact assessments is greatly determined by the precision of the elevation data (Najjar et al. 2000; Kleinosky et al. 2007; Poulter and Halpin 2008; Titus and Wang 2008; Gesch 2009; The Nature Conservancy 2010). Existing studies based on currently available elevation data do not provide the optimal degree of confidence for local decision making. There are important technical considerations that need to be incorporated for improving sea level rise impact assessments (Titus and Wang 2008).

Satellite remote sensing can provide operational digital elevation models (DEMs) through radar- based satellite images (UNESCO 2010). SLR vulnerability maps and assessments has been generated globally by the application of DEMs (Titus and Richman 2001; Dawson et al. 2005; Marfai and King 2008; Hinkel et al. 2010; Cooper et al. 2013a). Nevertheless, such assessments of the impact of SLR have generally dealt with a lack of elevation data by interpolating between contour intervals (Dawson et al. 2005). Perhaps the best use of coarser elevation datasets like global DEMs with horizontal grid spacing 30 arc seconds (GTOPO30), DEM of Shuttle Radar Topography Mission (SRTM), Cartostat, Advanced Spaceborne Thermal Emission and Reflection Radiometer (ASTER), and others is to give a more general description of low elevation coastal zones. In contrast, Light Detection and Ranging (LiDAR) data favors to create more accurate SLR vulnerability maps to take suitable decisions while planning for adaptation (Webster et al. 2004; Zhang et al. 2011; Cooper et al. 2013a, b).

Nevertheless, generating high-resolution SLR vulnerability maps from LiDAR are limited by lack of a common error standard and discrepancies between vertical data (Cooper et al. 2013b). However, the accuracy of any satellite-based DEM can be evaluated by comparing its elevation information with that derived from topographic maps, or comparing satellite imagery-based DEM elevation values of specific targets with those derived from Differential Global Positioning System (DGPS) surveys (Bhakar et al. 2010). However, the virtues of geomatics technology like DGPS have not been deservingly exploited in SLR impact studies (Ramasamy et al. 2010). In this context, this study addresses the importance of the accuracy of the coastal elevation data to assess the impacts of SLR to plan suitable coastal adaptation strategies with the Pichavaram mangrove region in the east coast of Tamil Nadu, India as a case study.

2 Pichavaram Coastal Mangrove Region

The study area Pichavaram coastal mangroves (Fig. 1) are present between Vellar-Coleroon estuarine region (11°30′14.69 N, 79°46′38.14 E; 11°21′41.40 N, 79°49′51.24 E) and it is situated on the east coast of Tamil Nadu, India (Khan et al. 2012a). It is placed at about 225 km southward of Chennai and 5 km north east of Chidambaram, Cuddalore district, Tamil Nadu (Planning Commission 2008; Khan et al. 2012b). The total study area considered for the present study is stretching about 16.5 km and covering total area of 6,541 ha. Mangroves, sand dunes and mud flats are the major coastal natural resources within the study area along with aquaculture farms, agricultural lands and other coastal vegetations. The total area of the Pichavaram mangrove forest is about 1,470 ha (Selvam et al. 2003). The forest scattered across 51 islets with an area from 10 m^2 to 2 km^2 and estranged by convoluted water streams that connect both the estuaries (Kathiresan 2000; Khan et al. 2014a). The large parts of the coastal zone of the study area are low-lying with gentle slope resulting large inundation, thus increasing the vulnerability of the region. The southern parts of the Cuddalore district near Chidambaram town in the vicinity of the Kolladam (Coleroon) river indicate larger areas under threat due to multiple hazards, because of lower mean elevation (Mahendra et al. 2011; Khan et al. 2012b).

3 Methodological Approaches

The knowledge of Land Use and Land Cover (LULC) changes is very important in understanding coastal natural resources, their utilization, conservation and management (Arunachalam et al. 2011). To study LULC of the study area satellite image such as IRS P6- LISS-4 has been procured from National Remote Sensing Centre (NRSC), Department of Space, Government of India and 1:50,000 Toposheet No. 58 M/15 from Survey of India (SOI). Study area boundaries were extracted from the topographic maps obtained from SOI by manual digitizing methods. Standard LULC procedure such as image processing, image classification, accuracy assessment and post classification analysis has been done using ERDAS Imagine and ARC GIS software (Fig. 2). LULC are classified based on ISRO-Space Application Centre Coastal LULC classification (Nayak et al. 1992).

Similarly, an assessment has been performed to compare DEM of ASTER 90 m, SRTM 90 m and DGPS for the selected study site of 16.5 km length with total area of 6,541 ha. ASTER 90 m and SRTM 90 m dataset was obtained from the publically available website of Earth Remote Sensing Data Analysis Centre (ERSDAC) and the DGPS data was obtained based on real-time ground level elevation measurement by DGPS surveying process. For this purpose, a DGPS measurement survey based on a ground elevation (above mean sea level) was done for the entire study area at a 100 m interval, using a Trimble R3 Digital Field Book (Version 6.0). ERDAS Imagine software was used to generate the DEM from all three source data (ASTER 90 m, SRTM 90 m and DGPS). Contours were then generated from the DEM at an interval of 0.25 and 0.5 m SLR (mid estimate of IPCC-AR5) using Arc-GIS. Area of inundation was quantified based on inundation model that works on the principle of equilibrium method or bath tub method (Khan 2013). The quantified area of inundation for ASTER and SRTM were compared with that of DGPS (Fig. 2).

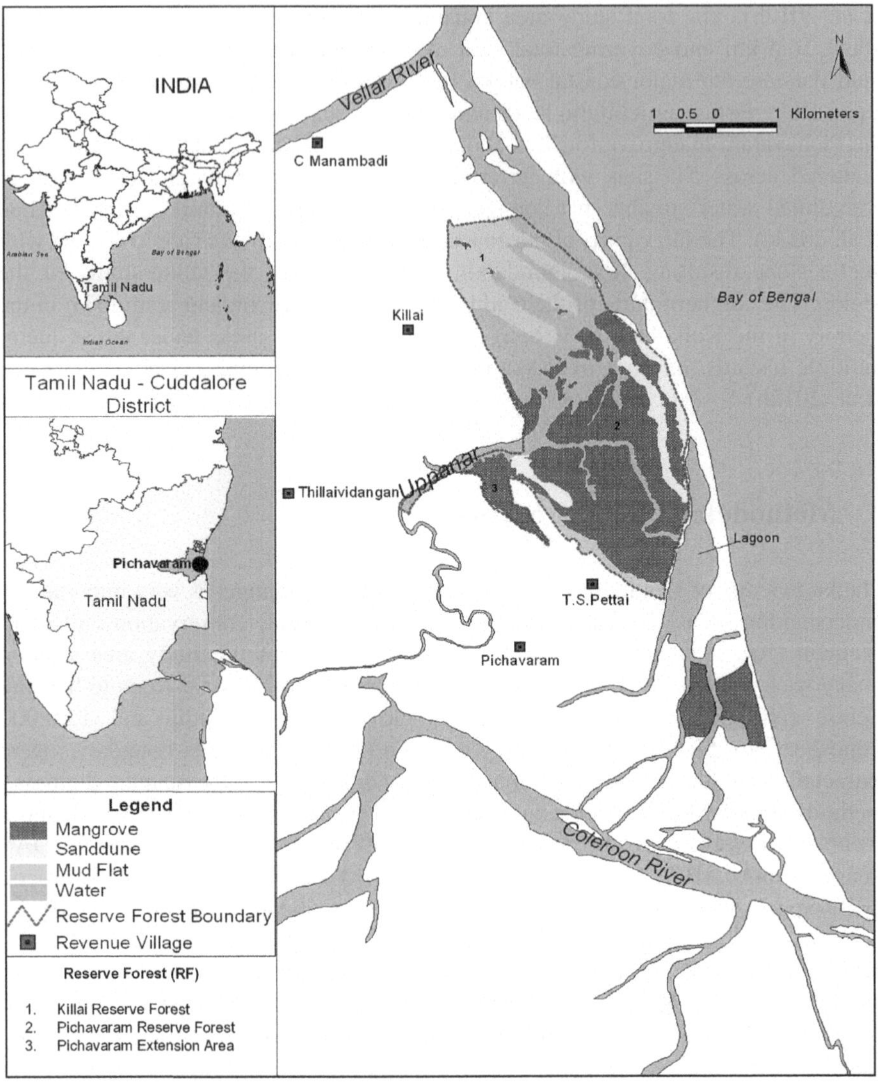

Fig. 1 Pichavaram coastal mangrove region, Tamil Nadu, India. *Source* Elaborated by the authors

4 Land Use and Land Cover

Land Use and Land Cover (LULC) maps are regarded to be fundamental for the purpose of coastal management planning (Cicin-Sain and Knecht 1998; Mumby et al. 1999; Stevens and Connolly 2004; Crossland et al. 2005). There are some new and important considerations that enter into planning and decision-making with

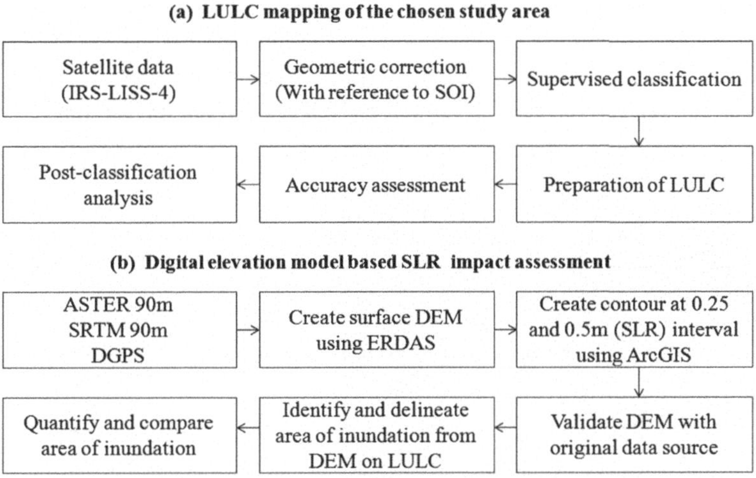

Fig. 2 Schematic representation of the methodological approach. *Source* Elaborated by the authors

respect to climate change (Tobey et al. 2010). These considerations include the need for analyzing the coastal LULC from the lens of climate change by understanding the pattern of LULC changes. This offers valuable information on the dynamics of LULC of the study area to adaptation planners to plan suitable and effective coastal adaptation policies to changing climate in the future (Khan 2013). The analysis of LULC of this study area has been classified into 14 classes that have occupied an area of 6,541 ha. It has been identified as agricultural land, aquaculture farm, beach, habitation, habitation with vegetation, man-made forest, mangrove dense, mangrove sparse, marsh/salt marsh, mud flat/tidal flat, open/vacant land, other vegetation, sandy area/dune and tanks respectively. In addition, nearly 20 hamlets (10 farming hamlets, 9 fishing hamlets and 1 combination of both) are spread across the region (Fig. 3). Thus, the outcome from these assessment could be used for the appraisal of coastal adaptation planning policies that could be applied in order to offset some of the possible impacts of climate change (such as SLR) and associated hazards (Boateng 2012).

5 An Assessment on Comparison of DEM Based SLR Impact Analysis

The valuable coastal habitats such as mangroves and other coastal wetlands has lost its area, and the projected future predication reveals more reduction of area will occur mainly due to inundation. In addition inundation also threats human safety and shoreline protection from coastal disasters (Gilman 2004; Gilman et al. 2007).

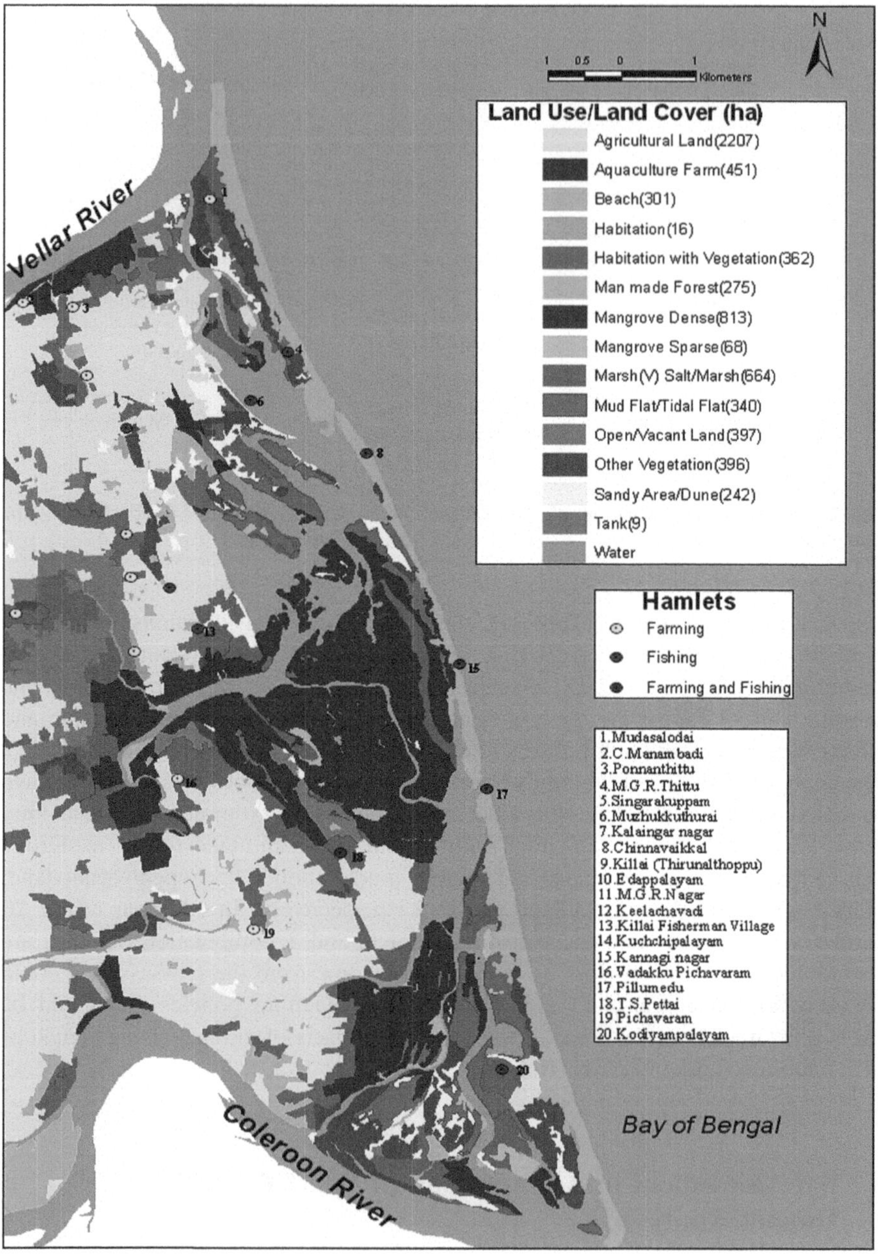

Fig. 3 LULC of the study area. *Source* Elaborated by the authors

Elevation of the coastal lands from the mean sea level plays a key role in estimating the area of inundation to the rising sea-levels. It is observed that the elevation of the study area based on ASTER 90 m has an error of 20 m, where as 10 m error for SRTRM 90 and less than a meter error for DGPS. It has been also noted that the area of inundation by 0.25 m SLR for ASTER 90 m and SRTM 90 m shows an alarming difference when compared with on-ground survey data by DGPS. From the total LULC of 6,541 ha of the study area, the area of inundation for a predicted impact of 0.25 m SLR using ASTER 90 m data was 74 ha, and for SRTM 90 m data was about 510 ha. However, the area of inundation from DGPS was 2,286 ha for the same study area (Fig. 4). Likewise, it has been also noted similarly that the area of inundation by 0.5 m SLR for ASTER 90 m and SRTM 90 m shows an alarming difference when compared with on-ground survey data by DGPS. From the total LULC of 6,541 ha of the study area, the area of inundation for a predicted impact of 0.5 m SLR using ASTER 90 m data was 100 ha, and for SRTM 90 m data was about 546 ha. However, the area of inundation from DGPS was 2,618 ha for the same study area (Fig. 5).

The findings of the study raise a concern that the accuracy of elevation data for SLR impact assessment studies plays a key role in planning adaptation strategies to SLR on ground. It has been observed (Table 1) that the sub-meter level accuracy of elevation data such as DGPS based elevation data are essential for planning adaptation strategies to a smaller study area and at the local level. In particular, even in data-sparse areas, open-access global DEMs are extensively used to model floods in order to meet the scarcity of high-accuracy DEM. Nevertheless, better flood estimations are provided by high-accuracy DEMs and can be deemed a 'must-have' for any flood model (Hawker et al. 2018). It is clear that currently available global

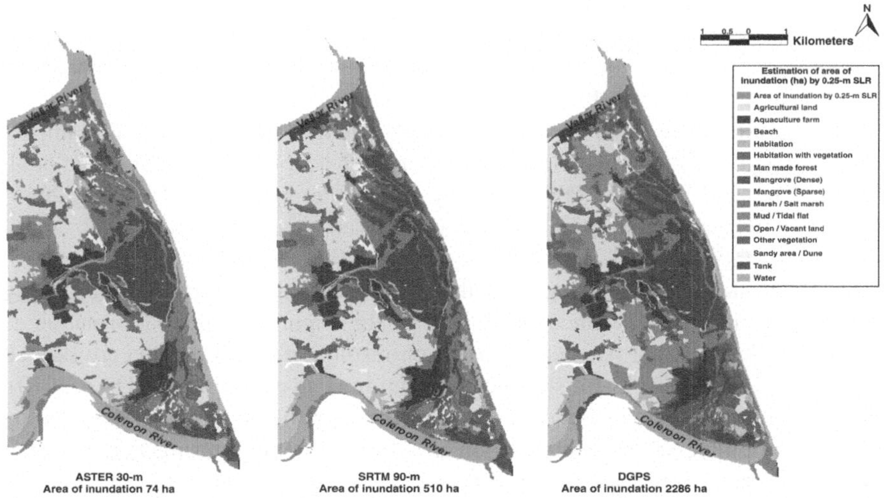

Fig. 4 Comparison of ASTER 90 m, SRTM 90 m and DGPS based DEM data for quantifying area of inundation by 0.25 m SLR. *Source* Elaborated by the authors

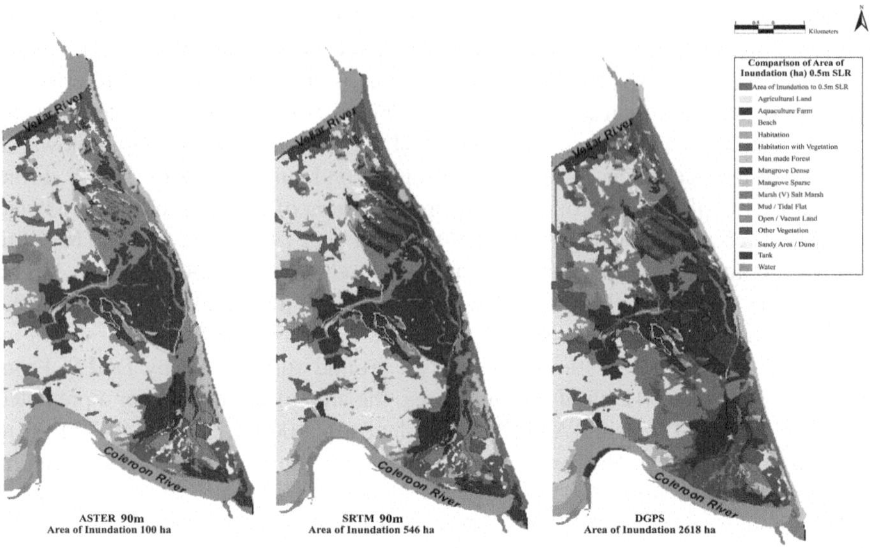

Fig. 5 Comparison of ASTER 90 m, SRTM 90 m and DGPS based DEM data for quantifying area of inundation by 0.5 m SLR. *Source* Elaborated by the authors

DEMs cannot be used to accurately simulate or predict local scale processes or impacts thereof. Even after considerable preprocessing to remove significant biases (due to vegetation and other physical structures) and to reduce inherent vertical errors, publicly available global DEMs still suffer from inaccuracies oftentimes orders of magnitude greater than length scales of the processes that are simulated (Schumann and Bates 2018). Generally, the data with the highest vertical accuracy will be the most accurate; however other factors such as the age of the data, the resolution of the data, and the availability of the data should also be considered in the evaluation process (FEMA 2016).

Many studies have explored the value of using more sophisticated coastal impact models and higher resolution elevation data in SLR adaptation planning (Zhu et al. 2015). Bales and Wagner 2009; Coveney and Fotheringham 2011; Zhang 2011; Fraile-Jurado and Ojeda-Zújar 2012; Sampson et al. 2016; Wolff et al. 2016; Mogensen and Rogers 2018; Paprotny et al. 2018; Vousdoukas et al. 2018; Reimann et al. 2018 has recognized that in inundation assessments, the quality (spatial resolution and vertical accuracy) of the input elevation information is important (Gesch 2018; Amante 2019). High-resolution elevation data are needed to produce credible results for site-specific planning and assessment (NOAA 2010). Whereas, from the above mentioned assessment it may be important to note that by providing not more accurate information as a result of compromising with available data may create serious possibility of developing adaptive responses and strategies to rising sea level that are maladaptive (IPCC 2014; Scheraga et al. 2003; Satterthwaite et al. 2009; Pittock 2011). Hence, systematic collection of

Table 1 Elevation values of the sample points of the study area based on the DEM of ASTER, SRTM and DGPS

S. no	Longitude and latitude	Elevation (in meters)		
		ASTER 90 m	SRTM 90 m	DGPS real-time
1	79°46′52.9692″ E 11°29′40.3511″ N	3	2	−1
2	79°45′36.3459″ E 11°28′47.7625″ N	7	0	5.3
3	79°46′46.1422″ E 11°27′57.2056″ N	10	7	−0.4
4	79°47′58.4141″ E 11°27′17.9559″ N	14	6	4.8
5	79°46′34.8949″ E 11°26′57.2004″ N	18	1	−1.1
6	79°46′52.9951″ E 11°26′91.4532″ N	26	2	3.48
7	79°48′27.2645″ E 11°26′41.2536″ N	9	1	8.0
8	79°48′2.91981″ E 11°25′49.3657″ N	13	2	−1.6
9	79°46′15.7522″ E 11°25′10.8505″ N	17	5	6.5
10	79°47′32.4417″E 11°24′16.0557″ N	11	1	3.34
11	79°49′10.6345″ E 11°23′22.5007″ N	16	1	−0.6
12	79°48′25.7483″ E 11°22′27.9349″ N	5	4	−1.2
13	79°46′11.2553″ E 11°28′01.3138″ N	10	1	−1.6
14	79°46′36.4908″ E 11°27′29.8516″ N	11	2	5.7
15	79°47′51.0316″ E 11°27′34.3007″ N	11	6	0.69

Source Elaborated by the authors

high-quality elevation data will improve the ability for detailed assessment (Titus and Wang 2008) and to avoid maladaptation by poor planning of elevation-dependent coastal adaptation strategies for inundation to rising sea levels.

6 Conclusions

SLR is a global issue, but the distribution and impact of SLR is not evenly distributed due to differences in ocean-atmospheric interactions, and the elevation and geological makeup of the land (Miller 2019; IPCC 2019). Low-elevated coastal regions are at high risk to SLR inundation (Khan et al. 2012a). Taking into account SLR impacts, responses and the future uncertainty, there is a need for no regret strategies to avoid maladaptation. However, this paper has advocated for the accuracy of the elevation data for planning suitable adaptation strategies to the risk of climate change induced SLR. The present work achieved it objective by assessing the impacts of SLR on Pichavaram mangrove regions using various DEMs of SRTM, ASTER and DGPS. Importantly, it acknowledges the importance of real-time on-ground survey data using geomatics techniques like DGPS. However, generating elevation data covering an entire coastline using the method of real-time on-ground elevation measurement may be impractical and prohibitively expensive. For this purpose, there is a new approach suggested by this study and it is, instead of depending solely on remote sensing elevation data, ground surveys can be used to supplement or monitor their accuracy, even the most reliable from LiDAR or other technologies (The Arlington Group 2013) wherever possible. LiDAR is the best source of terrain data for that particular application and that, although very useful (Costa et al. 2019). To assist researchers to take suitable adaptation decisions, LiDAR plays a significant role by generating more accurate SLR vulnerability maps (Cooper et al. 2013b). Unfortunately, the availability of LiDAR DEM for smaller study areas is uncertain. Thus, the inclusive collection of more precise information required is hardly available and not often owned by the decision-makers accountable for management within the coastal regions (Ricketts 1992; O'Regan 1996; Khan et al. 2014b). Nevertheless, a "hybrid DEM" with combination of elevation data from extensive ground survey and remote sensing will furnish much more accurate and reliable information, despite daunting challenges for generating widespread real-time ground survey data. Ultimately, the most accurate assessments of vulnerability to rising seas, especially for smaller areas, will require development and public release of improved coastal area elevation datasets building directly off of new high resolution observations increasingly collected by satellites today (Kulp and Strauss 2018; Kulp and Strauss 2019). One of the important limitations of this study is the unavailability of high resolution data set such as LIDAR for this study area. Therefore, it is important to know because high-resolution data are likely to not be available in some data-poor coastal areas and running more complicated models is relatively time-consuming, expensive and needs additional expert knowledge (Zhu et al. 2015). In addition to challenges of conducting DGPS survey on ground, the limitations also include noise and data gaps of SRTM and ASTER data sets. It is clear that currently available global DEMs cannot be used to accurately simulate or predict local scale processes or impacts thereof. Even after considerable preprocessing to remove significant biases and to reduce inherent vertical errors, publicly available global DEMs still suffer from inaccuracies (Schumann and Bates 2018). The information garnered from this study emphasizes the importance of the elevation based coastal land use planning and to sustainably manage and adapt the coastal natural resources to the impacts of SLR. Thus, the sustainable land-use planning improves the coastal resilience, is a more effective solution in coastal regions that have experienced dramatic land-use changes caused by development activities (Kim et al. 2017).

References

Amante CJ (2019) Uncertain seas: probabilistic modeling of future coastal flood zones. Int J Geogr Inf Sci 33(11):2188–2217

Arunachalam S, Maharani K, Chidambaram S, Prasanna MV, Manivel M, Thivya C (2011) A study on the land use pattern change along the coastal region of Nagapattinam, Tamil Nadu. Int J Geomatics Geosci 1(2):700–720

Bales JD, Wagner CR (2009) Sources of uncertainty in flood inundation maps. J Flood Risk Manage 2:139–147

Bhakar R, Srivastav SK, Punia M (2010) Assessment of the relative accuracy of ASTER and SRTM digital elevation models along irrigation channel banks of Indira Gandhi canal project area, Rajasthan. J Water Land-Use Manage 10(1–2):1–11

Boateng I (2012) GIS assessment of coastal vulnerability to climate change and coastal adaptation planning in Vietnam. J Coastal Res 16(1):25–36

Cicin-Sain B, Knecht RW (1998) Integrated coastal and ocean management: concepts and practices. Island Press, Washington DC, USA

Commission Planning (2008) Report on visit to Pichavaram in Tamil Nadu: a wetland included under National Wetland Conservation and Management Programme of the Ministry of Environment and Forests. Planning Commission, Government of India, New Delhi, India

Cooper HM, Chen Q, Fletcher CH, Barbee MM (2013a) Assessing vulnerability due to sea-level rise in Maui, Hawai'i using LiDAR remote sensing and GIS. Clim Change 116:547–563

Cooper HM, Fletcher CH, Chen Q, Barbee MM (2013b) Sea level rise vulnerability mapping for adaptation decisions using LiDAR DEMs. Prog Phys Geogr 37:745–756

Costa RT, Mazzoli P, Bagli S (2019) Limitations posed by free DEMs in watershed studies: the case of river Tanaro in Italy. Front Earth Sci 7(141):1–16

Coveney S, Fotheringham AS (2011) The impact of DEM data source on prediction of flooding and erosion risk due to sea-level rise. Int J Geogr Inf Sci 25:1191–1211

Crossland CJ, Baird D, Ducrotoy J, Lindeboom H (2005) Chapter 1: The coastal zone—a domain of global interactions. In: Crossland CJ, Kremer HH, Lindeboom H, Marshall C, Le Tissier MDA (eds) Coastal fluxes in the anthropocene: the land-ocean interactions in the coastal zone project of the international geosphere-biosphere programme. Springer, Berlin, Heidelberg, pp 1–37

Dawson RJ, Hall JW, Bates PD, Nicholls RJ (2005) Quantified analysis of the probability of flooding in the Thames Estuary under imaginable worst-case sea level rise scenarios. Int J Water Resour Dev 21:577–591

FEMA (2016). Guidance for flood risk analysis and mapping: elevation guidance. Federal Emergency Management Agency (FEMA). https://www.fema.gov/media-library-data/1469794589266-f404b39e73fa7a1c5ffe4447636634d4/Elevation_Guidance_May_2016.pdf

Fraile-Jurado P, Ojeda-Zújar J (2012) The importance of the vertical accuracy of digital elevation models in gauging inundation by sea level rise along the Valdelagrana beach and marshes (Bay of Cádiz, SW Spain). Geo-Mar Lett 33:225–230

Gesch DB (2009) Analysis of Lidar elevation data for improved identification and delineation of lands vulnerable to sea-level rise. J Coastal Res SI(53):49–58

Gesch DB (2012) Elevation uncertainty in coastal inundation hazard assessments. In: Cheval S (ed) Natural disasters. Intech Press, Croatia, pp 121–140. http://cdn.intechopen.com/pdfs-wm/30671.pdf

Gesch DB (2018) Best practices for elevation-based assessments of sea-level rise and coastal flooding exposure. Front Earth Sci 6(230):1–19

Gilman E (2004) Assessing and managing shoreline response to projected relative sea level rise and climate change. UNESCO World Heritage Central Pacific Project Planning Workshop, Kiribati Island, Republic of Kiribati

Gilman E, Ellison J, Coleman R (2007) Assessment of mangrove response to projected relative sea-level rise and recent historical reconstruction of shoreline position. Environ Monit Assess 124(1–3):105–130

Hawker L, Bates P, Neal J, Rougier J (2018) Perspectives on digital elevation model (DEM) simulation for flood modeling in the absence of a high-accuracy open access global DEM. Front Earth Sci 6(233):1–9

Hinkel J, Nicholls RJ, Vafeidis AT, Tol RSJ, Avagianou T (2010) Assessing risk of and adaptation to sea-level rise in the European Union: an application of DIVA. Mitig Adapt Strat Glob Change 15:703–719

IPCC (2014) Adaptation needs and options. In: Climate change 2014: impacts and adaptation and vulnerability, intergovernmental panel on climate change working group II. Cambridge University Press, Cambridge and New York

IPCC (2019) Chapter 4: Sea level rise and implications for low lying islands, coasts and communities. In: Pörtner HO, Roberts DC, Masson-Delmotte V, Zhai P, Tignor M, Poloczanska E, Mintenbeck K, Alegría A, Nicolai M, Okem A, Petzold J, Rama B, Weyer NM (eds) IPCC special report on the ocean and cryosphere in a changing climate. The Intergovernmental Panel on Climate Change (in press). https://report.ipcc.ch/srocc/pdf/SROCC_FinalDraft_Chapter4.pdf

Kathiresan K (2000) A review of studies on Pichavaram mangrove, southeast India. Hydrobiologia 430:185–186

Khan AS (2013) Climate change induced sea level rise and its impacts on the Tamil Nadu coast, India: framing ecosystem and community based adaptation strategies. Phd thesis. Anna University, Chennai, pp 220–226

Khan AS, Ramachandran A, Usha N, Punitha P, Selvam V (2012a) Predicted impact of sea-level rise at Vellar-Coleroon estuarine region of Tamil Nadu Coast, in India: mainstreaming adaptation as a coastal zone management option. Ocean Coast Manag 69:327–339

Khan AS, Ramachandran A, Usha N, Arul Aram I, Selvam V (2012b) Rising sea and threatened mangroves: a case study on stakeholders' engagement in climate change communication and non-formal education. Int J Sustain Dev World Ecol 19(4):330–338

Khan AS, Ramachandran A, Palanivelu K (2014a) Climate Portfolio of Pichavaram mangrove region of Tamil Nadu coasts, India: an add-on information for adaptation policy planning. J Integr Environ Sci 11(3–4):173–186

Khan AS, Ramachandran A, Palanivelu K, Selvam V, Punitha S (2014b) Impact of sea level rise on mangrove ecosystem and its dependent fishing communities in the coastal regions of Cauvery delta: a message for policy planners to frame suitable anticipatory adaptation strategies (abstract id. PA21A-12). American Geophysical Union, Fall Meeting 2014, San Francisco, USA. https://ui.adsabs.harvard.edu/abs/2014AGUFMPA21A..12A/abstract

Kim M, You S, Chon J, Lee J (2017) Sustainable land-use planning to improve the coastal resilience of the social-ecological landscape. Sustainability 9(7):1086

Kleinosky LR, Yarnal B, Fisher A (2007) Vulnerability of Hampton Roads, Virginia to storm-surge flooding and sea-level rise. Nat Hazards 40:43–70

Kulp SA, Strauss BH (2018) CoastalDEM: a global coastal digital elevation model improved from SRTM using a neural network. Remote Sens Environ 206:231–239

Kulp SA, Strauss BH (2019) New elevation data triple estimates of global vulnerability to sea-level rise and coastal flooding. Nat Commun 10(4844):1–15

Mahendra RS, Mohanty PC, Bisoyi H, Kumar TS, Nayak S (2011) Assessment and management of coastal multi-hazard vulnerability along the Cuddalore, Villupuram, East coast of India using geospatial techniques. Ocean Coast Manage 54(4):302–311

Marfai M, King L (2008) Potential vulnerability implications of coastal inundation due to sea level rise for the coastal zone of Semarang city, Indonesia. Environ Geol 54:1235–1245

Miller T (2019) Regional effects of sea level rise. Earth Systems and Environmental Sciences. https://www.sciencedirect.com/topics/earth-and-planetary-sciences/sea-level-rise

Mogensen LA, Rogers K (2018) Validation and comparison of a model of the effect of sea-level rise on coastal wetlands. Sci Rep 8:1369. https://doi.org/10.1038/s41598-018-19695-2

Mumby PJ, Green EP, Edwards AJ, Clark CD (1999) The cost effectiveness of remote sensing for tropical coastal resources assessment and management. J Environ Manage 55:57–166

Najjar RG, Walker HA, Anderson PJ, Barron EJ, Bord RJ, Gibson JR, Kennedy VS, Knight CG, Megonigal JP, O'Connor RE, Polsky CD, Psuty NP, Richards BA, Soenson LG, Steel EM, Swanson RS (2000) The potential impacts of climate change on the Mid-Atlantic coastal region. Climate Res 14:219–233

Nayak SR, Bahuguna A, Shaikh MG, Chauhan HB, Rao RS, Arya AS (1992) Coastal environment, scientific note, RSAM/SAC/COM/SN/11/92. Space Applications Centre, Ahmedabad

NOAA (2010) Technical considerations for use of geospatial data in sea level change mapping and assessment. NOAA technical report NOS 2010-01. National Oceanic and Atmospheric Administration (NOAA), Silver Spring

O'Regan PR (1996) The use of contemporary information technologies for coastal research and management: a review. J Coastal Res 12:192–204

Paprotny D, Morales-Nápoles O, Vousdoukas MI, Jonkman SN, Nikulin G (2018) Accuracy of pan-European coastal flood mapping. J Flood Risk Manage e12459. https://doi.org/10.1111/jfr3.12459

Pittock J (2011) National climate change policies and sustainable water management: conflicts and synergies. Ecol Soc 16(2):25

Poulter B, Halpin PN (2008) Raster modeling of coastal flooding from sea-level rise. Int J Geogr Inf Sci 22:167–182

Ramasamy SM, Kumanan CJ, Saravanavel J, Rajawat AS, Tamilarasan V, Ajay (2010) Geomatics based analysis of predicted sea level rise and its impacts in parts of Tamil Nadu Coast, India. J Indian Soc Remote Sens 38(4):640–653

Reimann L, Vafeidis AT, Brown S, Hinkel J, Tol RSJ (2018) Mediterranean UNESCO world heritage at risk from coastal flooding and erosion due to sea-level rise. Nat Commun 9 (4161):1–11

Ricketts PJ (1992) Current approaches in geographic information systems for coastal management. Mar Pollut Bull 25:82–87

Sampson CC, Smith AM, Bates PD, Neal JC, Trigg MA (2016) Perspectives on open access high resolution digital elevation models to produce global flood hazard layers. Front Earth Sci 3:85. https://doi.org/10.3389/feart.2015.00085

Satterthwaite D, Huq S, Pelling M, Reid H, Lankao PR (2009) Adapting to climate change in urban areas: the possibilities and constraints in low- and middle-income nations. Earthscan, London, pp 112. http://pubs.iied.org/pdfs/10549IIED.pdf

Scheraga JD, Ebi KL, Furlow J, Moreno AR (2003) From science to policy: developing responses to climate change. In: McMichael AJ, Campbell-Lendrum DH, Corvalan CF, Ebi KL, Githeko AK, Scheraga SD, Woodward A (eds) Climate change and human health: risk and responses. World Health Organization, World Meteorological Organization and United Nations Environment Programme, Copenhagen, pp 237–266. http://www.who.int/globalchange/publications/climchange.pdf

Schumann GJP, Bates PD (2018) The need for a high-accuracy, open-access global DEM. Front Earth Sci 6(225):1–5

Selvam V, Ravichandran K, Gnanappazham L, Navamuniyammal M (2003) Assessment of community-based restoration of Pichavaram mangrove wetland using remote sensing data. Curr Sci India 85(6):794–797

Stevens T, Connolly RM (2004) Testing the utility of abiotic surrogates for marine habitat mapping at scales relevant to management. Biol Cons 119(3):351–362

The Arlington Group (2013) Sea level rise adaptation primer-a toolkit to build adaptive capacity on Canada's south coasts. https://www2.gov.bc.ca/assets/gov/environment/climate-change/adaptation/resources/slr-primer.pdf

The Nature Conservancy (2010) Developing a regional framework for assessing coastal vulnerability to sea level rise in Southern New England-part 1. A case study from the nature conservancy. https://coast.noaa.gov/data/digitalcoast/pdf/slr-newengland-casestudy.pdf

Titus JG, Richman C (2001) Maps of lands vulnerable to sea level rise: modeled elevations along the US Atlantic and Gulf coasts. Clim Res 18:205–228

Titus JG, Wang JEM (2008) Maps of lands close to sea level along the Middle Atlantic coast of the United States: an elevation data set to use while waiting for LiDAR. In: Titus JG, Strange EM (eds) Background documents supporting climate change science program synthesis and assessment product 4.1. United States Environmental Protection Agency, Washington DC, pp 1–44. http://papers.risingsea.net/federal_reports/Titus_and_Strange_EPA_section1_1_Titus_and_Wang_may2008.pdf

Tobey J, Rubinoff P, Robadue D, Ricci G, Volk R, Furlow J, Anderson G (2010) Practicing coastal adaptation to climate change: lessons from integrated coastal management. Coast Manage 3:317–335

UNESCO (2010) Application of satellite remote sensing to support water resources management in Africa: results from the TIGER initiative. Division of Water Sciences, United Nations Educational, Scientific and Cultural Organization, Paris, France. http://unesdoc.unesco.org/images/0018/001880/188045m.pdf

UNU (2009) Sea level rise and the vulnerability of coastal peoples responding to the local challenges of global climate change in the 21st century. UNU Institute for Environment and Human Security (UNU-EHS), Bonn

Vousdoukas MI, Bouziotas D, Giardino A, Bouwer LM, Mentaschi L, Voukouvalas E (2018) Understanding epistemic uncertainty in largescale coastal flood risk assessment for present and future climates. Nat Hazards Earth Syst Sci 18:2127–2142

Webster TL, Forbes DL, Dickie S, Shreenan R (2004) Using topographic LiDAR to map flood risk from storm-surge events for Charlottetown, Prince Edward Island, Canada. Can J Remote Sens 30:64–76

WII (2015) Strengthening capacity for sustainable and participatory management of coastal and marine protected areas. Ministry of Environment, Forest and Climate Change, Government of India, Wildlife Institute of India. New Delhi, India. https://wii.gov.in/WII-GIZ_Marine_Programme

Wolff C, Vafeidis AT, Lincke D, Marasmi C, Hinkel J (2016) Effects of scale and input data on assessing the future impacts of coastal flooding: an application of DIVA for the Emilia-Romagna coast. Front Marine Sci 3:41. https://doi.org/10.3389/fmars.2016.00041

Zhang K (2011) Analysis of non-linear inundation from sea-level rise using LIDAR data: a case study for South Florida. Clim Change 106:537–565

Zhang K, Dittmar J, Ross M, Bergh C (2011) Assessment of sea-level rise impacts on human population and real property in the Florida keys. Clim Change 107:129–146

Zhu M, Hoctor T, Volk M, Frank K, Linhoss A (2015) The conservation value of elevation data accuracy and model sophistication in reserve design under sea-level rise. Ecol Evol 5 (19):4376–4388

Planning for Sustainable Coastlines in the United States in an Era of Climate Change: An Examination of Major Policy Barriers

Chad J. McGuire, Michael Goodman, and Devon Lynch

1 Introduction

The United States provides an intriguing case study of the dynamics associated with sustainable coastal development in an age of sea-level rise. Because of its diverse geography, dynamic development history, and complicated political frameworks, there is a somewhat unique opportunity to consider how past and current policies have impacted the ability to accommodate the effects of climate change in many coastal regions. For example, low-lying and heavily developed coastal areas of the United States have difficult decisions to make relative to balancing the competing interests of development and maintaining the ecosystem integrity of coastal watersheds (Sweet and Park 2014). Meanwhile, more options exist in less developed and higher elevation coastal areas (Buchanan et al. 2016).

While there is diversity in the physical characteristics and historical development of coastal areas throughout the United States, there are consistent themes of policy evolution at the national level impacting local coastal development decisions. These themes can be examined using a policy evolution analytical framework to better understand the dynamic between existing human expectations relative to coastal land use, how those expectations have been influenced by pro-development policies, and how the resulting mix of expectations and supportive policy effect the ecological integrity of low-lying coastal regions. These themes are similar to issues being experienced by many coastal nations where existing policy frameworks impede the ability to reorient policy in a way that sustainability balances coastal

C. J. McGuire (✉) · M. Goodman
Department of Public Policy, University of Massachusetts Dartmouth, 285 Old Westport Road, Dartmouth, MA, USA
e-mail: cmcguire@umassd.edu

D. Lynch
Departments of Economics and Public Policy, University of Massachusetts Dartmouth, 285 Old Westport Road, Dartmouth, MA, USA

W. Leal Filho et al. (eds.), *Sustainability in Natural Resources Management and Land Planning*, World Sustainability Series,
https://doi.org/10.1007/978-3-030-76624-5_23

development against the increasing risks brought by climate change (Nicholls 2015).

This chapter provides an examination of these larger themes of national policy evolution relative to coastal development. It attempts to identify the relationship between policy development, evolution, and the social institutions that become reliant on the resulting policy environment created. It then juxtaposes this overarching policy environment against the kinds of policies needed to counteract the effects of coastal climate change, particularly sea-level rise and the resulting loss of natural coastal attributes. Changes to existing policy required to harmonize coastal expectations with the reality of coastal erosion are explored. Finally, some observations and recommendations are made in how to think about altering existing expectations are discussed. The goal is to provide an exemplar of how to balance the dual considerations of development and ecosystem integrity in high demand but low-lying coastal areas subject to the negative impacts of climate change.

2 Materials and Methods

The research approach for this chapter utilizes a policy evolution and diffusion analytical framework to identify and analyze major national policies that influence coastal development decision-making. Policy evolution and diffusion is utilized to understand the dynamic nature of policy development and implementation, including analyzing discontinuities between the purposes and intentions of policies with their impact and effect (Bauler 2012; Luers 2005; Tews 2005) This combines a conceptual framework about risk assessment and public attitudes towards climate change risks with the diffusion of national policies that incentivize and support coastal development. These policies are then analyzed to determine the impact they have on the assessment of risk relative, either directly or indirectly, through their implementation.

A review of the most updated literature on coastal policy, specifically pro-development policies, in light of climate change was conducted. Major policies supportive of lowering perceptions of risk in coastal areas in light of objective evidence of increasing risk were identified and analyzed using the conceptual framework for understanding the dynamics of risk perception highlighted by McGuire (2015).

3 Overview of Coastal Development and Climate Change in the US

US counties abutting shorelines hold approximately fifty million housing units, with developed properties worth at least $1.4 trillion USD within one-eighth of a mile of the coastline. It is estimated that damage from climate change, including sea-level rise and increased frequency and intensity of storm events, will cause billions of dollars of property damage annually by 2050, with the Atlantic and Gulf coasts facing disproportionate risks (Neumann et al. 2015). In addition, based on low versus high estimates of future sea-level rise, it is estimated that between $66 and $106 billion USD of current real estate value (2020 dollars) will be below sea level by 2050; and between $238 and $507 billion USD by 2100 (Houser et al. 2015).

The physical assets at-stake in coastal areas of the US is only one dimension of consideration. Coastal regions also represent a significant amount of economic output and total employment relative to their size. An economic, population and geographic size comparison of US coastal areas is provided in Table 1.

Coastal infrastructure—such as roads, bridges, utility conduits, sewer systems, etc.—is also at-risk (Kunz et al. 2013). Currently there are more than 60,000 miles of roads and bridges in coastal floodplains that are vulnerable to coastal storms (USDOT 2008). And that vulnerability has both direct and indirect economic impacts. For example, indirect costs include the more expected, like loss of business, and the less expected like adverse sociopsychological impacts including disaster-related depression, anxiety, and even post-traumatic stress disorder (Shen and Aydin 2014).

Cumulatively, coastal property and infrastructure losses can have cascading effects, with the potential to disrupt the economic output and stability of these areas. Focusing on real property development, increased effects of climate change can influence demand for real estate development as well as access to financing. These effects can impact associated loan servicing, insuring, and financial mortgage-backed securities (Ouzad and Kahn 2019). Reduced demand for coastal real estate can lead to drops in market values. And this, in-turn, can be exasperated

Table 1 Economic, population, and geographic comparison of US coastal areas (adapted from National Ocean Economics Program 2016)

Region	Employment		GDP		Population		% land area
	Millions	% of US	Trillions (USD)	% of US	Millions	% of US	
United States	134.0		$16.7		316.5		
All coastal states	109.2	81.5	$13.9	83.7	257.9	81.5	57.0
Coastal zone counties	56.2	42.0	$8.0	48.0	133.2	42.1	19.6

through changes in how coastal hazards risks are insured against loss, and also how risky coastal development is financed (discussed later in this chapter). And this can all have negative public financing impacts. For example, less coastal real estate at lower valuations will lead to lower real estate tax revenues, which are a main source of funding for basic public services in many of these communities (Sapat and Esnard 2017).

All told, the impacts of climate change on coastal areas of the US can be substantial. This is because about half of the total US economic output occurs in coastal counties, which represent about twenty percent of the total US landmass (Table 1). Currently, however, these coastal areas have experienced, and continue to experience, strong demand. And that demand has been supported by a framework of historical and current policies. The following section examines the policy evolution of coastal development in the US in more detail.

4 History of Coastal Development in the US

The history of coastal development in the United States has been marked by, at least, three major factors. First, there is the demographic reality: most of the US population leaves near the coastline (Table 1). Second, this preference for coastal living has worked hand-in-hand with historical policies supporting coastal development. And third, the demographic reality of, and supportive policies of, coastal development all occurred in an American system that prioritizes private property rights. These overarching factors provide the conditions upon which policy has evolved in the US in a pro-development manner that, even in the face of increasing natural hazards today, helps to define a path dependence supportive of continued coastal development.

As previously mentioned, the federal government has developed a number of infrastructure policies that aided state and local communities in developing coastal infrastructure. European settlement of the US occurred mainly by ships, allowing settlement of coastal regions. Shipping continued as an important means of commerce, both nationally and internationally. And that historical reality continues today, as most international commerce is conducted through maritime shipping and the use of coastal seaports, handling approximately ninety-nine percent of overseas trade (Olsen 2015).

This historical reality provides a rationale for coastal development and its associated economic activity. But that reality is based in large part on an assumption that coastlines, while always dynamic, are not subject to large-scale substantial and irreversible change. In essence, the presumption that continues along coastal areas of the US is that these areas are net *revenue* centers: where the benefits—economic and otherwise—exceed the costs. It is this presumption that is at issue when the reality of climate change is considered. For example, estimates of adapting one hundred major commercial coastal seaports for experienced and anticipated sea level rise is between \$56–112 billion USD (Becker et al. 2017).

The question is whether the costs of climate change and the increasing hazards it presents for many coastal areas, particularly low-lying areas in the US, is being fully considered. In other words, is current US policy considering the emerging hazard of climate change? In this calculus, coastal areas may be reinterpreted as *cost* centers rather than revenue centers. And incorporating these costs must likely include an evaluation of not only current costs, but also future anticipated costs. Using the commercial seaport example, additional research has been conducted to determine these longer-term costs in a way that allows for a better determination of net value for continued investment or, potentially, divestment (Becker et al. 2018). The willingness of policy instruments to incorporate the emerging realities of climate change will be vital for the development of sustainable coastlines. Currently there is strong evidence this is simply not the case in major federal US coastal policy.

There are three existing federal policies of particular note worth further examination and focused on coastal development. The first has to do with insuring against the risk of losses associated with coastal hazards. The second has to do with direct government payments to those who suffer natural disaster losses along coastal areas. The third is more general, attempting to create incentives for private citizen home ownership, which then feeds into the number of properties influenced by the first and second policies. All three are important in understanding the foundations of coastal development policy that continue today. They are equally important in considering their influence not only on incentivizing coastal development, but also the impact such incentives have on coastal ecosystem protection in an era of sea-level rise. Before describing and analyzing these policies in greater detail, a short framing section of the dynamic conflict between coastal development and ecosystem protection follows.

5 Coastal Development and Ecosystem Protection: A Dynamic in Conflict

One way to understand the current state of coastal policy in the US relative to incorporating climate change is to consider how well current policy protects existing natural coastal habitat. Is there a clear preference in policy to protect natural coastal attributes, allowing them to move landward as sea levels rise? Or is there rather an inclination to protect coastal development, including at the expense of natural coastal attributes? There are many different ways to examine this question. Clearly, policies that are anti-coastal development would tend to favor the protection of existing coastal attributes. And policies that are pro-coastal development, including the protection of existing development, would tend to diminish the priority of natural coastal attributes. For example, policies favoring hard armoring techniques to protect existing or planned coastal development tend to result in the destruction of natural coastal attributes as approaching seas erode those attributes (Brandon 2016).

It is important to note there are current policies in the US that help to protect coastal attributes. There are few at the national level, but they tend to exist more at the state and local level. For example, states like Florida, which rely on their sandy beaches and seaside amenities for tourism, prioritize the importance of coastal attributes. Other coastal states are beginning to understand the importance of barrier islands and similar coastal attributes as natural forms of protection against storm surges and similar hazards impacting development landward. And, in some cases, there is recognition of the ecological importance of coastal zones, including and excluding direct economic benefits (Ariza et al. 2014).

But the question is how effective are these more local policies, particularly in comparison to overarching national policies that help to incentivize development and redevelopment, including in low-lying, risky coastal areas. The three federal policies mentioned above—insuring against losses, direct financial assistance post-disaster, and financially incentivizing coastal development—are focused on here because they create path dependence reinforcement: they engender a set of behaviors and patterns aimed at recreating existing expectations irrespective of new information. In the parlance of climate change along the coastlines, they reinforce past decisions while discounting the increasing risks posed by climate change. In doing so, they artificially increase the value of coastal development at the expense of natural coastal attributes. A closer examination of these national policies follows.

6 Key National Coastal Climate Change Policies in the United States

The three national policies described above can be separated into two programmatic areas for further examination; one deals with insuring against flood risk and providing compensation when coastal disasters occur, while the other deals more generally with the financial mechanisms involved in coastal development in the US. An examination of these three policies can provide insights into the dynamics of the "institution" of coastal zone management. The first two can be combined because they both are responsive policies aimed at mitigating the impacts of natural hazards. The third focuses on capital markets associated with incentivizing coastal home ownership. It is a policy that is not dependent on natural hazards, as the first two. But it is related, as it underlies the total coastal housing stock available for public flood insurance and disaster relief.

6.1 Insuring and Compensating Against Coastal Damage

The first programmatic national policy deals with insuring against flood risk and providing compensation when coastal damage occurs due to natural disaster. The two federal policies in the United States that compromise this programmatic approach to risk of damage along coastlines are the National Flood Insurance Program (NFIP) and federal disaster assistance. As Knowles and Kunreuther (2014) point out, while federal disaster relief has existed in the United States since its beginnings as a nation, national flood insurance began much later in response to escalating federal costs borne by greater financial damage via increasingly intense coastal storm events. These two national policies have helped to shape a social-institutional context of coastal management by influencing the expectations of those who live along the coast. The underlying principles of institutional development from these two policies are observed through a historical examination of their co-evolution.

Today, in the United States provides public flood insurance at the national level due to a mix of historical antecedents and political necessity. As summarized by Knowles and Kunreuther (2014), federal disaster relief was the main national approach to remediate the financial and logistical harms of flood disasters in the US prior to 1968. At this time, flooding due to "natural" phenomenon was seen as something similar to human-induced disasters caused by war. During the cold war, the US federal government was busy planning for harms caused by nuclear fallout. Natural disasters provided an analog to the kinds of anticipated harms associated with nuclear war. And since fault could not be attributed to civilians harmed by the acts of war, they were equally not attributed to the harms caused by natural disaster. As such, it was socially and politically acceptable for the federal government to compensate those harmed by, and help remediate the harm caused by, natural disasters (Dauber 2013).

The issue with national relief being given to those who lived along flood-prone areas was that it made the risk of living in these areas acceptable. While people understood events like hurricanes and coastal flooding would occur at somewhat regular intervals, they also understood that federal disaster relief provided a ready-made relief in the form of federal financial and other assistance. The federal government also understood this unintended consequence. They saw how federal disaster assistance created incentives to both develop and live along coastal areas by reducing the implied risk. As a result, the US government developed and adopted a model of public flood insurance. The concept behind the model was that now, unlike under federal disaster assistance, individuals and coastal communities would have to being to internalize the risks of developing in their areas by mandating flood insurance and cost-sharing (Anderson 1974). Communities most at-risk would pay the highest premiums, while lower-risk communities would pay lower premiums, all borne by individual policies made to individual coastal homeowners. The moral hazard created by federal disaster relief providing de facto, zero premium hazard insurance would be reduced through mandated public flood insurance and penalties for non-compliance, most notably the loss of availability of federal disaster assistance for the coastal community to supplement costs not covered under public flood insurance (CRS 2013).

These original goals for coastal flood insurance and disaster relief policy have never been fully implemented for the following reasons. At the beginning, national flood insurance was implemented as a wholly voluntary program (Kunreuther et al. 1978). Coastal municipalities and individuals were not mandated to carry flood insurance, even in areas with the highest risks of flooding. And because there was no tie-in between adoption of national flood insurance and the availability of federal disaster relief, most deemed voluntary flood insurance as an unnecessary expense (Michel-Kerjan 2010). Realizing that voluntary measures hindered national flood insurance adoption, the US government began making flood insurance mandatory for the most at-risk coastal communities. Making sure to create incentives at both the individual and community level, review of flood risk was done at both the individual parcel and at the community level. For the individual, failure to receive and maintain flood insurance could result in losses not covered by federal funds. For the community, failure to implement flood resilient community practices could lose the community certification which could result in the loss of federal disaster assistance funding.

While these changes to the system have increased the number of national flood insurance policies in existence, the program remains heavily subsidized. Decades of coastal development when risks were either unknown or discounted, along with increasing risks due to climate-induced sea-level rise, have led to aggregating risks that cannot be easily insured because of a lack of spreading that risk among lower-risk homes. Thus, the premiums paid by individual coastal homeowners, are substantially below what is required to maintain a liquid fund for accruing losses. As of 2012 there were approximately 5.5 million policies providing US$1.28 trillion in property coverage as of December 2012 (Kunreuther et al. 2013). Essentially all of these policies have premiums that are well below the actual risks of loss, meaning the system of national insurance is not actuarily sound without additional federal funding to subsidize the system (PCI 2011).

Due to a decade of substantial coastal storms in the 2000s, the national flood insurance program faced substantial deficits. By 2012 it had to borrow, through taxpayer funding, some US$30 billion to cover programmatic loss shortfalls (CRS 2013). Due to the burgeoning debt, the US Congress passed a federal law in 2012 to reform the national flood insurance program. The reform, like the one passed decades before creating the national flood insurance program, was to further remove federal subsidies for living in risky coastal regions. The law updated its definition of risky coastal areas, began to phase out the subsidy in premiums paid over five years, and began to exclude from insurability a number of coastal structures including second homes, businesses, and any structure suffering repetitive losses (GAO 2013). Unfortunately, the United States had, by this time, become habituated to the inherent subsidies associated with coastal development. Coastal homeowners, coastal communities, developers, real estate interests, and others banded together to oppose the recently passed reforms. The result was an essential repeal of almost all of the noted updates identified in 2014 (Knowles and Kunreuther, 2014). Since 2012, the United States has seen some of most catastrophic and expensive coastal storms in its history. For example, in 2017–18 Hurricanes Harvey, Irma, and Maria devastated different areas of the US coastline. Over US$100 billion has already been spent on disaster relief as of 2020, approximately 50% of the total amount spent on 949 declared federal disasters between 2004 and 2019 (CRS 2019).

6.2 Incentivizing Coastal Home Ownership

Economic research has recently emerged exploring the dynamics between coastal real estate prices and climate risk (Bakkensen and Barrage 2017; Ortega and Taspinar 2018; Zhang and Leonard 2019). That research shows that coastal home valuations are robust, even in light of emerging evidence of increasing coastal risk. In trying to understand the price-to-risk conundrum, Ouzad and Kahn (2019) focus their attention on how the mortgage industry responds to evidence of increasing coastal risks. Their findings indicate risk is not being internalized into coastal property lending. There are a number of factors contributing to this conclusion. For example, the NFIP and natural disaster payouts (described in the previous section) lead to reductions in mortgage debt; this was found by Gallagher and Hartley's (2017) analysis of post-Hurricane Katrina damaged coastal properties. Kousky (2018) added to Gallagher and Hartley's findings by noting that since 2006 (post-Hurricane Katrina), both the total number of NFIP policies and the total amount of dollar value under NFIP insurance has declined substantially. This leaves potential greater losses for mortgage lenders.

As a result of the aforementioned, one would expect lending standards to become more stringent for increasingly risky coastal properties, and the cost of financing would be higher, coinciding with the increased risk of loss and lack of insurance (NFIP or otherwise) to cover the loss. But Ouzad and Kahn (2019) have found the exact opposite in their research. They analyzed fifteen "billion-dollar" events: major coastal storms in the US that have caused at least $1 billion USD in losses as estimated by the US National Oceanic and Atmospheric Association (NOAA). They find that lending in low-lying coastal areas actually *increases* after a major coastal disaster. And the main reason seems to be securitization; private originators of loans are able to pass on the risk of the loan defaulting to one of two Government Sponsored Enterprises (GSEs) in the US: Fannie Mae and Freddie Mac.

Fannie Mae and Freddie Mac were created in conjunction with the federal government to expand home ownership to more Americans. They have specific lending standards that are generally more lenient than private lending. And importantly, they do not currently consider flood risk as a factor in either making loans or determining the interest rate charged on the loan. They do, however, have an upper limit on total loan amounts, referred to as a conforming loan limit. This amount varies by region of the US depending on average home prices for that region. What Ouzad and Kahn have been able to discern is that in each of the fifteen billion-dollar loss events reviewed, the rate of mortgage originations increased significantly from rates prior to the disaster, often for periods of years. And importantly, denial rates for non-conforming loans (where the risk of default is held by the originator) increased, while the denial rates for conforming loans (loans sold to GSEs) decreased.

Based on the findings presented by Ouzad and Kahn, there is a problem of adverse selection in the coastal mortgage industry. To summarize: private banks and other mortgage originators are increasing lending in risky coastal areas at an increasing rate following a disaster. But the increase in lending is only for GSE-conforming loans, which are immediately "transferred" to the GSE. Conversely, these same private banks and mortgage originators *decrease* lending of non-conforming mortgages after disaster. Such loans cannot be transferred to GSEs, and thus the risk of holding the loan would remain with the private bank. The increased lending reflects the financial rewards of receiving mortgage originating and servicing fees for the originators. This is in-line with the research of Elenev et al. (2016) showing that underpriced government mortgage guarantees lead to more and riskier mortgage originations.

The sum of this process leads to overcapitalization, and thus mispricing, in the coastal real estate market. There is a significant literature examining the US housing market's equilibrium pricing of natural disaster risk (Bakkensen and Barrage 2017; Ortega and Taspinar 2018; Zhang and Leonard 2019). But as Ouzad and Kahn (2019) identify, origination fees that do not carry climate risk are insufficient to reflect mortgage risk. And this is particularly true where the originator immediately passes any risk onto a GSEs balance sheet. This cumulatively leads to unhedged and unanticipated systemic risk; a distortion that feeds back into the dynamics of public insurance and compensation against coastal damage described earlier. They cumulatively create policy intervention that prevents an objective and balanced approach towards sustainable coastal use in the US in an era of climate change.

7 Evolving Policy Needs: Coastal Policy Evolution for the Future

Best practices for coastal climate change adaptation espouse a general framework that includes the following considerations. First, in an era of climate change, the risks to coastlines must be fully understood and appreciated. Second, understanding that for many coastlines adaptation planning will need to include loss of coastal land due to sea level rise and erosion, management practices need to be deployed that ensure the ability of coastal features to migrate inland (McGuire 2017). This second point incorporates a deeper understanding of the benefits of coastlines to include not only direct economic benefits, but also the more indirect economic benefits such as maintaining natural coastal features and ecological integrity (Titus et al. 2009). In sum, establishing and maintaining a buffer between a retreating coastline and the built environment ensures the safety of humans and the integrity of the coastline.

The policy instruments described in this paper have collectively acted as barriers to adaptation strategies based on climate change in the following ways. First, the US has developed a dependency on subsidizing coastal development, mainly by discounting the risks associated with living along the coastline. And this risk discounting has developed into a lowered perception of coastal hazard risk by those living, or considering living, in coastal regions. As a result, new information about increased coastal risks contradicts this low perception of risk. In addition, the existing policy environment has created clear counter-incentives for adapting to climate change in the coastal zone. While best practices in adaptation call for land use planning that limits development along the coastline and allows for the migration of coastal zones landward, existing policies favor continued development through publicly financed risk-shifting mechanisms.

There is ample evidence being uncovered on the effects of subsidized national flood insurance, federal disaster relief, and real estate financing. Reviews of subsidized insurance and disaster relief have identified important insights in how these programs attenuate the inverse relationship between actual risk and risk perception (McGuire 2015), influence how long homeowners keep flood insurance policies active under voluntary scenarios (Michel-Kerjan et al. 2012), and even showing limited policy tenure under mandatory insurance scenarios (Kriesel and Landry 2004). In all cases, the uncertainty of a loss event occurring, coupled with the low flood premiums, seems to reinforce a belief that the risk of coastal loss is lower than the actual risk, even when presented with accurate information regarding the risk of loss. All of this is further reflected in the emerging work on coastal mortgage financing as described earlier in this chapter.

If coastal adaptation is about accommodating the effects of climate change, in particular more intense coastal storms and increased erosion exacerbated by sea level rise, then the current state of coastal development and financing policy in the US is running counter to this reasonable definition of accommodation. Beyond the population demographics showing a clear preference for living along the coastline, current policy environment is paving a path for coastal development in spite of climate change. And this does not bode well for the protection or maintenance of coastal features, particularly those in low-lying, developed, disaster-prone areas.

8 Conclusions and Recommendations

Public policies are created and evolve within a context that is dynamic and subject to multiple influences over time (Jones and Baumgartner 2005). The examples of national flood insurance, disaster relief, and coastal development financing provided in this paper show how a collection of policies can combine to create a policy environment that is inhospitable to coastal climate change adaptation strategies.

Each of these policies have their own history and rationales outside of a climate change context. Most exist in a responsive setting, where policy has been created to deal with an acute issue presented. Disaster relief developed under the umbrella of US federalism, where the federal government could aid state governments under duress due to natural disasters. National flood insurance emerged in response to the conditions created by federal disaster assistance. Coastal development flourished in an era where the risks of development could be shifted to the federal government. Without private insurers willing to step into the void, the federal government developed a plan to incrementally begin shifting the risks of coastal development back onto the states through insurance requirements. Finally, the financial mortgage origination entities described earlier, the GSEs, were created to expand home ownership opportunities to a wider range of citizens, including those who might not otherwise qualify under wholly private standards.

When viewed as a whole, these policy mechanisms create both literal and constructive barriers to coastal adaptation strategies in an era of climate change. Heavily subsidized coastal living engenders a sense of entitlement to live along the coastline. Indeed, when disasters strike, the political response carries a clear message of rebuilding; the concept of "resiliency" in this context becomes one of building bigger and stronger rather than accommodating the coastline for climate change. And even when faced with options of how to be resilient along the coastline, financial mechanisms in place can create perverse incentives to increase liquidity, engendering development and redevelopment in risky coastal areas.

What is not known from this analysis is the potential for policy development and response as conditions along the coastline worsen over time. It is entirely possible that the described policies will evolve as the effects of climate change reveal themselves along the coastlines over time. The question will be whether or not this evolution will happen in a sufficiently proactive manner to protect against the worst-case outcomes. As Knowles and Kunreuther (2014) discussed, efforts in the United States to rationalize the heavily subsidized national flood insurance program after Hurricane Sandy in 2012 ultimately failed due to public backlash. This experience reinforces the need to be proactive in revising the unintended consequences of policy well before the effects have fully realized (McGuire 2015).

The key, then, is being able to see how existing policy instruments can, individually and collectively, create and then reinforce a set of public expectations that run counter to coastal adaptation planning in an era of climate change. These policies can exist in any environment: they are certainly not unique to the United States. By seeing how existing policy structures influence the socio-political process, one can properly identify the barriers, whether literal or conceptual, to implementing meaningful coastal adaptation planning. The prognosis can then be crafted. For example, the inherent subsidies created in national flood insurance and disaster relief must be removed so the public can properly gauge coastal risk. And, certainly, GSEs much incorporate climate risk into their lending standards, removing the adverse selection problem. Of course, even with a clear prescription in-hand, the process of moving from thought to action is difficult, particularly when dealing with entrenched policies. Helping the public understand and appreciate the objective risks of climate change would certainly be helpful. Thus, while this chapter may not be able to establish an exact guide to fixing existing policy barriers that run counter to sustainable coastal policy in the US in an era of climate change, it provides a framework for identifying these barriers and placing their impact into a larger context that can be replicated in various settings.

References

Anderson DR (1974) The National Flood Insurance Program: problems and potential. J Risk Insur 41(4):579–599

Ariza E, Lindeman KC, Mozumder P, Suman DO (2014) Beach management in Florida: assessing stakeholder perceptions on governance. Ocean Coast Manag 96:82–93

Bakkensen LA, Barrage L (2017) Flood risk belief heterogeneity and coastal home price dynamics: going under water? Bureau of Economic Research working paper series: 23854. https://www.nber.org/papers/w23854

Bauler T (2012) An analytical framework to discuss the usability of (environmental) indicators for policy. Ecol Ind 17:38–45

Becker A, Hippe A, Mclean EL (2017) Cost and materials required to retrofit US seaports in response to sea level rise: a thought exercise for climate response. J Marine Sci Eng 5(3):44

Becker A, Ng AKY, McEvoy D, Mullett J (2018) Implications of climate change for shipping: ports and supply chains. WIREs Clim Change 9(2):

Brandon TO (2016) Nationwide permit 13, shoreline armoring, and the important role of the U.S. Army Corps of Engineers in coastal climate change adaptation. Environ Law 46(3):538–576

Buchanan MK, Kopp RE, Oppenheimer M, Tebaldi C (2016) Allowances for evolving coastal flood risk under uncertain local sea-level rise. Clim Change 137:347–362

CRS (2013) The National Flood Insurance Program: status and remaining issues for congress. Congressional Research Service, Washington, DC

CRS (2019) The disaster relief fund: overview and issues. Congressional Research Service, Washington, DC

Dauber ML (2013) The sympathetic state: disaster relief and the origins of the American welfare state. Chicago University Press, Chicago

Elenev V, Landvoight T, Van Nieuwerburgh S (2016) Phasing out the GSEs. J Monetary Econ 81:111–132

Gallagher J, Hartley D (2017) Household finance after a natural disaster: the case of Hurricane Katrina. Am Econ J Econ Policy 9(3):199–228

GAO (2013) Flood insurance: more information needed on subsidized properties. United States Government Accountability Office, Washington, DC

Houser T, Hsiang S, Kopp R, Larsen K, Delgado M et al (2015) Economic risks of climate change: an American prospectus. Columbia University Press, New York

Jones BD, Baumgartner FR (2005) The politics of attention: how government prioritizes problems. Chicago University Press, Chicago

Knowles SG, Kunreuther HC (2014) Troubled waters: The National Flood Insurance Program in historical perspective. J Policy Hist 26(3):327–353

Kousky C (2018) Financing flood losses: a discussion of the National Flood Insurance Program. Risk Manage Insur Rev 21(1):11–32

Kriesel W, Landry C (2004) Participation in the National Flood Insurance Program: an empirical analysis for coastal properties. J Risk Insur 71(3):405–420

Kunreuther HC, Ginsberg R, Miller L, Sagi P, Slovic P et al (1978) Disaster insurance protection: Public policy lessons. Wiley, New York

Kunreuther HC, Michel-Kerjan E, Pauly M (2013) Making America more resilient towards natural disasters: a call for action. Environ Sci Policy Sustain Dev 54(4):15–23

Kunz M, Muhr B, Kunk-Plapp T, Daniell JE, Khazai B et al (2013) Investigation of superstorm Sandy 2012 in a multi-disciplinary approach. Nat Hazards Earth Syst Sci 13(10):2579–2598

Michel-Kerjan E, Lemoyne de Forges, S, Kunreuther H (2012) Policy tenure under the U.S. National Flood Insurance Program. Risk Anal 32(4):644–658

Luers AL (2005) The surface of vulnerability: an analytical framework for examining environmental change. Glob Environ Change 15(3):214–223

McGuire CJ (2015) The role of risk perception in building sustainable policy instruments: a case study of public coastal flood insurance in the United States. Interdiscip Environ Rev 16 (2,3,4):232–252

McGuire CJ (2017) Adapting to sea level rise in the coastal zone: law and policy considerations. Routledge, New York

Michel-Kerjan E (2010) Catastrophe economics: the National Flood Insurance Program. J Econ Perspect 24(4):165–186

National Ocean Economics Program (2016) State of the U.S. ocean and coastal economies: 2016 update. Middlebury Institute of International Studies at Monterey, Monterey

Neumann JE, Emanuel K, Ravela S, Ludwig L, Kirshen P, Bosma K, Martinich J (2015) Joint effects of storm surge and sea-level rise on US coasts: new economic estimates of impacts, adaptation, and benefits of mitigation policy. Clim Change 129:337–349

Nicholls RJ (2015) Planning for the impacts of sea level rise. Oceanography 24(2):144–157

Olsen JR (2015) Adapting infrastructure and civil engineering practice to a changing climate. ASCE, Reston

Ortega F, Taspinar S (2018) Rising sea levels and sinking property values: Hurricane Sandy and New York's housing market. J Urban Econ 106:81–100

Ouzad A, Kahn ME (2019) Mortgage finance in the face of rising climate risk. National Bureau of Economic Research working paper series: 26322. https://www.nber.org/papers/w26322

PCI (2011) True market-risk rates for flood insurance. Property Casualty Insurers Association of America. https://www.pciaa.net/pciwebsite/common/page/attachment/13821

Sapat A, Esnard AM (2017) Coming home after disaster: multiple dimensions of housing recovery. Routledge, New York

Shen G, Aydin SG (2014) Highway freight transportation disruptions under and extreme environmental event: the case of Hurricane Katrina. Int J Environ Sci Technol 11(8):2387–2402

Sweet WV, Park J (2014) From the extreme to the mean: acceleration and tipping points of coastal inundation from sea level risk. Earth's Future 2(12):579–600

Tews K (2005) The diffusion of environmental policy innovations: Cornerstones of the analytical framework. Environ Policy Governance 15(2):63–79

Titus JG, Hudgens DE, Trescott DL et al (2009) State and local governments plan for development of most land vulnerable to rising sea level along the US Atlantic coast. Environ Res Lett 4 (4):04408

USDOT (2008) Highways in the coastal environment, 2nd edn. United States Department of Transportation Federal Highway Administration, Washington, DC

Zhang L, Leonard T (2019) Flood hazards impact on neighborhood house prices. J Real Estate Financ Econ 58(4):656–674

Handbook of Wildlife Fires: Monitoring, Control and Management Under a Changing Climate

Droughts and Wildlife Fires Formation Due to Stratosphere-Troposphere Interactions

Aliaksandr Krasouski, Siarhei Zenchanka, and Tsimafei Schlender

1 Introduction

Considering forest fires and climate change in the 21st century, Flannigan et al. (2006) makes the conclusion that "fire is the major stand-renewing disturbance in the circumboreal forest. Weather and climate are the most important factors influencing fire activity and these factors are changing due to human-caused climate change". But this article, as has many others, considers climate change to be a result of greenhouse gases impact only. At the seminar "Global Expert Workshop on Fire and Climate Change" (Robinne 2018), experts discussed the complex relationships between forest fires, climate change, land use, and response measures in such emergencies. The concept of forest fires has several definitions and criteria for determining the level of fire. All known definitions of fires, their nature, impact and behavior, such as large, very large, and extremely large fire, megafires and extreme wildfire events [EWE] were considered by Tedim et al. (2018). Regarding the definition of an "extreme wildfire event", many authors disagree. Bowman et al. (2017) identifies the criteria to be EWE; exposure and emissivity of fire, Lannom et al. (2014) considers the burning area, its duration, area of the burnt territory and distance to residential areas. Some work (McRae and Sharples 2011; Sampson et al. 2000; Weber and Dold 2006) associates it with atmospheric factors. Increasingly, the main causes of extreme fires

A. Krasouski
Belarusian State University, 4, Nezalezhnosti Av, 220030 Minsk, Belarus
e-mail: krasovsky@bsu.by

S. Zenchanka (✉)
Minsk Branch of Plekhanov Russian University of Economics,
40, Radialnaya Str, 220070 Minsk, Belarus
e-mail: zench@tut.by

T. Schlender
Faculty of Geography and Geoinformatics, Belarusian State University,
4, Nezalezhnosti Av, 220030 Minsk, Belarus

W. Leal Filho et al. (eds.), *Sustainability in Natural Resources Management and Land Planning*, World Sustainability Series,
https://doi.org/10.1007/978-3-030-76624-5_24

are meteorological droughts caused by various atmospheric processes. According to changes in the global trend of fire hazardous weather (Jolly et al. 2015), the number of fire hazardous regions has significantly increased from the period 1996–2013, compared with the period 1979–1993. Droughts are accompanied by low moisture content, dry thunderstorms and abnormally high surface temperatures. In the subtropical regions of Africa, Australia and North America, frequent forest fires occur due to dry and hot atmospheric conditions (Global Wildfire Information System n.d.). Experts call these zones, as well as tropical forests of South Asia and the Mediterranean region, the most fire hazardous areas in the world. The zones of boreal forests in dry periods are characterized by the occurrence of underground and peat fires.

Forest fires rarely occur in the polar regions of the Arctic tundra and adjacent taiga; however, due to the inaccessibility of the territory and the stability of anticyclones, it is difficult or impossible to extinguish fires in these areas. The causes of forest fires in the polar latitudes are complex—the rapid warming of the Arctic, abnormal weather conditions, soils and the human factor all contribute. For example, in the southern part of Western Siberia thunderstorms are the cause of forest fires in 26% of cases, and of burnt forest areas 47% of the time (Gorbatenko et al. 2020). Due to global warming, as well as the difference in insolation of the hemispheres (Berger 1988), the polar regions of both hemispheres warm faster in spring than other areas of low latitudes (IPCC 2013, 2014). In this regard, the importance of monitoring and forecasting forest fires in the taiga of Canada and Siberia during April–May (Flannigan et al. 2013; Ponomarev et al. 2016), and the forests of southern Australia from November–December (Virgilio et al. 2019) is increasing. The modeling of the stratospheric-tropospheric relationships suggests that in a significant number of cases the long-term localization of high altitude fire events is the result of stratospheric-tropospheric impacts (Krasouski et al. 2014).

In addition to surface weather conditions contributing to the fire hazard of wildlife, anomalous phenomena in the stratosphere, causing specific tropospheric changes in weather and climate, are also identified in the literature. For example, Lim et al. (2018, 2019) suggests the cause of fires in Australia in August 2019 to be a weakening of the polar vortex at the South Pole and the formation of Sudden stratospheric warming [SSW], which was only the third warming in the Southern hemisphere during the entire observation. This SSW was accompanied by a shift of westerly winds to the north, thereby creating colder and wetter conditions in southern New Zealand and western Tasmania. For southern Australia, this displacement of the winds to the equator led to warmer and drier conditions, or the negative phase of the Southern Annual Mode [SAM]. At the same time, the SSW period coincided with the positive phase of the Indian Ocean Dipole [IOD] and El Nino-Southern Oscillation [ENSO] (Min et al. 2013; Perkins et al. 2015).

The formation of SSW in the circumpolar region can be caused by variations of tropospheric planetary waves caused by the Madden-Julian Oscillation [MJO] (Garfinkel et al. 2012), ENSO (Domeisen et al. 2019a, b), and from the Quasi-Biennial Oscillation [QBO] in the tropical stratosphere (Holton and Tan 1980). Based on the fact that the atmosphere is a single whole both in the vertical dimension—from the earth's surface to the mesopause, and in the horizontal dimension on latitude and longitude, any effect on its integrity is reflected in other parts of its shell (Manney and Coy 2014). The same view was presented by Pedatella et al. (2014, 2018) where the influence of SSW on both the upper and lower layers of the atmosphere was considered. Recent studies have also shown the mutual influence of the state of the ozone layer and climate change associated with greenhouse gases (Hegglin et al. 2015; IPCC/TEAP 2005; Karoly 2003).

This paper analyzes, in accordance with satellite, ground-based observations and reanalysis of several cases of EWE when the SSW phenomena or high total ozone (TO) values in one hemisphere affects atmospheric events in the other hemisphere. This is caused by displacing planetary atmospheric cells, high-altitude planetary fronts, and changing the height of the tropopause; which causes abnormal hot and dry atmospheric conditions in areas where a high probability of extreme forest fires is formed.

2 Terms and Definitions

Climate change, forest fires, and other natural phenomena are defined by atmospheric processes. To better understand these processes and their interactions it is important to define different atmospheric phenomena used in this paper:

Antarctic polar vortex—is persistent westerly stratospheric circulation during winter in the Southern Hemisphere, which is most intense between latitudes 60° and 70° S and increases with height up to the stratopause (WMO 1992).

The Arctic polar vortex—is persistent westerly stratospheric circulation during winter in the Northern Hemisphere; it is most intense between latitudes 60° and 70° N and increases with height up to the stratopause. Though of stationary front strength, its maximum speeds are less than half those of the Antarctic stratospheric vortex (WMO 1992).

Extreme wildfire events (EWE)—the adoption of 90th, 95th and 99th percentile of the burned area is considered as the threshold for an EWE. This procedure is not related to a specific value but to a relative context and thus could be accepted worldwide (Lannom et al. 2014).

Hadley cells—is a meridional circulation model first proposed by G. Hadley (in 1735) which explains the trade winds. The circulation in each hemisphere consists of a low-level equatorward movement of air from about 30° latitude to the equator, rising air near the equator, poleward flow aloft from the equator to 30° and descending motion near 30° (WMO 1992).

Stationary front—is a flat tubular, quasi-horizontal current of air generally near the tropopause, whose axis is along a line of maximum speed and which is characterized by great speeds and strong vertical and horizontal wind shears. There are polar and subtropical frontal stationary fronts as the main strong wind zones in the upper troposphere which divides the Polar, Ferrel and Hadley atmospheric cells (WMO 1992).

Potential vorticity—is vorticity which a column of air between two adjacent isentropic surfaces would have if it were brought to an arbitrary "standard" latitude and then stretched or shrunk to an arbitrary "standard" thickness; it is a conservative air mass property used for identifying a mass of air and tracing its movement (WMO 1992).

PVU—has become accepted to define $1.0 \times 10^{-6} m^2 s^{-1} K kg^{-1}$ as one potential vorticity unit (1 PVU) (Hoskins et al. (1985).

Stratopause—is the top of the inversion layer in the upper stratosphere at about 50–55 km (WMO 1992).

Stratosphere—is the region of the atmosphere, situated between the tropopause (near 10–15 km) and the stratopause (near 50–55 km), in which the temperature generally increases with height (WMO 1992).

Sudden stratospheric warming (SSW)—is the temporary or permanent breakdown of the antarctic or arctic stratospheric vortex, in late winter or early spring, due to a rapid rise in temperature of the polar stratosphere (up to about 50 K in a few days) (WMO 1992).

The Dobson Unit (DU)—is a unit of ozone amount equal to 10^{-2} mm depth at standard temperature and pressure (WMO 1992).

Total ozone (TO)—is the amount of ozone in an entire atmospheric column. It may vary from between 2 and 6 mm thickness at standard temperature and pressure (WMO 1992).

Tropopause—is the boundary between the troposphere and the stratosphere, where an abrupt change in lapse rate usually occurs. It is defined as the lowest level at which the lapse rate decreases to 2 °C km^{-1} or less, provided that the average lapse rate between this level and all higher levels within 2 km does not exceed 2 °C km^{-1}. Occasionally, a second tropopause may be found if the lapse rate above the first tropopause exceeds 3 °C km^{-1} (WMO 1992).

Troposphere—is the lower part of the terrestrial atmosphere, extending from the surface up to a height varying from about 9 km at the poles to about 17 km at the equator, in which the temperature decreases fairly uniformly with height (WMO 1992).

3 Data and Methods

In this work ground-based observational data, satellite data, and reanalysis data were used. The data of surface temperature anomalies were taken from the archive of the Hydrometeorological center of the Russian Federation (https://meteoinfo.ru), ECMWF ERA-5 reanalysis data (Hersbach et al. 2019) and the Australian Weather Bureau (http://www.bom.gov.au). ERA-5 provides a numerical description of the climate for the period 1979–2020, combining the calculation data of the Integrated Forecasting System [IFS] model with satellite and ground-based observations.

ERA-5 has a time interval of 1 and 3 h, horizontal resolution of 31 and 62 km and 137 vertical levels to a height of 0.01 hPa (about 80 km). Satellite data on TO and tropopause height were obtained thanks to the AIRS tool (https://airs.jpl.nasa.gov/).

The TO and tropopause heights (respectively, the position of stationary atmospheric fronts) in the regions of the Southern and Northern Hemisphere, limited to 800 and 1,500 meridians for 2019 and 2020, were compared. In the Northern Hemisphere it is a region of Siberia, in the Southern Hemisphere it is an area of the Southern Ocean near Australia.

To study the influence of the position of the planetary subtropical high-altitude frontal jets (the southern boundary of the intratropical convergence zone) on the spread of Australian fires in January 2014, the altitudinal frontal zones and other auxiliary maps (TO, dynamic tropopause) were constructed according to the Global Forecasting System [GFS] (Environmental Modeling Center 2003). GFS is a numerical weather forecast model that was developed by the National Centers for Environmental Prediction [NCEP]. Horizontal resolution is 13, 28 and 70 km, vertical—64 levels to a height of 0.2 hPa. The forecast is calculated for 16 days in advance every 6 h. To distinguish the areas of high-altitude fronts (for a height of 100–150 hPa) as a gradient of the geopotential height along the meridian, the specialized program "Tropopause frontal zones" [TrFZ] (Mitskevich and Partasenok 2014) was created in the Python3 programming language based on the method described by Hewson (1998).

The main features of this method are the simultaneous analysis of changes (anomalies) inTO, tropopause altitude and surface air temperature in a specific part of the globe and its predicted lead time for a day or month, and comparison of the behavior of solar-terrestrial links in the northern and southern hemisphere.

4 SSW and Forest Fires in Australia, 2014

From late December 2013 to mid-January 2014 the surface air temperature was observed to be about 40 °C in southern and southeastern Australia. In the last week of December 2013 (Fig. 1) the anomalies in the maximum surface air temperature exceeded the norm by 10 °C in southern Australia and by 8 °C in eastern. For January 2014, the anomalies amounted to 3–4 °C from the norm in the southeastern part of Australia (Fig. 1). Negative anomalies were observed in the northern and northwestern part of the mainland. Such dry and hot atmospheric conditions contributed to the spread of forest fires. The state of Victoria suffered the most: 16 residential buildings burned down and 50 thousand hectares of forest burned in the Grampian National Park. The situation was aggravated by the widespread drought in Australia, even in places with a more or less favorable climate (https://www.abc.net.au/news/2014-01-16/bushfire-threat-as-australia-south-east-suffers-heatwave/5204250?nw=0).

Using the method of determining frontal altitude zones as a gradient of geopotential heights (Mitskevich and Partasenok 2014), it was found that on 19th

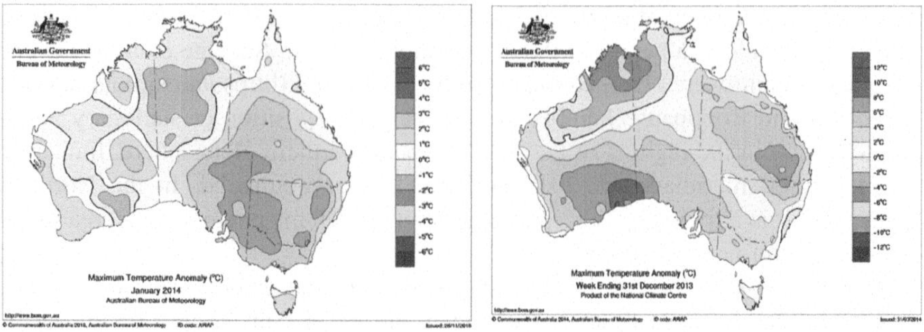

Fig. 1 Anomalies in Australia's maximum surface air temperature in late December 2013 (left) and January 2014 (right) according to the Australian Weather Bureau (http://www.bom.gov.au/)

January 2014 a subtropical stationary front at an altitude of 100 hPa of the Southern Hemisphere (Fig. 2) was located south of its normal position, namely, at the latitude of Tasmania. As a result, the entire territory of Australia was located within the Hadley cell (Dias and Bradley 2004), or the intratropical convergence zone, which led to an increase in surface air temperature. Such changes in the atmospheric circulation were reflected in the anomalies of the maximum surface air temperature, which on January 19th 2014 (Fig. 2) were in the southeastern and western parts of Australia over 10 °C, and in the northern and central parts—below 12 °C.

This case of forest fires in Australia at the end of 2013 and the beginning of 2014 was considered under the application of the model of the ozone mechanism of Climate change (Krasouski et al. 2014), in order to study hazardous meteorological phenomena such as drought. The climatic norm for this period assumes the autumn

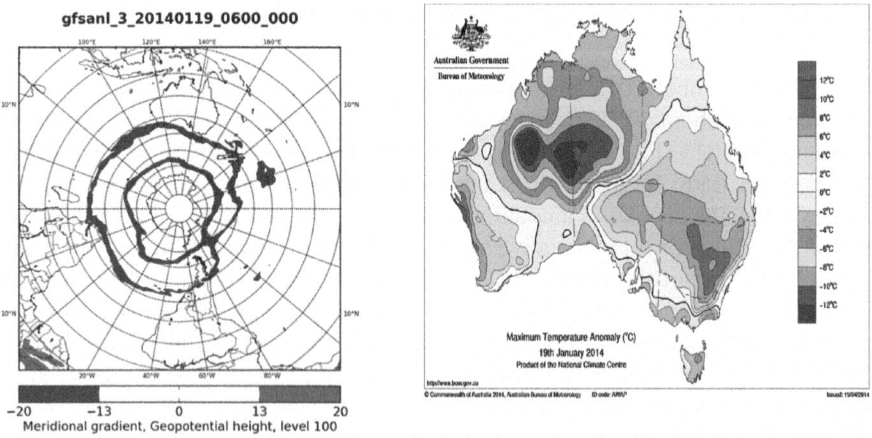

Fig. 2 The position of the high-altitude frontal zones at 100 hPa for the Southern Hemisphere, gpm/deg, left (https://nomads.ncep.noaa.gov/), and maximum temperature anomalies, °C, right (http://www.bom.gov.au/) on January 19, 2014 for the territory of Australia

minimum of TO and the maximum shift of all frontal zones to the north. Over the territory of Eastern Siberia on November 11 (Fig. 3), a pronounced local maximum of TO (over 440 DU) and a sharp increase in temperature in the lower stratosphere (by 25–30 °C for several days) was observed. This pattern of TO distribution caused a corresponding local decrease in the dynamical tropopause (2 PVU) (Kunz et al. 2011) over Eastern Siberia and the subsequent shift of the subtropical stationary front at an altitude of 150 hPa from the Northern Hemisphere to the south. Over southern Australia at this time the dynamical tropopause had a height similar to extratropical latitudes.

During November and December 2013, TO increased over Eastern Siberia and the Far East (over 520 DU). Based on observations of the dynamical tropopause and the position behind the high-altitude frontal zones, a shift to the south of the polar and subtropical frontal jets of the Northern Hemisphere, and an extrusion of the zone of intratropical convergence into the Southern Hemisphere was observed. By the beginning of January 2014 (Fig. 4), a resultant southward shift of the frontal zones of the Southern Hemisphere occurred. Thus, in two months over Australia the subtropical stationary front at an altitude of 150 hPa in the Southern Hemisphere

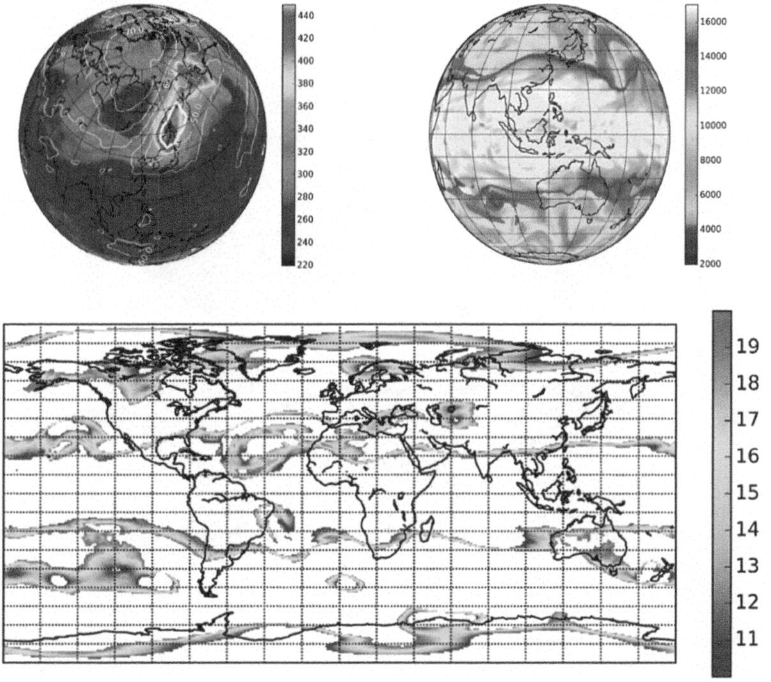

Fig. 3 TO (DU) and temperature at a height of 30 hPa, °C (left) and the geopotential height of the dynamical tropopause 2 PVU, gpm (right) on November 11, 2013, masked gradient of geopotential height (gpm /deg) at 150 hPa (bottom) on November 5, 2013 according to GFS (https://nomads.ncep.noaa.gov/)

shifted from the northern part of the mainland to the Tasmanian region. Ultimately, such changes in the upper atmosphere led to abnormal heat which contributed to intense forest fires in Australia.

5 SSW and Forest Fires in Siberia, 2019

Fires in Eastern Siberia began in July 2019. According to Greenpeace (2019), by the beginning of August 2019 forest burning areas reached record levels:—4 million hectares, with more than 13 million hectares of burned forests, and as a result, the amount of CO^2 emissions into the atmosphere was 166 million tons (https://www.euronews.com/2019/08/06/siberian-wildfires-engulf-area-almost-the-size-belgium-as-states-of-emergency-are-declared).

Anomalies in the average surface temperatures of the summer months in those parts of Siberia where extreme fires were burning, were almost 10 °C higher than the long-term average temperature for the period between 1981 and 2010 (Fig. 5). Temperatures in Alaska beat record highs, reaching 32 °C in July. The hottest for Siberia were June and August.

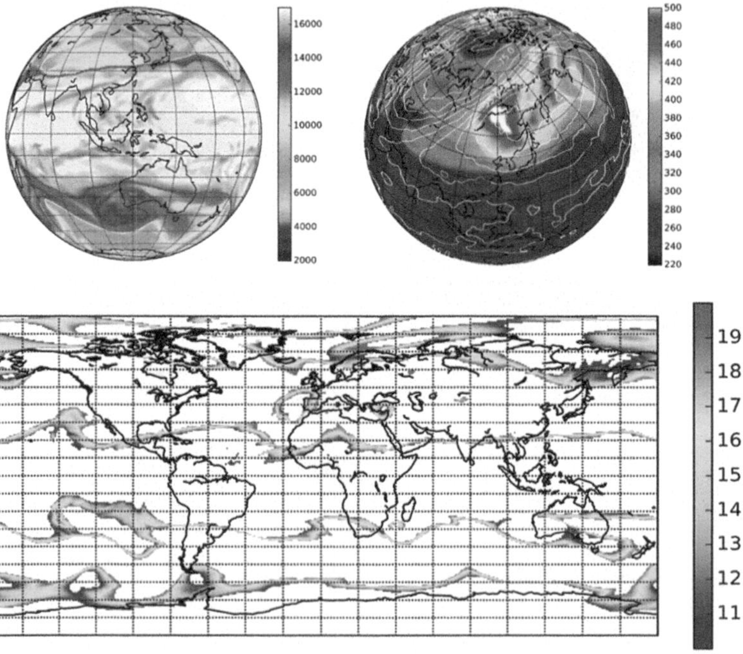

Fig. 4 TO (DU) and temperature at a height of 30 hPa, °C, (left) and the geopotential height of the dynamical tropopause 2 PVU, gpm, (right), masked gradient of geopotential height, gpm/deg, at 150 hPa (bottom) on January 10 2014 according to GFS (https://nomads.ncep.noaa.gov/)

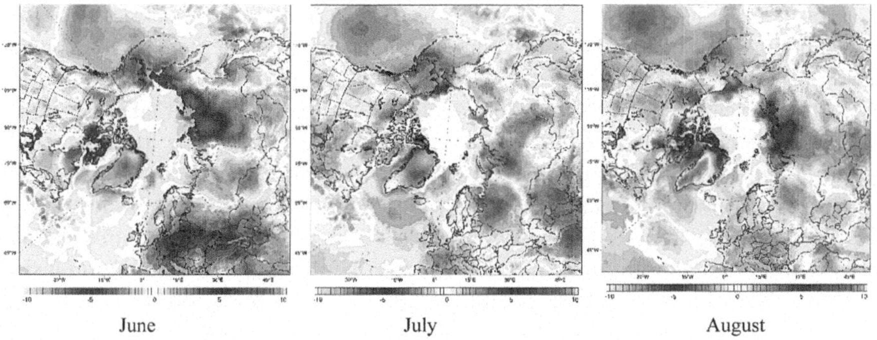

Fig. 5 Monthly average anomalies of surface air temperature (2 m) in the Arctic region in the summer (June, July, August) 2019 according to ERA5 reanalysis (https://cds.climate.copernicus.eu/)

Ground-based data of the surface air temperature anomalies of the Oymyakon, Yakutsk and Chita meteorological stations, which are located in Siberia and the Far East, showed hot and dry periods in summer (Fig. 6). Figure 6 shows that for the more western stations (Chita and Yakutsk) an abnormal heat which started back in June and reached surface air temperature deviations of +5° – +6 °C. In July, it was replaced by periods of a negative anomaly of −5 °C, and in August at all stations an increase in air temperature of +4 °C from the climatic norm for the period between 1981 and 2010 was observed. At the more eastern station Oymyakon, which is the center of the cold for the Northern Hemisphere, a positive stable anomaly of surface air temperature was observed only from early August to +5 °C.

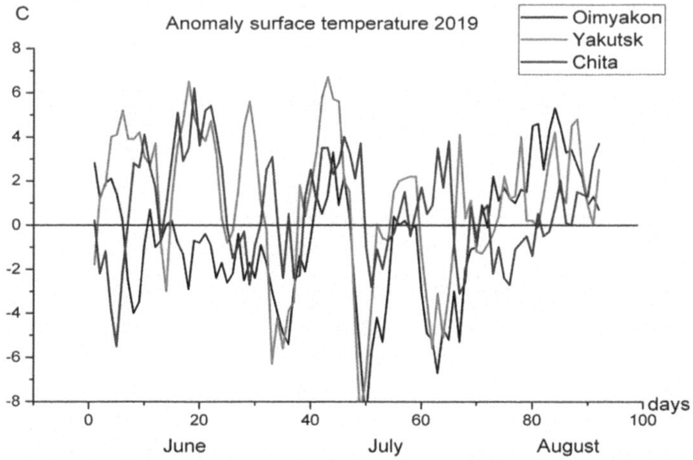

Fig. 6 Graph of surface air temperature anomalies for Oymyakon, Yakutsk and Chita stations for the period June–August 2019. according to Russian Hydrometeorological Center (https://meteoinfo.ru)

From the point of view of the ozone mechanism model (Krasouski et al. 2020), this situation of severe and extensive fires in Siberia in July–September may be due to the presence of high values of TO or SSW in the Southern Hemisphere during that time of year. Analysis shows that stratospheric ozone creation takes place just near stratopause; the high ozone concentration is transporting due to gravitational lowering and diffusion of ozone through the all stratosphere, and the height of the tropopause is lowered directly under high TO values. Namely, the presence of high TO values in the Southern hemisphere causes a local shift of the subtropical stationary front and an intratropical convergence zone to the north, resulting in a shift of Hadley's atmospheric tropical cell to the north of its normal position within the zone of the same longitude. As a result, in Northern hemisphere tropospheric processes, which are characteristic of equatorial and tropical latitudes, shifts north and from dry and hot weather conditions accompanied by strong convective heating of the Earth's surface, may eventually cause forest and peat fires in Siberia.

In fact, in May–June 2019 (Fig. 7 a, b) positive anomalies of TO (up to 360–370 DU) began to form more often in the Southern Ocean area. Already from mid-July to the end of August (Fig. 7c, d), a significant increase inTO (from 370 to 500 DU) was observed and stratospheric warming formed, which led to the displacement of the Antarctic polar vortex. All these events of increased TO and warming in the stratosphere led to the shift of atmospheric circulation cells north along the longitudes 80°–150°, which can be observed by the uncharacteristic high height of tropopause (hPa 190–200) in southern Siberia (Fig. 7). It is important to note that the high TO values in the temperate latitudes of the southern hemisphere correspond to the low TO values in the temperate latitudes of the northern hemisphere at one longitude and vice versa, depending on the season.

All positive surface temperature anomalies (over 6 °C) at the polar stations Oymyakon, Yakutsk and Chita of the Northern Hemisphere from the end of May to the beginning of September coincide with an increase inTO (up to 400 DU) in the opposite region of the Southern Hemisphere—the southern Indian Ocean and Southern Ocean (Fig. 7). But the most significant temperature increase at all stations in mid-August can be attributed to a significant increase inTO (up to 500 DU) in the Southern Hemisphere at around 90°–120° East longitude.

6 SSW and Forest Fires in Siberia, 2020

By mid-April 2020, more than 800 thousand hectares of forest were burning in Siberia and the Far East. Forests burned not only in Eastern Siberia, but also partially in the Western area of Siberia as well. Abnormal dry and hot weather in the Asian part of Russia has been established since mid-April, according to the Russian Hydrometeorological Center (https://meteoinfo.ru/novosti/17093-letnyaya-zhara-v-sibiri). At the same time, March was the second warmest month in the history of observation on a global scale (https://www.ncei.noaa.gov/news/global-climate-202003). According to the Copernican Atmosphere Monitoring Service [CAMS]

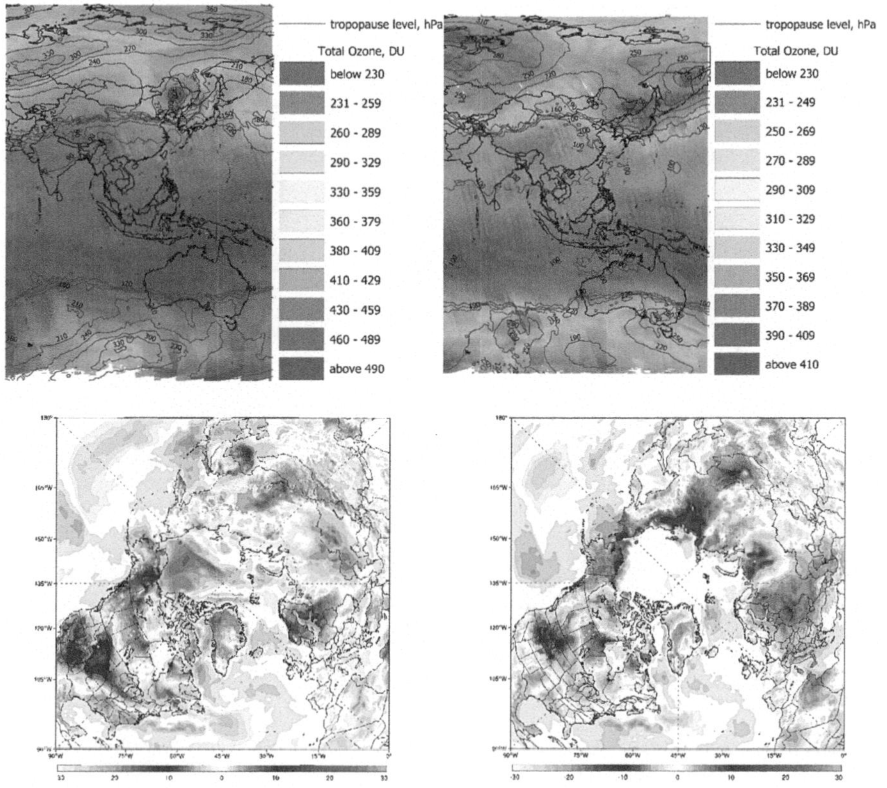

A B

Fig. 7 TO (DU, color fill, according to AIRS data) and tropopause height (hPa, contours according to AIRS, https://airs.jpl.nasa.gov/), top, surface temperature anomalies (°C, according to ERA5, https://cds.climate.copernicus.eu/), bottom, for the Arctic: **a** May 21, 2019, **b** June 21, 2019, **c** July 12, 2019, **d** August 31, 2019

(https://atmosphere.copernicus.eu/cams-tracks-record-breaking-arctic-ozone-hole), a record depletion of the ozone was observed over the Arctic, which is associated with a change in the Arctic polar vortex and very low temperatures. The last time the same strong depletion of the ozone layer was observed in the Arctic was in spring 2011 (https://public.wmo.int/en/media/news/arctic-ozone-depletion-reached-record-level).

Similar events such as those that occurred in 2019 may be associated with atmospheric processes in the Southern Hemisphere, namely in the Southern Ocean near Australia. Figure 8, according to the data of the AIRS satellite, shows how high TO (360 DU) values in the Southern Hemisphere and, as a result, an increase of the tropopause (up to 190–200 hPa) in the high latitude zone of the Northern

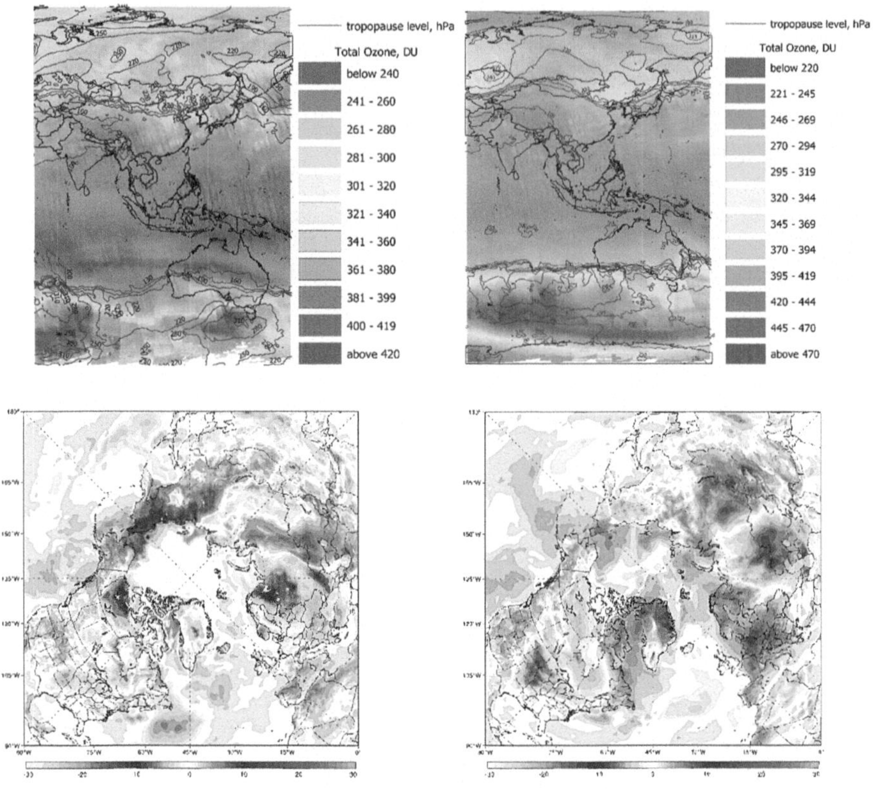

C D

Fig. 7 (continued)

Hemisphere led to significant surface air temperature anomalies (10–15 °C) in Siberia and the Far East on April 28 and May 5, 2020 (Fig. 8a, b).

7 Summary and Discussion

As a result, it is possible to conclude that the two hemispheres have a certain link in the distribution of TO. The maxima (minima) of TO in the Southern Hemisphere are characterized by the minima (maxima) of TO in the Northern Hemisphere at a similar longitude. An increase inTO leads to the lowering of the tropopause locally (in height) and to rise globally (in longitude). Such changes in the height of the tropopause disrupts the structure of global altitudinal frontal zones. The offset of the polar and subtropical frontal jets stream moves atmospheric global cells with a high tropopause height to areas where it is usually lower, which ultimately leads to

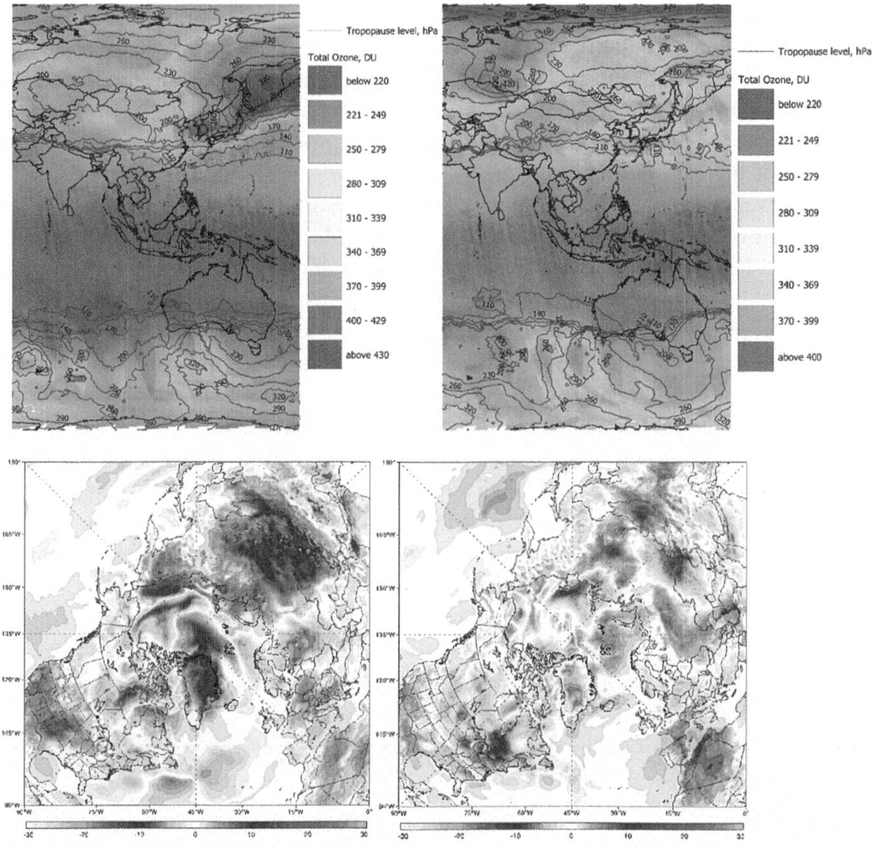

A B

Fig. 8 TO (DU, color fill, according to AIRS) and tropopause height (hPa, black contours, according to AIRS, https://airs.jpl.nasa.gov/), top, surface temperature anomalies (°C, according to ERA5, https://cds.climate.copernicus.eu/), bottom, for the Arctic: **a** April 28, 2020, **b** May 5, 2020

abnormal meteorological phenomena, such as droughts, thunderstorms, dust storms, etc. Opposite behavior of atmospheric events of the two hemispheres were also distinguished by Dzerdzeyevskiy (1967). Similar findings were reported by Morozova et al. (2019), where in almost 69% of cases processes in the Northern and Southern hemispheres were interrelated. That is, similar processes (blocking) develop simultaneously (or almost simultaneously) at the same longitudes—taking into account the differences in the seasons.

The fact that there is a symmetry between the hemispheres in the distribution of TO in the lower stratosphere and atmospheric formations in the troposphere is not surprising. In addition, in high latitudes of the Northern and Southern Hemispheres, feedback is observed in antiphase in fluctuations in the temperature of the oceans

(Scafetta 2012). Berger (1988) notes that, according to Milankovich, the distribution of insolation is equally important in the summer hemispheres; respectively, there will be heating of the surface, expansion of the atmosphere and the rise of the tropopause. In the winter hemisphere, where insolation is low, the tropopause lowers. These are natural climatic and atmospheric processes; fluctuations under the influence of the Sun (Chizhevsky 1976) (Fig. 9).

The presence of high temperature gradients of various sizes and signs leads to the division of atmospheric layers into (Zdunkowski and Bott 2004) "stable" and "unstable". The troposphere and mesosphere are unstable layers (the troposphere is a conditionally unstable layer, but the mesosphere is an absolute unstable layer). Between them there is an absolute stable layer—the stratosphere, which is limited by inversion zones—the tropopause and stratopause. Thus, the tropopause and stratopause are fundamentally different in their importance in the formation of solar-terrestrial connections. While the tropopause is an "inversion" and is driven by competing "up-down" processes (Tsyganenko and Sitnov 2005; Snyder et al. 1963; Tassev et al. 2003; Verronen et al. 2011), the stratopause demonstrates complete instability—any process of adding energy in this layer leads to a local uplift of the entire mesosphere layer upwards (Sinnhuber et al. 2012; Sisir Kumar Mitra 1952; Manney and Coy 2014; Zvyagintsev et al. 2004).

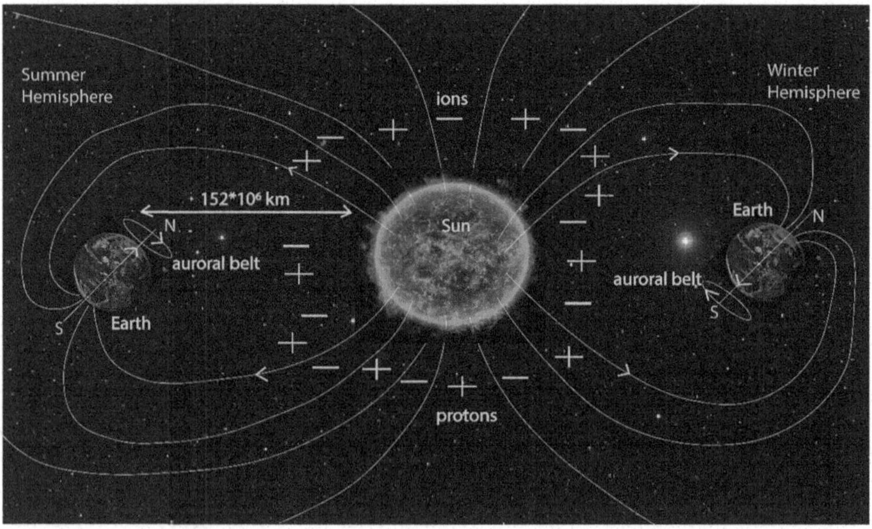

Fig. 9 Sun-earth interactions in summer and winter hemisphere [authors picture in accordance with Berger (1988) and Chizhevsky (1976)]

8 Conclusion

As a result, it can be argued that the two processes affect the changes in the tropopause's height and maintenance content, and accordingly, the fires that arise due to the movement of high-altitude frontal zones and atmospheric cells. The first is a bottom-up process, where the height of the tropopause rises due to vertical convective flows and the rise of ozone-depleting substances (H_2O). It is characteristic of the summer hemisphere. And the second top-down process, when solar protons act on the ionosphere and ozone formation in the stratopause region, and its subsequent gravitational deposition in a stable layer into the lower stratosphere. This process is characteristic of the winter hemisphere. These are the ways the Sun controls the Earth's climate.

The most important result of the presented cases is identifying unusual behavior of SSW in the Southern Hemisphere. In recent years the SSWs have been observed in the Southern Hemisphere, which appear very rarely there and appeared relatively recently, from the beginning of the 21st century (Sunkara et al. 2017; Yamazaki et al. 2020). The appearance of the SSW over the ocean surface is especially unexpected, it has never been observed before either in the Northern or Southern Hemispheres. There is no physical basis for an SSW to be created over the ocean. Usually, such phenomena form over land, as presented above (Sect. 4).

Nevertheless, the SSWs and positive TO anomalies were formed over the Southern Ocean near Australia in 2019 and 2020. In the Northern Hemisphere, on the same meridians in the corresponding regions of Eurasia, forest fires have been observed in areas close to the Arctic Circle, where they had never been before. As a result of this shift of atmospheric circulation cells in the meridional direction, hurricanes, rainstorms, landslides and river floods can form in tropical regions. As mentioned above, behind a perfectly stable layer of the stratosphere is a completely unstable layer of the mesosphere, which, moreover, is a charged electrical layer. In the stratopause region, the formation of ozone can also have an electrical nature and is determined by the height of the ionization layer belt. If the mesosphere layer in the stratopause region is heated, this will lead to an uplift of the positive ionization layer, a decrease in the amount of NOx in the stratopause region, and a local increase in the concentration of stratospheric ozone (Bruchkouski et al. 2018). Then, going down, due to gravity the stratospheric ozone contributes to the local lowering of the tropopause and the shift of high-altitude stationary atmospheric fronts (Krasouski et al. 2020). This suggests that the formation of wildfires in the Southern Hemisphere in the Australian region is artificial.

Acknowledgements The authors gratefully acknowledge the support of the ECMWF ERA-5 reanalysis data; the NASA's Aqua satellite for support AIRS data, the Australian Bureau of Meteorology and Russian Hydrometcenter. We thank Yaroslav Mitskevich for providing Figs. 2–4 and creating program TrFZ (Tropopause frontal zones), and anonymous reviewers.

References

Berger A (1988) Milankovitch theory and climate. Rev Geophys 26(4):624–657. https://doi.org/10.1029/RG026i004p00624

Bowman DMJS et al (2017) Human exposure and sensitivity to globally extreme wildfire events. Nat Ecol Evol 1(3):58. https://doi.org/10.1038/s41559-016-0058

Bruchkouski I, Krasouski A, Turishev L et al (2018) Aerosol profile measurements in the coastal zone of Antarctica: instrumentation and preliminary results, vol. 10786. In: Proceedings of SPIE, pp 107860 M-1–107860 M-9

Chizhevsky AL (1976) The terrestrial echo of solar storms. Moscow, 366 p

Dias HF, Bradley RS (eds) (2004) The hadley circulation: present, past and future. Adv Global Change Res 21. Springer, Dordrecht. https://doi.org/10.1007/978-1-4020-2944-8

Domeisen DIV et al (2019a) The role of the stratosphere in subseasonal to seasonal prediction: 1. Predictability of the stratosphere. J Geophys Res Atmos 125(2). Available at: https://agupubs.onlinelibrary.wiley.com/pb-assets/journal-banners/21698996-1586276389343.jpg. Accessed 10 Aug 2020

Domeisen DIV, Butler AH Charlton-Perez AJ et al (2019b) The role of the stratosphere in subseasonal to seasonal prediction: 2. Predictability arising from stratosphere-troposphere coupling. J Geophys Res Atmos 125(2). https://doi.org/10.1029/2019JD030923

Dzerdzeyevskiy BL (1967) Comparison of the main patterns of atmospheric circulation over the Northern and Southern hemispheres. Newslett Soviet Antarct Expedition 65:58–68

Environmental Modeling Center (2003) The GFS atmospheric model. NCEP office note 442, U. S. Department of Commerce, 14 p

Flannigan M, Amiro BD, Logan KA et al (2006) Forest fires and climate change in 21st centure. Mitig Adapt Strat Glob Change 11(4):847–859

Flannigan M, Cantin A, de Groot W et al (2013) Global wildland fire season severity in the 21st century. For Ecol Manag 294:54–61

Garfinkel CI, Feldstein SB, Waugh DW et al (2012) Observed connection between stratospheric sudden warming sand the Madden-Julian Oscillation. Geophys Res Lett 39:L18807. https://doi.org/10.1029/2012GL053144

Global Wildfire Information System (n.d.) Available at: https://gwis.jrc.ec.europa.eu/. Accessed 10 Aug 2020

Gorbatenko VP, Volkova MA, Nosyreva OV, Zhuravlev GG, Kuzhevskaia IV (2020) Influence of climatic conditions on Western Siberian Forest. In: Fires, predicting, monitoring, and assessing forest fire dangers and risks, chapter 12, pp. 269–293. https://doi.org/10.4018/978-1-7998-1867-0.ch012

Greenpeace (2019) Massive forest fires in Siberia is a climate emergency. Available at: https://www.greenpeace.org/international/press-release/23660/massive-forest-fires-in-siberia-is-a-climate-emergency. Accessed 10 Aug 2020

Hegglin MI, Fahey DW, McFarland M et al (2015) Twenty questions and answers about the ozone layer: 2014 update, scientific assessment of ozone depletion: 2014. World Meteorological Organization, Geneva, Switzerland, 84 p

Hersbach H et al (2019) Global reanalysis: goodbye ERA-Interim, hello ERA5. ECMWF Newslett Meteorol 59:17–24. https://doi.org/10.21957/vf291hehd7. Available at: https://www.ecmwf.int/node/19027. Accessed 10 Aug 2020

Hewson T (1998) Objective fronts. Meteorol Appl 5:37–65. https://doi.org/10.1017/S1350482798000553

Holton JR, Tan Y-C (1980) Influence of the equatorial quasi-biennial oscillation on the global circulation at 50 mb. J Atmos Sci 37:2200–2208

Hoskins BJ, Mcintyre ME, Robertson ALW (1985) On the use and significance of isentropic potential vorticity maps. Quart. J. R. Meteor. Soc. 111:877–946

IPCC (2013) Climate change 2013: the physical science basis. contribution of working group I to the fifth assessment report of the intergovernmental panel on climate change. Cambridge University Press, Cambridge, United Kingdom and New York, NY, USA, 1535

IPCC (2014) Climate change 2014: synthesis report. contribution of working groups I, II and III to the fifth assessment report of the intergovernmental panel on climate change [Core writing team, Pachauri RK, Meyer LA (eds)]. IPCC, Geneva, Switzerland, 151 p

IPCC/TEAP (2005) Special report safeguarding the ozone layer and the global climate system: issues related to hydrofluorocarbons and perfluorocarbons. In: Metz B, Kuijpers L, Solomon S, Andersen SO, Davidson O, Pons J, de Jager D, Kestin T, Manning M, Meyer L (eds). Cambridge University Press, UK, pp 478

Jolly W, Cochrane MA, Freeborn PH et al (2015) Climate-induced variations in global wildfire danger from 1979 to 2013. Nat Commun 6(1):7537. https://doi.org/10.1038/ncomms8537

Karoly DJ (2003) Ozone and climate change. Science 302(643):236–237. https://doi.org/10.1126/science.1090851

Krasouski A et al (2014) Observational and modelling studies of the short-term climate influences of the ozone mechanism. In: SPARC 2014 general assembly, Queenstown, New Zealand

Krasouski A et al (2020) Ozone layer holes, regional climate change and possible ways for their forecasting. In: Leal Filho W, Nagy G, Borga M, Chavez P, Magnuszewsk A (eds) Climate change, hazards and adaptation options: handling the impacts of a changing climate. Springer International Publishing, Switzerland, pp 311–331

Kunz A, Konopka P, Müller R, Pan LL (2011) Dynamical tropopause based on isentropic potential vorticity gradients. J Geophys Res 116:D01110. https://doi.org/10.1029/2010JD014343.331

Lannom KO, Tinkham WT, Smith AMS et al (2014) Defining extreme wildland fires using geospatial and ancillary metrics. Int J Wildl Fire 23:322–337

Lim E-P Hendon HH, Thompson DWJ (2018) On the seasonal evolution and impacts of stratosphere-troposphere coupling in the Southern Hemisphere. J Geophys Res Atmos 123. https://doi.org/10.1029/2018JD029321

Lim E-P, Hendon HH, Boschat G et al (2019) Australian hot and dry extremes induced by weakenings of the stratospheric polar vortex. Nat Geosci 12:896–901. https://doi.org/10.1038/s41561-019-0456-x

Manney G, Coy L (2014) Satellite observations of extreme events in the polar middle atmosphere. In: SPARC 2014 general assembly, Queenstown, New Zealand

McRae R, Sharples J (2011) A conceptual framework for assessing the risk posed by extreme bushfires. Aust J Emerg Manag 26:47–53

Min S-K, Cai W, Whetton P (2013) Influence of climate variability on seasonal extremes over Australia. J Geophys Res Atmos 118:643–654

Mitra SK (1952) The upper atmosphere. Asiatic Soc 713 p

Mitskevich Y, Partasenok I (2014) Simulation of high-altitude frontal zones in the tropopause and their role in the formation of the synoptic situation and climatic variability in the Northern Hemisphere and Europe region. In: 21st century challenges in regional climate modelling, Lund, Sweden, 16–19 June, pp 257–258

Morozova S, Polyanskaya E, Kononova N et al (2019) Differences in the formation of blocking processes between the Northern and Southern Hemispheres. IOP Conf Ser Earth Environ Sci 386: https://doi.org/10.1088/1755-1315/386/1/012005

Pedatella NM, Raeder K, Anderson JL, Liu H-L (2014) Ensemble data assimilation in the whole atmosphere community climate model. J Geophys Res 119:9793–9809. https://doi.org/10.1002/2014JD021776

Pedatella NM, Liu H-L, Marsh DR et al (2018) Analysis and hindcast experiments of the 2009 sudden stratospheric warming in WACCMX + DART. J Geophys Res Space Phys 123:3131–3153. https://doi.org/10.1002/2017JA025107

Perkins SE, Argüeso D, White CJ (2015) Relationships between climate variability, soil moisture, and Australian heatwaves. J Geophys Res Atmos 120:8144–8164

Ponomarev E, Kharuk V, Ranson KJ (2016) Wildfires dynamics in Siberian Larch forests. Forests 7:125. https://doi.org/10.3390/f7060125

Robinne F et al (2018) International Union of Forest Research Organizations, 2018. Global fire challenges in a warming world. In: Robinne F-N, Burns J, Kant P, de Groot B, Flannigan MD, Kleine M, Wotton DM (eds) Occasional paper no. 32. IUFRO, Vienna

Sampson RN, Atkinson RD, Lewis JW (2000) Mapping wildfire hazards and risks. CRC Press, Boca Raton

Scafetta N (2012) A shared frequency set between the historical mid-latitude aurora records and the global surface temperature. J Atmos Sol Ter Phys 74:145–163

Sinnhuber M, Nieder H, Wieters N (2012) Energetic particle precipitation and the chemistry of the mesosphere/lower thermosphere. Surv Geophys 33:1281–1334. https://doi.org/10.1007/s10712-012-9201-3. Accessed 26 Aug 2020

Snyder CW, Neugebauer M, Rao UR (1963) The solar wind velocity and its correlation with cosmic-ray variations and with solar and geomagnetic activity. J Geophys Res 68:6361–6370

Sunkara E et al (2017) Do minor sudden stratospheric warmings in the Southern Hemisphere (SH) impact coupling between stratosphere and mesosphere–lower thermosphere (MLT) like major warmings? Earth Planets Space 69:119. https://doi.org/10.1186/s40623-017-0704-5

Tassev Y, Velinov P, Mateev L, Tomova D (2003) A comparison between effects of solar proton events and of geomagnetic storms on the ozone profiles. Adv Space Res 31(9):2163–2168. https://doi.org/10.1016/S0273-1177(03)00126-1

Tedim F, Leone V, Amraoui M et al (2018) Defining extreme wildfire events: difficulties, challenges, and impacts. Fire 1(9). https://doi.org/10.3390/fire1010009

Tsyganenko NA Sitnov MI (2005) Modeling the dynamics of the inner magnetosphere during strong geomagnetic storms. J Geophys Res 110. https://doi.org/10.1029/2004JA010798

Verronen PT, Santee ML, Manney GL et al (2011) Nitric acid enhancements in the mesosphere during the January 2005 and December 2006 solar proton events. J Geophys Res 116. https://doi.org/10.1029/2011jd016075

Virgilio G, Evans J, Blake S et al (2019) Climate change increases the potential for extreme wildfires. Geophys Res Lett 46. https://doi.org/10.1029/2019GL083699

Weber RO, Dold JW (2006) Linking landscape fires and local meteorology—a short review. JSME Int J Ser B 49(3):590–593

WMO (1992) World Meteorological Organization, international meteorological vocabulary. WMO/OMM/BMO, no. 182

Yamazaki Y, Matthias V, Miyoshi Y et al (2020) September 2019 Antarctic sudden stratospheric warming: quasi-6-day wave burst and ionospheric effects. Geophys Res Lett 47: e2019GL086577. https://doi.org/10.1029/2019GL086577)

Zdunkowski W, Bott A (2004) Thermodynamics of the atmosphere. A course in theoretical meteorology. Cambridge University Press, New York, 252 p

Zvyagintsev AM, Kruchenitsky GM, Chernikov AA (2004) Long-term ozone changes in the troposphere. In: Proceedings of XX quadrennial ozone symposium. Kos, Greece, pp 925–926

Aliaksandr Krasouski PhD is an associate professor of Faculty of Geography and Geoinformatics, Belarusian State University. He was a founder and first director of NOMREC BSU. Main research area(s): Total and surface ozone measurement, UV investigation, ozone-climate change investigations.

Siarhei Zenchanka PhD Minsk Branch of Plekhanov Russian University of Economics, the Department of management, accounting and finance, professor. Main research area(s): Environmental management, Sustainable Development, Education for Sustainable Development.

Tsimafei Schlender PhD student of Faculty of Geography and Geoinformatics, researcher NOMREC, Belarusian State University. Main research area(s): Total and surface ozone measurement.

Wildfires in Paraguay: Environmental and Human Impacts

Genaro Coronel, Max Pastén, Norman Breuer, Aramí Celeste, Larissa Rejalaga, Federico Monte Domecq, and Gustavo J. Nagy

1 Introduction

Climate change already observed in Paraguay, drought events and biomass burning in the Amazon region increase the risks of forest fires in the region (de Oliveira Alves et al 2015). The greenhouse gas (GHG) emissions into the atmosphere alter the climate variables precipitation and temperature that will likely cause more severe droughts; the changes produce conditions that may increase the likelihood of wildfires. In turn, increased forest fires produced release large amounts of GHG into the atmosphere, which accelerates global warming (Nobre et al. 2016). The fires cause large-scale forest mortality in drier places like the Chaco (Brando et al. 2014) and also expand to humid areas as climate and land-use change (Le Page et al. 2017). The geographic distribution of air pollution during the periods of forest fires was analysed with data from the Ozone Monitoring Instrument (OMI/NASA) and the HYSPLIT model from the National Oceanic and Atmosphere Agency (NOAA). High average concentrations of toxic nitrogen dioxide (NO_2) were found over various regions of Paraguay, mainly in August, September and October (Recalde and Coronel 2019).

G. Coronel (✉) · M. Pastén · A. Celeste · G. J. Nagy
Master Degree Programme: Global Change and Climate Risks, Facultad Politécnica (FPUNA), Universidad Nacional de Asunción, San Lorenzo, Paraguay

N. Breuer
CTA – FCyT - Universidad Católica "Nuestra Señora de La Asunción", Asunción, Paraguay

L. Rejalaga
Facultad de Ciencias Agrarias, Universidad Nacional de Asunción (UNA)-Paraguay, San Lorenzo, Paraguay

F. M. Domecq
Asociación Paraguaya de Recursos Hídricos, Asunción, Paraguay

G. J. Nagy
Facultad de Ciencias, Universidad de La Republica (UdelaR), Montevideo, Uruguay

W. Leal Filho et al. (eds.), *Sustainability in Natural Resources Management and Land Planning*, World Sustainability Series,
https://doi.org/10.1007/978-3-030-76624-5_25

A better understanding of the fire regimes in the country could help direct efforts to increase the capacity in the region to prevent accidental forest fires and their negative impacts on ecosystems, socio-economic well-being, biodiversity and health (Nepstad et al. 2008). A system of delicate and deep interconnections between the intense evaporation of the warm waters of the northern tropical Atlantic Ocean (ZCAS), the trade winds, the Amazon forests and the Andes mountain range, cause rainfall to reach the la Plata Basin. However, due to deforestation in dry months, forests would cease to perform their functions as a climate regulator (Marengo et al. 2007). The forest fire season in the Amazon rainforest in 2019 saw an increase in fires in the Amazon rainforest and the Amazon biome within Brazil, Bolivia, Paraguay and Peru during that year's Amazon tropical dry season (Wikipedia 2019).

Fire control is a modelling element for all forest ecosystems on Earth, and the humans use as a cleaning agent to create open land that allows agriculture. In Paraguay, during the second half of the 20th century, this practice was encouraged by an agrarian statute that sought to expand the agricultural area and the non-productive occupation of forest lands. This practice is currently prohibited by law.

Fire is indirectly responsible for desertification processes that are causing social, economic and political disorders in the planet. The evolution of forest fires depends mainly on how they originate, which can be caused by the action of atmospheric phenomena, or accidental or intentional anthropogenic origin. Due to the damaging effects of fire, preventing them is a priority, so more regional and national investigations should be conducted.

Paraguay is a landlocked country located in the Rio de la Plata basin, in the centre of South America. It has an area of 406,752 km^2, between parallels $-19°18'$ and $-27°03'$ latitude, and meridians $-54°15'$ and $-62°38'$ longitudes. Forest fires are considered a risk in Paraguay, are cyclical and recurrent, increasing in intensity from July to October annually (Fig. 1). The environmental vulnerability to these fires is related to the existing intrinsic predisposition of the climate and climate scenarios in the country, which are reflected in human losses or socio-economic damage. In recent times, there has been a more significant increase in the activity of forest fires globally and the countries of the region (Moreno et al. 2020), the number of outbreaks of fires, duration, extent and severity has increased. Ecosystems vulnerable to fire are increasing, and intense droughts prolong devastating fires, the Paraguayan Chaco is going through one of the worst droughts in 50 years (La Nación newspaper 2020).

The most vulnerable by air pollution caused by forest fires are children, the elderly and pregnant women in urban and rural sectors far from the sources of fire, observed by the HYSPLIT model (Stein et al. 2015), developed by the Air Resources Laboratory of NOAA (Crinó et al. 2014). During droughts, the fire season lengthens and the number of days with fire danger increases in the agricultural-livestock production areas. Better knowledge about fires will increase resilience in the regions allowing the population and resources to be better protected. The fires that occurred in the Paraguayan Chaco from August to September 2019 drew attention to its possible contamination effects throughout the country. The General Directorate of Health Surveillance activated the epidemiological alert

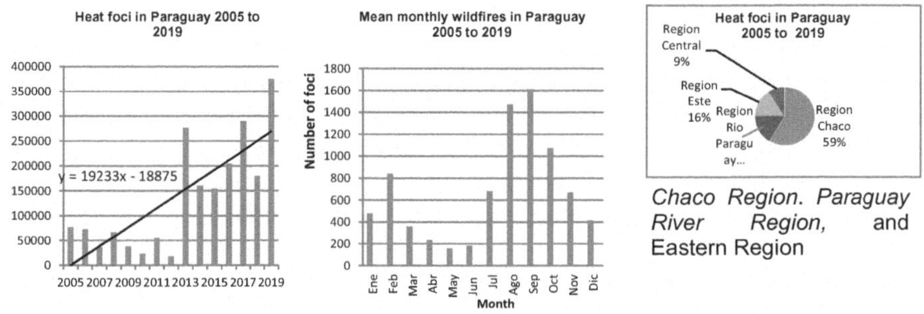

Fig. 1 The graphs show wildfire outbreak trends, months with the most fires, and the regions with the most fires, during the years 2005 to 2019. *Source* Own elaboration with data from the INPE

of risks and damage to health as a result of forest fires registered at the country level (La Nación newspaper 2019). Forest fires are among the main environmental and development problems, mainly due to their effects on health and climate change (Vázquez et al. 2007)

Climate change and forest fires in recent times have gained global interest, going from impacts to adaptation, which leads to new challenges to education and research. The purpose of this work is to present the geographic and temporal evolution of forest fires in the country and to create awareness for adaptation in the national and regional community from an interdisciplinary point of view. "Forest Fires in Paraguay: Environmental and Human Impacts" is a collaborative effort of an academic group from across the spectrum of natural sciences focused on improving the vision of forest fires for education and resilience. Exposure to air pollution is a significant cause of hospital admissions due to respiratory diseases (Machin et al. 2019). This interdisciplinary study on forest fires, climate threats, environment, the health of populations and sustainability that alters the well-being of the population, synthesises knowledge and suggests possible solutions, including stakeholder commitment. For the success of this work, national and civil organisations must play an essential role in controlling forest fires, such as putting more budgetary pressure on national governments, taking care of and managing donations appropriately. Promote the growing global consensus that the problem of forest fires be included in budget planning and proactive preparation for vulnerability reduction. The success of these wishes will depend on local community mobilisation and the connection between governments and citizens, especially in alliance with higher education institutions to influence.

One desired outcome of this study is for national and civil organisations to play a more prominent role in reducing the risk of disasters from forest fires. Several factors motivate the participation of civil society and other stakeholders, one factor is increased budgetary pressure on governments, partly due to a change in donor funding that predisposes them to share the cost of risks. Another factor is the growing global consensus that the problem of forest fires must include planning and proactive preparation and vulnerability reduction. Developing roles include community and local mobilisation and the connection between governments and citizens, especially in alliance with higher education institutions in order to influence national government policy. The knowledge and results generated from the investigations in the universities should reach the institutions in charge of risk management and those responsible for policy formulation. This knowledge can also help introduce innovative local governance practices, in particular by trying to link research, adaptation to climate change and the provision of development services.

2 Methods, Objectives, and Organisation

This chapter is based on a non-systematic literature review of the understanding of vulnerabilities in critical sectors related to forest fires.

For a description of the 2019 wildfire season in Paraguay, up-to-date information was found on the newspapers.

The temporal evolution of hotspots is investigated using data available in "Banco de Datos de Queimadas" (Forest fires database) from the Brazilian Space Research Institute (INPE).

The study responds to the need to provide information and data in a synthesised manner to the public and to government officials to develop adaptation actions and plans, prioritising these actions according to national and regional needs.

The chapter aims to provide an integrated vision of forest fires in the Western (Chaco) and Eastern (Humid) regions of Paraguay, to provide information on the climatic situations to which the population could be exposed and to what degree we can expect changes in future climate scenarios.

The chapter is organised as follows: Introduction, Methods objectives and organisation. Characterisation of the most fire region its geography and economic and productive aspects, Ecosystems function and service in the region, climate and scenarios, and trends in Paraguay. The vulnerability of the agriculture, livestock sector to wildfire. Region resilience, Summary: priority planning and preparation actions for adaptation to reduce vulnerability to wildfires in Paraguay. Conclusions. References.

3 Characterisation of the Region Where Most Wildfires Occur, Its Geography, and Economic and Productive Aspects

Wildfire outbreak data is from the Brazilian Institute for Space Research (INPE/Burn database) that processing more than 200 images per day to detect vegetation fires. Figure 1 records the fires in the regions of Paraguay, the months with the most outbreaks and trends (INPE 2020), Fig. 1.

Fig. 2 The regions of Paraguay; Western region (Chaco) and the humid Eastern region

During the winter of 2019, a large number of heat sources were detected in the country that alarmed the citizens. In Paraguay, there are two very different scenarios for forest fires, which differ from each other geographically, environmentally and socioeconomically (Fig. 2).

The Paraguayan Chaco that includes the Departments of Alto Paraguay, Boquerón and Presidente Hayes, some 59% of the focuses of fires of the years 2005-2019 were concentrated in the Western region of the country. The Paraguayan Pantanal is located in this region and reported the first sources of fire in the week of August 14, 2019. These ecosystems have cyclical fires and, despite the vast affected area, the damage is expected to be less than in other ecosystems due to their resistance to fire. In contrast, the other region has fewer ecoregions, such as the Chaco Cerrado that were severely affected as of mid-August. Fire outbreaks in the northern part of the country were aggravated by adverse weather conditions (high temperatures, winds of more than 20 km/h and low humidity); The National Emergency Secretariat (Secretaria de Emergencia Nacional 2020) reported that about 325,000 ha were affected by the fires. Northern Chaco and the rest of the territory showed patterns associated with heat sources and the change in land use.

Additionally, due to strong winds from the north, fires that originated across the border in Bolivia entered Paraguayan territory, affecting private properties whose owners mobilised to combat them. This region has characteristics associated with extreme climatic events such as drought, high temperatures, and wind that make wildfires very difficult to combat and extinguish them due to their large extent. This Chaco fires were a major environmental catastrophe, and they called attention to the need for a detailed assessment and the adoption of measures that drive restoration processes.

In these ecosystems the fires are cyclical, and despite the vast affected area, the damage is expected to be less than in other ecosystems due to their resistance to fire. Forest fires usually have several origins and can be caused by natural or spontaneous combustion, lightning, or generated by traditional practices of burning fields. In economic terms, these claims of intentional ignition are more relevant because of their consequences for biodiversity and productive systems. Wildfires that occur in the forests, which affects several areas of the region, have severe consequences on natural resources, as well as for the livestock sector directly and for a prolonged period.

In the Eastern Region of Paraguay, wildfire outbreaks generally affect livestock grazing and extensive agricultural areas, in Bella Vista Norte of the Amambay department there are sugar cane plantations, which regularly burn both intentionally and no intentionally. In both Eastern and Western Region, wildfires affect animal and human health due to smoke inhalation and poor air quality.

The calculation of environmental, monetary services is complicated and depends on the characteristics of each type of ecosystem, timber products and the market value of carbon sequestration, among others. In the case of forest fires, emissions from gases are a cost that local, national, regional, and the global community bear.

Forest and grassland fires had a substantial impact on the Chaco economy in 2019; forest fires not only cause damage to ecosystems and biodiversity but also have enormous social and economic impacts on the surrounding communities.

From January to August of 2019, some regions of South America were affected by increases in the number of forest fires; Brazil had 75,336; Venezuela follows with 26,491; Bolivia with 17,154: Peru with 5,681; finally Paraguay, with 9,861 wildfires.

In March–May of 2020, the total number of heat sources reached 652,917 active sources, some countries with more active sources detected with the VIIRS 375 m sensor are Venezuela with 278,266 sources, followed by Colombia 121,636, Brazil 78,548, Argentina 7,125 and in fifth place Paraguay 55,478. In Paraguay, it is not common to see so (Fig. 3) many active heat sources in March, the high peaks are observed from August to October. Such behaviour may be due to the existence of much combustible material, the performance of prescribed burns, climatic (dry), economic, social and cultural factors, among other factors (INPE 2020).

The first months of 2020 were anomalous; the President Hayes department suffered 21143 fire outbreaks, Boquerón 9057, Alto Paraguay 2961, Misiones 2482, Ñeembucú 5356, Paraguarí 2482 and Cordillera 1569 outbreaks.

The occurrence of the El Niño events (October to March) in the Paraguayan Chaco has an essential influence on wildfires (Pastén 2006) with decreased precipitation in general during La Niña events. During the cold ENSO phase, this, decrease in rains contributes to prolonged droughts such as those that occurred during 1999–2000 (Grassi et al. 2005). This water deficit, in turn, causes droughts and significant reductions in water in forests. Illegal burning of fields and crops complicates this situation and increases the risk of forest fires. High risk exists from the spread of forest fires in Brazil, Bolivia, Paraguay and Peru.

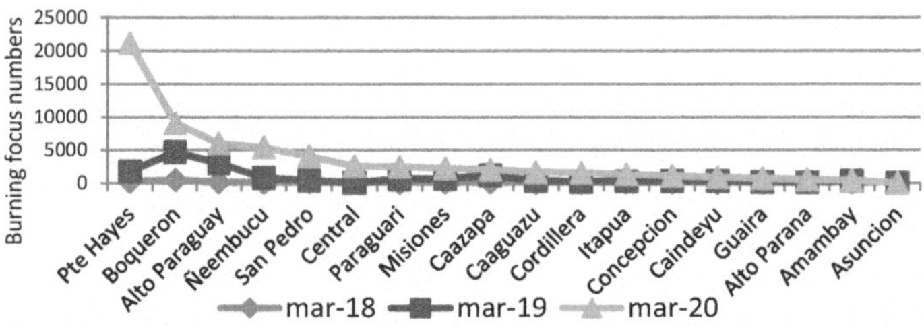

Fig. 3 The focus of burning in the regions of Paraguay, March 2018, 2019 and 2020 (NPP VIIRS 375 m INPE, 2020)

4 Ecosystems Function and Services in the Region

In urban ecosystems, rapid urbanisation turns cities into centres of environmental problems, and these urban areas are increasingly engines of innovation and socio-economic development. They must adapt to local and regional socio-environmental challenges; overpopulation makes urban areas increasingly key to sustainability. Balanced growth and development is part of the solution at local and regional scales. Therefore, it necessary to take into account the ecosystem services of cities and consider them as urban ecological systems (Leal Filho et al. 2020). Increasing knowledge would be the basis for improving practices in the management of urban ecosystems, and being able to respond to the growing need to develop urban areas and implement plans for their sustainability and resilience, which ultimately leads to the human well-being of the population.

The objective of studying urban ecosystems is to promote adoption and provide guidance for ecosystem approaches in planning and management of urban ecosystems, to improve the ecosystem services provided by urban ecosystems, and human well-being. It also aims to build a network of scientists, professionals, and policy and decision-makers to advance our ecological understanding of urban areas, to better plan and design urban ecosystems and to increase their sustainability and resilience.

Fire is a natural element for the functioning of numerous forest ecosystems; it is one of the natural elements that influence plant communities over time and plays an essential role in the health of particular ecosystems. The established dynamics between human action and fires and the higher frequency of the El Niño phenomenon have given rise to a situation in which fires are a significant threat to many forests and diversity (Nasi et al. 2001). Uncontrolled fires that originate without direct or indirect human intervention exert a significant influence on certain types of vegetation (Pritchett 1991). Forest fires can be unwarranted and uncontrolled in which large amounts of combustible material are burned and spread rapidly, often destroying everything in their path. The effects of fires are associated with the intensity, recurrence, and duration of a forest fire. The impacts of fires on a forest ecosystem involve alterations of the three main components: soil, water, vegetation and fauna,

Chile's wildfires affect the country on various fronts, generating three main types of problem: Environmental problems, Social problems, Economic problems (Ubeda and Sarricolea 2016) pointed out that fires cause soil degradation, mainly after the event, the level of degradation will depend on the topography of the site, the intensity of the erosive process, the regeneration rate of the post-fire vegetation cover, the recurrence, intensity and duration of the fire (Caon et al. 2014). Gonzalez (2017) pointed out that the main impacts occur at the level of the physical, chemical and biological, and soil productivity (Garcia-Orenes et al. 2017). Wildfires may produce several changes in the short and long term in the landscape and the soil system (Zavala et al. 2014). After fires, the nutrient content increases in the first ten centimetres of the surface, due to ash deposition, mineralisation of nutrients and formation of stable structures. This process was the basis of indigenous slash and burn agriculture and forest fires can modify the hydrological cycle at the level of the river basin. The climatic condition of the area is decisive on the probability of a fire occurring. Wildfire seasonality explains the fire regime, while the modification of the structure and composition of the vegetation due to forest fires can affect ecosystem services.

In forests, fire can have devastating effects on forest species of vertebrates and invertebrates, not only because it causes death, but also because it causes more lasting indirect effects such as stress and disappearance of habitats, territories, shelter and food. The disappearance of organisms of great importance for forest ecosystems, such as invertebrates, pollinators and decomposers, can significantly delay the recovery rate of the forest (Boer 1989).

Wildlife may be affected both negatively and positively. With the destruction of forests, many species of insects, mammals and birds that survive a fire lose their habitat and are forced to move. Fire plays a vital role in maintaining the health of particular ecosystems. However, in addition to climate change, human use of fire is now a threat to forests and their biodiversity; therefore, human action becomes vital (Meijaard et al. 2008). In Chaco fires almost always occur due to human intervention implies that the adverse effects of fires will be much higher than the positive effects. Despite the positive and negative effects of a fire and the discussion surrounding it, it can be generalised that fires and burns have impacts on different aspects of the environment such as air quality, wildlife, soil, and soil biota.

5 Climate and Scenarios, Extreme Events, and Climate Trends for Paraguay

One of the problems that arise at the global and local levels is the impacts of climate change on society, the economy and the environment; we use Global Climate Models to analyse this, evaluating changes in the temperature and precipitation at the regional and local level. The Regional Climate Models (R.C.M.s) reducing climate characteristics from global to regional and report in detail on a particular region. The high resolution of the R.C.M.s better simulates the current climate in its interaction with the terrain, more closely predicts future climate change, small area representation, better simulates and predicts extreme weather events, generates more detailed data for analysis and impact studies at the local and regional levels.

Climate scenarios for Paraguay were carried out with the Eta-20 km model for South America (Chou et al. 2014; Dereczynski et al. 2020), the results indicate that the country would be affected by climate change, with the most significant increases in temperature for the period 2031–2050 in the department of Amambay (+3.1 °C). The scenario shows that rainfall will decrease in the first three decades and then increase in the last decade of the period considered. Precipitation shows a decrease during the first three decades and in the 2041–2050 decade increases slightly but very close to its typical values for both scenarios. The highest increases in temperature from 2031 to 2050 are expected in the department of Amambay (+3.1 °C) (northeast of the Eastern Region). The departments of the country that expect an increase in temperature and decrease in precipitation may be more severely affected during La Niña years, because this phenomenon produces a decrease in rainfall in much of the country. The same can occur in those regions where temperature and precipitation are expected to increase since in El Niño more precipitation than average occurs in much of the country. The climate scenarios for the Chaco were carried out by Max Pasten with the Climate Scenarios Technical Team of "Centro de Conocimiento Del Gran Chaco Americano y Cono Sur" (SERIE CLIMA 2017).

The South America Low-level Jet (SALLJ) is a climatological feature with a critical role in the spatial and temporal distribution of precipitation in the region. Previous studies focused on the mechanisms and variability of the SALLJ in the central Andes (i.e., southeast Peru, Bolivia, and Paraguay), this study show that the SALLJ in the northern branch exceeds 10 ms^{-1} during September–February and the frequency can be as high as in the central region. When the SALLJ is active, wind speed is vital in the spread of fires along with rainfall. The frequency and intensity of the SALLJ in the northern Andes have substantially increased in the last 39 years (Jones 2019).

6 The Vulnerability of the Agriculture and Livestock Sector to Wildfire

The primary division that characterises agriculture in Paraguay is one between commercial and family agriculture. Wildfire from intentional and nonintentional burning of pastures may spread to sugar cane fields and damage other row crops, including limited-resource family farms. Of recent concern is the expansion of eucalyptus plantations for biomass production (Manrique et al. 2020), and silvopastoral system–which combine tree plantations with cattle grazing in Eastern Paraguay, and the new practice of forest grazing in the Chaco Region. Under the right conditions, these systems are susceptible to wildfire and may serve as a source affecting other systems.

The agricultural sector contributes 30.4% of Gross Domestic Product and 35% when agribusiness is included. It generates 40% of national exports and occupies nearly 40% of the country's workforce (World Bank 2014). Agriculture faces many types of production risk, many of which are agroclimatic in nature, and the most important of which is drought. The available literature does not mention vulnerability to wildfire as a significant risk in row crop agriculture, and livestock production, on the other hand, is susceptible to wildfire risk. With the rise of agroforestry and forest plantations on former livestock grasslands, wildfire risk to pastures that sustain beef production is likely to rise. However, it is essential to note that the sharp distinction between livestock and forest systems is being blurred by new practices (including forest grazing), and the reclassification of palm savannahs, in the Western Region of Paraguay, that are natural fire ecosystems, as forestlands.

In Paraguay, the bulk of agricultural production is concentrated in three crops, soy, maise and wheat. The gross value of the production of these three products represents about 76% of total agriculture in Paraguay, with soybeans alone accounting for 46%. Irrigated rice is also growing in importance at the commercial scale. Occupying a smaller area, but not less important from the socio-economic point of view for family farming, are cassava, beans, white corn, cotton, sesame, vegetables and fruit trees. Family farming mainly using family labour and is composed of production units under 50 ha in size in the Eastern Region and 500 ha in the Western or Chaco Region; family farmers represent 91% of all the farms. The geographical distribution of family farming is very heterogeneous; however, it is mainly concentrated in the departments of San Pedro, Caaguazú, Caazapá, Paraguarí, Guairá and Cordillera, in the Eastern Region of the country. In general, they coexist in the same areas as commercial holdings. The area cultivated by family farming has remained constant or decreased for some products, during the last decade (SENACSA 2019). Some crops on family farms, especially sugar cane, are affected by local wildfires in small 2–10 ha patches that may appear unimportant even in aggregate but can be devastating to a family farm with low capital and low technology.

Currently, more than 123,000 livestock producers exist with a stock of nearly 14 million head of cattle (Ministerio de Hacienda 2018). Today, livestock is a crucial sector of the Paraguayan economy, especially for its contribution to exports (beef), which have increased sharply, reaching 1.5 billion USD in 2018 (Ministerio de Agricultura y Ganadería 2013). Many traditional ranchers still burn their natural grasslands in the lower (humid Chaco) and the natural grasslands of the Oriental Region in the austral winter months. When drought, high temperatures and high winds create unfavourable conditions, they can spread over broad areas, and usually do so every year. Many ranchers burn only small fields after first rains and with favourable or no wind using controlled backfires. In the northern or dry Chaco, burning pastures is not customary. A law regulates wildfire use for this purpose, Law 4014/2010 that regulates this activity. However, the Ministerio Del Ambiente y Desarrollo Sostenible (MADES) has little human and financial resources to enforce it.

The expansion of soybeans was mainly at the expense of forest areas and pastures Currently it also expands on lands traditionally occupied by family agriculture, leased or sold by peasant farmers, mainly in the departments of Caaguazú, San Pedro and Caazapá (Fraisse et al. 2008). Commercial agriculture and livestock are growing in the area, whereas family farming is stagnant or decreasing. Risk evaluations focus on the value chains of soybeans, corn, wheat, rice, cassava, sesame, cotton, sugar cane and vegetables, in addition to livestock.

Agroclimatic risks, along with pests and diseases, affect grain and oilseed production in Paraguay. The main risk factor for soybeans is summer drought (mainly in January), whose impact is exacerbated by high temperatures and enhanced in producing areas with soils with less water retention capacity (more sandy soils in San Pedro and Canindeyú). Weeds and other pests and diseases systematically affect grain production in Paraguay. Their intensity varies depending on climate conditions and crop management. La Niña years usually exacerbate seasonal drought and are responsible for losses as high as 38% (World Bank 2014). In the case of second planting (zafriña), maise, early frosts, and late summer drought are the leading causes of yield variations. The occurrence of early frosts may threaten and limit the sowing of the second maise if the soybean harvest falls behind. Finally, excess rainfall in the harvest season of the second maise (June) is frequent at harvest time.

For livestock production, drought is a significant risk. Drought occurs in all regions of Paraguay, affecting the Western region or Chaco more drastically from June to September. In this period of the year, rainfall is scarce, and wildfires tend to occur. Droughts occur at apparent semi-cyclical intervals of 4–5 years, with extreme periods every ten years (Pezzoli et al. 2016). Production losses include reductions in pregnancy rate, 30%; rate birth rate, 25%; weight loss, 20%; and decrease in slaughter rate, 10%. Wildfires effects are immediate, as pasture burns along with forest reserves, and hay. Cattle often need to be physically transported to rental properties or sent to slaughter underweight.

In Paraguay, prolonged drought, heat and the burning of grasslands aided by strong winds, spread flames, giving rise to intense forest fires on reserves and commercial ranches. Since wetlands and streams are almost dry, wildlife must migrate or perish. With such a large sector of the economy facing production risk, it is imperative to assume the challenge of developing a management plan for these situations, to reduce their negative impacts (World Bank 2014; Food and Agricultural Organization 2017).

7 The Adaptive Capacity of the Region

Law N° 4014 (Fire Prevention and Control, 5/12/2009) establish suitable regulations to prevent and control rural, forest, vegetation and interface fires, in Paraguay. The law also seeks to reduce the uncontrolled burning of pastures, forests, scrublands. It forbids burning fallow, natural fields, sawdust or any other cereal, of legumes or type of flammable organic material that could generate any of the fires defined in this law. However, the law has not been regulated to date. Besides, the law enforcement fails due to lack of resources at the municipal, departmental, and Ministry level. The country's municipalities are the law enforcement application authority for wildfire in coordination with the Paraguayan Network of Prevention, Monitoring and Fire Control, which is a specialised unit that is coordinated by the National University de Asunción, together with public and private institutions related to the matter.

Forest ecosystem management: a need for the environment (Kimmins 1974), and must serve the needs and desires of the surrounding people (Lutz et al. 2001), Paraguay's current population of around seven million puts pressure on the country's forests. New paradigms must emerge as models for forest management, and forestry faces new social pressures, ideas, and concepts, as well as the ability to anticipate their long-term and medium-term social and environmental consequences. These forests and values will need to meet the current and future needs of both the local population and society (Franklin et al. 1997, 2002; Franklin 1997).

Forest ecosystems are tree-dominated landscape areas. They consist of biologically integrated communities of plants, animals, and microbes along with local soils and atmospheres with which they interact, insects, the disease may have recently altered forests, but continue as forests due to the biological and physical legacies of the previous forest. Forest ecosystem management is the management of forest ecosystem disorder processes and regimes to sustain desired ecosystem values and services. It is also the management of the human use of the forest and its interaction with it because men are part of forest ecosystems. An understanding of structure, function, and energy, nutrient, and information flows are paramount for this type of forest management.

The environmental impact of forest fires is not limited to biodiversity, the Eastern region and the Chaco, in particular, suffered dangerous wildfires in late austral winter and spring of 2019. The fires negatively impacted biodiversity may have increased desertification in the northwestern-most regions, and air and water quality negatively affected. Humans cause between 80 and 90% of fires. Biodiversity of the burned area changes its structure and composition. Wildfire negatively affects soil and water since both are related. The vegetable mantle disappears, and with it, the natural barrier that retains water. Wildfires emit greenhouse gases such as carbon dioxide into the atmosphere. To this must be added the economic costs derived from any forest fire. The ecosystem loses its appeal for both leisure and tourism activities. To all, this must be added the economic cost of regeneration. Land-use changes cause most forest fires. However, in some ecosystems such as cerrado and savannahs, wildfires play a vital role as they have a regulatory role in ecosystems and are crucial to biodiversity.

8 Summary, Planning and Preparation Actions for Adaptation to Reduce Vulnerability to Wildfires in Paraguay

Fire management is complex and involves many stakeholders; the ability to integrate external information is key to proactive planning and preparation for wildfire mitigation. Seasonal climate forecasts complement satellite data, and local populations, agricultural and livestock firms, insurance companies, businesses, health and transportation professionals, and others have an interest in fire management for adaptation. Institutional actors hold most of the decision-making power; they can incorporate climate forecasts and climate change scenarios into participatory planning and preparation processes that are central to current risk management approaches. They should also communicate monitoring, alerts, and adaption information to citizens ahead of time and in real-time. Stakeholders must evolve from being passive users of scientific information to being protagonists in a participatory and iterative consultation process for readiness. Interaction among public officials, scientists, and stakeholders requires stable and long-term funding and institutional incentives for preparation rather than reactive action. Interdisciplinary research agendas, long-term programs, sustained partnerships, and continuous outreach and training will be crucial to the adaptation to wildfire in a changing climate.

Some recommendations that may be of interest to local actors are the following.

- Consolidate a Valuation System for economic losses from forest fires in the entire region
- Involve and integrate the community of the Paraguayan Network for Fire Prevention and Control, under art. 4 of Law 4014/2010
- The application of fire prevention should not hinder economic development. A focus on positive effects on the environment is a better approach,
- The use of satellite monitoring techniques should increase, due to the climatic conditions of the Chaco, because of its low population density and difficult to access environment.

9 Conclusions

Paraguay has two great regions: Western region (Chaco) and the humid Eastern region divided by the river that bears his name, with three departments in Chaco and fifteen in the Eastern region. Burning in Paraguay, in the study period, showed a positive trend, given by the equation $y = 19233 \times -18875$.

The highest monthly averages in the study period are observed from July to October.

The largest area of fires occurs in El Chaco Humedo (Wet Chaco) (Western Region adjacent to the Paraguay River) where the vegetation is exuberant. Since 2005 the regions most affected by burns are (i) Chaco (59%), (ii) Eastern regions, and (iii) the adjacent areas to the Paraguay River (16% each). The region with the fewest fires is the Central department that surrounds Asuncion (Capital) with 9%.

During the first months of the year, the occurrence of fires is relatively low, except for the year 2020, in March the department of Presidente Hayes was exceptional where 21,143 outbreaks occurred, and in Boquerón with 9,057 sources of burning, as shown in Fig. 3.

References

Boer C (1989) Investigations of the steps needed to rehabilitate the areas of East Kalimantan seriously affected by fire: effects of the forest fires of 1982/83 in East Kalimantan towards wildlife. F.R. report no. 7. Deutsche Forest Service/ITTO/GTZ Samarinda, Indonesia

Brando PM, Balch JK, Nepstad DC, Morton DC, Putz FE, Coe MT, Silvério D, Macedo MN, Davidson EA, Nóbrega CC, Alencar A, Soares-Filho BS (2014) Abrupt increases in Amazonian tree mortality due to drought–fire interactions. Proc Natl Acad Sci USA 111:6347–6352

Caon L, Vallejo VR, Ritsema CJ, Geissen V (2014) Effects of wildfire on soil nutrients in Mediterranean ecosystems. Earth Sci Rev 139:47–58

Chou SC, Lyra A, Mourão C, Dereczynski C, Pilotto I, Gomes J, Campos D (2014) Assessment of climate change over South America under RCP 4.5 and 8.5 downscaling scenarios. Am J Clim Change 3(05):512

Crinó E, Coronel G, Piacentini RD (2014) Events of air pollution on Asuncion (Paraguay) and Clorinda (Argentina) produced by NO_2 gas and aerosol emitted during intense biomass burning. ANALES AFA 25(1):13–17

de Oliveira Nilmara, Alves N, Brito J, Caumo S, Arana A, de Souza Hacon S, Artaxo P, Hillamo R, Teinilä K, Batistuzzo de Medeiros SR, de Castro Vasconcellos P (2015) Biomass burning in the Amazon region: aerosol source apportionment and associated health risk assessment. Atmos Environ 120:277–285

Dereczynski, Chan Chou S, Lyra A, Sondermann M, Regoto P, Tavares P, Chagas D, Gomes JL, Carneiro Rodrigues D, Skansic MM (2020) Downscaling of climate extremes over South America–part I: model evaluation in the reference climate. Weather Clim Extremes 29:100273. https://doi.org/10.1016/j.wace.2020.100273. Elsevier B.V

Food and Agricultural Organization (2017) Gestión integral del riesgo de desastres en el sector agrícola y la seguridad alimentaria en los países del CAS. Análisis de capacidades técnicas e institucionales. Paraguay. FAO-IICA-CAS, Santiago de Chile. ISBN 978-92-5-130048-0

Fraisse C, Cabrera V, Breuer NE, Báez J, Quispe J (2008) El Niño—southern oscillation influences on soybean yields in Eastern Paraguay. Int J Climatol 1399–1407

Franklin JF (1997) Ecosystem management: an overview. In: Boyce MS, Haney A (eds) Ecosystem management: applications for sustainable forest and wildlife resources, p 21

Franklin JF, Berg DR, Thornberg DA, Tappeiner J (1997) Alternative silviculture approaches to timber harvesting: variable retention harvesting systems. In: Kohm KA, Franklin JF (eds) Creating forestry for the 21st century: the science of ecosystem management, pp 11–139

Franklin JF, Spies TA, Van Pelt R, Carey AB, Thornburgh DA, Berg DR, Lindenmayer DB, Harmon ME, Keeton WS, Shaw DC, Bible K, Chen J (2002) Disturbances and structural development of natural forest ecosystems with silvicultural implications using douglas-fir forests as an example. For Ecol Manage 155:399–423

Garcia-Orenes F, Arcenegui V, Chrenková K, Mataix-Solera J, Moltó J, Jara-Navarro AB, Torres MP (2017) Effects of salvage logging on soil properties and vegetation recovery in a fire-affected Mediterranean forest: a two-year monitoring research. Sci Total Environ 1057–1065

González PU (2017) Impacto de los incendios forestales en suelo, agua, vegetación y fauna. BIBLIOTECA DEL CONGRESO NACIONAL DE CHILE| SUP 11507

Grassi B, Max PA, Armoa J (2005) Un análisis del comportamiento de la precipitación en el Paraguay. Informe final del proyecto. Universidad Nacional de Asunción, San Lorenzo

INPE, Instituto Nacional de Pesquisas Espaciais (2020) Banco de Dados de queimadas. http://queimadas.dgi.inpe.br/queimadas/portal

Jones C (2019) Recent changes in the South America low-level jet. Npj Clim Atmos Sci 2(1). https://doi.org/10.1038/s41612-019-0077-5

Kimmins JP (1974) Sustained yields, timber mining, and the concept of ecological rotation: a British Columbian view. For Chronicle 50:27–31

La Nación Newspaper (2019) Alerta por riesgo y daños a la salud a causa de incendios forestales. https://www.lanacion.com.py/pais/2019/09/16/alerta-por-riesgo-y-danos-a-la-salud-ante-incendios-forestales/

La Nación newspaper (2020) El Chaco paraguayo pasa una de las peores sequías en 50 años. https://www.lanacion.com.py/tag/sequia/

Le Page Y, Morton D, Hartin C, Bond-Lamberty B, Pereira JMC, Hurtt G, Asrar G (2017) Synergy between land use and climate change increases future fire risk in Amazon forests. Earth Syst Dyn 8:1237–1246

Leal Filho W, Barbir J, Nagy GJ, Sima M, Kalbus A, Paletta A, Villamizar A, Martínez R, Azeiteiro UM, Pereira MJ, Mussetta P, Ivars J, Salgueirinho Osório de Andrade Guerra JB, Da Silva Neira S, Moncada S, Galdies C, Klavins M, Nikolova M, Gogu RC, Balogun A-L, Bouredji A, Bonoli A (2020) Reviewing the role of ecosystems services in the sustainability of the urban environment: a multi-country analysis. J Cleaner Prod. https://doi.org/10.1016/j.jclepro.2020.121338

Lutz W, Sanderson W, Scherbov S (2001) The end of world population growth. Nature 412:543–543

Machin AB, Nascimento LF, Mantovani K, Machin EB (2019) Effects of exposure to fine particulate matter in elderly hospitalisations due to respiratory diseases in the South of the Brazilian Amazon. Braz J Med Biol Res 52(2):e8130. https://doi.org/10.1590/1414-431X20188130, ISSN 1414-431X

Manrique SM, Torreiro Villarino Y, Contreras Rodríguez ML, Sánchez Hervás JM, Garrido SM, Curbelo Alonso AJ (2020) Recursos, tecnologías, transferencia y políticas: Una mirada desde múltiples perspectivas y dimensiones a los sistemas de bioenergía en Iberoamérica, 27. CYTED (Programa Iberoamericano de Ciencia y Tecnología para el Desarrollo), Madrid, España

Marengo JA, Nobre CA, Salati E, Ambrizzi T (2007) Caracterização do clima atual e definição das alteraçoes climáticas para o território brasileiro ao longo do século XXI. Ministério do Meio Ambiente, SBF/DBIO, Brasília

Meijaard DR, Nasi ER, Gustafson L (2008) Biodiversity conservation in S.E. Asian timber concessions and overview of the implementation of guidelines and recommendations. Ecol Soc 13(1): 25

Ministerio de Agricultura y Ganadería (2013) Registro Nacional de Agricultura Familiar. Base de Datos, San Lorenzo, Paraguay

Ministerio de Hacienda (2018) Reporte de Comercio Exterior 2018. https://www.economia.gov.py/application/files/6315/4824/5481/RCE_Diciembre_-_2018.pdf

Moreno JM, Laguna-Defior C, Aldunce P, Barros V, Bilbao B, Bustamante M, Calvo Buendía E, Camilloni I, Cardona Arboleda OD, Cortés J, Delgado GC, Marengo JA, Mena C, Mendo J, Moreno AR, Oswald Spring U, Poveda G, Scarano FR, Taboada MA, Vicuña S (2020) Resumen para los responsables de políticas. En: Adaptación frente a los riesgos del cambio

climático en los países iberoamericanos–Informe RIOCCADAPT. Spring (eds). McGraw-Hill, Madrid España
Nasi R, Meijaard DR, Applegate EG, Moore P (2001) Los Incendios forestales y la diversidad biológica. Unasylva. http://www.fao.org/3/y3582s/y3582s08.htm
Nepstad DC, Stickler CM, Soares Filho B, Merry F (2008) Interactions among Amazon land use, forests and climate: prospects for a near-term forest tipping point. Philoso Trans Royal Soc London Biol S.C.I 363(1498):1737–1746
Nobre CA, Sampaio G, Borma LS, Castilla-Rubio JC, Silva JS, Cardoso M (2016) Land-use and climate change risks in the Amazon and the need of a novel sustainable development paradigm. Proc Natl Acad Sci USA 113:10759–10768
Pastén M (2006) Cambio climático en Paraguay Escenarios futuros. II Congreso Paraguayo de Ciencia del Suelo/V Simposio Paraguayo de Manejo y Conservación de Suelos/17 a 18 Agosto 2007/ Hotel Savoy- Encarnación
Pezzoli A, Ponte E, Tiepolo M (2016) Vulnerability and resilience to drought in the Chaco Paraguay—9783110480795. De Gruyter Online at 09 Oct 2016
Pritchett WL (1991) Suelos forestales. Propiedades, conservación y mejoramiento. Editorial LIMUSA México, D. F
Programa de Desarrollo de las Naciones Unidas (2006) Uso y Gobernabilidad del Agua en el Paraguay. Available at https://www.py.undp.org/content/paraguay/es/home/library/poverty/uso-y-gobernabilidad-del-agua-en-paraguay.html
Recalde C, Coronel G (2019) Contaminante Dióxido de Nitrógeno en la Atmósfera. Análisis sobre Regiones de Paraguay con datos de satelites. Campus Universitario UNA, San Lorenzo. Julio 2019. Private communication
Secretaria de Emergencia Nacional (2020) Fire in Paraguay 2019–2020. Asunción, Paraguay. https://disasterscharter.org/web/guest/activations/-/article/fire-in-paraguay-activation-621
Secretariat of the Convention on Biological Diversity (2001) Impacts of human-caused fires on biodiversity and ecosystem functioning, and their causes in tropical, temperate and boreal forest biomes. SCBD, Montreal, p 42
SENACSA-Servicio Nacional de Calidad y Salud Animal (2019) Plan Estratégico Institucional 2013–2018. Asunción, Paraguay
SERIE CLIMA 1 (2017) Evaluación de Vulnerabilidad e impacto al cambio climático en el Gran Chaco Americano, EQUIPO TÉCNICO de Escenarios Climáticos, Centro de Conocimiento para el Gran Chaco Americano y Cono Sur. Programa de las Naciones Unidas para el Medio Ambiente – PNUMA. Portal Regional para la Transferencia de Tecnología Acción frente al Cambio Climático en América Latina y el Caribe – REGATTA. Available at http://www.granchacoyconosur.net/, ISBN: 978-99967-859-7-9
Stein AF, Draxler RR, Rolph GD, Stunder BJB, Cohen MD, Ngan F (2015) NOAA's HYSPLIT atmospheric transport and dispersion modeling system. Bull Amer Meteor Soc 96(12):2059–2077. https://doi.org/10.1175/BAMS-D-14-00110.1
Ubeda X, Sarricolea P (2016) Wildfires in Chile: a review. Global Planet Change 146:152–161 (Elsevier B.V)
Vázquez C, Bobadilla ML, Villalba S, Coluchi N (2007) Epidemia de Infección por Hantavirus en el Chaco Paraguayo. Año 2001. In: 6° Congreso Paraguayo de Infectología 2007 Asuncion 2007
Wikipedia (2019) Amazon rainforest wildfire. Available at https://en.wikipedia.org/wiki/2019_Amazon_reinforest_wildfires
World Bank (2014) Paraguay—strategy of alliance for Work Bank at years 2015–18 Uso y Gobernabilidad del agua en Paraguay. Asunción, Paraguay. Available at https://www.py.undp.org/content/paraguay/es/home/library/poverty/uso-y-gobernabilidad-del-agua-en-paraguay.html
Zavala L, Celis R, Jordán A (2014) How wildfires affect soil properties. Cuadernos de Investigación Geográfica 40:311. http://dx.doi.org/10.18172/cig.2522

Post-fire Regeneration of the Palm *Mauritia flexuosa* in Vichada, Orinoco Region of Colombia

Gabriela Huidobro, Rodrigo Bernal, and Sven Wagner

1 Introduction

Fire is a disturbance factor in the gallery forests and palm swamps of the Orinoco region of Colombia, due to the recurrent burnings of the surrounding savannas. Most fires in these savannas are of anthropogenic origin, due to management practices (Armenteras et al. 2005; Roberts 2008). Since fire is used as a cost-effective land-management tool, savannas are usually burned once a year in the dry season. After it rains, the savanna grasses resprout and can be used for cattle grazing (Hoyos 1991; Higgins et al. 2000; Robinne et al. 2018). Sometimes these burning practices get out of control if they are not associated with an adequate planning. This often leads to fires spreading through the savannas, triggering land degradation.

The dendritic pattern of gallery forests and associated streams makes up ca. 16% of the savanna landscape in the Orinoco region (Crawford 1996; Veneklaas et al. 2005), which is evident when looking at satellite imagery (Fig. 1). Gallery forests in these savannas are usually a natural barrier to fire, as they act as edges, mainly due to their associated humidity (Armenteras et al. 2005; Veneklaas et al. 2005). With the stream ecosystem in the middle and the savanna grassland ecosystem on the sides, these gallery forests are highly prone to fragmentation (Veneklaas et al. 2005). However, the response to habitat loss and fragmentation is not equally documented for all taxa (Browne and Karubian 2016). For example, the conse-

G. Huidobro (✉) · S. Wagner
Faculty of Environmental Sciences, Technische Universität Dresden, Pienner Str. 8, 01737 Tharandt, Germany
e-mail: gabriela.huidobro@tu-dresden.de

S. Wagner
e-mail: sven.wagner@tu-dresden.de

R. Bernal
Reserva Natural Guadualito, Montenegro, Quindío, Colombia

W. Leal Filho et al. (eds.), *Sustainability in Natural Resources Management and Land Planning*, World Sustainability Series,
https://doi.org/10.1007/978-3-030-76624-5_26

quences of burning practices on palm trees (Arecaceae), a significant element of the Orinoco landscape, are not yet well studied.

The dioecious palm *Mauritia flexuosa* (Arecaceae) is one of the most abundant and widely distributed palm trees in South American lowlands (Rull and Montoya 2014). It is also one of the top 25 hyperdominant trees in the Amazonia, with an estimated population of 1.57 billion individuals (Ter Steege et al. 2013). In the Orinoco region it is found in gallery forests or in palm swamp aggregations (González and Rial 2011). Its natural regeneration is influenced by the interaction of various biotic and abiotic factors (Vormisto et al. 2004; Costa et al. 2009; Khorsand Rosa et al. 2014; Galeano et al. 2015; Isaza et al. 2017), including fire, which is often considered as a disturbance factor (Fig. 2) (e.g. Rull and Montoya 2014).

The natural regeneration dynamic of plants plays an essential role in both the maintenance and recovery of populations after disturbance (Cintra and Horna 1997; Chazdon and Guariguata 2016). Although studies of palm communities in fragmented landscapes suggest that life stages may be affected differently (Scariot 1999; Baez and Balslev 2007; Wang 2008), no studies relating the regeneration stage to fire disturbance have been conducted for *M. flexuosa*. Moreover, fire has been considered a cause of biodiversity loss (Cochrane 2009), and a driver of climate change (IPCC 2014). The savanna biome is maintained by fire but, simultaneously, fire is also a major source of global greenhouse gas (GHG) emissions (Van der Werf et al. 2010). Therefore, understanding the role of fire in the regeneration of *M. flexuosa* in the savannas of the Orinoco region is vital to optimize the role of this swamp palm in reducing GHG emissions.

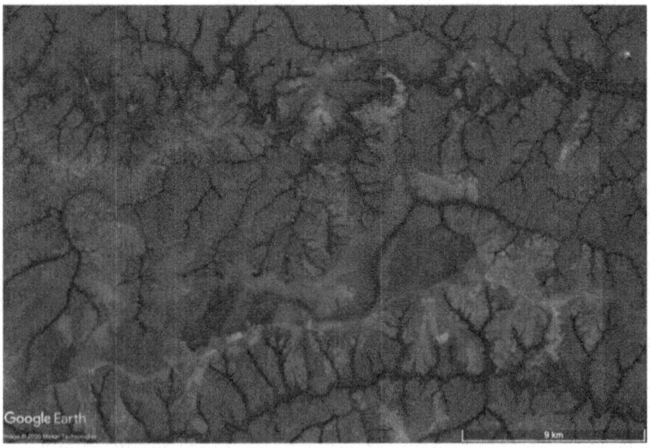

Fig. 1 The dendritic drainage pattern to which the gallery forests are associated within the savanna landscape in the Orinoco region of Colombia

Fig. 2 External view of the tip of a *Mauritia flexuosa* palm swamp impacted by the savanna fires in Vichada, Orinoco region of Colombia

2 Aims and Objectives

This chapter evaluates how fire frequencies impact the regeneration of *M. flexuosa*, by comparing seedling and sapling density in palm swamps with different time since last burn in the department of Vichada, Orinoco region of Colombia. It attempts to give recommendations for fire management in the savannas of the region.

3 Materials and Methods

The information was obtained through on-site mapping of all seedlings and saplings along 50 × 20 m transects established across palm swamps. The mother palms located inside and in the vicinity of the transects were likewise mapped in the Euclidian plane. Whenever possible, the first two vertices of the transect were established 15 m outside of the palm swamp. For each sampling site, the dates when the last fire occurred (time since last burn) were obtained from satellite imagery of the VIIRS sensor and corroborated with the information provided by the landlords. Three burning categories were established: *burning class 1*: fire event 0–2 years before the study; *burning class 2*: fire event 2–4 years before the study and; *burning class 3*: no fire event at least in the last 4 years. It was not possible to have a control group, as in the study area no palm swamps were found that had never been burned, or at least not burned in a period longer than 6 years, except in the case of forest plantations, where adjacent gallery forests are normally isolated by firebreaks, thus preventing fire from spreading. Environmental factors along the transects were mapped in grids of 10 × 1 m and classified in four different levels of habitat (Fig. 3) (*palm swamp*, *seasonal savanna*, *open savanna* and *other vegetation* as areas covered by ferns or by *Caraipa llanorum* tree formations) to analyze its effect on seedlings density.

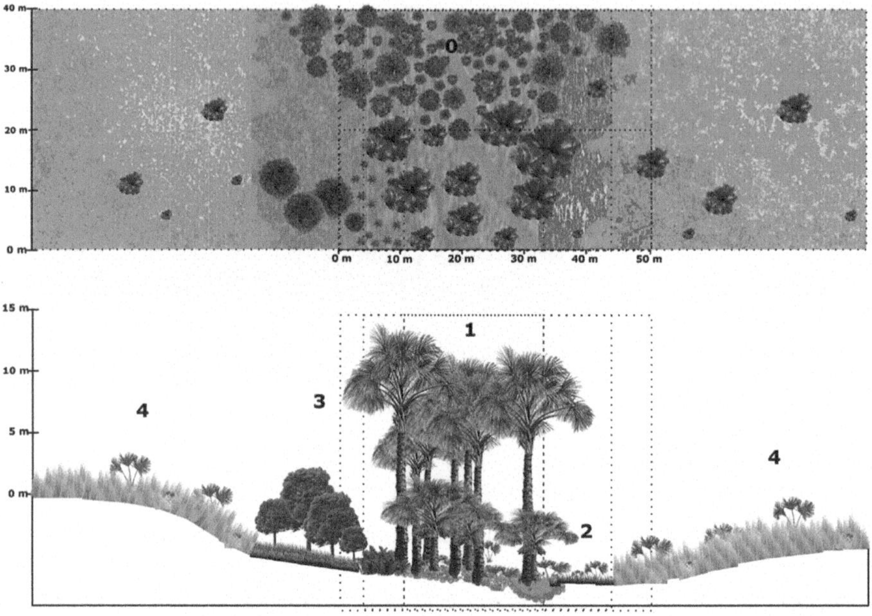

Fig. 3 Habitats of *Mauritia flexuosa* in the savannas of Vichada, Colombia. The horizontal profile is showing the gallery forest (0) connected to the palm swamp and the transect of study in a window of 20 × 50 m. In the vertical profile: (1) Palm swamp. (2) Seasonal savanna (short grasses). (3) Other vegetation (*Caraipa llanorum* and ferns). (4) Open savanna

Mapped information on the spatial position of the mother palms, the seedlings and saplings, was processed via Point Patterns Analysis. This is a set of techniques to analyze spatial point data. It allows to describe the distribution of points in the established transects and it models the relationship between the distribution of points and the covariates, by defining that relation mathematically (Baddeley and Turner 2005). Modelling is done by exploring the variability in point intensity as a function of a covariate, using a generalized linear mixed model. Seedlings, saplings and mother palms were represented by points on the Euclidean plane, whereas the explanatory data, such as the environmental variables and burning levels, were regarded as covariates. As we took into account the interaction between the regenerating individuals, we assumed the Gibbs process to represent and appropriate null model. The following exponential function shows the intensity estimation by m^2 and represents the modelling approach:

$$\text{intensity} = \exp(\beta 0 + \beta 1\text{distmin} + \beta 2\text{habitat} + \beta 3\text{burning} + S(r))$$

with "distmin" as the distance to the next potential mother palm, and

$$habitat = \begin{cases} 0 = & Seasonal\,savanna\,4\,years\,ago \\ 1 = & Open\,savanna \\ 2 = & Palm\,swamp \\ 3 = & Other\,vegetation \end{cases}$$

and

$$burning = \begin{cases} 0 = & more\,than\,4\,years\,ago \\ 1 = & 2-4\,years\,ago \\ 2 = & 0-2\,years\,ago \end{cases}$$

and $S(r)$ as the interaction term.

For further details of point pattern models, see Illian et al. (2008) or Baddeley et al. (2016). To model point patterns in R software version 3.6.0 (R Development Core Team 2014) the Spatstat package (version 1.60-1) was applied.

4 Post-fire Regeneration Response to Different Burning Categories

The generated model allowed us to analyze differences in seedling intensity values, involving habitat and burning covariates. Enabling us to determine the post-fire relation between seedling density and fire occurrence, and the habitat where the highest natural regeneration of *M. flexuosa* is likely to be found. Furthermore, the data provided a descriptive understanding of seedling distribution in relation to potential mother palms (Fig. 4), and a further insight into the significance of parental distance for seedling recruitment (Fig. 5).

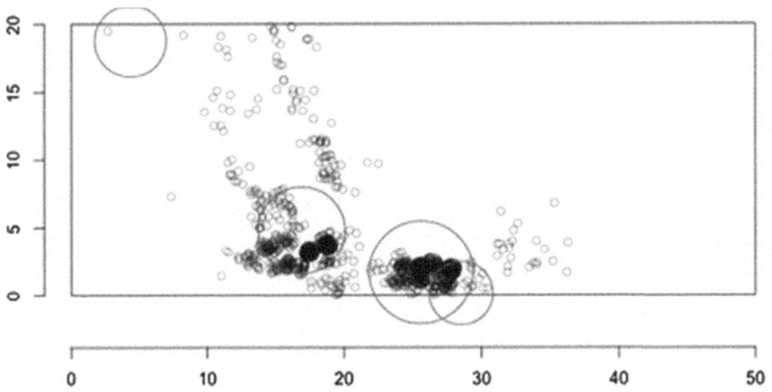

Fig. 4 Spatial distribution of seedlings (smaller circumferences) and mother palms (larger circumferences) of *Mauritia flexuosa* in Vichada, Colombia, for a particular transect. The intensely colored areas indicate seedlings clusters

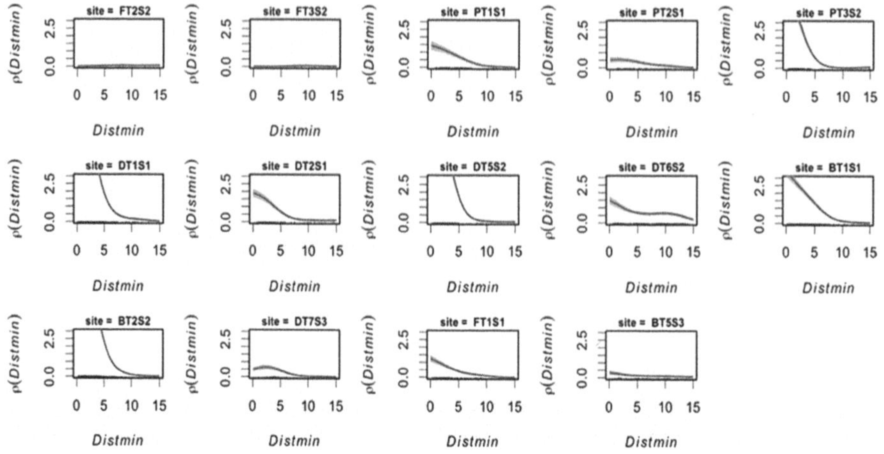

Fig. 5 Estimated seedling intensity of *Mauritia flexuosa* in Vichada, Colombia, as a function of distance to the next mother palm for each of the transects. The transect data suggest that the intensities are increasing the closer the seedlings are positioned to a mother palm. The effect seems to diminish in distances larger than 10 m

The estimated model parameters for seedlings are presented in Table 1. The value column indicates the parameters for the given variables. “Burning class 3”, “seasonal savanna” and “no distance” to the next potential mother palm are the reference conditions, which were taken to evaluate the significance of seedling intensity for the other categories.

Burning class 2 (2–4 years since last burn) had a significantly different parameter value ($p < 0.05$) than *burning class* 3 (no fire event at least in the last 4 years) and a higher intensity than *burning class 1* (0–2 years). The different parameters lead to a highest seedling intensity estimate in the transects with a time since last burn from 2 to 4 years (Fig. 6). As the formula is purely additive, *burning class 2* had the highest intensity values for all of the four habitats (Table 2).

Table 1 Values of the fitting-model for seedlings of *Mauritia flexuosa* in palm swamps in the Orinoco region of Colombia

	Value	Std. error	Freedom degrees	t-value	*p*-value
Intercept	−2.044	0.29	44,480	−7.02	0.0000
Minimum distance	−0.172	0.01	44,480	−27.97	0.0000
Open savanna	0.888	0.20	44,480	4.52	0.0000
Palm swamp	0.843	0.19	44,480	4.35	0.0000
Other vegetation	0.344	0.27	44,,480	1.26	0.2181
¨Burning class 2”	0.721	0.31	11	2.45	0.0320
¨Burning class 1”	0.325	0.30	11	1.11	0.2932
Interaction	4.474	0.05	44,480	95.51	0.0000

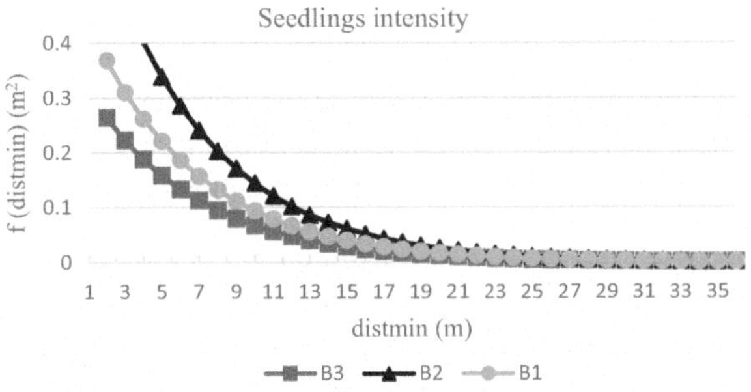

Fig. 6 Seedlings intensity of *Mauritia flexuosa* in function of the distance to the potential mother palm, for three burning categories in "Open savanna", disregarding the interaction term. The effect of the distance to the mothers is negative (−0.17), meaning that with an increasing distance to the mothers, the seedling intensity is decreasing

Table 2 Seedling intensity estimates of *Mauritia flexuosa* in palm swamps in the Orinoco region of Colombia, in relation to habitat levels and burning categories for a given distance of 15 m from the potential mother (individuals/m^2) disregarding interaction between seedlings

	Seasonal savanna	Open savanna	Palm swamp	Other vegetation
Burning 1	0.014	0.033	0.032	0.019
Burning 2	0.020	0.049	0.047	0.028
Burning 3	0.010	0.024	0.023	0.014

A reasonable explanation for the higher number of seedlings in transects burned 2–4 years earlier is that fertile palm individuals had more time to recover and produce infructescences. Although fructification in this species does not occur annually at the individual level, it does at the population level (González 1987; Urrego 1987). In the case of the seasonal and open savanna, burning can facilitate plant succession and even contribute to soil enrichment (Roberts 2008). Burning of natural vegetation is a practice that facilitates the mobilization of nutrients found in biomass and makes them return to the soil (González-Vila and Almendros 2011), creating suitable conditions for seeds to germinate.

Fire may also trigger seed regeneration directly or indirectly (Keeley and Fotheringham 2000; Silva 2013). Directly, by inducing the germination of dormant soil-stored seed banks or as a requirement for seedling recruitment in the case of "fire-dependent species". Indirectly, by initiating seedling recruitment by opening gaps in closed vegetation. Therefore, suitable conditions for regeneration may occur under certain fire frequencies. However, none of the aforementioned situations are applicable to *M. flexuosa*. Firstly, the seeds of this palm are recalcitrant and intolerant to desiccation (Silva et al. 2014; Veloso et al. 2016), and palms usually do not form seed banks, but seedlings banks (Dickie et al. 1992). Secondly, the pioneering role of the palm in the savanna rules out the need for gaps in the vegetation (González 1987). The purported pioneering role of the palm, however, needs further evaluation. Our observations of current and past satellite imagery of the palm swamps studied in this research, as well as of other *M. flexuosa* palm swamps in the region, show that the extent of the branches in the dendritic pattern of the gallery forests has hardly changed at all in the past 50 years. This suggests that the palm swamp communities of *M. flexuosa* may be persistent communities in a steady interaction with fire.

For all four habitats, *burning class 2* had higher seedling intensity than *burning class 1,* although the differences were not significant. The latter category represents a recent fire disturbance (0–2 years), which leaves a short time for the palm swamp to recover and to generate suitable conditions for regeneration. In these transects, some of the seeds on the ground were completely burned, as well as the infructescences that were still hanging on the palm.

Burning class 3, the category with the longest time since the last burn, had the lowest seedling densities in all four habitats. The transects under this category were in some cases inside forest plantations, therefore isolated by firebreaks, which did not allow fire disturbance during that period. Moreover, the savanna grasses, when unburnt, may grow up to 1.5 m height, resulting in areas of closed vegetation or even colonization by shrubby grassland species.

Although the low seedling density on these long-unburnt sites may be due to the growth of pre-existing seedlings into the next size category, it is also remarkable that no significant seedling recruitment occurred. This finding contrasts with observations made by González (2016), who stated that when the swamp graminoid grasses remain unburned for a period of more than six years, the area is gradually colonized by *M. flexuosa,* due to the dispersal of fruits and seeds from the nearby swamps, leading to the regeneration. We believe that the opposite situation takes place here, instead—a longer period without burnings leads to a large accumulation of grasses and other vegetation, thus increasing the intensity of any new fires, and consequently affecting fertile adults at the edge of the gallery forest, to a point where they cannot recover from the disturbance.

5 Post-fire Sapling Establishment

The generated generalised linear model was also applied to saplings, the next age category found along the transects. The results of the fitting model for saplings (Table 3), show that the distance to the next mother palm is also significant in saplings ($p = 0.0001$). Meaning that, as with seedlings, an increasing distance to the next mother palm significantly reduces their density. Unlike seedlings, whose density increased in plots burned 2–4 years before the study, neither *burning category 1* nor *burning category 2* had any significant effect of fire on sapling intensity ($p > 0.05$), both compared to the *burning category 3*.

"Palm swamp" was the only habitat which had a significant effect on seedling density, when compared to seasonal savanna ($p < 0.05$), and the only habitat with a positive parameter estimate (1.454). In contrast, "open savanna" and "other

Table 3 Values of the fitting-model for saplings of *Mauritia flexuosa* in palm swamps in the Orinoco region of Colombia

	Value	Std. error	Freedom degrees	t-value	*p*-value
Intercept	−3.703	0.75	13,756	−4.96	0.0000
Minimum distance	−0.108	0.03	13,756	−3.85	0.0001
Open savanna	−0.332	0.63	13,756	−0.53	0.5957
Palm swamp	1.454	0.55	13,756	2.62	0.0087
Other vegetation	−0.186	0.80	13,756	−0.23	0.8160
Burning factor "1"	−0.584	0.67	11	−0.88	0.3993
Burning factor "2"	−0.839	0.70	11	−1.19	0.2592
Interaction	4.091	0.30	13,756	13.52	0.0000

vegetation" had negative parameter values (−0.332 and −0.186 respectively), and had no significant effect of fire on sapling density ($p = 0.5957$ and 0.8160). Thus, whereas seedling density benefits from both "palm swamp" and "open savanna", regardless of fire conditions, sapling density benefits only from "palm swamps".

We also calculated sapling intensity in relation to both habitat level and burning category. As shown in Table 4, there were no relevant differences among burning classes, except for the category with the longest time since the last burn in "palm swamp", which had the highest value (0.021). This result is consistent with the value shown in Table 3, where "palm swamp" is the only habitat presenting a positive value. These findings, again, do not support a pioneer role of the palm, pointing also to a permanent nature of the *Mauritia flexuosa* swamps.

6 Field Observations on the Resprouting Capacity of the Species

The survival of *M. flexuosa* individuals to a blaze depends on the fire type. Fires can spread along the ground surface, beneath it, or even above it, if the vegetation structure allows it. These are called surface fires, ground fires and crown fires, respectively (Cochrane 2009). After a surface fire, individuals of *Mauritia* can survive, whereas the highest lethality is associated with underground fires (González and Rial 2011). Moreover, the effects of a fire depend on its intensity, duration, and extent, as well as season in which it occurs and the amount of time since previous fire(s) (frequency) (Cochrane 2009).

The single apical meristem at the upper end of palm stems divides and grows continuously, meaning there is no rest period (González 1987). This growth of palms at the expense of a single apical meristem makes them susceptible to mortality, if the meristem is destroyed (Ponce et al. 1999). For *M. flexuosa,* our field observations evidenced the resprouting ability of individuals after a fire event (Fig. 7). Not only adult individuals, but also seedlings, saplings and juveniles

Table 4 Sapling intensity estimates of *Mauritia flexuosa* palm swamps in the Orinoco region of Colombia, in relation to habitat levels and burning categories for a given distance of 15 m disregarding the interaction term (individuals/m^2)

	Seasonal savanna	Open savanna	Palm swamp	Other vegetation
Burning 1	0.003	0.002	0.012	0.002
Burning 2	0.002	0.002	0.009	0.002
Burning 3	0.005	0.003	0.021	0.004

produced new leaves if the meristem was not affected by fire. Within days after it had rained, the growth of green tissue could be observed. These findings appear to support the hypothesis of a potential pyrophilus nature of this palm (Montoya et al. 2011).

Besides, factors as available fuel, and humidity beneath the forest canopy that keeps fuel moist, play a role. Fuel moisture is the regulator of fire in these systems (Ryan 1991). Our research suggests that the humidity kept on the leaf sheath (petiole base) is of major importance for the survival of *M. flexuosa*, as the leaf sheaths and any moisture accumulated surround the apical meristem allow resprouting after the fire event.

7 Problems Posed by Poor Burning Practices

Fire has been used traditionally in agriculture and animal husbandry as a tool to remove dry vegetation quickly, for the control of pest and weeds, and mainly to stimulate the regrowth of grasses and add nutrients to the soil (Rippstein et al. 2001; Hernández-Valencia and López-Hernández 2002; López-Hernández et al. 2008; Robinne et al. 2018). Nevertheless, despite the ecological benefits of fire, these

Fig. 7 Seedling and saplings of *Mauritia flexuosa* resprouting after a fire event

ancestral savanna management practices are not associated with management programs, and some of these recurrent burns get out of control, thus facilitating the spread of fire along the extensive savannas, and affecting palm swamps and gallery forests, which are usually considered as a natural barrier to fire (Armenteras et al. 2005).

Not just the lack of planning but large-scale burnings during the dry season can increase the risk of wildfires. The outcome of burnings in the savannas depends on the magnitude of annual precipitation and its seasonal concentration. In the savannas of the Orinoco region, the average annual precipitation is between 1,000 and 2,000 mm, concentrated in a rainy season of ca. 6–7 months (Silva 2013). In these savannas with frequent burns (annual or biannual), fires occur mainly during the dry period (between December and March), when rainfall is at its lowest, and they are strongly associated with socioeconomic factors, such as the expansion of the agricultural and livestock frontier (Hernández-Valencia and López-Hernández 2002). These uncontrolled burnings during the dry season undermine the basis of biological wealth, affecting the balance of natural ecosystems, and reducing their resilience (Roberts 2008).

Human-induced fires not only affect the structure and composition of ecosystems, but they also alter important ecological processes, such as productivity, and the water and nutrient cycles of these ecosystems (Medina et al. 1978; Cochrane 2009). However, with limited fuel availability in the savanna ecosystem, these fires last a short time, thus they seldom cause further damage (Robinne et al. 2018).

In the Eastern Plains of Colombia, burning is a common practice in agricultural management. Local inhabitants in Vichada, mentioned fire as a useful tool, and they recognize its use as a necessity since it helps them to prepare the land for cultivation, guaranteeing food for livestock and preventing future fires. Additionally, they pointed out that the lack of care when using fire is one of the leading causes of uncontrolled fires (Camacho et al. 2016).

8 Role of Fire in the Dynamics of Palm Swamps

Fires can be part of a natural process that characterizes some forest ecosystems (Nasi et al. 2002; WWF 2017). The gallery forests and the natural savannas of the Orinoco region, for example, have an association with fire that induces the colonization of pre-existing forests by savanna vegetation, generating a new flammable ecosystem, intrinsically related to the dynamics of fire in natural conditions. Savanna and gallery forest ecosystems are frequently considered as a sign of a positive relationship between fire and savanna vegetation (Beerling and Osborne 2006; Beckage et al. 2011).

Gallery forests have core zones, which are rarely affected by fire, and peripheral zones that suffer fire incursions, patchily distributed in time and space. In peripheral areas, fires play a role comparable to that of gaps in the continuous forest canopy, as a variety of new light regimes are induced, but they also create a specific class of microhabitats to which a subset of tree species is specialized (Kellman and Meave 1997). The fuels in the gallery forests can reach an ignitable state late in the dry season, but fire entry is probably precluded by the tendency of savanna fires to occur earlier in the dry season and by discontinuities in fuels at the savanna/forest contact (Biddulph and Kellman 1998).

The post-fire regeneration and survival of the palm swamps associated with the gallery forests not only depend on fire frequency and burning season, but on many other factors too. The model generated in this study provided information on seedling intensity regarding the fertile adults, and an evaluation of the relation between habitat and regeneration after a fire event, by categories of time since the last burn. The finding that the highest regeneration intensity occurs in the category "between 2 and 4 years" suggests how frequency impacts the regeneration of the species and ecosystems within the savanna landscape. The reduction of grasses in the savannas through a managed fire frequency could reduce the intensity of the next fire event, thus allowing the palm individuals to survive. A similar situation was found in the Chaco region of Argentina, where savannas of the grass *Elionurus muticus* play a key role in the fire regime of the ecosystem, by maintained productivity under a normal or medium fire frequency (one fire every 3–4 years) (González et al. 2001).

Knowledge on the fire regime of a savanna is vital to understand the role of fire in the gallery forest and in palm swamp ecosystems. Such knowledge may include a description of fire frequency and intensity, the nature of fire disturbance and the effects of fire on vegetation (Agee 1996; Cochrane 2009). This information is needed for the conservation and management of the savanna, and for the development of long-term fire management plans at local and regional scales (Bravo et al. 2001). However, more information is needed on the effect of the fire regime in the Orinoco region, to generate an adequate prescription of fire as a tool for the sustainable management of the palm swamp ecosystem.

9 Policy Inconsistency and Its Role

Although fire governance has historically advocated fire suppression, in the new fire reality this is no longer an option. The key to wildfire disaster risk reduction now lies in learning how to live with fire (Robinne et al. 2018). In many countries, the main strategy has been to implement a "zero fire policy" for all situations (Mutch et al. 1999). However, as recognition of the ecological role of fire in some ecosystems has grown, as well as our understanding of how fire suppression may increase long-term fire risk, many countries have begun to pay greater attention to integrating the use of fire into their strategic management approach (Robinne et al. 2018).

Despite the increase of wildfires (Armenteras et al. 2019), Colombia does not have a national fire policy, although there is a regulatory framework based on fire suppression. The country has enacted regulations that recognize fire as a threat and emphasize risk reduction, emergency preparedness and disaster response. Social co-responsibility against wildfires was proposed, although the forms of use and perceptions of fire held by different communities are not recognized (Meza et al. 2019). Therefore, the results of the scientific community or local experiences for specific ecosystems have not yet been taken into account.

The practice of rural burning is prohibited by Colombian law, except for controlled burning in rural areas for agricultural activities, such as soil preparation. Among other provisions, the regulation defines the minimum protection distances for controlled burnings in rural areas for agricultural activities. A protection distance of 100 m is required around urban perimeter, water sources, and limits of forest reserves and conservation areas; a restriction distance of 30 m to the banks of water currents, and an additional 15 m around protective vegetation. In areas of 25 ha or more, the regulation defines burning schedules, procedures for burning practices and fire management, permitted atmospheric emissions, and monitoring of the impact of burning on the environment and communities. It also establishes that in dry times and under special climate conditions, as El Niño, any controlled burns must be suspended (Meza et al. 2019).

The preceding aspects of Colombian fire legislation reveal that the country is not in line with global challenges in terms of fire management, even though a potentially vicious cycle linking fire and climate change is emerging, showing a trend of increasing frequency and intensity of uncontrolled fires (Robinne et al. 2018). Although there are no normative guidelines at the global level that cover the diversity of ecological, socioeconomic and cultural conditions, there are at least general policy principles that are common to most situations. These include promoting a balance between fire management, prevention and control, protecting fire-sensitive ecosystems, managing fire-dependent ecosystems, and recovering areas affected by fires (FAO 2006; Robinne et al. 2018). Comprehensive and responsible fire management can be achieved through the existence and proper implementation of a legal framework endorsed by policies that provide clarity on the use of fire (FAO 2006). Those policies should be based on the integration of all the relevant biological, ecological, technological, economic, social, cultural and environmental aspects that affect the use of fire and the presence of forest fires (FAO 2006; Gutiérrez Navarro et al. 2017).

Furthermore, policies and practices that fail to take into account the needs and concerns of local communities, as well as their traditional knowledge, are unlikely to lead to desirable outcomes. Awareness and community engagement thus need to focus not solely on information provision, but also on interactive and cooperative efforts (Robinne et al. 2018). Solutions to fire risks lie in holistic environmental management and governance.

10 Fire as a Climate Change Participant

Variations in climate conditions lead to more extreme weather events, which means more fires since extreme weather drives fire activity (CBD 2019). Many climate models have projected an overall increase in temperature worldwide and a drying trend in many subtropical and mid-latitude regions. In many areas, the frequency of fires is expected to increase, along with the length of the fire seasons, the fire intensity, the areas burned and the fuel consumption (i.e., more emissions). These, in turn, may cause a marked loss of soil organic carbon that might not be compensated by vegetation regrowth. The predicted changes pose a challenge for fire management, as high-intensity fires will occur outside the "traditional fire season" (Liu et al. 2014; Robinne et al. 2018).

International scientific collaborations have demonstrated that coexistence with changing activity of fire is necessary, and that, at a global level, societies must adapt to climate change and maintain healthy, resilient and safe natural and cultural landscapes for future generations (Robinne et al. 2018). The role of fire has been recognized from a multisectoral perspective that favors the balance of the socio-cultural, environmental and economic dimensions of fire management (Mutch et al. 1999; Gutiérrez Navarro et al. 2017; FAO 2006; Robinne et al. 2018).

Intensely burning fires that occur under dry and warm conditions release more carbon than fires that occur under cool, moist conditions. Savanna fires contribute 65% of the total of fire GHG emissions annually, followed by tropical forests (15%) (Van der Werf in Robinne et al. 2018). However, savannas also regrow annually and thus take back almost all of the CO_2 they emit within a year (Robinne et al. 2018). When this ecological disturbance occurs on the edge of gallery forests, fire supress the ecotone; consequently, not only does the release of GHG take place, but it also induces a gradual selection of pyrophilic species, with strategies of colonization of open areas (Beckage et al. 2009).

While fire suppression can delay burns, it cannot mitigate fires over the long term, as some wildfires will continue to exceed the capacity of suppression, especially if there is an increase in fire-prone weather severity. Fire suppression may even have adverse long-term consequences by increasing fuel loads leading to more extreme future fires (Donovan and Brown 2007). Modeling work done in Europe estimated a potential increase of average annual burned areas under the "no adaptation" scenario to be within the range of 120–270% by the 2090s, relative to the 2000s (Robinne et al. 2018). Thus, improvements in fire suppression might reduce the impact of climate change.

The environmental consequences of burning savannas in the Orinoco region, as an extended practice, represent a serious challenge, as it is a practice that affects ecological processes at landscape scales and involves other important ecosystems. This phenomenon is in itself an important object of research and pertinent for its direct implications in global phenomena related to GHG emissions and the search for alternatives to capture atmospheric carbon (Higgins et al. 2007). Increased fire activity is not just an outcome of the changing climate, but it is also a participant in the change (Liu et al. 2014).

11 Conclusions

Fire is recognized as a necessary tool for ecosystem management and agricultural production, as it maintains the structure and composition of grasslands for feeding local fauna and livestock. In the Orinoco region, burning is an important factor in plant succession processes and in soil enrichment. However, anthropogenic fires are increasingly frequent (annual) and extensive (uncontrolled), not only resulting in the release of GHG but affecting the balance of natural ecosystems and even reducing their resilience.

The effects of a fire event on the ecosystem depend on several factors (intensity, duration, season, and time elapsed since previous fires, among others), but when it comes to evaluating the role of fire it should be taken into account that not all the ecological niches within an environment are fire adapted.

In our study we evaluated the post-fire response of *M. flexuosa* in palm swamps located at the tips of the gallery forests. The results indicated that the regeneration responded positively to fire disturbance when the time since last burn was between 2–4 years. Therefore, this period might be an adequate fire frequency for the palm swamps in Vichada, Orinoco region. The management of the fuel in these savannas through a controlled fire frequency could reduce the intensity of the next fire event, thus allowing the palm individuals to survive.

Additional information is needed on the effects of the fire regime in the Orinoco region, in order to contribute to long-term fire management plans at local and regional scale, as well as to an adequate prescription of fire as a tool for the sustainable management of the savanna landscape. Solutions to fire risks lie in comprehensive environmental management and in policies that take into account the specific needs of the ecosystem under current climate change scenarios. Ultimately, adaptation to fire can only be achieved with political willingness to address the challenge.

12 Recommendations

The results of this research give us an idea of the frequency ranges that could be applied if the prescribed burns were practiced promoting the regeneration of *M. flexuosa*. However, further studies must be performed on the regimes for this ecosystem and a law to be passed. This would allow fire management to be a real option and therefore incorporated into the conservation programs of the region.

In planning for long-term prescribed burns, not only the frequency of fires plays a role. Prescribed burning early in the dry season has been found to be more effective in reducing GHG emissions than burning late in the dry season. Therefore, understanding the role of fire in the regeneration of *M. flexuosa* in the Orinoco savannas is essential in order to optimize the role of this swamp palm in the reduction of GHG emissions. The challenge for proper fire management lies in attaining a balance in the frequencies of burning practices and burning methods in the savanna.

Uncontrolled and excessive burning practices can directly affect the ecosystems and ecological cycles of the natural landscape of the Orinoco region. Annual burning increases the productivity of savanna grasslands in the short-term, but this repeated frequency can cause an imbalance in the ecosystem that affects natural formations such as palm swamps. On the other hand, the high burning frequencies lead to fuel accumulation in the grasslands, which makes them prone to burn more intensely. The suppression of fires in the savannas does not ensure the protection of the gallery forests and the associated palm swamps. In the case of forest plantations, where palm swamps were found to have the longest time since the last burn, it is advisable that the use of firebreaks go hand in hand with integrated fire management.

On the day of a prescribed fire, some variables should be considered, such as wind direction, lignified plant material, atmospheric pressure, and temperature, among others. Even in the rainy season, nutrient leakage must be avoided. The list of factors to be considered is long. Therefore, it is fundamental to promote and provide institutional support for local capacity building in activities that can mitigate local fire risk and ensure proper management. The integration of available information and traditional fire knowledge is key to adapt to local changes in fire activity, using known techniques for the reduction of hazardous fuel loads, prescribed long-term burning and sustainable landscape management practices.

References

Agee JK (1996) Fire ecology of Pacific Northwest Forests. Island Press, Washington, DC

Armenteras D, Romero M, Galindo G (2005) Vegetation fire in the savannas of the Llanos Orientales of Colombia. World Resour Rev 17(4):531–543

Armenteras D, Schneider L, Dávalos LM (2019) Fires in protected areas reveal unforeseen costs of Colombian peace. Nat Ecol Evol 3(1):20–23

Baddeley A, Rubak E, Turner R (2016) Spatial point patterns—methodology and applications with R. CRC Press, Boca Raton

Baddeley A, Turner R (2005) Spatstat: an R package for analyzing spatial point patterns. J Stat Softw 12(6):1–42

Baez S, Balslev H (2007) Edge effects on palm diversity in rain forest fragments in western Ecuador. Biodivers Conserv 16(7):2201–2211

Beckage B, Platt WJ, Gross LJ (2009) Vegetation, fire, and feedbacks: a disturbance-mediated model of savannas. Am Nat 174(6):805–818

Beckage B, Gross LJ, Platt WJ (2011) Grass feedbacks on fire stabilize savannas. Ecol Model 222 (14):2227–2233

Beerling DJ, Osborne CP (2006) The origin of the savanna biome. Glob Change Biol 12 (11):2023–2031

Biddulph J, Kellman M (1998) Fuels and fire at savanna-gallery forest boundaries in southeastern Venezuela. J Trop Ecol 14(4):445–461

Bravo S, Kunst C, Gimenez A, Moglia G (2001) Fire regime of a *Elionorus muticus* Spreng. savanna, western Chaco region, Argentina. Int J Wildland Fire 10(1):65–72

Browne L, Karubian J (2016) Diversity of palm communities at different spatial scales in a recently fragmented tropical landscape. Bot J Linn Soc 182(2):451–464

Camacho A, Huidobro G, Victorino A (2016) Notibita 6-Tierra, fuego, agua... vida en Bita. Instituto de Investigación de Recursos Biológicos Alexander von Humboldt. Bogotá

CBD (2019) Voluntary guidelines for the design and effective implementation of ecosystem-based approaches to climate change adaptation and disaster risk reduction (no. 93)

Chazdon RL, Guariguata MR (2016) Natural regeneration as a tool for large-scale forest restoration in the tropics: prospects and challenges. Biotropica 48(6):716–730

Cintra R, Horna V (1997) Seed and seedling survival of the palm *Astrocaryum murumuru* and the legume tree *Dipteryx micrantha* in gaps in Amazonian forest. J Trop Ecol 13(2):257–277

Cochrane MA (2009) Fire in the tropics. In: Tropical fire ecology. Springer, Berlin, Heidelberg

Costa FR, Guillaumet JL, Lima AP, Pereira OS (2009) Gradients within gradients: the mesoscale distribution patterns of palms in a central Amazonian forest. J Veg Sci 20(1):69–78

Crawford E (1996) Land use dynamics in the Colombian Llanos. Internal report. CIAT, Cali

Development Core Team R (2014) R: a language and environment for statistical computing. R Foundation for Statistical Computing. Vienna

Dickie JB, Balick MJ, Linington IM (1992) Experimental investigations into the feasibility of *ex situ* preservation of palm seeds; an alternative strategy for biological conservation of this economically important plant family. Biodivers Conserv 1(2):112–119

Donovan GH, Brown TC (2007) Be careful what you wish for: the legacy of Smokey Bear. Front Ecol Environ 5(2):73–79

FAO (2006) Fire management: voluntary guidelines. Principles and strategic actions. Fire management working paper 17. FAO, Rome

Galeano A, Urrego LE, Sánchez M, Peñuela MC (2015) Environmental drivers for regeneration of *Mauritia flexuosa* L.f. in Colombian Amazonian swamp forest. Aquat Bot 123:47–53

González V (1987) Los morichales de los llanos Orientales. Un enfoque ecológico. Ediciones Corpoven. Caracas

González V (2016) Los palmares de pantano de *Mauritia flexuosa* en Suramérica: una revisión. In: Lasso CA, Colonnello G, Moraes M (eds) Morichales, Canaguchales y Otros Palmares Inundables de Suramérica. Parte II: Colombia, Venezuela, Brasil, Perú, Bolivia, Paraguay, Uruguay y Argentina. IAvH, Bogotá, pp 45–83

González V, Rial A (2011) Las comunidades de morichal en los Llanos Orientales de Colombia, Venezuela y el delta del Orinoco: Impactos de la actividad humana sobre su integridad y funcionamiento. In: Lasso CA, Rial B, Matallana CL, Ramírez W, Señaris JC, Díaz-Pulido A, Corzo G, Machado A (eds) Biodiversidad de la cuenca del Orinoco. II. Áreas prioritarias para la conservación y uso sostenible. IAvH, Bogotá, pp 125–146

González CC, Studdert GA, Kunst C, Albanesi A (2001) Comportamiento de algunas propiedades del suelo en una sábana del Chaco Semiárido Occidental bajo distintas frecuencias de fuego. Ciencia del Suelo 19(2):92–100

González-Vila FJ, Almendros G (2011) El controvertido efecto de los incendios. El fuego en el ciclo del carbono en los ecosistemas mediterráneos. Revista Mètode 70:82–87

Gutiérrez Navarro A, García Barrios LE, Parra Vázquez M, Rosset P (2017) De la supresión al manejo del fuego en la Reserva de la Biosfera La Sepultura, Chiapas: perspectivas campesinas. Región y sociedad 29(70):31–70

Hernández-Valencia I, López-Hernández D (2002) Pérdida de nutrimentos por la quema de la vegetación en una sabana de Trachypogon. Revista de Biología Tropical 50(3–4):1013–1019

Higgins SI, Bond WJ, Trollope WS (2000) Fire, resprouting and variability: a recipe for grass–tree coexistence in savanna. J Ecol 88(2):213–229

Higgins SI, Bond WJ, February EC, Bronn A, Euston-Brown DI, Enslin B, Scheiter S (2007) Effects of four decades of fire manipulation on woody vegetation structure in savanna. Ecology 88(5):1119–1125

Hoyos P (1991) Manejo de la quema de la sabana nativa en la altillanura de Los Llanos Orientales de Colombia. Pasturas Tropicales 13:49–51

Illian J, Penttinen A, Stoyan H, Stoyan D (2008) Statistical analysis and modelling of spatial point patterns. Wiley, Chichester

IPCC (2014) Summary for policymakers. In: Field CB, Barros VR, Dokken DJ, Mach KJ, Mastrandrea MD, Bilir TE, Chatterjee M, Ebi KL, Estrada YO, Genova RC, Girma B, Kissel ES, Levy AN, MacCracken S, Mastrandrea PR, White LL (eds) Climate change 2014: impacts, adaptation, and vulnerability. part A: global and sectoral aspects. Contribution of working group II ,to the fifth assessment report of the intergovernmental panel on climate change. Cambridge University Press, Cambridge, New York

Isaza C, Bernal R, Galeano G, Martorell C (2017) Demography of *Euterpe precatoria* and *Mauritia flexuosa* in the Amazon: application of integral projection models for their harvest. Biotropica 49(5):653–664

Keeley JE, Fotheringham CJ (2000) Role of fire in regeneration from seed. Seeds Ecol Regeneration Plant Communities 2:311–330

Kellman M, Meave J (1997) Fire in the tropical gallery forests of Belize. J Biogeogr 24(1):23–34

Khorsand Rosa R, Barbosa RI, Koptur S (2014) Which factors explain reproductive output of *Mauritia flexuosa* (Arecaceae) in forest and savanna habitats of northern Amazonia? Int J Plant Sci 175(3):307–318

Liu Y, Goodrick S, Heilman W (2014) Wildland fire emissions, carbon, and climate: Wildfire–climate interactions. For Ecol Manage 317:80–96

López-Hernández D, Hernández-Valencia I, Güerere I (2008) Cambios en parámetros físicos, químicos y biológicos en el suelo de una sabana protegida de quema y pastoreo durante veinticinco años. Bioagro 20(3):151–158

Medina E, Mendoza A, Montes R (1978) Nutrient balance and organic matter production in the *Trachypogon* savannas. Trop Agric 55(3):243–253

Meza MC, González TM, Armenteras D (2019) Propuesta para el manejo integral del fuego en Colombia. Ediprint, Bogotá

Montoya E, Rull V, Stansell ND, Abbott MB, Nogué S, Bird BW, Díaz WA (2011) Forest–savanna–morichal dynamics in relation to fire and human occupation in the southern Gran Sabana (SE Venezuela) during the last millennia. Quatern Res 76(3):335–344

Mutch RW, Lee B, Perkins JH (1999) Public policies affecting forest fires in the Americas and the Caribbean. FAO Forestry Paper, pp 65–108

Nasi R, Dennis R, Meijaard E, Applegate G, Moore P (2002) Forest fire and biological diversity. UNASYLVA-FAO 36–40

Ponce ME, Brandín J, Ponce MA, González V (1999) Germinación y establecimiento de plántulas de *Mauritia flexuosa* L.f. (Arecaceae) en los llanos sur-orientales del Estado Guarico, Venezuela. Acta Botanica Venezuelica 167–183

Rippstein G, Escobar G, Motta F (2001) Agroecología y biodiversidad de las sabanas en los Llanos Orientales de Colombia. CIAT, Cali

Roberts G (2008) Policies and instruments for the adaptation of forests and the forest sector to impacts on climate change as indicated in United Nations framework convention on climate change national reports. Occasional paper no.22. IUFRO, Vienna

Robinne FN, Burns J, Kant P, Flannigan M, Kleine M, de Groot B, Wotton DM (2018) Global fire challenges in a warming world. Occasional paper no.32. IUFRO, Vienna

Rull V, Montoya E (2014) *Mauritia flexuosa* palm swamp communities: natural or human-made? A palynological study of the Gran Sabana Region (northern South America) within a neotropical context. Quatern Sci Rev 99:17–33

Ryan KC (1991) Vegetation and wildland fire: implications of global climate change. Environ Int 17(2–3):169–178

Scariot A (1999) Forest fragmentation effects on palm diversity in central Amazonia. J Ecol 87 (1):66–76

Silva JF (2013) El fuego y los árboles de la sabana. In: Medina E, Huber O (eds) Recorriendo el paisaje vegetal de Venezuela. Instituto Venezolano de Investigaciones Científicas (IVIC), Caracas, pp 157–151

Silva RS, Ribeiro LM, Mercadante-Simões MO, Nunes YRF, Lopes PSN (2014) Seed structure and germination in buriti (*Mauritia flexuosa*), the swamp palm. Flora-Morphol Distrib Funct Ecol Plants 209(11):674–685

Ter Steege HT, Pitman NC, Sabatier D, Baraloto C, Salomão RP, Guevara JE, Monteagudo A et al (2013) Hyperdominance in the Amazonian tree flora. Science, 342(6156)

Urrego LE (1987) Estudio preliminar de la fenología de la canangucha (*Mauritia flexuosa* L.f.). Colombia amazonica 2(2):57–81

Van der Werf GR, Randerson JT, Giglio L, Collatz GJ, Mu M, Kasibhatla PS, van Leeuwen TT (2010) Global fire emissions and the contribution of deforestation, savanna, forest, agricultural, and peat fires (1997–2009). Atmos Chem Phys 10(23):11707–11735

Veloso VHS, Ribeiro LM, Mercadante-Simões MO, Nunes YRF (2016) Cytological aspects of recalcitrance in dormant seeds of *Mauritia flexuosa* (Arecaceae). Acta Physiol Plant 38(7):1–11

Veneklaas EJ, Fajardo A, Obregón S, Lozano J (2005) Gallery forest types and their environmental correlates in a Colombian savanna landscape. Ecography 28(2):236–252

Vormisto J, Tuomisto H, Oksanen J (2004) Palm distribution patterns in Amazonian rainforests: what is the role of topographic variation? J Veg Sci 15(4):485–494

Wang YH (2008) Palm community structure and land cover changes in the San Juan Biological Corridor, Costa Rica. Biotropica 40(1):44–54

WWF (2017) Study forests Ablaze. Causes and effects of global forest fires. WWF Deutschland, Berlin

Climate Downscaling for Fire Management

Yongqiang Liu, Scott Goodrick, and John A. Stanturf

1 Introduction

Fires occur and re-occur on almost a third of the global landmass at different return intervals, with burned area averaging 4.5 million km^2 annually (Giglio et al. 2013; IUFRO 2018). Wildfires shape forest structure and composition; forests are dependent on, sensitive to, independent of, or influenced by fire (Myers 2006). Despite recent decreases in the annual area burned, catastrophic wildfires apparently are increasing in frequency globally (FAO 2001). For example, bushfires in Australia burned more than 16 million ha during the 2019–20 fire season, leading to losses of over 3,000 houses and 33 lives (Richards et al. 2020). Over 3 million ha were burned by wildfires during 2019 in Siberia (https://en.wikipedia.org/wiki/2019_Siberia_wildfires). The Amazon rainforest burned about 142,000 ha during the 2019 dry season with the highest fire count since 2012 (https://www.globalfiredata.org/updates.html). Extremely large fires occurred in the western United States (US) with burned areas of about 1.2 and 1 million ha in 2017 and 2018, respectively. The increased frequency of catastrophic wildfires, especially mega-fires (Stephens et al. 2014; Williams 2004; Williams and Hyde 2009) are due to multiple causes, including extreme weather events such as extended drought (Acácio et al. 2009; Dai 2011; González et al. 2018; Liu et al. 2010b). Climate is a primary driver for fire activity. Climate change that results in drier, warmer con-

Y. Liu (✉) · S. Goodrick
Center for Forest Disturbance Science, USDA Forest Service, 320 Green St., Athens, GA 30602, USA
e-mail: yongqiang.liu@usda.gov

S. Goodrick
e-mail: scott.l.goodrick@usda.gov

J. A. Stanturf
Estonian University of Life Sciences, Kreutzwaldi 5, 51014 Tartu, Estonia
e-mail: johnalvin.stanturf@emu.ee

W. Leal Filho et al. (eds.), *Sustainability in Natural Resources Management and Land Planning*, World Sustainability Series,
https://doi.org/10.1007/978-3-030-76624-5_27

ditions potentially could increase fire occurrence and intensify fire behavior. These changes may already be occurring (Abatzoglou and Williams 2016; Flannigan et al. 2009; Reinhard et al. 2005; Westerling et al. 2006). Human factors also play central roles by increasing fuel loads due to years of suppression activity (DeWilde and Chapin 2006; O'Brien et al. 2017) and increasing wildland-urban interface (Hammer et al. 2009; Radeloff et al. 2005; Stewart et al. 2007). High intensity fires that occur outside the traditional fire season will exceed existing fire management capacities (IUFRO 2018; Jolly et al. 2015).

Climate information is necessary for understanding and predicting fire regimes, seasonal and inter-annual variability, and future wildfire trends. Fire risks are assessed daily for individual forests using tools such as the US National Fire Danger Rating System (NFDRS) (Deeming et al. 1977) and the Canada Forest Fire Danger Rating System (CFFDRS) (Van Wagner 1987). Weather conditions including temperature, precipitation, humidity, and wind are the major parameters used to assess fire occurrence, spread, and intensity. At broader spatial and long temporal scales, seasonal fire activity is predicted for geographical regions using dynamical and empirical planning tools. Future trends of fire regimes are projected mainly based on the projected climate change. Atmospheric variability, especially extremes such as droughts, is a major predictor.

Besides predicting wildfire activity, another response to the challenge of increased wildfire occurrence is to reduce hazardous fuel loads by prescribed burning (Bradstock et al. 2012; Fernandes et al. 2013; Tolhurst and McCarthy 2016; Wiedinmyer and Hurteau 2010). Frequent burning under mild fire weather conditions has been a common fire management practice in the southern US for decades (Fernandes and Botelho 2003; Waldrop and Goodrick 2012) and has been advocated in other regions of the US and as "indigenous fire treatments" in Australia (Eriksen and Hankins 2014; Gott 2005; Vigilante et al. 2009). Prescribed burning is not a panacea, however; managing smoke from intentional burning is critical to securing public acceptance (Hardy 2001). Additionally, climate change that alters fire weather may lengthen fire seasons (Liu et al. 2010a) and reduce the amount of time that burning may be conducted without significantly increasing the risk of escape (e.g., Sun 2006; Wonkka et al. 2015) or problems with smoke (McKenzie et al. 2014; O'Neill et al. 2017), thereby taxing management resources. The ability to project future climate conditions will assist managers in planning and efficiently allocating resources.

Climate change is projected mainly using global climate models (GCMs). Despite advances in complexity and the ability to resolve global climate, GCMs have limited ability to simulate regional climate. Significantly, most wildfires occur at spatial scales up to tens of kilometers in area, resulting in a significant size mismatch with GCM output. Nevertheless, the climate and fire communities have made great strides in developing and applying climate downscaling techniques to overcome the spatial deficiencies of GCMs. These techniques use either a dynamical approach by running regional climate models (RCMs) driven by GCMs or from measurements, or a statistical approach applying statistical tools to obtain relationships between historical meteorological conditions at local sites and global

GCM grids, for high-resolution (meters to tens of kilometers) climate information. Both approaches and their applications in various regions of the world have been documented and reviewed, for example, by Abatzoglou and Brown (2012), USAID (2014), Xue et al. (2014), Xu et al. (2019), and illustrated in recent publications by Kitoh et al. (2016), Li et al. (2019a), and Takhsha et al. (2018) for dynamical downscaling and Gebrechorkos et al. (2019), Jiang et al. (2018), and Navarro-Racines et al. (2020) for statistical downscaling. The downscaled climate provided necessary information for assessing the impacts of climate change on fires in various regions of the world, including Africa (e.g., Strydom & Savage 2017), Asia (e.g., Li et al. 2019b), Australia (e.g., Dowdy), Europe (e.g., Dupuy et al. 2020; Faggian 2018; Lozano et al. 2017), North America (e.g., Stambaugh et al. 2018), and South America (e.g., Silva et al. 2016).

The climate modeling community is advancing global climate change projections through implementing the Coupled Model Intercomparison Projects, Phase 6 (CMIP6). The CMIP6 projections use the Intergovernmental Panel on Climate Change (IPCC) new "Shared Socioeconomic Pathways" (SSPs) emission scenarios (Eyring et al. 2016; Meehl et al. 2016) and will be a part of the upcoming 2021 IPCC Sixth Assessment Report (AR6). Efforts have been underway to downscale the CMIP6 projections for impact assessment in various fields including wildfires. Here we introduce fire managers and researchers to climate downscaling techniques in this active field and their significance for fire management. First, we introduce dynamical and statistical downscaling approaches and compare their strengths and weaknesses. Second, available downscaling tools and products are described. Finally, some examples of actual fire applications are illustrated. The information provided in this paper is expected to provide managers with a basis for selecting the high-resolution climate products from both current resources and the coming new downscaling products of the CMIP6 projections, which are needed to achieve the specific fire management goals and to understand the connections of wildfires (including the recent large and mega-fires) to climate change.

2 Methodology

Each of the three topics of downscaling approaches, tools and products, and fire applications are presented with separate descriptions of dynamical and statistical downscaling (Fig. 1). The two essential tools for dynamical downscaling (that is, GCMs and RCMs) and their differences are first described. The GCM description focuses on the future global climate change projections made by GCMs through CMIP. The RCM description focuses on some classical models developed in various regions of the world. The methods and or models for statistical downscaling are then described. Finally, the strengths and weaknesses are compared between dynamical and statistical downscaling. The dynamical downscaling tools are described using the WCRP Coordinated Regional Climate Downscaling Experiment (CORDEX) (https://cordex.org) and its North America region as an

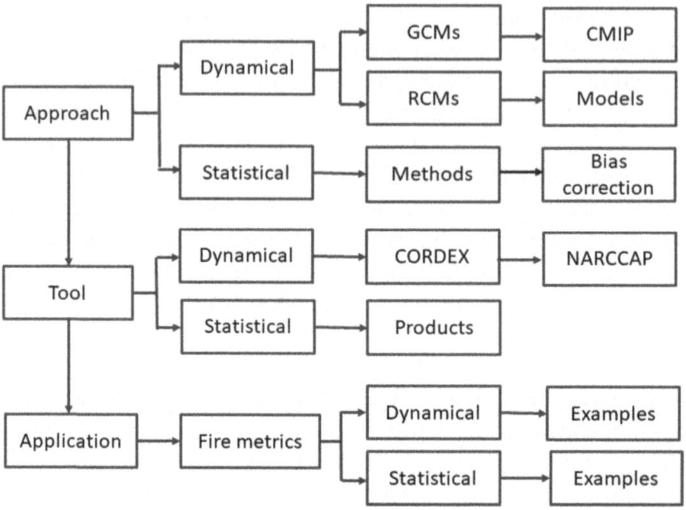

Fig. 1 A schematic diagram of topics (from top to bottom on left) and related issues (from left to right)

example. The statistical downscaling tools include a few major global and regional products. For fire applications, the need for downscaling climate for calculation of a metrics of fire weather indices is described, followed by examples of fire risk evaluation and projections under different downscaled regional climate change scenarios.

The information used to describe the above topics and issues was obtained from a number of sources. The GCMs and RCMs were based on our knowledge of climate modeling. The CMIPs and statistical downscaling methods were obtained from the literature. The downscaling tools were from online information. The fire applications were from literature, including studies by the authors.

3 Downscaling Approaches

3.1 Dynamical Downscaling

3.1.1 Approach

Dynamical downscaling utilizes RCMs with boundary conditions provided by GCMs or measurements to produce high-resolution regional climate change scenarios. A GCM is comprised of equations to describe atmospheric dynamics based on momentum, heat energy, mass, and moisture conservation and atmospheric physics to represent radiation, turbulence, cloud and precipitation, etc. In addition,

GCMs also include other climate system components (ocean, ice, soil, vegetation, etc.) and gas and particle emissions. GCMs simulate and predict global climate, seasonal and inter-annual variability, and long-term change.

One of the important applications of GCMs is to project climate change due to the increasing atmospheric greenhouse gas concentrations, including carbon dioxide. A useful source for GCM projections is CMIP output. The CMIP provides a framework for coordinated climate change experiments for the major GCMs all around the world. The third (CMIP3) and fifth (CMIP5) phases of the project provided the major information for the scientific assessment of IPCC AR4 and AR5, respectively (Semenov and Stratonovitch 2010; Hausfather et al. 2020). About 25 GCMs participated in CMIP3 for the the Special Report on Emission Scenarios (SRES) emission scenarios (Nakicenovic et al. 2000). These scenarios distinct between strong economic values and strong environmental values and between increasing globalization and increasing regionalization. About 26 research centers participated in CMIP5 (Taylor et al. 2012) with some centers providing more than one GCM. The climate change projections were made based on the Representative Concentration Pathways (RCPs) (Van Vuuren et al. 2011). The pathways describe four possible climate futures, their likelihoods depending on the amount of greenhouse gases to be emitted in the years to come. The four RCPs (RCP2.6, 4.5, 6.0, and 8.5) are named after a possible range of radiative forcing values in the year 2100 relative to pre-industrial values (+2.6, +4.5, +6.0, and +8.5 W/m^2, respectively). About 100 GCMs are participating in CMIP6 for the emission scenarios defined in 8 SSPs, 4 of which correspond to the 4 RCPs.

Each GCM has a horizontal resolution, usually of hundreds of kilometers (Fig. 2). Global processes such as the El Niño and Southern Oscillation (ENSO) and synoptic systems such as cyclones and fronts can be identified at this resolution. However, regional climate and mesoscale processes such as convective storms are largely missed. The effects of local and regional forcing such as terrain, land cover variability, and aerosols emitted from local or regional natural and anthropogenic sources are often not well represented in GCMs.

RCMs are developed based on limited-area meteorological models (LAMs) to simulate regional climate and variability. LAMs are a primary tool for numerical

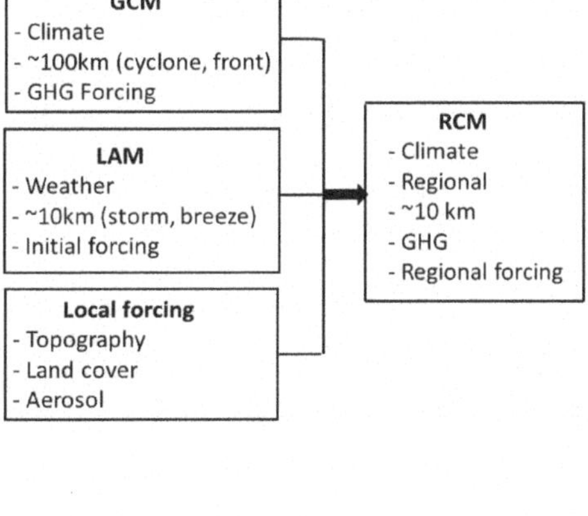

Fig. 2 Different features and relations among global climate models (GCMs), limited area models (LAMs), and regional climate models (RCMs)

weather simulation and prediction for a specific geographic region. They have similar components of dynamics and physics to GCMs, but with resolutions of tens of kilometers or higher and usually run for a short period such as a few days. They can identify storms and local circulations such as seas breezes. Initial conditions are very important in addition to internal dynamics, while lateral conditions and physical processes such as radiation and land-surface processes are less important.

An LAM is expanded into a RCM by including more detailed physical schemes for local and regional properties. Because RCMs are run over a longer period (months and years), initial conditions become less important while lateral conditions and forcing on the ground and inside the atmosphere become more important. The lateral boundary conditions of meteorological variables could be provided by a GCM, which makes RCMs a useful tool for climate downscaling of GCM simulations and projections. A RCM runs over the same or parts of the simulation period of GCM with frequently provided boundary conditions (e.g., once a day or a few hours).

The regional climate modeling technique was first developed in the US National Center for Atmospheric Research (NCAR) (Dickinson et al. 1989; Giorgi and Bates 1989) based on the standard NCAR/Penn State Mesoscale Model Version 4 (MM4) (Anthes et al. 1987). It has grown tremendously over the last three decades (Giorgi 2019) and become a powerful tool for climate downscaling (Tapiador et al. 2020). Table 1 lists some RCMs developed in the early time.

Table 1 Early developed regional climate models

RCM	Full name	Developer	Significance	Reference
CRCM	Canadian Regional Climate Model	UQAM	Regional version of Canadian GCM	Caya and Laprise (1999)
MM5	Mesoscale Model Version 5	PSU/ NCAR	One of the widely used early US mesoscale models	Grell et al. (1994)
PRECIS	Providing REgional Climates for Impacts Studies	Hadley Centre	Easily applied to any area of the globe	Jones et al. (2004)
RAMS	Regional atmospheric modeling System	CSU	One of the widely used early US mesoscale models	Liston and Pielke (2000)
RegCM3	Regional Climate Model3	NCAR	First RCM	Giorgi and Bates (1989)
RSM	Regional Spectral Model	NCEP	Spectral model	Juang and Kanamitsu (1994)
WRF	Weather Research and Forecasting	NCAR/ NOAA	For both research and operational forecasting	Liang et al. (2006)

UQAM, PSU, and CSU represent Université du Québec à Montréal, Pennsylvania State University, and Colorado State University

3.1.2 Evaluation and Bias Corrections

Evaluation is a process to assess a model's performance by comparing simulated historical averages, variability, and extremes of atmospheric as well as other climate system components to observed values. Although evaluation of a RCM is part of its development, its performance is dependent on the simulation region and time, and scheme and parameter setting. Thus, when more than one RCM is used for climate downscaling, it is useful to compare the performances of all models in the ensemble. Potential issues can be found and may need to be addressed before the model is used for climate downscaling. The evaluation process can also provide information on the uncertainty in the downscaled climate results.

RCMs can produce large biases in simulating present climate conditions. There are two ways to remove or at least reduce the biases. One way is to improve the models before they are used to predict regional climate (Jin et al. 2011). Xu and Yang (2012) used adjusted GCM climatology according to regional re-analysis to drive a RCM. Their results showed that the downscaled climatological means and extremes events were better than the downscaling without bias correction.

The other way to address bias is to correct the output of projected climate. For example, McGinnis et al. (2012) applied statistical bias correction to monthly mean surface air temperature and precipitation data from the North American Regional Climate Change Assessment Program (NARCCAP) regional climate change scenarios (Mearns et al. 2012) and examined the characteristics of the distributions of these fields on a regional basis. Appropriate statistical distributions (Gaussian for temperature and Gamma for precipitation) were used to fit RCM output for the monthly average of each climatic region. Quantile mapping (QM) was used in bias correlation through creating a transfer function to adjust the distribution of the simulated current climate so that they will match the cumulative distribution function (CDF) of observed climate. The same transfer function was then applied to output from the future climate projections.

3.2 Statistical Downscaling

3.2.1 Approach

Statistical downscaling is a technique developed based on the relationships between GCM predicted climate and historical weather observation data to produce high-resolution spatial and temporal climate at the regional level. Denoting GCM global and regional meteorological variables as X and Y with subscripts 0 and 1 for present and future conditions, the data for three of the four variables, X_0, X_1, and Y_0, are available. Y_1 is what statistical downscaling will provide. The procedure for statistical downscaling is as follows: (1) Building a relationship F between X_0 and Y_0 using a statistical method. (2) Correct X_0 based on Y_0 (optional). (3) Assume that F and correction are also applicable to future condition, thereby obtaining Y_1 using F and X_1.

3.2.2 Statistical Methods

The methods of statistical downscaling can be categorized into different types. One categorization (Wilby and Wigley 1997) classifies all methods into three types: (1) Transfer functions that relate large-area climate to local weather, or from upper air to the ground, (2) Weather typing that relates large-scale circulation systems or patterns to local weather, and (3) Weather generator that relates parameters from large- to local-scale. The transfer function type can be further classified as scaling methods and regression-based approaches (Schoof 2013).

1. *Transfer functions*

Transfer functions are developed using statistical techniques such as multiple linear regression, spatial pattern tools [e.g., empirical orthogonal functions (EOF), canonical correlation analysis (CCA), and singular value decomposition (SVD)], and artificial neural networks (ANN). These functions transfer information on large-scale meteorological variables provided by GCMs (predictors) to regional climate (responses) measured by direct meteorological variables such as temperature and precipitation. Transfer functions are computation efficient and therefore easily produce multiple scenarios for ensemble analysis. Large amounts of data and good correlations between predictor and response variables are the keys for successful applications.

Regressions develop linear relationships between individual variables. For example, Vasiliades et al. (2009) used multiple regressions between GCM variables and observed precipitation to downscale monthly precipitation values in order to estimate a meteorological drought index, the Standardized Precipitation Index (SPI) at multiple timescales. The validation indicated the accuracy of the methodology and the uncertainties propagated by the downscaling procedure.

The spatial pattern tools are similar to regressions but for relationships between spatial patterns of predictor and response fields. For example, SVD is a technique to identify coupled spatial patterns with the maximum temporal covariance between a variable from GCM and a variable from historical data. Uvo et al. (2001) developed linear regression models based on SVD analysis to downscale atmospheric variables to estimate average rainfall in a region in Japan. The results revealed that the Bai-u front was responsible for the majority of summer rainfall, the strong circulation pattern associated with autumn rainfall, and the strong influence of orographic lifting creating a pronounced east-west gradient across the local region.

The ANN approach can be used to build non-linear relationships between meteorological variables in a large network and variables within the network (Snell 2000). Trigo and Palutikof (1999) compared performances of linear models and non-linear ANNs in downscaling the site temperature at Coimbra, Portugal from large-scale atmospheric variables. Their results showed that even a simple configuration of a 2-layer non-linear neural network significantly improved on the performance of a linear model. Knowledge of complex spatial patterns and ANN are essential for downscaling with these tools.

2. *Weather typing*

Weather typing relates large-scale weather or circulation patterns to regional climate variations. Weather or circulation patterns can be selected based on the synoptic or statistical analyses of atmospheric systems responsible for local weather. Spatial analysis also can be used for weather typing. For example, the frequency distributions of regional climate variables can be derived by weighting the local climate conditions with the relative frequencies of the weather classes. Climate change then can be estimated by determining the change of the frequency of weather classes. Weather typing is based on understanding the physical processes of synoptic or planetary atmospheric systems and their connections with local climate. Knowledge of such systems is needed to apply the technique but may or may not exist for a specific location.

Abatzoglou and Brown (2012) used a weather type model, the Multivariate Adapted Constructed Analogs (MACA) that identifies commonality between the synoptic-scale field from a GCM and a catalog of observed synoptic-scale fields from observations to download weather for complex wildfire analyses in areas with diverse terrain. They found the model out-performed results obtained from direct interpolation of re-analysis. The MACA method was also found to be superior to the bias correction and spatial downscaling method for fire applications due to the improved spatial representation of analogs over interpolation and the MACA's multivariate approach that maintains the physical relationships among variables.

3. *Weather generators*

Weather generators are stochastic parameter models that generate synthetic time series of weather data based on historical weather at a location (Wilks 1999). Precipitation often is a primary weather field to be generated. There are two ways to create weather generators. One (Richardson 1981) first models wet day probability using a Markov procedure and then uses a frequency distribution to estimate rainfall amount. The other (Racsko et al. 1991) finds dry and wet day series and then estimates rainfall amount. The other variables such as temperature are then computed based on correlation relations under dry or wet conditions. Weather generators can create long-term series with less need for computation resources. However, the accuracy is likely low and depends on parameter specification.

The data produced by a weather generator can then be used to downscale a GCM. For example, Chen et al. (2006) used a first-order Markov chain as a weather generator to simulate wet day probability, the Gamma distribution function to describe variation of wet-day precipitation amounts, and statistical downscaling to transfer large-scale GCM (in both space and time) future precipitation series to station/local scales. The results showed the capacity of the technique to reproduce the observed mean daily amount and model parameters of the daily precipitation series at station/local scales in the study region.

3.2.3 Bias Correction

One feature of statistical downscaling is the use of large amounts of meteorological observations, Y_0. Bias correction uses this advantage in obtaining the differences between Y_0 and the present climate, X_0 simulated by GCM. The differences or biases are assumed to exist as well between the downscaled future local climate and the GCM projected climate. The downscaling was improved by applying the differences (bias correction) that could be obtained based on the cumulative distribution function or quantile instead of the original variable values.

Selection of the downscaling and bias correction methods depends on many factors, including the meteorological variables, seasons, regions, and time frequency (Maurer and Hidalgo 2008). Wood et al. (2004) found that the bias-correction and spatial disaggregation (BCSD) method was able to obtain the main features of the observed hydrometeorology based on the simulated climate sin the western US, while linear interpolation led to unacceptably biased hydrologic simulations and the spatial disaggregation result was not useful for hydrologic simulation purposes without bias-correction. However, performance of the bias correction method relative to other methods varies, depending on specific situations. Goodess et al. (2012) found that it was not possible to identify a consistently superior method for statistical downscaling of temperature and precipitation in the majority of 22 cases with a focus on extreme conditions in European regions. They therefore recommended the use of a range of the better statistical downscaling methods for the construction of scenarios of extremes.

3.2.4 Extremes

Two problems could arise when the statistical methods described above are used to download weather and climate extremes. First, extremes occur in the very far tail of both magnitude and frequency distributions and they may not meet the conditions required for the statistical distributions. There might be less reliable relationships connecting the extremes to their predictors. Second, the statistical downscaling technique is based on statistical relationships between large-scale fields or circulations and local weather and bias corrections of GCM projections using observations. They usually need a large number of sample data. Extremes, however, are rare events and there are much fewer observations for extremes than normal values. In addition, rainfall extremes often occur under circulation anomalies at much larger spatial scale and longer duration. The downscaling techniques for normal or mean conditions may only consider circulation at more local area and shorter periods.

Recently new statistical tools have been applied to solve these problems. Extreme value theory (EVT) is one of these widely used tools. Different from regular statistical tools that describe mean, variance and skewness (the first, second, and third moment of a probability distribution function), EVT describes the tail (fourth moment). EVT uses the generalized extreme value (GEV) distribution to model the frequency of extremes such as a series of maxima or minima. The GEV is a cumulative exponential distribution function with location, scale, and shape parameters and depending on the parameters is a general form for one of the three special cases of Gumbel, Fréchet and Weibull distributions. Using the generalized Pareto distribution (GPD) the EVT can model the magnitude of extremes such as the abnormal values exceeding a high threshold or very high quantiles.

There have been an increasing number of EVT applications to statistical downscaling. Friederichs (2010) assessed the potential of using EVT for statistical downscaling of the re-analysis of daily extreme precipitation events to German weather stations. The results from two approaches (a Poisson point process representation with non-stationary parameters using defined constant threshold and parameters, and the peak-over-threshold representation using the GPD with defined threshold) indicated that both approaches led to reliable estimates of high quantiles. It also showed that the EVT application produced less uncertain quantile estimates than the censored quantile regression. Another study by Mannshardt-Shamseldin et al. (2010) fit the GEV distribution to the right tail of the distribution of both rain gauge and gridded events and developed regressions to connect them. They found that return values computed from rain gauge data were extremely higher than those computed from gridded data and that it was possible to project future changes in precipitation extremes at the point-location level based on results from climate models.

3.3 *Comparisons of Two Approaches*

3.3.1 Strengths and Limitations

The major strengths and limitations of the dynamical and statistical downscaling approaches are compared in Table 2. Dynamical downscaling is physically consistent between RCMs and GCMs and among meteorological elements to be downloaded. The RCMs have almost the same components of dynamics and physics as GCMs. In contrast, statistical downscaling has to develop separate relationships for each of the variables and therefore the physical relationship between variable may be weak or lacking.

Dynamical downscaling provides a wide range of variables with high frequency, while statistical downscaling mostly provides temperature and precipitation although humidity and wind are also available from some datasets. Dynamical downscaling products can provide daily or even hourly temperature and precipitation. Daily data is required in calculating some fire weather indices such as the Keetch-Byram Drought Index (KBDI) (Keetch and Byram 1968). Hourly data is useful for estimating some atmospheric and fuel properties whose values at certain

Table 2 Comparisons of the strengths and limitations with dynamical and statistical downscaling approaches

Feature	Dynamical	Statistical
	Strength	Weakness
Consistence	Physically consistent	Inconsistent among variables
Variable	Multiple variables	Limited variables
Forcing	Local interactions	Not included
Vertical	Yes	No
Data need		Observations, sparse in some areas
Assumption		Empirical relations applied to future
	Weakness	Strength
Computation	Need in large capacity	Efficient
Resolution	Lower (but more frequent)	Higher (but lower frequency)
Scenario	Limited number	Multiple no. for assembling
Bias	Passed over from GCMs	Control with observations
	Common features	
Extreme	Included	EVT used to downscale extremes
Dependence	Model dependent	Method dependent

periods of a day are important to fire behavior modeling and fire management. For example, prescribed burning is conducted mostly between 9 am and 3 pm when wind, fuel moisture, and atmospheric stability during this period are key for burn prescriptions. Wind and humidity are useful for estimating the Fosburg fire weather index and many NFDRS indices (Deening et al. 1977). Dynamical downscaling also provides many energy and water fluxes and soil temperature and moisture conditions that are needed for fuel condition and change modeling.

RCMs include many physical interactions and feedbacks important for projections of climate change, which are largely missed in GCMs due to coarse grid spacing. The snow-radiation-warming feedback is an example. The summits of mountains are often covered by snow; due to global warming, part of that snow is melting. This leads to reduced albedo and more absorption of solar radiation. As a result, warming becomes more significant. With finer spatial resolution, RCMs are better than GCMs to represent topography and therefore the feedbacks. Dynamical downscaling provides data at multiple atmospheric layers, which are needed, for example, for calculation of the Haines Index (Haines 1988; Winkler et al. 2007).

Dynamical downscaling requires present and future, large-scale climate projections from GCMs. Besides GCM outputs, statistical downscaling also needs a large amount of observational data to build empirical relationships with the present climate simulated by a GCM. Such relationships may not exist for a particular variable at a specific location. Also, it is especially difficult to develop such relationships in complex terrain such as mountains where data are spatially sparse and may lack representativeness. Another major limitation with statistical downscaling is that the relationships developed using the current data are assumed applicable in the future, which may be not a valid assumption.

One of the major strengths with statistical downscaling is that fewer computation resources are needed in comparison with dynamical downscaling. Because of this, statistical downscaling for a specific region can be conducted using multiple GCMs, emission scenarios, and statistical methods, importantly allowing inter-comparisons and uncertainty analyses. More importantly, it allows obtaining ensemble results, thereby permitting use of higher spatial resolution for statistical downscaling, although this is limited by the resolution of observations. The resolution of statistical downscaling could be as high as meters, while the spatial resolution of dynamical downscaling is usually at tens of kilometers, or a few kilometers if using multiple-nesting techniques. Further, it is relatively easy to update statistical downscaling products using the most recent GCM projections. For example, there are already many downscaling studies using the CMIP5 GCM projections (Brekke et al. 2013). These advantages of statistical downscaling are very difficult to do with dynamical downscaling.

Both downscaling approaches can produce large errors. Those for dynamical downscaling may arise from the errors passed over from GCMs to RCMs and from internal dynamics of a particular RCM. One strength with statistical downscaling, however, is the bias correction procedure using observational data. For statistical downscaling, errors may arise from inaccuracy in observations and empirical relationships.

Extreme events present a challenge for both approaches. Because RCMs include all physical processes needed to simulate mesoscale weather systems and better represent topography and other local forcing for convective precipitation, dynamical downscaling is able to directly produce weather and climate extremes. Many studies, however, have so far indicated limited capacity of RCM simulations of weather and climate extremes. The regular tools useful for statistical downscaling, on the other hand, are unable to deal with the extreme issue. However, the development and application of the extreme value theory may provide a useful solution to this problem.

The regional climate change scenarios depend on which RCM or method is used and on the selection of GCMs and emission scenarios. Specific RCMs and statistical methods have been compared in various geographic regions. Haylock et al. (2006) found that the differences in the future changes of the seven downscaled precipitation indices in England among six statistical and two dynamical downscaling models were no smaller than the differences between the emission scenarios for a single model, emphasizing the importance of including multiple types of downscaling techniques, global prediction models, and emission scenarios. Tryhorna and DeGaetanoa (2011) found that two statistical models (BCSD and the Statistical DownScaling model-SDSM) and a regional climate model (HadRM3) simulated reasonable estimates of extreme rainfall across the northeastern US. The SDSM matched observed extreme climatology the best, while HadRM3 tended to overestimate mean precipitation and extremes. Ayar et al. (2015) showed that the occurrence and intensity of rain were better simulated by stochastic and re-sampling-based statistical downscaling, whereas spatial and temporal variability were better modeled by RCMs and re-sampling-based statistical downscaling. Tang et al. (2016) found that WRF driven by two GCMs tended to underestimate the surface temperature over most of China. Both approaches required further work to improve their ability to downscale precipitation. Zhang et al. (2020) found that PRECIS driven by 20 GCMs better represented temperatures in China than the GCMs alone. Kreienkamp et al. (2019) found that the grid cell bias for yearly values was mostly less than 0.1 °C for temperature and 10% for precipitation totals.

3.3.2 Hybrid Approach

The two approaches of dynamical and statistical downscaling have been used together in some modeling efforts. This hybrid approach aimed to maximize advantages while minimizing the limitations of each approach. For example, dynamical downscaling could be used first to get physically based climate at moderate-resolutions of tens of kilometers. The data can then be used with observations using statistical models to obtain climate at high-resolution of kilometers or higher. Yoon et al. (2012) indicated that a hybrid forecast system has the potential to maximize prediction skill. Walton et al. (2015) described another method for hybrid downscaling. Dynamical downscaling was first applied to five GCMs among the entire 32 CMIP5 GCMs. A simple statistical model was then built relating the GCM input and the dynamically downscaled output. This statistical model was used to approximate the warming patterns of the remaining GCMs.

4 Climate Change Downscaling Tools and Products

4.1 Dynamical Downscaling

4.1.1 CORDEX

CORDEX was initiated in 2009 to evaluate and benchmark model performance and design a set of experiments to produce climate projections for use in impact and adaptation studies (Giorgi et al. 2009; Giorgi and Gutowski 2015). One of the goals of CORDEX is to produce coordinated sets of regional downscaled projections worldwide. The domains of the current 14 regions cover essentially all land areas of the globe at a grid spacing of about 50 km. Dynamical downscaling has been conducted using a large number of RCMs under the forcing from CMIP3 and CMIP5 outputs. The downscaled climate data are provided through the Earth System Grid Federation (ESGF), a geographically distributed archiving system. CORDEX also provides statistical downscaling products.

4.1.2 NARCCAP

NARCCAP (Mearns et al. 2012) is one of the CORDEX regions covering North America. The GCMs used were the Community Climate System Model (CCSM), the Coupled Global Climate Model, version 3 (CGCM3), the Geophysical Fluid Dynamics Laboratory climate model (GFDL), and the Hadley Centre Climate Model, version 3 (HadCM3) for the SRES A2 emissions scenario. The GCM-RCM combinations are listed in Table 3. Simulations were conducted for the present period of 1971–2000 and the future period of 2041–2070 at a spatial resolution of 50 km. The NARCCAP projections provide 3-hourly, 2-D surface meteorological variables (air temperature, precipitation, humidity, winds, pressure), fluxes (radiation, heat, water, momentum) and soil variables, daily 2-D variables (maximum and minimum air temperature, 10-m wind speeds and sea ice), and 3-hourly 3-D atmospheric variables at multiple vertical levels. Constant parameters such as land-cover type, latitude and longitude of grid points, capacity of soil water, surface altitude, root depth etc. are also included. The RCMs were evaluated by running these models during 1980–2004 with the boundary conditions provided by the NCEP/DOE re-analysis data.

Using the projections driven by CCSM3 as an example, the GCM projected strong winter warming in northern Central Canada by about 7 °C. The large warming area extended southward to the Great Lakes by about 5 °C and further to the Deep South by about 4 °C. There is a small isolated warming area by about 4 °C in the Intermountain Region of the western US. Three RCMs projected a similar warming pattern, with a comparable strong warming with WRF, weaker in the Deep South with CRCM, and weaker in both the Great Lakes and Deep South with MM5. The summer warming projected by CCSM3 was the largest in the western US centered on the Intermountain Region by about 5 °C, and extended north towards southwest Canada and northeast towards the Northern Plains. However, warming was only about 1.5 °C in some areas in the South and Southeast US. The RCM projections were much different. The warming center in the western US was weaker, but warming in the eastern US was larger with WRF. The warming in Canada was much weaker. Two other RCMs go to opposite extremes. CRCM projected strong warming by about 5 °C across the continental US, while weaker warming by 2.5 °C was projected by MM5 for the southern US with a value of 3.0 °C in the Southern Plains.

Table 3 RCM-GCM combinations used in the NARCCAP dynamical downscaling simulations (https://www.narccap.ucar.edu/about/plan.html#rcm-gcm)

	GFDL	HADCM3	CGCM3	CCSM
RegCM3	X		X	
ECPC	X	X		
PRECIS	X	X		
CRCM			X	X
WRF			X	X
MM5		X		X

Projections of precipitation were also different between RCMs and the driven GCM. The CCSM3 projected drying summer in the Pacific Coast of the US and southern Canada by 20–50%, and wetting in Southwest US by more than 60% and in the Northern Rocky Mountains by about 40%. The RCM projections were overwhelmingly drier in the US. Although the Pacific Coast remained dry and the Northern Rockies wet, the Southwest turned to weak wetting or even strong drying, together with drying in most of the eastern US.

4.2 *Statistical Downscaling*

In comparison with dynamical downscaling, statistical downscaling applies multiple methods, covers greater ranges of spatial resolution, and demands much less computational resources. As a result, a larger number of the tools and products for statistical downscaling are available. Many of them are online that provides more power for user involvement and visualization. The major features of a number of tools and products for statistical downscaling summarized below are obtained from their websites. Some of these tools and products also come with dynamical downscaling.

(1) *Climate Wizard* (http://www.climatewizard.org)

This tool provides global maps of past and projected future temperature and precipitation trends, based on user-selected climate models and emissions scenarios, time steps, and regional boundaries (e.g., country or state). There is an associated tool of Climate Wizard Custom Extremes, which outputs include 23 temperature and precipitation variables at daily time steps, summary statistics, figures of statistical trends, and optional GIS data downloads. Resolution of outputs varies, ranging from 4 to 50 km depending on the datasets used. The first version of Climate Wizard Basic launched in 2009 using CMIP3. There are plans to update the database for both tools using CMIP5.

(2) *WorldClim* (https://www.worldclim.org)

WorldClim is comprised of a set of global historical and future climate conditions. The spatial resolution is about 1 km. The data can be used for spatial GIS mapping and numerical modeling. The data are provided for use in research and related activities; and some specialized skill and knowledge is needed to use them. More easily available data for the general public will be available soon.

(3) *Earth Exchange Downscaled Climate Projections (NEX-DCP30)* (https://www.nccs.nasa.gov/services/data-collections/land-based-products/nex-dcp30)

NEX-DCP30 was developed by NASA. It applied a statistical technique to downscale maximum and minimum air temperature and precipitation from 33 of the CMIP5 climate models to a very fine, 800-m grid over the contiguous United States

(CONUS). The full dataset covers the historical period (1950-2005) and 21st century (2006-2099) under the four RCPs.

(4) *National Climate Change Viewer (NCCV)* (https://www2.usgs.gov/landresources/lcs/nccv.asp)

The NCCV was provided by US Geological Survey. It includes the historical and future climate projections from 30 of the downscaled models for RCP4.5 and RCP8.5 from NEX-DCP30. The NCCV allows users to visualize projected changes in climate (maximum and minimum air temperature and precipitation) and the water balance (snow water equivalent, runoff, soil water storage and evaporative deficit) for any US state, county and USGS Hydrologic Units (HUCs, a hierarchical classification of hydrologic drainage basins).

(5) *MACA Datasets* (https://climate.northwestknowledge.net/MACA)

The datasets were obtained using the MACA method to downscale the model output from 20 CMIP5 GCMs for historical GCM forcing (1950–2005) and RCP 4.5 and RCP8.5 (2006–2100) from the native resolution of the GCMs to either 4 km or ~6 km in CONUS.

(6) *1/8 degree-CONUS Daily Downscaled Climate Projections* (https://cida.usgs.gov/thredds/catalog.html)

This dataset was provided by USGS using an advanced statistical downscaling method, asynchronous regional regression model. The method combines high-resolution observations with outputs from 16 different GCMs based on four AR3 emission scenarios to generate the most comprehensive dataset of daily temperature and precipitation projections. The gridded dataset covers CONUS, southern Canada and northern Mexico at one-eighth degree resolution and Alaska at one-half degree resolution. The data period is January 1, 1960–December 31, 2099.

(7) *ENSEMBLES Regional Scenario Web Portal* (https://crudata.uea.ac.uk/projects/ensembles/ScenariosPortal)

The ENSEMBLES project is coordinated by the Hadley Centre for Climate Prediction and Research at the UK Met Office. It is one of the CORDEX regions, covering Europe. The ENSEMBLES portal provides statistical downscaling techniques and simulations, which allows users to choose a statistical downscaling method. Downscaling is performed through selecting predictors, and predictand, and downscaling method.

5 Fire Applications of Climate Downscaling

5.1 Climate Downscaling Needs for Fire Applications

While not intended as an exhaustive review, this section describes some applications of downscaling to investigations of changes in wildland fire potential. How one utilizes downscaled climate information to evaluate future wildland fire conditions depends on the scale of the analysis and the weather data needed to describe fire conditions. At the global scale, changes in fire potential can be directly addressed from GCM data without downscaling (Liu et al. 2010a). Analyses at the continental scale can also utilize GCM data directly, for example for Canada by Wotton et al. (2010) with the realization that there may be issues in areas of complex terrain due to the large grid size. As the scale of analysis becomes more refined, downscaling becomes more important and the choice of downscaling method will be driven by the complexity of weather data required for describing fire conditions for a given analysis as not all methods will supply the required information. For some descriptions of fire conditions, precipitation and temperature data may be sufficient as is the case for applications using an index such as the KBDI (Liu et al. 2013). Other descriptions can also involve more detailed indices such as those comprising a fire danger rating system which will require a broader range of weather inputs, often collected at a specific time. Table 4 provides a list of metrics used to describe fire conditions in a number of downscaling studies along with the weather data required for their calculation.

Table 4 Fire metrics used in downscaling studies and weather inputs required for calculation

Fire metric	Weather inputs
KBDI	Daily surface high temperature, daily and annual average precipitation
Fosberg Fire Weather Index	Surface temperature, relative humidity, wind speed
Haines Index	Temperature and dew point depression at two atmospheric pressure levels which are determined by the elevation. Elevation: 950 & 850 hPa (low), 850 & 700 hPa (mid), 700 & 500 hPa (high)
US NFDRS Indices	Surface temperature and relative humidity (daily max/min and 1,300 local time), surface wind speed, and precipitation amount and duration for the previous 24-h from the observation time
Canadian Fire Weather Index	Maximum temperature, relative humidity, wind speed at 10 m above ground, and precipitation
McArthur Forest Fire Danger Index	Surface temperature, relative humidity, wind speed and precipitation

5.2 Applications of Dynamically Downscaled Climate

Dynamic downscaling has been employed for fire studies. Liu et al. (2013) estimated fire potential for the contiguous US using KBDI and a modified form of the Fosberg Fire Weather Index (mFFWI; Goodrick 2002). The study used a NARCCAP dynamical downscaled regional climate change scenario from the HadCM3 GCM with the PRECIS. Findings were expected increased fire potential across the Southwest, Rocky Mountains, northern Great Plains, Southeast, and Pacific coastal regions driven mainly by future warming trends. Rocco et al. (2014) draw upon the climate projections of Liu et al. (2013) for their study of climate change impacts on fire regimes and key ecosystem services in Rocky Mountain, US forests. Rather than relying solely upon a metric such as KBDI for their assessment, they employed a conceptual model of fuel dynamics and fire regimes that relied on current knowledge of climate-fire regime relations combined with literature on future climates to allow for the projection of how climate change is likely to alter both short- and long-term fire regimes in each of the four forest types studied within the region.

Bedel et al. (2013) conducted a similar study to Liu et al. (2013) targeted at the southeastern US but used a different set of GCM/RCM pairings available from NARCCAP. In addition to looking at changes in KBDI, Bedel et al. (2013) also explored changes in the Haines Index that is often related to the potential for large, plume-driven wildfires to grow and exhibit unpredictable behavior (Haines 1988; Winkler et al. 2007). Calculation of the Haines Index involves temperature and specific humidity from different levels in the atmosphere. For the low elevation version of the index appropriate for the southeastern US, that data came from the 950 and 850 hPa pressure levels. The ability to utilize more than just surface data is an advantage of dynamically downscaled over statistical methods.

Other studies have used dynamical downscaling and focused on specific countries and regions. For example, Carvalho et al. (2011) examined the impact of climate change on fire danger at the regional scale using Portugal as a case study. They used the MUGCM with MM5. Canadian Fire Weather Index (FWI) System components were estimated from the MM5 outputs, at 10 km resolution. They found higher FWI values in 2050 at the beginning of summer and increased total area burned. Fox-Hughes et al. (2014) used data from six dynamically downscaled CMIP3 climate models for 1961–2100 to estimate daily values of the McArthur Forest Fire Danger Index (MFFDI) at 10 km resolution over Tasmania, Australia. The MFFDI was developed in the 1960s by CSIRO to measure the degree of danger of fire in Australian forests. Model projections showed a broad increase in fire danger across Tasmania, but with substantial regional and seasonal variation. Days of elevated fire danger increased in frequency during the simulated twenty-first century. They concluded that dynamically downscaled model data were useful projections of future fire danger for landscape managers and the community.

5.3 Applications of Statistically Downscaled Climate

Abatzoglou and Brown (2012) used the MACA statistically downscaled data to analyze fire conditions in the western CONUS (west of 104° W longitude) with the complex meteorology inherent to its heterogeneous landscape. The desired description of fire conditions focused on two daily fire danger metrics widely used in operational fire management, the Energy Release Component (ERC) and the Fosberg Fire Weather Index (FFWI; Fosberg 1978). The ERC is a widely used metric from the US NFDRS (Deeming et al. 1977) that provides a cumulative drying index that responds on multiple time scales tied to fuel size classes based on previous daily temperature, precipitation and humidity. In contrast, the FFWI is not tied to fuel conditions and reflects short-term, weather-driven impacts on wildfire potential that are a function of temperature, wind speed and relative humidity.

The MACA methodology and the more general constructed analog (CA) approach have been used in other fire related studies. Abatzoglou and Kolden (2011) used MACA data in their analysis of the role of climate change and invasive annual grasses on wildfire potential in the deserts of the western United States. Stavros et al. (2014) and Barbero et al. (2015) employed the MACA data in their assessments of the potential for very large fires in the western US and the CONUS, respectively. In these assessments, the MACA data and derived fire danger indices were used as inputs to regional relationships predicting changes in potential for very large wildfires, adding another layer to the description of future fire conditions.

In an examination of climate change and population growth impacts on California wildfires, Westerling et al. (2011) developed daily climate data from six realizations of future climate by downscaling to a 1/8-degree grid usings the CA method following the work of Maurer and Hidalgo (2008). The monthly probability of large (>200 and >8,500 ha) wildfires was modeled using generalized linear models driven by land-surface characteristics, human population, and climate on a 1/8-degree grid. Most scenarios signaled increases in wildfire burned area across much of California, although the range of outcomes was large and increases with time.

6 Concluding Remarks

Climate change under various emission scenarios are projected by GCMs with a typical resolution of a few hundreds of kilometers. The size of most wildfires, on the other hand, is less than tens of kilometers. Downscaling fills the space gap between GCM output and fire applications. Dynamical downscaling is especially valuable for assessing the impacts of climate change on fires at regional and continental scales, in mountains where RCMs are able to capture amplified climate change due to topography-atmosphere interactions and feedbacks, and for estimating fire indices that need atmospheric conditions of multiple variables, at hourly scale, and at atmospheric levels above the ground. Statistical downscaling is especially valuable for assessing the impacts of climate change on fires at local scale, with long and dense meteorological observations, for ensemble analyses, and active involvement of users due to relatively easy applications of the models and low requirements of computing resources.

The CIMP6 GCMs are providing updated global climate change projections under the new SSPs emission scenarios. The corresponding downscaling products are becoming available from, for example, WorldClim (https://www.worldclim.org/data/cmip6/cmip6climate.html) and EURO-CORDEX (Jacob et al. 2020). The CMIP6 projections provide improved simulations of historical climate and higher spatial resolutions with many GCMs, outputs from more GCMs and scenarios for future, and considerations of socioeconomic factors. The downscaled regional and local climate change scenarios of CMIP6 global projections are expected to provide more accurate and climate information and more ensemble products for wildland fire managers to improve decision support. In addition, there is an increasing number of Earth System Models (ESMs) with the CMIP6 projections. ESMs project not only atmospheric but also vegetation conditions, which are important for fires. The vegetation condition themselves are expected to change in future due to climate change, which would change future wildland fires. Thus, the climate change scenarios from CMIP6 would be specifically valuable for assessing the impacts of forest fuel change on wildland fires under changing climate.

Acknowledgments This work was supported by a US Forest Service Southern Research Station-US Department of Energy Oak Ridge Institute for Science and Education (ORISE) project and by the ASTRA project "Value-chain based bio-economy". The authors do not have conflicts of interest.

References

Abatzoglou JT, Brown TJ (2012) A comparison of statistical downscaling methods suited for wildfire applications. Int J Climatol 32:772–780. https://doi.org/10.1002/joc.2312

Abatzoglou JT, Kolden CA (2011) Climate change in western US deserts: potential for increased wildfire and invasive annual grasses. Rangeland Ecol Manage 64(5):471–478. https://doi.org/10.2111/REM-D-09-00151.1

Abatzoglou JT, Williams AP (2016) Impact of anthropogenic climate change on wildfire across western US forests. Proc Natl Acad Sci 113:11770–11775

Acácio V, Holmgren M, Rego F, Moreira F, Mohren GM (2009) Are drought and wildfires turning Mediterranean cork oak forests into persistent shrublands? Agrofor Syst 76:389–400

Anthes RA, Hsie EY, Kuo YH (1987) Description of the Penn State/NCAR mesoscale model version 4 (MM4). NCAR technical note TN-282-STR, Boulder, CO, US. National Center for Atmospheric Research, 67 pp

Ayar PV, Vrac M, Bastin S, Carreau J, Déqué M, Gallardo C (2015) Intercomparison of statistical and dynamical downscaling models under the EURO-and MED-CORDEX initiative framework: present climate evaluations. Clim Dyn 46:1301–1329

Barbero R, Abatzoglou JT, Larkin NK, Kolden CA, Stocks B (2015) Climate change presents increased potential for very large fires in the contiguous United States. Int J Wildland Fire 24 (7):892–899. https://doi.org/10.1071/WF15083

Bedel AP, Mote TL, Goodrick SL (2013) Climate change and associated fire potential for the south-eastern United States in the 21st century. Int J Wildland Fire 22:1034–1043. https://doi.org/10.1071/WF13018

Bradstock RA, Boer MM, Cary GJ, Price OF, Williams RJ, Barrett D et al (2012) Modelling the potential for prescribed burning to mitigate carbon emissions from wildfires in fire-prone forests of Australia. Int J Wildland Fire 21:629–639

Brekke L, Thrasher BL, Maurer EP, Pruitt T (2013) Downscaled CMIP3 and CMIP5 climate projections, p 47. http://gdo-dcp.ucllnl.org/downscaled_cmip_projections/

Carvalho AC, Carvalho A, Martins H, Marques C, Rocha A, Borrego C, Viegas DX, Miranda AI (2011) Fire weather risk assessment under climate change using a dynamical downscaling approach. Environ Model Softw 26:1123–1133

Caya D, Laprise R (1999) A semi-implicit, semi-Lagrangian regional climate model: the Canadian RCM. Mon Weather Rev 127:341–362

Chen YD, Chen X, Xu C-Y, Shao Q (2006) Downscaling of daily precipitation with a stochastic weather generator for the subtropical region in South China. Hydrol Earth Syst Sci Discuss 3:1145–1183

Dai A (2011) Drought under global warming: a review. Wiley Interdisc Rev Clim Change 2:45–65

DeWilde L, Chapin FS (2006) Human impacts on the fire regime of interior Alaska: interactions among fuels, ignition sources, and fire suppression. Ecosystems 9:1342–1353

Deeming JE, Burgan RE, Cohen JD (1977) The National Fire-Danger Rating System-1978. Gen. Tech. Rep. INT-39. Ogden, UT: U.S. Department of Agriculture, Forest Service, Intermountain Forest and Range Experiment Station. 63 p. https://www.fs.fed.us/rm/pubs_int/int_gtr169.pdf

Dickinson RE, Errico RM, Giorgi F, Bates GT (1989) A regional climate model for the western United States. Clim Change 15:383–422

Dupuy J, Fargeon H, Martin-St. Paul N, Pimont F, Ruffault J, Guijarro M, Hernando C, Madrigal J, Fernandes P (2020) Climate change impact on future wildfire danger and activity in southern Europe: a review. Ann Forest Sci 77(35)

Eriksen C, Hankins DL (2014) The retention, revival, and subjugation of indigenous fire knowledge through agency fire fighting in Eastern Australia and California. Soc Nat Resour 27:1288–1303

Eyring V, Bony S, Meehl GA, Senior CA, Stevens B, Stouffer RJ, Taylor KE (2016) Overview of the coupled model intercomparison project phase 6 (CMIP6) experimental design and organization. Geosci Model Dev (Online) 9

FAO (2001) Global forest fire assessment 1990–2000. Food and Agriculture Organization of United Nations, Rome

Faggian P (2018) Estimating fire danger over Italy in the next decade. Euro-Mediterr J Environ Integr 3:13–15. https://doi.org/10.1007/s41207-018-0053-1

Fernandes PM, Botelho HS (2003) A review of prescribed burning effectiveness in fire hazard reduction. Int J Wildland Fire 12:117–128

Fernandes PM et al (2013) Prescribed burning in southern Europe: developing fire management in a dynamic landscape. Front Ecol Environ 11. https://doi.org/10.1890/120298

Flannigan M, Stocks B, Turetsky M, Wotton M (2009) Impacts of climate change on fire activity and fire management in the circumboreal forest. Glob Change Biol 15:549–560

Fosberg MA (1978) Weather in wildland fire management: the fire weather index. In: Proceedings of conference on Sierra Nevada meteorology, June 19–21, Lake Tahoe, CA, pp 1–4

Fox-Hughes P, Harris R, Lee G, Grose M, Bindoff N (2014) Future fire danger climatology for Tasmania, Australia, using a dynamically downscaled regional climate model. Int J Wildland Fire 23:309–321. https://doi.org/10.1071/WF13126

Friederichs P (2010) Statistical downscaling of extreme precipitation events using extreme value theory. Extremes 109–132

Gebrechorkos SH, Hülsmann S, Bernhofer C (2019) Statistically downscaled climate dataset for East Africa. Sci Data 6:31. https://doi.org/10.1038/s41597-019-0038-1
Giglio L, Randerson JT, van der Werf GR (2013) Analysis of daily, monthly, and annual burned area using the fourth-generation global fire emissions database (GFED4). J Geophys Res Biogeosci 118:317–328. https://doi.org/10.1002/jgrg.20042
Giorgi F (2019) Thirty years of regional climate modeling: where are we and where are we going next? J Geophys Res Atmospheres 124:5696–5723. https://doi.org/10.1029/2018JD030094
Giorgi F, Bates GT (1989) The climatological skill of a regional model over complex terrain. Mon Weather Rev 117:2325–2347
Giorgi F, Gutowski WL (2015) Regional dynamical downscaling and the CORDEX initiative. Annu Rev Environ Resour 40:467–490
Giorgi F, Jones C, Asrar G (2009) Addressing climate information needs at the regional level: the CORDEX framework. WMO Bull 58(3):175–183
González ME, Gómez-González S, Lara A, Garreaud R, Díaz-Hormazábal I (2018) The 2010–2015 megadrought and its influence on the fire regime in central and south-central Chile. Ecosphere 9:
Goodess CM, Anagnostopoulou C, Bárdossy A, Frei C, Harpham C, Haylock MR et al (2012) An intercomparison of statistical downscaling methods for Europe and European regions—assessing their performance with respect to extreme temperature and precipitation events, vol 11. Climatic Research Unit Research Publication, University of East Anglia, pp 69. www.cru.uea.ac.uk/publications/crurp/
Goodrick SL (2002) Modification of the Fosberg fire weather index to include drought. Int J Wildland Fire 11:205–211
Gott B (2005) Aboriginal fire management in south-eastern Australia: aims and frequency. J Biogeography 1203–1208. https://doi.org/10.1111/j.1365-2699.2004.01233.x
Grell GA, Dudhia J, Stauffer DR (1994) A description of the fifth generation Penn State/NCAR mesoscale model (MM5). 398 + STR, Boulder, CO, US. National Center for Atmospheric Research
Haines DA (1988) A lower atmosphere severity index for wildland fires. Natl Weather Digest 13:23–27
Hammer RB, Stewart SI, Radeloff VC (2009) Demographic trends, the wildland–urban interface, and wildfire management. Soc Nat Resour 22:777–782
Hardy CC (2001) Smoke management guide for prescribed and wildland fire. National Wildlife Coordinating Group. https://www.nwcg.gov/
Hausfather Z, Drake HF, Abbott T, Schmidt GA (2020) Evaluating the performance of past climate model projections. Geophys Res Lett 47:e2019GL085378
Haylock MR, Cawley GC, Harpham C, Wilby RL, Goodess CM (2006) Downscaling heavy precipitation over the United Kingdom: a comparison of dynamical and statistical methods and their future scenarios. Int J Climatol 26(10):1397–1415
IUFRO (2018) Global fire challenges in a warming world. International Union of Forest Research Organizations, Vienna
Jacob D, Teichmann C, Sobolowski S, Katragkou E, Anders I, Belda M et al (2020) Regional climate downscaling over Europe: perspectives from the EURO-CORDEX community. Reg Environ Change 20:51. https://doi.org/10.1007/s10113-020-01606-9
Jiang Y, Kim JB, Still CJ, Kerns BK, Kline JB, Cunningham PG (2018) Analysis: inter-comparison of multiple statistically downscaled climate datasets for the Pacific Northwest, USA. Scientific Data 5:180016. https://doi.org/10.1038/sdata.2018.16
Jin JM, Wang SY, Gillies RR (2011) An improved dynamical downscaling for the Western United States, climate change—research and technology for adaptation and mitigation. In: Blanco J, Kheradmand H (eds) InTech, London, pp 23–38. ISBN: 978-953-307-621-8, https://doi.org/10.5772/22991
Jolly WM, Cochrane MA, Freeborn PH, Holden ZA, Brown TJ, Williamson GJ, Bowman DM (2015) Climate-induced variations in global wildfire danger from 1979 to 2013. Nat Commun 6:7537

Jones RG, Noguer M, Hassell DC, Hudson D, Wilson SS, Jenkins GJ, Mitchell JFB (2004) Generating high resolution climate change scenarios using PRECIS. Met Office Hadley Centre, Exeter, UK, p 40

Juang H, Kanamitsu M (1994) The NMC nested regional spectral model. Mon Wea Rev 122:3–26

Keetch JJ, Byram GM (1968) A drought index for forest fire control. US Department of Agriculture, Forest Service, Southeastern Forest Experiment Station, Asheville, NC

Kitoh A, Ose T, Takayabu I (2016) Dynamical downscaling for climate projection with high-resolution MRI AGCM-RCM. J Meteor Soc Japan 94A:1–16

Kreienkamp F, Paxian A, Früh B, Lorenz P, Matulla C (2019) Evaluation of the empirical-statistical downscaling method EPISODES. Clim Dyn 52:991–1026. https://doi.org/10.1007/s00382-018-4276-2

Li D, Feng J, Xu Z, Yin B, Shi H, Qi J (2019b) Statistical bias correction for simulated wind speeds over CORDEX-East Asia. Earth Space Sci 6:200–211. https://doi.org/10.1029/2018EA000493

Li Y, Li Z, Zhang Z, Chen L, Kurkute S, Scaff L, Pan X (2019a) High-resolution regional climate modeling and projection over western Canada using a weather research forecasting model with a pseudo-global warming approach. Hydrol Earth Syst Sci 23(4635–4659):2019. https://doi.org/10.5194/hess-23-4635-2019

Liang X-Z, Xu M, Choi HI, Kunkel KE, Rontu L, Geleyn J-F, Müller MD, Joseph E, Wang JXL (2006) Development of the regional climate-weather research and forecasting model (CWRF): treatment of subgrid topography effects. In: Proceedings of the 7th annual WRF user's workshop, Boulder, CO, 19–22 June, 5 pp

Liston GE, Pielke RA (2000) A climate version of the Regional Atmospheric Modeling System. Theor Appl Climatol 66: 29–47

Liu Y-Q, Goodrick SL, Stanturf JA (2013) Future US wildfire potential trends projected using a dynamically downscaled climate change scenario. For Ecol Manage 294:120–135. https://doi.org/10.1016/J.FORECO.2012.06.049

Liu Y-Q, Stanturf J, Goodrick S (2010a) Trends in global wildfire potential in a changing climate. For Ecol Manage 259:685–697

Liu Y-Q, Stanturf J, Goodrick S (2010b) Wildfire potential evaluation during a drought event with a regional climate model and NDVI. Ecol Inf 5:418–428

Lozano OM, Salis M, Ager AA, Arca B, Alcasena FJ, Monteiro AT, Finney MA, Del Giudice L, Scoccimarro E, Spano D (2017) Assessing climate change impacts on wildfire exposure in Mediterranean areas. Risk Anal 37(10):1898–1916. https://doi.org/10.1111/risa.12739

Mannshardt-Shamseldin EC, Smith RL, Sain SR, Mearns LO, Cooley D (2010) Downscaling extremes: a comparison of extreme value distributions in point-source and gridded precipitation data. Ann Appl Stat 4(1):484–502. https://doi.org/10.1214/09-AOAS287

Maurer EP, Hidalgo HG (2008) Utility of daily vs. monthly large-scale climate data: an intercomparison of two statistical downscaling methods. Hydrol Earth Syst Sci 12:551–563. http://www.hydrol-earth-syst-sci.net/12/551/2008/

McGinnis SA, Mearns LO, Nychka DW, Thompson JA (2012) Regional characteristics of statistically bias-corrected NARCCAP data. 2012 AGU Meeting, San Francisco, December 2012

McKenzie D, Shankar U, Keane RE, Stavros EN, Heilman WE, Fox DG, Riebau AC (2014) Smoke consequences of new wildfire regimes driven by climate change. Earth's Future 2:35–59

Mearns LO, Arritt R, Biner S, Bukovsky MS, McGinnis S, Sain S, Caya D et al (2012) The North American regional climate change assessment program: overview of phase I results. Bull Amer Met Soc (September issue):1337–1362

Meehl GA, Senior CA, Stouffer RJ, Taylor KE (2016) Overview of the coupled model intercomparison project phase 6 (CMIP6) experimental design and organization. Geoscientific Model Dev 9:1937

Myers RL (2006) Living with fire: sustaining ecosystems & livelihoods through integrated fire management. Nature Conservancy, Global Fire Initiative, 36 p. https://www.frames.gov/catalog/701

Nakicenovic N, Alcamo J, Grubler A, Riahi K, Roehrl R, Rogner H-H, Victor N (2000) Special report on emissions scenarios (SRES), a special report of working group III of the intergovernmental panel on climate change. Cambridge University Press, London

Navarro-Racines C, Tarapues J, Thornton P, Jarvis A, Ramirez-Villegas J (2020) High-resolution and bias-corrected CMIP5 projections for climate change impact assessments. Sci Data 7:7. https://doi.org/10.1038/s41597-019-0343-8

O'Brien JJ, Hiers JK, Loudermilk EL (2017) The role of fuels for understanding fire behavior and fire effects. In: Kirkman LK, Jack SB (eds) Ecological restoration and management of longleaf pine forests. CRC Press, Boca Raton, pp 107–122

O'Neill S, Urbanski S, Goodrick S, Larkin S (2017) Smoke plumes: emissions and effects. Fire Manage Today 75:10–15

Racsko P, Szeidl L, Semenov M (1991) A serial approach to local stochastic weather models. Ecol Model 57:27–41

Radeloff VC, Hammer RB, Stewart SI, Fried JS, Holcomb SS, McKeefry JF (2005) The wildland–urban interface in the United States. Ecol Appl 15:799–805

Reinhard M, Rebetez M, Schlaepfer R (2005) Recent climate change: rethinking drought in the context of forest fire research in Ticino, South of Switzerland. Theor Appl Climatol 82:17–25

Richards L, Brew N, Smith L (2020) 2019–20 Australian bushfires—frequently asked questions: a quick guide. Research paper series, 2019–20, Australia Department of Parliamentary Services. https://parlinfo.aph.gov.au/parlInfo/download/library/prspub/7234762/upload_binary/7234762.pdf

Richardson CW (1981) Stochastic simulation of daily precipitation, temperature, and solar radiation. Water Resour Res 17:82–190

Rocco ME, Brown PM, MacDonald LH, Carrico CM (2014) Climate change impacts on fire regimes and key ecosystem services in rocky mountain forests. For Ecol Manage 327:290–305

Schoof JT (2013) Statistical downscaling in climatology. Geogr Compass 7:249–265. https://doi.org/10.1111/gec3.12036

Semenov MA, Stratonovitch P (2010) Use of multi-model ensembles from global climate models for assessment of climate change impacts. Climate Res 41:1–14

Silva P, Bastos A, DaCamara CC, Libonati R (2016) Future projections of fire occurrence in Brazil using EC-earth climate model. Rev Bras Meteorol 31(3). https://doi.org/10.1590/0102-778631320150142

Snell SE (2000) Spatial interpolation of surface air temperatures using artificial neural networks: evaluating their use for downscaling GCMs. J Clim 13:886–895

Stambaugh MC, Guyette RP, Stroh ED, Struckhoff MA, Whittier JB (2018) Future southcentral US wildfire probability due to climate change. Clim Change 147:617–631

Stavros EN, Abatzoglou JT, Larkin NK, McKenzie D, Steel EA (2014) Climate and very large wildland fires in the contiguous western USA. Int J Wildland Fire 23(7):899–914. https://doi.org/10.1071/WF13169

Stephens SL, Burrows N, Buyantuyev A, Gray RW, Keane RE, Kubian R, Liu S et al (2014) Temperate and boreal forest mega-fires: characteristics and challenges. Front Ecol Environ 12:115–122

Stewart SI, Radeloff VC, Hammer RB, Hawbaker TJ (2007) Defining the wildland–urban interface. J Forest 105:201–207

Strydom S, Savage MJ (2017) Potential impacts of climate change on wildfire dynamics in the midlands of KwaZulu-Natal, South Africa. Clim Change 143:385–397

Sun C (2006) State statutory reforms and retention of prescribed fire liability laws on US forest land. Forest Policy Econ 9:392–402

Takhsha M, Nikiéma O, Lucas-Picher P, Laprise R, Hernández-Díaz L, Wingeret K (2018) Dynamical downscaling with the fifth-generation Canadian regional climate model (crcm5) over the Cordex Arctic domain: effect of large-scale spectral nudging and of empirical correction of sea-surface temperature. Clim Dyn 51:161–186. https://doi.org/10.1007/s00382-017-3912-6

Tang J, Niu X, Wang S, Gao H, Wang X, Wu J (2016) Statistical downscaling and dynamical downscaling of regional climate in China: present climate evaluations and future climate projections. J Geophys Res Atmos 121:2110–2129. https://doi.org/10.1002/2015JD023977

Tapiador FJ, Navarro A, Moreno R, Sánchez RL, García-Ortega E (2020) Regional climate models: 30 years of dynamical downscaling. Atmos Res 235: https://doi.org/10.1016/j.atmosres.2019.104785

Taylor KE, Stouffer RJ, Meehl GA (2012) An overview of CMIP5 and the experiment design. Bull Amer Meteor Soc 93:485–498. https://doi.org/10.1175/BAMS-D-11-00094.1

Tolhurst KG, McCarthy G (2016) Effect of prescribed burning on wildfire severity: a landscape-scale case study from the 2003 fires in Victoria. Aust Forest 1–14. https://doi.org/10.1080/00049158.2015.1127197

Trigo RM, Palutikof JP (1999) Simulation of daily temperatures for climate change scenarios over Portugal: a neural network model approach. Clim Res 13:45–59

Tryhorna L, DeGaetanoa A (2011) A comparison of techniques for downscaling extreme precipitation over the Northeastern United States. Int J Climatol 31:1975–1989. https://doi.org/10.1002/joc.2208

USAID (2014) A review of downscaling methods for climate change projections, September 2014. http://www.ciesin.org/documents/Downscaling_CLEARED_000.pdf

Uvo CB, Olsson J, Morita O, Jinno K, Kawamura A, Nishiyama K, Koreeda N, Nakashima T (2001) Statistical atmospheric downscaling for rainfall estimation in Kyushu Island, Japan. Hydrol Earth Syst Sci 5(2):259–271

Van Wagner CE (1987) Development and structure of the Canadian forest fire weather Index system, vol 35. Can For Serv, For Tech Rep, Ottawa, 37 p

Vasiliades L, Loukas A, Patsonas G (2009) Evaluation of a statistical downscaling procedure for the estimation of climate change impacts on droughts. Nat Hazards Earth Syst Sci 9:879–894

Vigilante T, Murphy BP, Bowman DM (2009) Aboriginal fire use in Australian tropical savannas: ecological effects and management lessons. Trop Fire Ecol 143–167

Van Vuuren DP, Edmonds J, Kainuma M, Riahi K, Thomson A, Hibbard K, Hurtt GC, Kram T, Krey V, Lamarque J-F (2011) The representative concentration pathways: an overview. Clim Change 109:5

Waldrop TA, Goodrick SL (2012) Introduction to prescribed fires in Southern ecosystems science update SRS-054, vol 54. US Department of Agriculture Forest Service, Southern Research Station, Asheville, NC, 80 p, pp 1–80

Walton DB, Sun F, Hall A, Capps SB (2015) A hybrid dynamical–statistical downscaling technique, part I: development and validation of the technique. J Clim 28(12):4597–4617

Westerling AL, Bryant BP, Preisler HK, Holmes TP, Hidalgo HG, Das T, Shrestha SR (2011) Climate change and growth scenarios for California wildfire. Clim Change 109(Suppl 1):S445–S463. https://doi.org/10.1007/s10584-011-0329-9

Westerling AL, Hidalgo HG, Cayan DR, Swetnam TW (2006) Warming and earlier spring increase western US forest wildfire activity. Science 313:940–943

Wiedinmyer C, Hurteau MD (2010) Prescribed fire as a means of reducing forest carbon emissions in the western United States. Environ Sci Technol 44:1926–1932

Wilby RL, Wigley TML (1997) Downscaling general circulation model output: a review of methods and limitations. Prog Phys Geogr 21(530–548):1997

Wilks DS (1999) Multisite downscaling of daily precipitation with a stochastic weather generator. Clim Res 11:125–136

Williams J (2004) Managing fire-dependent ecosystems: we need a public lands policy debate. Fire Manage Today 64(2):6–11

Williams JT, Hyde A (2009) The mega-fire phenomenon: observations from a coarse-scale assessment with implications for foresters, land managers, and policymakers. In: Proceedings from the Society of American Foresters 89th national convention, Orlando, FL, 30 September–4 October 2009. Society of American Foresters, Bethesda

Winkler JA, Potter BE, Wilhelm DF, Shadbolt RP, Bian X, Piromsopa K (2007) A climatology of the Haines index for North America derived from NCEP/NCAR reanalysis fields. Int J Wildland Fire 16:139–152. https://doi.org/10.1071/WF06086

Wonkka CL, Rogers WE, Kreuter UP (2015) Legal barriers to effective ecosystem management: exploring linkages between liability, regulations, and prescribed fire. Ecol Appl 25:2382–2393. https://doi.org/10.1890/14-1791.1

Wood AW, Leung LR, Sridhar V, Lettenmaier DP (2004) Hydrologic implications of dynamical and statistical approaches to downscaling climate model outputs. Clim Change 62:189–216. https://doi.org/10.1023/B:CLIM.0000013685.99609.9e

Wotton BM, Nock CA, Flannigan MD (2010) Forest fire occurrence and climate change in Canada. Int J Wildland Fire 19:253–271

Xu Z, Han Y, Yang Z (2019) Dynamical downscaling of regional climate: a review of methods and limitations. Sci China Earth Sci 62:365–375

Xu ZF, Yang Z-L (2012) An improved dynamical downscaling method with GCM bias corrections and its validation with 30 years of climate simulations. J Clim 25:6271–6286. https://doi.org/10.1175/JCLI-D-12-00005.1

Xue YK, Janjic Z, Dudhia J, Vasic R, De Sales F (2014) A review on regional dynamical downscaling in intraseasonal to seasonal simulation/prediction and major factors that affect downscaling ability. Atmos Res 147:68–85

Yoon J-H, Leung LR, Correia J Jr (2012) Comparison of dynamically and statistically downscaled seasonal climate forecasts for the cold season over the United States. J Geophys Res 117: D21109. https://doi.org/10.1029/2012JD017650

Zhang L, Xu Y, Meng C, Li X, Liu H, Wang C (2020) Comparison of statistical and dynamic downscaling techniques in generating high-resolution temperatures in China from CMIP5 GCMs. J Appl Meteorol Climatol 59:207–235. https://doi.org/10.1175/JAMC-D-19-0048.1

The Monitoring, Control and Management of Wildlife Fires in Zimbabwe Under a Changing Climate: Challenges and Prospects for Doing It Right

Patience Mazanhi and Innocent Chirisa

1 Introduction

The frequency of wildfires has increased in many forest ecosystems. Corona et al. (2015) alludes that wildfires globally affect ecosystems in the Mediterranean as natural resources and biodiversity are facing rapid deterioration. However, despite the increasing frequency in wildfires, Dube (2013) denotes that fires will intensify as a result of climate changes and natural disasters that makes it important to review fire suppression strategies and find sustainable means of fire management. Together with climate change impact, the spreading frequency and conflagrations of wildfires have increased recently due to the increases in population growth among many other causes (IBR and World Bank 2018). Kaesin et al. (2012) allude that in the previous decades, climate change, coupled with soil degradation, desertification and the loss of biodiversity has become a great threat globally with its impact affecting ecosystems. The recent prognosis has it that climate change, together with wildfires will be top leads in the extinction of biodiversity with an estimated 20–30% of the natural species risking extinction as of 2050 (Kaesin et al. 2012). SCBD (2001) also highlights that forest fire conflagrations have increased in the previous decades globally. The trends on wildland fires require a revision of the management practices and approaches in fire management in order to minimise disastrous effects amidst a changing climate. Mutimukuru-Maravanyika (2010) states that the cen-

P. Mazanhi · I. Chirisa (✉)
Department of Demography Settlement and Development, University of Zimbabwe, Harare, Zimbabwe

I. Chirisa
Department of Urban and Regional Planning, University of the Free State, Bloemfontein, South Africa

W. Leal Filho et al. (eds.), *Sustainability in Natural Resources Management and Land Planning*, World Sustainability Series,
https://doi.org/10.1007/978-3-030-76624-5_28

tralized approaches to natural resources management have failed dismally and so is the participatory approach that has, instead of enhancing the management of resources, given disappointing and unexpected results towards the sustainable management of natural resources. Thus, this chapter seeks to explore the fire suppression approaches that were practiced in the ancient era and how relevant they are in managing wildfires presently. The chapter uses a desktop approach to explore literature relevant to the debates presented in the chapter and to concretise the arguments. A look to the ancient practices of environmental and wildlife fire management is relevant towards sustainable development.

2 Theoretical Framework

The success of the efforts to manage and control fires in the presence of a changing climatic environment requires the integration of various concepts, such as Institutionalism that helps regulate and formulate policies against fires. Some concepts, such as the prisoner's dilemma, common-pool resources, the tragedy of the commons and zoning of wildlife areas are critical in effecting the implementation of wildfire control measures. Chief of all being institutionalism, without institutionalism, it is difficult to coordinate human behaviour and manage the challenges that come on shared resources' management.

Institutionalism is explained by Schmidt (2014) as an embodiment of rules, structures and regularities within a political science structure or setup that influences the behaviour of humans and the management of common goods. According to Amenta and Ramsey (2009), institutionalists influence policies and decisions. This concept is important in the management of natural resources. Without an institutional setup, it may be difficult to manage the usage of natural resources and prevent the further deterioration of the environment for a common good.

Common pool resources is a theory or concept that focuses mainly on natural resources that is a result of human participation and lobbying to produce shared resources (Requier 2004). Commons are all the natural resources that include forests and water bodies that are shared by the community (Choe and Yun 2017). However, there is no extensive free usage of these commons due to the birth of capitalism that has seen many of these common resources being privatized leading to exclusion of other users despite the nature of the commons as non-excludable (Choe and Yun 2017).

The zoning concept brings in solutions to reduce conservation and development conflicts while maintaining culture (Rotich 2002). Zoning is defined by Harris and Hazen (2006) as an action that fosters the protection of the natural environment by setting aside sensitive areas and protecting their habitats. Such protection lessens conflicts of use by different people. It promotes sustainable utilisation of natural resources through discouraging some uses and developments that harm the environment (Eagles et al. 2002). Zoning is a very important tool that can be utilised to successfully monitor, control and manage wildlife zones (Rotich 2002). FAO

(2011) denotes that the most destructive impact on biodiversity and wildfires, apart from climate change, are as a result of human activities.

3 Literature Review

Wildfires have become a cause for concern in the management of natural resources and wildlife. Dube (2015) displays some of the damaging effects of wildfires to compose vegetation destruction, air pollution and wildlife destruction. In recent years, human activities and fuel consumption requirements have contributed to the out-bursts of forest fires across the globe. Mkhwananzi (2007) asserts that an estimate of 95% of forest fires has been as a result of human activities. This means that the successful management, control and monitoring of wildfires has to be a comprehensive and integrative approach that places the key players such as the community affront as a few cases result from natural causes (Nkomo and Sassi 2009). In Zimbabwe alone, wildfires destroyed approximately 950,905 ha of land in 2009, about 1 152,413 ha in 2010 and at least 1,320,325 in the year (EMA 2011). This is a real issue that needs proper strategic frameworks to deal with the enormous impact of wildfires both globally, regionally and locally if sustainability is to be achieved.

4 Monitoring, Control and Management of Wildlife Fires Under a Changing Climate

Wildfires, exacerbated by climate change conditions, have become a global challenge that requires immediate action in terms of policing, management and monitoring (Flannigan et al. 2009). The constantly changing climatic conditions have made wildfires settle inconveniently as drought effects tend to enhance fires. Global, regional and local efforts are imperative to the common problem of wildfires if biodiversity is to be protected. According to FAO (2011), a comprehensive approach to successfully fight the problem of veld fires should include, research (in-depth analysis of fire issues and exploration of possible options for positive change), risk reduction (preventative measures and channelling resources to the core causes), readiness (preparedness), response (giving proper responses against damaging fires) and recovery of what has been lost if the fire had affected livelihoods. FAO (2011) states that it is a blinded view that damaging fires are the danger that needs to be phased off but rather it is imperative to understand that the outbursts of fires are symptomatic and a signal to the underlying problems regarding management.

For the successful fight against fires, it is important to have supporting legislation and policy and enhanced community and institutional capacity. Fire has been in existence since time immemorial and their frequency is a cause for action and the need to move from technical management of fire but the addressing of underlying causes (FAO 2011). The use of the community based natural resource management (CBNRM) concept has become popular in recent years as this method brings together the states, communities and other stakeholders to a common goal to fight against the destruction of natural environments (Treue and Nathan 2007). FAO (2006) reiterates strategies that are required to enhance a collaborative approach to firefighting to include the development of common standards and international systems such as early warning mechanisms, the training and capacity building of important stakeholders and formulation of policies and frameworks for institutional capacity building and support. The common pool resources require common effort by every stakeholder in order to avoid the prisoner's dilemma and the careless neglect of wildfire management (Dube 2009). Also, Technology transfer and the use of satellites in the monitoring of wildfires have been of great use in most countries such as Australia. Luce et al. (2012) denote that the effort to control and manage fires are made complex and complicated by the rising populations, particularly in the United States. The collaborative management of fires has been successfully implemented in some parts of India with the government committed to greening and improving biodiversity. According to IBR and The World Bank (2018), the government of India policed, through its 'National Action Plan on Climate Change', to ensure an increase in forest cover on an estimated land size of 5 million hectares.

Climate change together with the natural fires has a bearing upon the livelihoods of people as they destroy the land to that many countries in the southern African region rely on for their living (Clover and Eriksen 2009). Dube (2015) highlights that fires tend to lower vegetal cover and cause air pollution that is detrimental to the health of people. One of the most common measures put in place by most African countries is the use of fireguards as a preventative means against wildfires. The management of wildfires in most African countries has been bottom-up, with the community playing a major role in suppressing the fires (Nyongesa and Vacik 2018). This is a relevant aspect of institutionalism that upgrades local people who experience the impacts on ground rather than have impositions through top-down approaches. Kenya as an example mostly makes use of traditional expertise on ecological management together with the Grass Fire Act (Cap 327). However, Nyongesa (2015) highlights that there is a neglect of following the regulations regarding the management of grass and bushfires.

Behaviour change in humanity is the immediate need to fight and channel all resources towards mitigating climate change rather than accommodating environment insensitive human activities that contribute to wildfires (Dube 2015). Kenya has embodied cultural beliefs that discourage forest destruction as they fight to preserve their sacred tree of Mukuyu (sycamore tree), Mukurwe and Mugumo trees. With no community member allowed to destroy the sacred trees, it also saves on the protection of other natural resources from destruction (Nyongesa and Vacik 2018). One of the effective approaches that can be applied in the management and control of wildfires is ecotourism that enhances the conservation and preservation of biodiversity (Acquah et al. 2017). This has multiple benefits as it does not only protect the environment but improves the economic livelihoods of the communities that benefit from the income generated (Ogato 2014). Poachers have become an issue in the management of wildfires as they carry out their hunting, with fire as a hunting tool and this has cost many tourist sites, such as resorts through fire destruction (Government of Kenya 2018). FAO (2012) highlights that about 350 million hectares of land is get affected yearly by wildfires and of this, approximately 250 million hectares consist of tropical rain forests.

5 Environmental Education and the Stewardship Ethic

Environmental stewardship has gained popularity in recent years, people have realised the need for environmental responsibility and protection (Arakawa et al. 2018). The application of environmental stewardship in the cultural norms of communities has proven to be one of the effective means to achieve environmental justice as argued by Worrell and Appleby (2000). Berry (2006) postulates that when environmental stewardship and education is embedded in the culture of communities, it empowers and capacitates communities and makes it easy for a bottom-up approach to environmental management. Stewardship practices that include greening mechanisms by combined community-based efforts have become more common and relied upon by most governments globally (Baker 2014). Successful environmental management is upon the realisation that institutions cannot work in isolation from thc community especially with regards to the common or shared resources and thus the top-down approach is rendered futile. According to Ramolini et al. (2012), stewardship practices to environmental management enhance the resilience of communities and brings them in unison with the government authorities to fight for a common cause.

Communities and individuals need to have a sense of ownership and responsibility. This would make even the enacted statutes easier followed as both the government and local people work towards achieving a common goal. For most of the developing countries in Africa, the reliance on stewardship is important as it covers government resource gaps through community participation (Arakawa et al. 2018). Most of the efforts that have been put in place in many countries anchored on environmental education to enhance community awareness to the dangers of not maintaining natural resources (Omoogun et al. 2016). The attainment of sustainable environment comes effectively in play when communities are well informed and practice environmental ethics that enable them to appreciate the sacredness of nature (Karatas 2014). Despite the various efforts that are being done to foster good environmental management and conservation practices, Omoogun et al. (2016) argue that many nations are still very far from attaining environmental justice.

A realisation of the need for common responsive action against wildfires enlarges the knowledge of communities and enables them to effectively fight wildfires. With the growing evidence of wildfire intensity amidst climate change imaoct, there is need for technology and information transfer. The information derived from the use of statistical models for projections and establishing the relationship between forest fires and the changing climate enables the projection of the likelihood of future wildfires (Mouillot and Field 2005). This enables the pro-active management, monitoring and control of wildfires before they even come to life. Krawchuk et al. (2009) support the notion highlighting the common use of such models globally and how effective they have aided in preparing and equipping against fires in the United States according to Westerling and Bryant (2008) and Canada's Boreal forest as highlighted by (Flannigan et al. 2005). Failure to act upon the damaging impact of wildfires can overwhelm natural resources and environmental managers as they intensify due to the exacerbating climate change (Halofsky et al. 2020). Luce et al. (2012) denotes that the efforts to design landscapes that are fire-smart are socially and technically challenging and, as such, there is a need for capacitating and increasing community awareness in managing fires.

It is indicated that regional efforts to manage fires tend to crumble due to insufficient training and the failure to prioritise wildfires as a critical issue in policy-making (Chinamatira et al. 2016). Globally, various departments exist that deal with other areas deemed important but no specific departments with a sole focus on fire management exist. However, with the engagement and practising of environmental stewardship, the management, control and monitoring of wildfires and environmental problems become an easy battle. Grassroots' efforts are what defines the stewardship ethic and ought to be initiated (Worrell and Appleby 2000).

6 Research Methodology

In order to support the research and arguments presented in the chapter, the chapter used a desktop approach and literature review to concretise the arguments put forward for environmental management and wildfire monitoring were used. The chapter made use of both global, regional and local scholarship and reports that are rich in wildland-fire-information. These have been visited to gain an appreciation of the nature of the problem of wildfires and how the efforts by other countries can be of use when applied in Zimbabwe. The chapter analyses the current efforts that are put in place towards the management of wildfires and how, with the moving of time, the old, culturally imbedded efficient practices have been lost along the way. The analysis is made from the pre-colonial, colonial and post-independence eras. Since most literature an defforts towards controlling wildfires are biased towards its suppression alone, this chapter focuses on the monitoring, control and management of wildfires in Zimbabwe under a changing climate. It tries to provided an ind-depth analysis of the extent to which the modern efforts are efficient and gives a comparison with the ancient fire management efforts. The evidences are presented in text, tables and skematic diagrams to fully provide a picture of the extent of the situation and how imperative is urgent action needed.

7 Analysis and Results

Climate change and wildfires seem to be interconnected and have simultaneous effects. Climate change and the degradation of biodiversity have become frequent and more rampant from the turn of the twenty-first century (UNISDIR 2012). Humanity and nature have not been spared of the effects of climate change, with humans becoming major contributors to bio-destruction (EEA 2012). Flannigan et al. (2009) reveals the efforts being put in place by fire agencies and managers to rectify and prevent frequent recurrences, severity and intensity of the fires. However, there is huge neglect by many nations of properly managing fires as in Canada, Australia and Russia, most of the wildfires conflagrate into unmanageable blazes as they are freely left to burn. This has indicated the existing battle against climate change and increasing frequency of wildfires. Many nations are embarking on mitigation approaches globally.

Climate change has become the topping agenda in many conferences globally as with its intensity and destruction, many facets such as bio-diversity are affected. The recent prognosis has it that climate change, together with wildfires will be top leads in the extinction of biodiversity with an estimated 20–30% of the natural species risking extinction as of 2050 (Kaesin et al. 2012). SCBD (2001) also highlights that forest fire conflagrations have, in the previous decades, increased globally. However, despite the immediate need to strategize on the best policies against wildfires in a changing climate, Flannigan et al. (2009) highlight the lack of an integrative approach to fire fighting and management as one of the contributing factors to the continued challenges of wildland fires. There is poor coordination and poor mobilisation of human and technical resources with the common goal of managing wildfires. There is a lack globally, of cooperation between international countries and this has led to poor or low capacity to fight wildfires.

8 Monitoring, Control and Management of Wildlife Fires Under a Changing Climate in Zimbabwe

The challenges of wildlife fires are slowly frequenting in Zimbabwe particularly as it is a Savannah climate. Nyamadzawo et al. (2013) argue that little efforts towards research on the major causes of wildfires in Zimbabwe have been done and this has affected proper planning for management and control of wildfires.

8.1 Colonial Institutions for Environmental Management

The issue of climate change and environmental protection goes long back and the efforts towards mitigating wildfires can be traced back to the colonial period where institutions and regulations for natural resource protection were put in place. The protection of natural resources in then Rhodesia, was enhanced by the natural resources board whose powers were postulated in the Natural Resources Act of 1975. According to Nickerson (1994), the natural resources board was mandated to monitor the utilisation of natural resources, recommend necessary legislative tools for resources management and also educating the public through awareness campaigns. The relevance of the protection of natural resources in Rhodesia has been weighed according to the protection, conservation and improvement of the natural resources as laid out in the natural resources Act (Mushuku and Ngwenya 2014). The natural resources board was also responsible for monitoring the efforts by landholder committees on the preservation of natural resources. This enhanced a bottom-up approach to the management of resources as landowners and communities at a local level were engaged in the management of natural resources (Nickerson 1994).

Nyamadzawo et al. (2013) highlights the efforts towards fire management that were functional in the colonial era that can be traced back to 1890. The regulations that were functioning to preserve natural resources are the Native Land Husbandry Act of 1951 and the Natural Resources Act (No. 9) of 1941 (Stocking 1978). Local farmers were given the responsibility to protect their farmlands from fires through the use of fire-guards and fencing off their farms as a way of restricting the movements of people and at the same time preventing outbursts of fires. According to Mudekwe (2007), the systems regarding fire management in Zimbabwe were borrowed from the colonial regime that ensured the suppression, detection and quick responses to fires. However, with the onset of the LRP of 2000, the system was done away with due to the realisation that it was not easy to prevent fires but rather it was efficient to formulate mitigating measures that are more sustainable to fire management.

8.2 Post-Colonial Institutions for Environmental Management: Change with Continuity?

The transition from the colonial to the post-colonial period brought with it some changes in the management of natural resources. Mapedza (2007) highlights that the colonial regulations and policies regarding environmental protection were harsh, inhuman and exclusionary to the indigenous people. The attainment of independence would have brought in some changes that would enable community participation in decision-making. However, Mapedza (2007) denotes that the institutions have remained insensitive and rather a continuity of the colonial regime as local people are regarded as exploiters. Of some important regulation, The Parks and Wildlife Act of 1996 embodies some regulations regarding fire management before the beginning of its season that includes the placement of boundaries or fireguards. Mudekwe (2007) also highlights the contribution of the Forestry Act (1996) that shunned smoking in any forestry. Some of the statutes that regulate wildfires, as highlighted earlier, include the Environmental Management Act (Chapter 20:27) together with theThe rural district councils through the Rural District Councils Act (Chapter 29:13) are also empowered to enact regulations that control fires and bylaws that enable the management of natural resources (Dube 2015).

Amongst the provisions for environmental management, the environmental management agency (EMA) stands constitutionally (Mangena 2014). The EMA is a board created by the government to facilitate the implementation of the environmental act (Environmental Management Act of 2000). The efforts by EMA to manage the environment are backed by the Act. The agency plays a crucial role in controlling human behaviour in the management of the environment through penalties for failure to comply and educational campaigns for human cooperation in management. The EMA was known as 'the Department of Natural Resources' (Mangena 2014). The agency plays various roles, such as taking audits on the environment, carrying out environmental impact assessments and approving when carried out by third parties (Mukwindidza 2008). This is crucial in protecting environmentally sensitive areas especially from destructive human developments on ecology. However, Mangena (2014) argues that most of the measures by the agency and the government to protect the environment are not considerate of non-human species and thus makeing it difficult to achieve a comprehensive and sustainable environment.

Richardson et al. (2006) reveal that there is a dearth of environmental law in most of the developing countries, Zimbabwe included. For Zimbabwe, it was a continuity of laws that were already in existence before independence. Various statutes were already in existence to fight against veld fires in the country. Nyamadzawo highlights that statutes, such as the Parks and Wildlife Act (Chapter 19:05) of 1996 and the Forest Act (Chapter 19:05) of 1996 were enacted to wedge against unprecedented fires. EMA (2007) highlights that various legislative and statutory mechanisms have been put in place by the government of Zimbabwe to hedge against wildfires. The SI 7 of 2007 requires that all landowners should ensure that they have preventative mechanisms of fire and to act in the suppression of fire in case of conflagrations. The instrument further apprehends that no one is permitted to set fire between July and October every year as it is the country's fire season.

Another important institution is Agritex that is a board focusing on agricultural practices that are climate-smart agriculture that do not worsen environmental degradation (Marambanyika et al. 2012). The department was officially initiated in 1980 as the Agricultural, Technical and Extension Services department (Hagmann et al. 2000). The department also offers training to farmers together with research on effective agricultural practices (Agritex 1994). However, Hagmann et al. (2000) argues that the board is not fully capacitated to assist all farmers as most farmers tend to rely mainly on non-governmental organisations. The department's role in environmental protection through farm work is relevant but has been affected dearly by financial challenges and lack of full governmental support. Apart from influencing technology to farmers (Hagmann et al. 2000), there is not much that the department has done towards environmental protection and particularly the management of wildfires.

Also, the Zimbabwe Parks and Wildlife Management Authority (Zimparks), is a board that focuses on the management of wildlife in national parks. It was in existence even before the attaining of independence by the country with its controlling powers driven by the Parks and Wildlife Act of 1975. The agency has made efforts to engage with the communities in the management of wildlife resources. One of the most popular programmes that the agency uses is the CAMPFIRE that has proven to be a relevant conservation means. The success of such a program is due to the acknowledgement of the community efforts in environmental preservation by giving them access to incentives and community benefits (Zimparks 2020). According to Frost and Bond (2007), the use of incentives in the management and protection of natural resources has been proven to be a successful measure that brings cooperation with the local communities. With the continued deterioration of the environment and natural resources through human activities, Zimparks becomes a relevant department that should play a crucial role in enabling resources management.

8.3 Wildlife Zoning in Zimbabwe

The zoning concept brings in solutions to reduce conservation and development conflicts while maintaining culture (Rotich 2002). Zoning protects vulnerable areas that need to be protected and the concept is advocated for by many organisations. It promotes sustainable utilisation of natural resources through discouraging some uses and developments that harm the environment (Eagles et al. 2002). Zoning is a very important tool that can be utilised to successfully monitor, control and manage wildlife zones as highlighted by Rotich (2002). Zoning in Zimbabwe has yielded positive results towards both the protection of the natural resources and community benefits. Mashuku and Ngwenya (2014) denote that as part of zoning practices in the country, people were relocated from sensitive areas to more humane conditions that were conducive for human habitation. The communities that were resettled in a bid to protect wildlife became the custodians of conservation benefits that today is termed, CAMPFIRE.

The redistribution of land from white commercial farmers has had an impact on the management of wildlife in Zimbabwe. According to Mashuku and Ngwenya (2014), the fast-track land resettlement programme did not only focus on the farming areas but also included the resettlement of people from wildlife dominated areas. The motive that was behind the resettlement program was to enhance the food security of communities initially. Positively, this also contributed to the protection of wildlife and natural resources. The move towards land resettlement also resulted in CAMPFIRE villages that were capacitated with ownership and conservation of wildlife zones as highlighted by Bong (1999). According Murphree (2011), this promoted good conservation practices as the CAMPFIRE beneficiaries benefited from accessing grazing land and wildlife as a community resource. The fast-track land resettlement program contributed to wildlife conservation through resettling some locals from environmentally sensitive areas, such as regions 4 and 5 as there were conflicts from human and wildlife interaction.

The most urgent development need in Zimbabwe is the inclusion of climate change in zoning, planning and decision-making to create climate-resilient developments. Jiang et al. (2017) allude that sustainable development measures that promote the protection of biodiversity and natural resources are of urgent need. A delay in meeting these challenges will lead to a complex scenario where it is going to be difficult for policy-makers to overhaul both climate change, wildfires and the ever-increasing population that is already becoming a developmental issue (Abunnarsr 2013). UNHABITAT (2016) emphasises on making climate a developmental issue as with its neglect, humanity is affected. Therefore, it is important for humanity to work towards resilient management and protection of wildlife. Zimbabwe makes use of the "National Climate Policy of 2017" that promotes evidenced-based methods and strategies for climate change mitigation (Government of Zimbabwe 2020). Climate change is a policy issue that needs a collaborative and integrative approach that involves all stakeholders in management and this requires good governance (Milan 2016).

8.4 Case Study: *Monitoring, Control and Management of Wildlife Fires Under a Changing Climate in Zimbabwe*

The successful management of wildlife in communal areas depends highly upon the cooperation of the farmers with the policy-makers. Mhuriro et al. (2017) highlight the existing conflicts between the communal farmers and the adjacent wildlife zones. According to Gandiwa et al. (2013), such conflicts have made it difficult to enhance the success of the CAMPFIRE programmes and thus, to maintain what already existed, there is need to ensure that the arising conflicts are addressed. Kahuni et al. (2014) highlight that most of the protected wild animals have become a danger to adjacent communities as they sometimes kill people and domesticated animals that affects the livelihoods of people. Gandiwa et al. (2013) provide workable examples of making habitat conservation successful through community engagement to include the exploration of other revenue generation and remuneration methods. This is because the revenue generated from the ordinary methods is not enough to be distributed to all community individuals. Therefore, to progress in wildlife management in most communal areas, it is important to ensure adequate community benefits and safety as a way of motivation.

Resettlement farms were a result of the government's efforts to achieve equity in the distribution of land through redistribution to local people (Njaya and Mazuru 2014). The redistribution of land was in such a way that local people would get about three lots (one for residential, one for grazing and the other for farm activities). The environment was impacted negatively as the vegetal cover was lowered as people cleared land for settlement and farming. Njaya and Mazuru (2014) postulate that the effects on the environment were also felt even in commercial farmlands as woodlands reduced. UNDP (2002) states that approximately woody vegetal cover lowered by 1.4% with cultivation land expanding by approximately 2%. Capacity utilisation for the management of natural resources for local farmers both in commercial and resettlement areas needs to be enhanced to successfully achieve environmental justice and protect natural resources. The white settlers, although they alienated black people, protected jealously their farmlands, natural resources and wildlife that the indigenous farmers today are very far from achieving. The present efforts are biased towards the suppression of fires and not more on the effective monitoring, management and control that saves to fight against even future conflagratons. It is imperative therefore that comprehensive approaches be adopted in order to effectively fight wildfires and minimise impacts in the prevailing climate change.

9 Discussion

Generally and progressively from the colonial to the present day, there is a deterioration in the application of local environmental management systems and a lack of better alternative methods to cover up (Njaya and Mazuru 2014). The progression from the old times has seen the loss of many vital practices and systems that would have been of more help today as climate change is worsening environmental deterioration. Table 1 summarises the NRM from the pre-colonial to the present day in Zimbabwe and the continuity and changes in NRM.

Manatsa and Chigwenya (2007) highlights that in all phases, there has been a failure to create institutions with a specific mandate to fight aginst wildfires and climate change. Government and NRM systems have changed from one to another, but little has changed regarding the management of natural resources but rather a downward spiral. There is a relaxation in the management of natural resources in the modern times as compared to the colonial times and this has cost the nation of vital natural resources thus placing it on a more vulnerable state against the continual climate change impact. It is as though the coming in of any government system brings its on NRM methods that are divergent of those that have been used from the past. This disintegration and lack of continuity has led to environmental degradation. The modern regime has to appreciate the indigenous efforts that were used in climate change and NRM. In as much as the colonial regime may be ‘hated’ by state to soeme extent, it is still mandatory that such differencs be put aside and borrow the efficient practices that kept the natural resources intact. This has become a crucial matter in the modern times where climate changed has taken its toll. Change with continuity is imminent for sustainability.

Table 1 NRM comparison in phases

Pre-colonial NRM	Colonial NRM	Post-colonial NRM
• Environmental Stewardship ethic was well imbedded (sacred reverence and holistic NRM) • Taboos and beliefs formed the customary laws • Cutting down of trees was prohibited with only dry wood allowed for collection • Strict protection of socially and religiously sacred areas eg forests • However: There was no • Exclusive communal use of resources • Environmental challenges were not rampant that era	• The Natural Resources Act of 1941 (coercive NRM with strict monitoring of natural resources) • The Land Husbandry Act of 1951 (arresting ecological deterioration, use of fireguards and fencing for fire protection) • 1896 Game Law promulgation (wildlife utilisation regulation) • 1929 Game and fish Act promulgation (led to the formation of the Matopo Reserve). However, it was top down-command-control NRM system • No commonly shared, no participation • Government local conflicts	• The land reform program led to the loss of many practices • SI 7 0f 2007 (all landowners to have fire preventative and suppression mechanisms of conflagrations • EMA Board, penalties for failureto comply and educational campaigns for human cooperation in management • Agritex (climate-smart agricultural practices • The Zimbabwe Parks and Wildlife Management Authority (Zimparks), (management of wildlife in national parks) • National Climate Policy of 2017 that promotes evidenced-based methods and strategies for climate change mitigaton • CAMPFIRE community based management of natural resources) • However: Land resettlement worsened environmental degradation and natural resources depletion

Source Manatsa and Chigwenya (2007)

The transition from the colonial to the post-colonial period brought with it some changes in the management of natural resources. Despite the harsh and exclusionary colonial rules on environmental management the white settlers appreciated well the importance of protecting natural resources and wildlife and as such were stricter (Mapedza 2007). However, the exclusionary policies that did not promote community participation in decision making were carried on after independence and this left behind the sense of responsibility towards the environment that the settlers had (Njaya and Mazuru 2014). Mapedza (2007) denotes that the institutions have remained insensitive and rather a continuity of the colonial regime as local people are regarded as exploiters. Literature has it that the most culprits in destroying the environment (Mkwananzi 2007), apart from the climate change impact, are humans and without much education and emphasized responsibility, sooner than later, the planet would have lost its life.

Therefore, to successfully manage, control and monitor of wildfires, there has to be a comprehensive and integrative approach that places the key players, such as the community affront as a few cases result from natural causes (Nkomo and Sassi 2009). This is a real issue that needs proper strategic frameworks to deal with the enormous impact of wildfires both globally, regionally and locally if sustainability is to be achieved. The experience of destructive wildfires is a global issue that has affected many countries be it the developing or the developed nations with anthropogenic land-use activities and "natural disturbances" being at the forefront of the causes (Nyongesa and Vacik 2018). According to Dube (2015), the involvement of all stakeholders is relevant in fighting against the destructive nature of fires.

10 Conclusion

The battle against climate change and wildlife fires does exist and needs immediate attention. The pre-colonial and the colonial regime had the environment at heart with strongly imbedded culture sensitive environmental and made efforts to protect the natural environment. Even though pre-colonial policies were exclusionary, they managed dearly to protect the natural resources that most indigenous people have failed dismally to carry on even after being granted the power, independence and land. The government that took place after independence took a biased stance that was leaning on equity in the distribution of resources and neglecting the importance of environmental protection. With the passing of generations and with those who used to hold the environmental protection principles passing away, most of the environmental principles were lost along the way. Indigenous principles need therefore to be traced back and combined together with the modern environmental management principles, may enable sustainability. A reactive approach in mitigating wildfire is the commonly used way, however, with climate change in the picture, more pro-active monitoring, management and control measures ought to be implemented. As most of the fire and environmental management practices were were lost along the way from the pre-colonial era, much work is now necessary to re-implant the ethics towards sustainable environmental management. Of late, the modern communities have lost the practices and only aware of fire suppression without futuristic mitigating and monitoring plans. Climate change is there to stay and humanity needs not to worsen the challenges by failing to manage the environment and by contributing to the destruction of natural resources.

11 Recommendations

- There is a need for community awareness to enforce environmental stewardship ethics for collaborative natural resources management.
- Community engagement and participation in wildlife and environmental management is imperative so that it becomes easier to even resuscitate the indigenous environmental management practices that were in place before.
- The chapter recommends the revisit of the former environmental management practices in order to appreciate how the environment was maintained, how fires were fought, monitored and controlled. This would inform the present and help in planning for the future.
- Conflict management between stakeholders is to be improved in order to enhance community participation in protecting common pool resources
- There is need to form solid institutions whose mandate is specifically to manage, control and monitor wildfires.
- The problem of wildfires is to be made a policy issue that has to be worked on in just the same urgent manner as with other policy issues such as housing challenges that are given high priority
- A revisit of ancient environmental ethics and practices is relevant and their application in the modern era in order to overcome the intensifying wildfires and climate change impact.

References

Abunnasr, YF (2013) Climate change adaptation: a green infrastructure planning framework for resilient urban regions. University of Massachusetts, Open Access Dissertations. https://doi.org/72510.7275/yg71-qy42

Acquah E et al (2017) The socio-cultural impact of ecotourism on park-adjacent communities in Ghana. Afr J Hosp Tour Leis 6:1–14

Amenta E, Ramsey KM (2009) Institutional theory. In: Handbook of politics: state and society in global perspective. Springer, California

Arakawa S, Sachdeva S, Sandas V (2018) Environmental stewardship. In: The engaged sustainability: pathways to community cohesion and cultivating meaningful engagement. Springer.

Baker D (2014) Environmental stewardship in New York City parks and natural areas: assessing barriers, creating opportunities and proposing a new way forward. Phd Dissertation, Yale School of Forestry and Environmental Studies.

Berry RJ (2006) Environmental stewardship: critical perspectives past and present. T&T Clark, London

Bond I (1999) CAMPFIRE as a vehiclw for sustainable rural development in the semi-arid communal lands of Zimbabwe: incentives for instituional change. DPhil Thesis, Department of Agricultural Economics and Extension. UZ Publishing, Harare.

Clover J, Eriksen S (2009) The effects of land tenure change on sustainability: human security and environmental change in Southern African Savannahs. Environ Sci Policy 12:53–70 http://Dx.Doi.Org/10.1016/J.Envsci.2008.10.012

Chinamatira L, Mtetwa S, Nyamadzawo G (2016) Causes of wildland fires, associated socio-economic impacts and challenges with policing, Chakari Resettlement Are, Kadoma, Zimbabwe. Fire Sci Rev 5(1)

Choe H, Yun S (2017) Revisiting the concept of common-pool resources: beyond Ostrom. Develop Soc 46(1):P113–129

Corona P et al (2015) Integrated forest management to prevent wildfire under mediterranean environments. Ann Socio-Cult Res Review Paper 39(1):1–22

Dube E (2015) Environmental challenges posed by Veld Fires in fragile regions: the case of the Bulilima and Mangwe districts in Southern Zimbabwe. MSU, Gweru

Dube OP (2009) Linking fire and climate: interactions with land use, vegetation and soil. Curr Opin Environ Sustain 1:161–169

Dube OP (2013) Challenges of wildland fire management in Botswana: towards a community inclusive fire management approach. J Weather Clim Extremes 1:26–41 (Elsevier)

EEA (2012) Urban adaptation to climate change in Europe: challenges and opportunities for cities together with supportive national and European policies, no 2. European Environment Agency

Eagles PFJ et al (2002) Sustainable tourism in protected areas. Guidelines for planning and management. IUCN Gland

EMA (Environmental Management Agency) (2007) Chapter 20:27 Environment Management Act (Environmental Impact Assessment & Ecosystems Protection) Regulations, 2007. Harare, Government of Zimbabwe

Environmental Management Agency (EMA) (2011) Are veld fires a threat to the environment. http://Www.Ema.Co.Zw/Index.Php/Ema-Newspaper-Articles/116-Veld-Fires

FAO (2006) Breed diversity in dryland ecosystems. In: Information Document 9, fourth session of the intergovernmental technical working group on animal genetic resources for food and agriculture. Rome, p 13

FAO (2011) Adapting to climate change: the ecosystem approach to fisheries and acquaculture in the near east and north Africa region. FAO Fisheries and Acquaculture Circular No. 1066, Rome. http://www.fao.org/docrep/014/i2146e00.html

FAO (2012) Forests, trees and people together in a living landscape: a key rural development committee on forestry paper COFO/2012/6.2. http://www.fao.org/docrep/meeting/026/me435e.pdf

Flannigan MD et al (2005) Future area burned in Canada. Clim Change 72:1–16. https://doi.org/10.1007/S10584-005-5935-Y

Flannigan MD et al (2009) The implication of the changing climate for global wildland fire. Int J Wildland Fire 18:483–507 CSIRO Publishing

Frost P, Bond I (2007) The CAMPFIRE programme in Zimbabwe: payments for wildlife services. Ecol Econ 65(4):776–787

Gandiwa L et al (2013) CAMPFIRE and human-wildlife comflicts in local communities bordering Northern Gonarezhou National Park, Zimbabwe. Ecol Soc 18(4)

Government of Kenya (2018) Mt Kenya ecosystem management plan 2010–2020. Www.Kws.Go.Ke/Sites/Default/Files/Parksresorces%3A/Mt.%20Kenya%20Ecosystem%20Management%20Plan%20%282010-2020%29.Pdf

Government of Zimbabwe (2020) Background report on national policy priorities, initiatives and institutions relevant for climate change capacity development in Zimbabwe. Government of Zimbabwe, Harare

Hagmann J, Murwira C, Connolly (2000) Learning together for participatory extension: a grind ytoo an approach developed in Zimbabwe. AGRITEX, Harare

Halofsky JE, Peterson DL, Harvey BI (2020) Changing wildlife, changing forests: the effects of climate change on fire regimes and vegetation in the pacific Northwest, USA. Assoc Fire Ecol Rev

Harris P, Hazen HD (2006) Power of maps: centre mapping for conservation. ACME E-J Crit Geographies 4(1):99–130

International Bank For Reconstruction (IBR)/The World Bank (2018) Strengthening forest fire management in India. Joint Report By IBR & World Bank, Washington DC

Jiang Y et.al (2017) A review of urban planning research for climate change. MDPI Article

Kahuni T, Zisadza-Gandiwa P, Mhlanga W, Libombo E, Mashapa C, Muboko N, Gandiwa E (2014) In: Mararike CG (ed) Land: an empowerment asset for Africa: the human factor perspective. UZ Publishing, Harare, pp 278–287

Kaesin E, Redmond I, Dudley N (2012) Wildlife in a changing climate. FAO Forestry Paper 16

Karatas A (2014) Environmental ethics education as a tool for the prevention of environmental problems in the community. Eur J Sustain Devel 3(4). Research Gate Publication

Krawchuk MA et al (2009) Global pyro geography: macro-scaled statistical models for understanding the current and future distribution of fire. PLoS One 4:E5102

Luce C (2012) Climate change, forests, fire, water and fish: building resilient landscapes, streams and managers. General Technical Reports RMRS-GTR-290. US department of agriculture, forest service, rocky mountain research station, Fort Collins, Co., 207 P

Manatsa D, Chigwenya A (2007) The history of natural resource management in Zimbabwe: a chronicle of how sustainable resource management has remained an elusive concept. J Sustain Devel Afr 9(2)

Mangena F (2014) Environmental policy, management and ethics in Zimbabwe, 2000–2008(1). J Pan-Afr Stud 6(10)

Mapedza E (2007) Forestry policy in colonial and post-colonial Zimbabwe: continuity with change. J Hist Geogr 33(4):833–835

Makwindidza, 2008. Makwindidza, E (2008) The implementation of environmental in Mutasa district of Zimbabwe. Unpublished MSc of Public Administration Degree. University of South Africa, Pretoria

Marambanyika T, Mutsiwegota C, Maringawa KCR (2012) Importance of community participation in sustainable utilisation of wetlands: case of chebvute in Zvishavane district, Zimbabwe. J Environ Sci Eng 13(1):832–844

Mhuriro P, Mashapa C, Mwakiwa E (2017) Determinants of communal farmers' willingness to pay for human-wildlife conflict management in the periphery of save-valley conservancy, southeastern Zimbabwe. J Animal Plant Sci 27(5):1678–1688

Milan BF (2016) Water security in an urbanized world: an equity perspective. Habitat International, Technische Universität Berlin. http://dx.doi.org/https://doi.org/10.14279/depositonce-5767

Mkhwananzi H (2007) Veld fire campaign in Lupane. Environment Africa, Lupane

Mouillot M, Field CB (2005) Fire history and the global carbon budget: a 1×1 fire history reconstruction for the 20th century. Glob Change Biol 11:398–420. https://doi.org/10.1111/J.1365-2486.2005.00920.X

Mudekwe J (2007) Management practices for the protection of forest reserves: the case of Kalahari sand teak forest reserves in western Zimbabwe. FAO, Rome

Murphree M (2011) Environmental management, African centre for disaster studies. North-West University, Potchefstroom

Mushuku A, Ngwenya TA (2014) Wildlife based land reform and its impact on household food security-a case from Zimbabwe. J Rural Dev 33(4):417–436

Mutimukuru-Maravanyika T (2010) Can we learn our way to sustainable management adaptive collaborative management in Mafungautsi state forest, Zimbabwe, Thesis. Wageningen University, Wageningen

Nickerson BJ (1994) The environmental laws of Zimbabwe: a unique approach to the management of the environment. BC 3rd World Law J 14(2)

Njaya T, Mazuru N (2014) Emerging New farming practices and their impact on the management of woodlands in A1 resettlement areas of Mashonaland central province in Zimbabwe. As Devel Policy Rev 2(1):1–19

Nkomo GV, Sassi M (2009) Impact of veld fires on land on smallholder farmers in cashel valley in Zimbabwe. NAF International working paper series no. 02/2009, International Research Network
Nyamadzawo G, Gwenzi W, Kanda A, Kundhlande A, Masona C (2013) Understanding the causes, socio-economic and environmental impacts, and management of veld fires in tropical Zimbabwe. Fire Sci Rev 2(1):1–13
Nyongesa KW (2015) Fire management if forests and national parks of Kenya: case studies at Kakamega
Nyongesa KW, Vacik H (2018) Fire management in mt Kenya: a case study of Gathiuru forest station. MDPI Article
Ogato GS (2014) Planning for sustainable tourism: challenges and opportunities for ecotourism development in Addis Ababa, Ethiopia. Am J Hum Ecol 2014(3):20–26
Omoogun A, Onnoghen UN, Egbonyi EE (2016) From environmental awareness to environmental responsibility towards a stewardship demarcation. J Educ Issues 2(2)
Parks and Wildlife act Chapter 20:14 (1996) The parks and wildlife act chapter 20:14. Government Printers, Harare. Http://Www.Environmental-Auditing.Org/Portals/0/Auditfiles/Zweng02ar_Sm_Wildlife.Pdf
Romolini et al., 2012. Romolini M, Brinkley K, Wolf LW (2012) What is urban environmental stewardship? Constructing a practitioner-derived framework. USDA Forest Service Reserch note PNW-RN, Research Gate Publlication
Requier-Desjardins D (2004) Produced common pool resources, collective action and sustainable local development: the case of food processing clusters. Research Gate
Richardson BJ, Mgheoji I, Botchway F (2006) Environmental law in post-colonial societies: aspirations, achievements and limitations. Research Gate
Rotich D (2002) Concept of zoning management in protected areas. J Environ Earth Sci 2(2) Kenya
Secretariat of the Convention on Biological Diversity (2001) Impacts of the human-caused fires on biodiversity and ecosystem functioning and their causes in tropical temperate and boreal forest biomes, Montreal, SCBD, 42p, SCBD Technical Series No. 5
Schimidt V (2014) Institutionalism. In: The encyclopaedia of political thought
Stocking MA (1978) Relationship of agricultural history and settlement to severe soil in Rhodesia. Zambezia 6(2):129–145
Treue T, Nathan I (2007) Community-based natural resource management. http://41.204.161.209/bitstream/handle/11295/42598/Technicalnotemarts2007CommunityBasedNaturalResourceManagement.pdf?sequence=1&isAllowed=y
UNHABITAT (2016) Addressing climate change in national urban policy: a policy guide for low carbon and climate resilient urban development. Cities and climate change initiative policy note, Nairobi
United Nations Development Programme, UNDP (2002) Evolving institutional framework for community-based natural resource management in Mozambique: a case study from Choa Highlands. Afr Stud Q 5(3). UNDP Interim Mission Report. UNDP, Harare.
UNISDR (2012) How to make cities more resilient: a handbook for local government leaders. A contribution to the global campaign 2010–2015 "Making cities resilient—my city is getting ready!" United Nations, Geneva.
Westerling AL, Bryant BP (2008) Climate change and wildfire in California. Clim Change 87:231–249. https://doi.org/10.1007/S10584-007-9363-Z
Worrell R, Appleby MC (2000) Stewardship of natural resources: definition, ethical and practical aspects. J Agric Environ Ethics 12(3):263–277

Monitoring Wildfires in Forest and Grassland Related to Sugarcane Burning with Geotechnologies

Luis Alberto Olvera-Vargas and Noé Aguilar-Rivera

1 Introduction

The importance of agricultural management, from the beginning to the end of the value chain, lies in the fact that these activities impact the environment and human health. Of the environmental impacts associated with agriculture, the most significant are deforestation and the growth of the agricultural frontier, water and soil contamination as the result of using agrochemicals, and crop burning (Youssef et al. 2019), the last of which is a very common practice associated with eliminating agricultural waste, cleaning the land, controlling weeds, pests, and insects and facilitating the absorption of nutrients. However, this activity generates air pollution and contributes to greenhouse gases because it emits black carbon, carbon dioxide, carbon monoxide, nitric oxide and other organic and inorganic compounds into the atmosphere (Santiago-De la Rosa et al. 2017). Depending on prevailing weather conditions, the products of biomass burning can have regional impacts due to the transport of smoke aerosols from agricultural to urban areas, which can affect the biogeochemical cycle and harm human health (Kalluri et al. 2020). The burning of grasslands by the slash-and-burn method for planting crops and of agro-industrial waste for harvesting are commonly related to the incidence of wildfires in forests, surrounding agricultural areas and rangelands, as well as increased temperatures,

L. A. Olvera-Vargas
Cátedra CONACyT-Centro de Investigación Y Asistencia en Tecnología Y Diseño del Estado de Jalisco, A.C. (CIATEJ), Av. Normalistas 800 Colinas de la Normal., 44270 Guadalajara, Jalisco, México
e-mail: lolvera@ciatej.mx

N. Aguilar-Rivera (✉)
Universidad Veracruzana, Facultad de Ciencias Biológicas Y Agropecuarias, Km. 1 Carretera Peñuela Amatlán de Los Reyes S/N. C.P., Córdoba, 94945 Veracruz, México
e-mail: naguilar@uv.mx

W. Leal Filho et al. (eds.), *Sustainability in Natural Resources Management and Land Planning*, World Sustainability Series,
https://doi.org/10.1007/978-3-030-76624-5_29

respiratory ailments and GHG emissions into the atmosphere (Adolf et al. 2018; Phairuang et al. 2017).

Rice, corn, sorghum, cotton, sugarcane and wheat are the crops that generate the most agro-industrial waste which, mainly in developing countries, is burned before, during and after the harvest; moreover, despite technological and machinery advances for harvesting, approximately 60% of these crops are still harvested manually, which leads to the burning of this waste as a generalized management practice (Ferreira et al. 2014). According to data from FAOSTAT (FAO 2017), agriculture produces emissions from various processes (manure management as fertilizer, rice cultivation, use of chemical fertilizers, manure composting, aerobic decomposition processes of crop residues, crop residue burning, savanna burning, use of waste as rural energy, etc.). These processes generate various environmental impacts and potentially GHGs. In 2017, 150 thousand Gg of methane (CH_4), 9020 Gg of nitrous oxide (N_2O) and 6 million Gg of carbon dioxide equivalent (CO_2eq) were emitted into the atmosphere worldwide. In the same year, the burning of agricultural residues and savanna alone accounted for the emission of 7 thousand Gg of CH_4, 500 Gg of N_2O and 318 thousand Gg of CO_2eq (FAO 2017; Tubiello et al. 2013).

Sugarcane, the feedstock for sugar, ethanol and bioproducts, normally accounts for the largest share of raw material costs and the main environmental and social impacts such as GHG emissions and job generation in rural areas. Therefore, the impacts of climate change on variables affecting sugarcane production, such as air temperature, rainfall, and atmospheric carbon dioxide (CO_2) concentration, are not fully known (Linnenluecke et al. 2020).

Chagas et al. (2016) concluded that the economic and environmental impacts of ethanol biorefineries with different sugarcane (*Saccharum* spp.) production technologies should be evaluated with a focus on harvesting systems, reduced tillage, controlled traffic farming to reduce soil compaction, and alternative crops for use in sugarcane rotation.

The sugarcane-based biorefinery is gaining the interest of the sugar industry worldwide as a way of maximizing the benefits of the whole sugar milling process. However, the environmental and economic implications of the existing and proposed systems must be assessed to ensure the transition to sustainability (Silalertruksa and Gheewala 2019).

In the particular case of the sugarcane agro-industry, actions towards sustainability are reported by Sahu (2018), who concludes that the production of sugarcane and its industrialization present technical, socioeconomic, commercial, health-related and social welfare problems that demand greater investment in research and development, where the impact of conventional technologies generated largely exceeds the resources invested.

According to Cardoso et al. (2019), when considering the current vertically-oriented production system (agricultural and industrial phases), manual harvesting provides greater job creation while mechanized sugarcane systems have better working conditions and workers earn a higher average income, especially in the agricultural phase. Environmental impact scenarios with mechanized harvesting without burning and trash recovery have the best comparative balance of environmental impacts because sugarcane harvesting systems, with and without burning, differ in relation to soil chemical and biological attributes. Sugarcane harvesting without burning favors soil attributes in relation to burning (Souza et al. 2012).

The tendency to eliminate the burning of crops and agricultural residues in the entire production process of sugarcane, and other crops, is on the agenda of international environmental agencies. For example, Brazil and India have implemented greencane management systems based on sustainable agriculture programs, including leaving field straw as soil protection against erosion, in addition to inhibiting the growth of weeds and improving the nutritional balance of the soil (Mugica-Alvarez et al. 2018; Leal et al. 2013).

2 Sugarcane Burning

Sugarcane is one of the crops that requires combustion during several of its processing stages, namely: burning is conducted before the harvest in order to facilitate manual harvesting and protect workers from insects and animals (Holder et al. 2017; França et al. 2012); straw and other debris are burned to decrease their total amount in the field and prepare the soil; and bagasse is used as fuel for the sugar mill (Mugica et al. 2015; Neto et al. 2011). The sugarcane part used as raw material to produce sugar as ethanol is the stem, which can be used with or without the practice of pre-harvest burning (França et al. 2014). Studies related to the emissions generated by sugarcane burning indicate that during the harvest the atmospheric concentrations of $PM_{2.5}$, PM_{10} and Black Carbon increase from 2 to 4 times, while polycyclic aromatic hydrocarbons (PAHs) increase up to 6 times (Mugica-Alvarez et al. 2018). To avoid dangerous climate change ($\Delta T > 2$ °C), global GHG emissions by 2050 must be: ½ current emissions level, or less than ¼ of projected 2050 "business-as-usual" emissions, with the biomass becoming much more valuable (including the possibility of negative GHG emissions when biomass is used with CO_2 capture and storage because, during burning, combustion is incomplete, with the formation of compounds that are not completely oxidized). Burning not only pollutes the environment and aggravates the situation of extreme heat at the time of harvest, but also affects soil fertility by decreasing the content of organic matter and nitrogen in the soil (Leal et al. 2013). This practice is the main cause of forest fires and contributes to the phenomenon of global warming.

Leite et al. (2018) conducted a review of work in PubMed, SciELO, Medline, and Lilacs data, from 1997 to 2017, and concluded that work in manual sugarcane cutting, especially when involving burning, exposes workers to various health risks, including respiratory, renal, cardiovascular, musculoskeletal, eye and skin problems. Crowe et al. (2013) reported that sugarcane workers had high levels of temperature stress within the burned fields and severe dehydration, and Silverira et al. (2013) found that sugarcane cutters develop a significant degree of genotoxicity due to inhalation of emissions from burnt sugarcane. It has also been determined that populations near sugarcane production areas may suffer health effects, since the burned material and the compounds can be transported downwind causing instantaneous exposure to the contaminants (Santiago-De La Rosa et al. 2018; Singh et al. 2018; Mugica-Alvarez et al. 2017). Reid et al. (2016), Du et al. (2018) and Ramos et al. (2019) reviewed the scientific literature to evaluate studies on the impact of smoke exposure from forest fires and crop burning in terms of mortality and respiratory, cardiovascular, mental and perinatal health problems. These reviewed papers document associations between exposure to biomass burning smoke and general respiratory health effects, specifically exacerbation of asthma, chronic obstructive pulmonary disease (COPD), an increased risk of respiratory infections, allergies and conjunctivitis, with mounting evidence supporting an association with all-cause mortality.

Burning of sugarcane residues is a common practice all over the world, especially in developing countries. This practice is allowed by state legislation in Louisiana, the United States, on the grounds that there is no scientific evidence of negative impacts (Ribeiro 2008). In Mexico, 92% of the sugarcane harvest is carried out with open burning (harvest), and this occurs between November and May, and although regulations that limit agricultural burning through a comprehensive law were promoted for prevention and comprehensive waste management, this activity is commonly carried out every season without any restrictions (Mugica-Alvarez et al. 2018; Santiago-De la Rosa et al. 2017; Cámara de Diputados 2015).

Compared to the consequences caused by the burning of cane fields, some benefits of this type of traditional harvesting system for sugar mills are:

1. Increases cutting force
2. Reduces transportation costs due to evaporation of water in the stalks during burning, maximizes the removal of foreign materials (soil, plant material, foliage, stones, logs, etc.), which would act as sponges during milling, capturing the juice extracted from the sugarcane, reducing the efficiency of juice extraction
3. Facilitates the production process because it is a precooked working material

In a green harvest about 20–25% of the material is waste, whereas in a burned harvest around 16% is waste. The most sustainable uses for these wastes are:

(1) Incorporation into the soil to improve its organic matter content
(2) Packing for use as livestock feed (preparation of rations and silage)
(3) Pickup from the field for energy generation (use of biomass)

Additionally, Cherubin et al. (2018) concluded with regard to GHG emissions that there is no consensus about the potential impact of harvesting crop residues. In general, harvesting crop residues decreases CO_2 and N_2O emissions from the decomposition process, but it has no significant effect on CH_4 emissions. Plant growth responses to soil and microclimate changes due to harvesting crop residues are site and crop specific.

Accordingly, the effects of burning or a green harvest are widespread, affecting processing, harvesting, growing, soil and the environment. The complex cause and effect pattern of these effects are summarized by Davies (1998):

Easier harvesting.
Ruptures cane stalk and destroys its protective wax coating.
Heat of fire destroys some sugar in stalks.
Environmental and soil damage.
Declining soil productivity over time.
Kills bees and beneficial insects.
Removes trash.
Rapid fall in sucrose content.
Rapid accumulation of impurities.
Lower cane yields.
Loss of biomass.
Reduced quantity of sugar production.
Increased sugar losses to molasses.
Crushing season extended.
Longer processing time.
Processing becomes slower, more costly and less effective.
More caustic cleaning required and chemical waste to dispose.
Ingress of sucrose devouring bacteria and yeasts.

Despite this monitoring of the atmospheric emissions generated by sugarcane burning reported by Flores-Jiménez et al. (2019), few studies have been conducted on the impact of crop burning on wildfire generation in Mexico, where such research is especially needed since it is a tropical country with dense vegetation such as evergreen and deciduous forests along with different tropical forests, which represent high availability of potential fuel (Prato and Huertas 2019). In addition, the harvest season (zafra) coincides with the dry season in the northern hemisphere, which influences the spread of fires to areas surrounding sugarcane fields (Mugica et al. 2015) where there is no trash recovery for energy purposes (Carvalho et al. 2017).

3 Using Geotechnologies to Monitor Wildfires in Forest and Grassland Areas Caused by Sugarcane Burning

Forest loss is being driven by several factors, the most important having to do with agriculture: growth of the agricultural frontier and wildfires caused by the uncontrolled burning of agricultural residues, so much so that 24% (±3) of forests are being lost for the first reason and 23% (±4) due to wildfires, where 89% of these fires were caused by the burning of agricultural residues (Curtis et al. 2018). Wildfires, of natural or anthropogenic origin, have negative impacts on the economy of the communities that suffer them, with reverberating effects at the regional, national and even international level (Rasch and McCaffrey 2019). In Mexico, losses of over US $337 million in wood and around US $39 million in reforestation costs have been reported (Bautista Vicente et al. 2014). Based on information from the National Forestry Commission (CONAFOR), fires were reported in 3929 km^2 in Mexico between 2010 and 2019. However, many wildfires are not documented because they occur in inhospitable areas and can only be detected from satellite sensors and geotechnologies (Yin et al. 2019; Bautista Vicente et al. 2014).

Thus, the use of geographic technologies, particularly remote sensing, has improved the monitoring and surveillance of wildfires. Yin et al. (2019) identified $PM_{2.5}$ contamination from aerosols from burning crop residues with MODIS imagery; Kalluri et al. (2020) used the same images, in addition to combining satellite products such as EOS Terra, NASA Aura, and CALIPSO, which include different spectral and spatial resolutions, and combined them to assess the spatial and temporal variation of atmospheric aerosols during the burning of agricultural crop residues. More specifically, França et al. (2014) showed that remote sensing maps help to measure emission and combustion factors for sugarcane, as well as to measure the increase in sugarcane area over time, and the replacement of forest areas using the slash-and-burn method to plant sugarcane. Phairuang et al. (2017) estimated the emission of air pollutants in areas in Thailand, showing that the areas with sugarcane and surrounding forest vegetation were the places with the most emissions in that country. Also, satellite products are used to generate models that identify the susceptibility to wildfire, associating spectral resolution values to hot-spots and the vicinity with agricultural areas, as carried out by Fernandez-Guisuraga et al. (2019).

In wildfires studies, the most widely used satellite images are those of the MODIS and Landsat sensors (MSS, TM, ETM + and OLI). Through a search of the databases of high impact international journals, in the last 5 years 4383 articles with reference to "Wildfire & Remote Sensing" have been published; of these, 46.4% used MODIS imagery to identify wildfires, their causes and consequences, while 40.3% used Landsat imagery for the same objectives.

The MODIS aboard both the Terra and Aqua satellites orbits around the earth from north to south, visualizing the Earth's surface almost every day, acquiring data in 36 spectral bands between wavelengths 0.41–15 μm (Remer et al. 2005). The images have been cataloged as instruments to understand the global dynamics and processes that occur on land, in the oceans and in the lower atmosphere, where MODIS develops and maintains algorithms for creating climate data records based on the Earth's atmospheric properties (aerosols, clouds, water vapor). From these, it obtains various products, such as MCD64A1 v006, which provides data on burned areas on a monthly basis (Giglio et al. 2015).

The Landsat program has been acquiring imagery since 1972, with the spatial and spectral resolutions provided improving with successive missions. Landsat 8 consists of two scientific instruments: the Operational Land Imager (OLI) and the Thermal Infrared Sensor (TIRS). These two sensors offer images with a spatial resolution of 30 m (visible, NIR, SWIR), 100 m (Thermal) and 15 m (panchromatic), covering wavelengths from 0.44 to 12.51 μm, divided into 11 bands (Markham et al. 2014). These images are widely used to detect and analyze forest fires, due to their high spatial resolution and because they contain the SWIR (Shortwave infrared) band that helps identify the spread of fires (Kato et al. 2018; Nguyen et al. 2018; Huo et al. 2014).

For the above, this chapter aims to highlight the distribution patterns of the burned areas, during the period from 2010 to 2020 in the sugarcane-growing areas of the Huasteca Potosina, Mexico, by monitoring their impact on adjoining forest and rainforest areas using high and low spatial resolution satellite images.

4 Materials and Methods

Two information inputs were used: satellite imagery and official databases on wildfires in Mexico; the latter was obtained from the databases administered by CONAFOR and the National Forest Information and Management System (SNIGF), both of which are institutions attached to the Secretariat of the Environment and Natural Resources (SEMARNAT). The information obtained corresponds to wildfires reported between 2010 and 2020, separated by year and Julian-days, and contains, among other attributes, data such as fire dates, affected area, cause of fire and geographical position. The information is from all over Mexico, divided by state and municipality where the fires were reported. Several filters were made to refine the information. First, only data corresponding to the state of San Luis Potosí was accessed and then the 20 municipalities belonging to the Huasteca Potosina were extracted, and then the information was filtered by cause (fires due to agricultural activity) and by type (fire by sugarcane burning); once this was done, the sum of the hectares was obtained with these filters. The data was reviewed and did not appear to be extreme, so it was organized and refined for the analysis. The database was saved in CSV format (comma-separated values) and imported as a *shapefile* format within ArcGIS 10.2 software, where it was edited as cartography.

For spatial monitoring of wildfires, MODIS imagery was downloaded from the LAADS DAAC (Level-1 and Atmosphere Archive & Distribution System / Distributed Active Archive Center) site, specifically the product MCD64A1 version 6 from Terra and Aqua, which provides information on burned areas on a monthly basis with a global grid of 500 m. The imagery uses an index of vegetation sensitive to burned areas derived from the composition of bands 5 and 7 of the infrared surface reflectance (Giglio et al. 2015). Twelve images corresponding to the months of the year, starting from January 2010 to March 2020, and totaling 123 images, were downloaded. These images cover a large part of central and northern Mexico (Fig. 1b), so a mask for the study area was generated to be able to cut out only the study area. This process was carried out in ArcGIS 10.2 software. The pixel values of each image indicate that all values greater than 0 are areas affected by fire. The images were organized, and the area affected by wildfires was quantified by month and year.

Information was also downloaded from the North American Drought Monitor (NADM) of the National Oceanic and Atmospheric Administration to have data on the drought index and temperature and precipitation anomalies, to associate them with the wildfire data and relate them to the cause (natural or anthropogenic). Information was obtained from the same dates (2010–2020); the data are organized by month and express the percentage of area that has a type of drought: Abnormally Dry (D0), Moderate Drought (D1), Severe Drought (D2), Extreme Drought (D3) and Exceptional Drought (D4). The information is registered at the country level; however, the cut was made only for the study area, to relate it to the Huasteca Potosina wildfires.

Landsat 8 imagery was downloaded from the GloVis platform of the USGS (United States Geological Survey), corresponding to two dates: February 14 to observe the areas before the burning of sugarcane and April 3, five days after the burning of crops and reports of burned forest areas. Both dates were for 2019, which was a year in which several fires with an environmental impact occurred in the study area (CONAFOR 2019). To fully cover the Huasteca Potosina region, three Landsat image scenes are required: 27/44 (Path/Row), 27/45 and 26/45 (Fig. 1a). The three scenes were joined through the creation of a mosaic, to be later cut by the boundaries of the municipalities in the study area. With the images, an index called Normalized Burn Rate (NBR) was generated, which is used to map the severity of burns in forest and agricultural ecosystems, considering the loss of soil and aerial organic matter (Cardil et al. 2019; Key and Benson 2005). The areas with NBR index values were extracted and quantified within a specific area. To know if the burned areas correspond to agricultural or forest areas, the geo-referenced pattern of sugarcane was used (SIAP 2019).

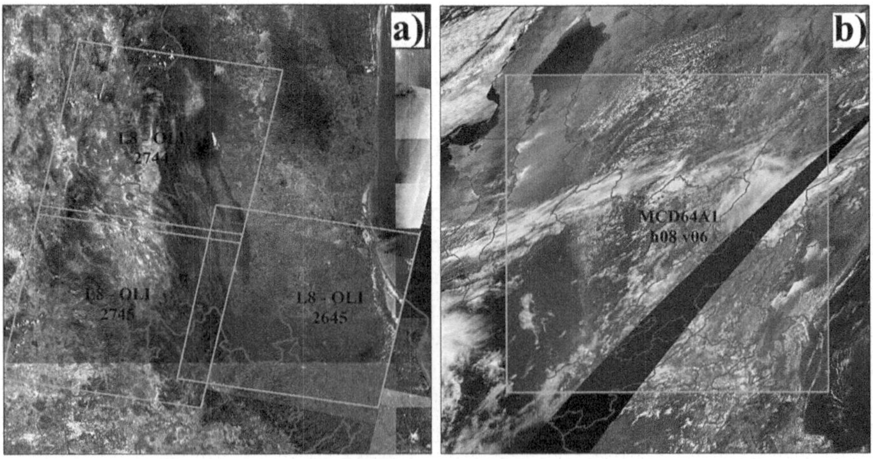

Fig. 1 Scene coverage used by sensor type: **a** Landsat OLI and **b** MODIS imagery

5 Sugarcane and Wildfires in the Huasteca Potosina

Sugarcane cultivation is important in Mexico, since 847,000 ha are cultivated in 264 municipalities in 15 states, producing 56.3 million tons and generating profits of more than US $2.2 billion (SIAP 2019). Approximately 0.5% of Mexico's Gross Domestic Product (GDP) corresponds to this activity, 400,000 families with permanent and temporary jobs depend on it, and indirectly 2.5 million Mexicans depend directly on the sugar agribusiness (Figueroa Rodríguez et al. 2015). However, the crop is no longer competitive, mainly due to the stagnation of the low sugarcane yield (between 60 and 70 ton/ha) and that of sucrose, the high fiber content in stems and the lack of advanced technology in all production processes (Aguilar-Rivera et al. 2012). These factors have negatively affected the sugar industry and farms, and have constantly eroded its sustainability and profitability, and it has also been seen as an unsustainable crop, due to the use of agrochemicals to control pests and diseases, which contaminate the soil and waters, and the burning of cultivated fields as a general practice (Aguilar-Rivera et al. 2019).

In the sugarcane field, 75% of the growers have a farm between 0.5 and 5 ha with a national average of 3.97 ha; 61% of the crop is rainfed, the most common varieties are Mex-69 290 and CP 72-2086, and average productivity is 65 t/ha^{-1} without the ability to add value to the crop.

Huasteca Potosina is a sugarcane area as an agro-industrial survival activity. It is in the east of San Luis Potosí state, and is the most boreal area of cultivation. Four of Mexico's 50 sugar mills (Plan de Ayala, Plan de San Luis, Alianza Popular and San Miguel del Naranjo) are located here. More than 100,000 ha are cultivated in this region, producing 7.4% of Mexico's total sugarcane yield (SIAP 2019). Sugarcane cultivation in the Huasteca Potosina region is very important, since it accounts for 26% of its total agricultural area and generates 56% of the profits from agriculture. In this region, up to 90% of the harvested sugarcane area is burnt (Aguilar et al. 2018), which generates up to 36% of the nitrogen monoxide (NO) and 37% of the carbon dioxide (CO_2) emissions in the state of San Luis Potosí (SEGAM 2013). Sugarcane is grown in 20 municipalities of the Huasteca region: Axtla Terrazas, Matlapa, Tamazunchale, Tampacán, San Martin Chalchicuautla, Tampamolón Corona, Coxcatlán, Huehuetlán, Tancanhuitz Santos, Xilitla, San Antonio, Tanquián de Escobedo, San Vicente Tancualayab, Aquismón, Ciudad de Maíz, Ciudad Valles, El Naranjo, Tamasopo, Tamuín and Tanlajás (Fig. 2). The sector provides survival, well-being and generation of rural employment for 13,578 producers and their families.

In addition, some sugarcane fields abut grassland areas which are used for livestock feeding and which are highly susceptible to wildfires during the dry season; these areas also overlap with other crops and forestland (Fig. 3).

According to CONAFOR reports, between January 2010 and March 2020, 197 wildfires were detected in the Huasteca Potosina, affecting 27,941 ha of forests and rainforests. In total, 78% of this area was damaged due to sugarcane field burning (CONAFOR 2019). Figure 4 shows the wildfire areas in the study period and the

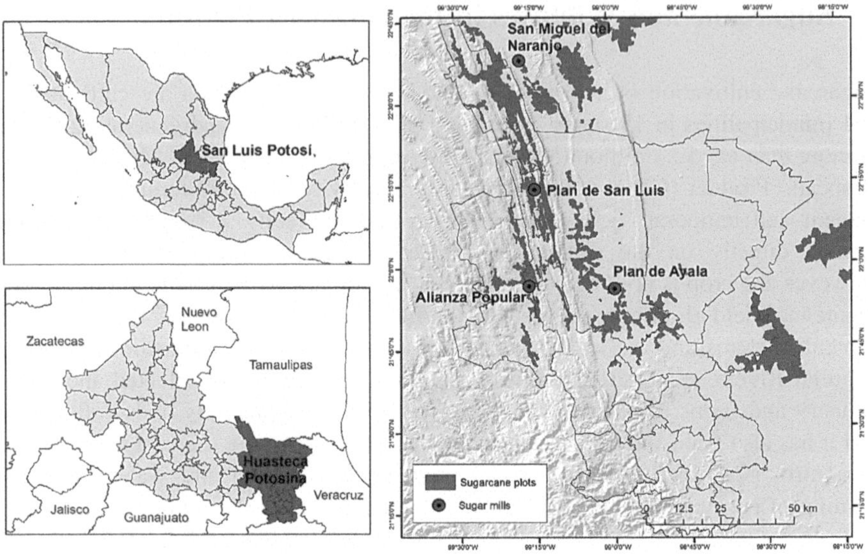

Fig. 2 Identifying sugarcane-growing areas in the Huasteca region, Mexico

proximity of the sugarcane plots to the forest and rainforest areas, particularly in the northwest of the Huasteca Potosina. As mentioned above, the months in which sugarcane is burned (harvest season) coincide with the driest season in the region (December–May). Distance is an important factor for the spread of agricultural fires to forest areas, but also landscape, since sugarcane plots are located in the mountain valleys, and in the transition from slopes to mountains and hills where grades increase and forests or rainforests begin (Flores-Jiménez et al. 2019).

The relationship between wildfires, soil type and drought is strong (Fig. 5), particularly because as the magnitude and duration of a drought increase, the soil's moisture retention capacity also increases the ignition capacity of the biomass present and in the case of forests and rainforests, already weakened by the presence of forest pests and reduced moisture content, the forest canopy is more susceptible to fire (Stephens et al. 2018). With this, the drought monitor was related to the CONAFOR fire reports, specifically with the wildfire area in the Huasteca Potosina (Fig. 1). It can be seen that there is little relationship between extreme drought conditions (D3) and exceptional ones (D4), since the drought that began in early 2011 and ended in mid-2012 generated few wildfires, compared to other years. However, due to its continuity over the years, abnormal (D0), moderate (D1) and severe (D2) droughts weaken the trees of forest ecosystems and leave adequate space for wildfires to start. This relationship between wildfires and sugarcane burning is evidenced by the dates when there is the most burned area, which corresponds in almost every year to the January–July period, being the most intense in April and May (Fig. 6).

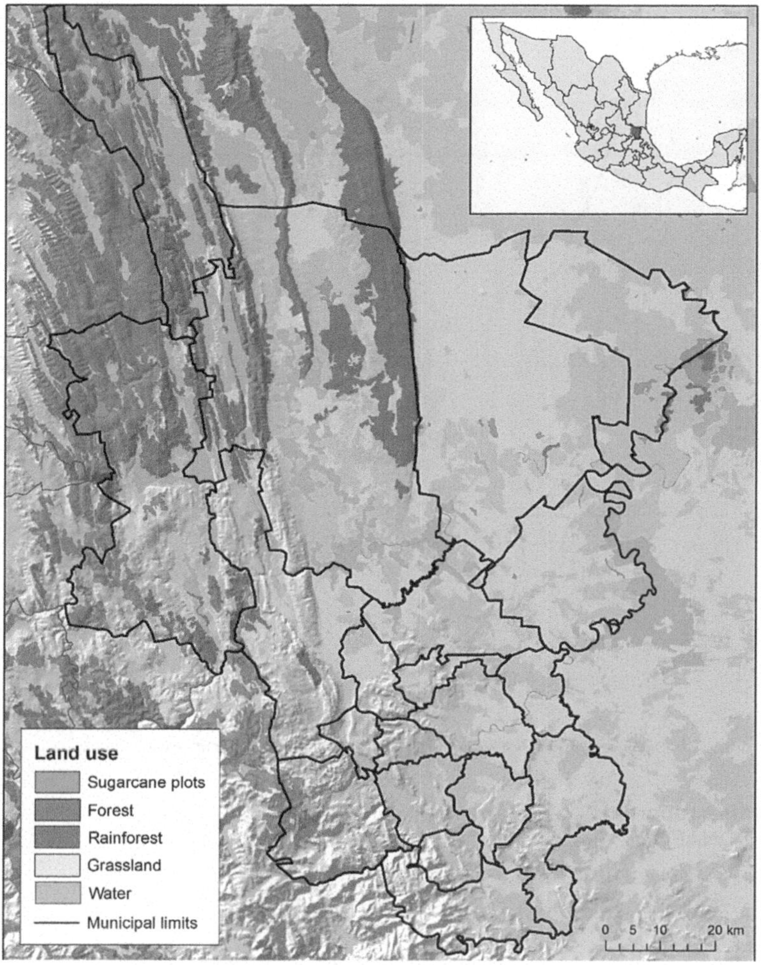

Fig. 3 Land use and sugarcane area in the Huasteca, Mexico

The fire monitoring information obtained with the MODIS images was very good; the timing of the fires (wildfires and sugarcane burns) was per month, since the MCD64A1 product gives data on areas burned monthly and with a spatial resolution of 500 m per pixel (Giglio et al. 2015). As shown in Fig. 8, fires occur cyclically each year, with the January–May period having the most burned areas. In the rainy season (between July and November), there are very few or no fires. In burned area, 2013 stands out since fires were recorded in more than 20,000 ha. CONAFOR for that year reported 11,200 ha affected by fires. Figure 8 also shows that in 2011 and 2019 the fires reached almost 10,000 ha. Between January 2010 and March 2020, the MODIS sensor calculated a burned area of 100,709 ha in Huasteca Potosina, equivalent to 10.2% of its total area (Fig. 7).

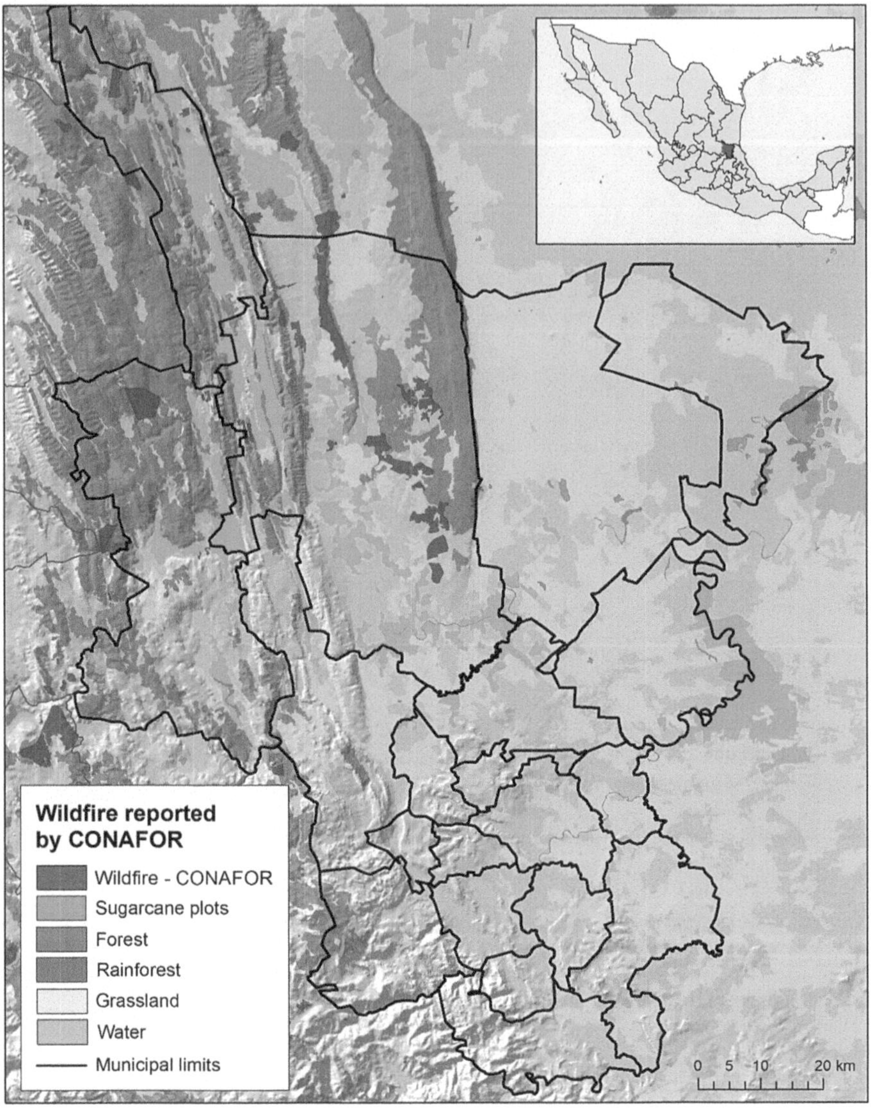

Fig. 4 Wildfires reported by CONAFOR between 2010 and 2020

Once the burned areas were calculated, the frequency of occurrence at the same site was obtained, corresponding to 7300 ha where there were more than 2 fires in the study period, and on 3500 ha more than 4 fires in the last 10 years. These areas all correspond to sites that are less than 500 m from dense sugarcane areas (Fig. 8). Regarding the more than 100,000 ha burned in the study period, 43% corresponded to wildfires, with 35,000 ha in rainforests and 8300 in forest areas. According to

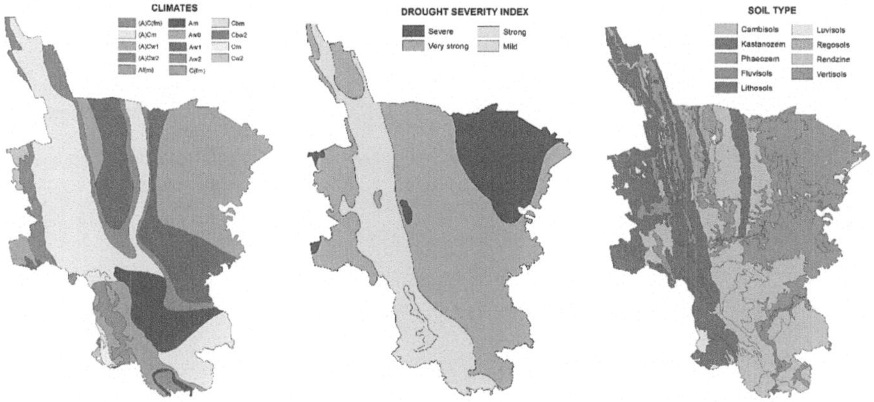

Fig. 5 Climates, soil types and drought severity index in the Huasteca region, Mexico

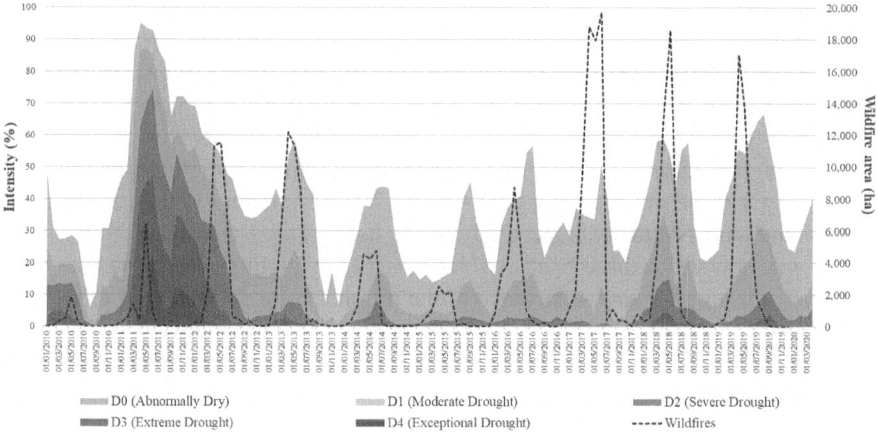

Fig. 6 Relationship between the drought monitor and wildfire frequency

some CONAFOR reports (2019), with this affected area, 46.6 tons of CO_2 per hectare are being released into the atmosphere, equivalent to 2 million tons in the last 10 years.

The evaluation of burned areas with the MODIS imagery resulted in a different assessment of the size of the total burned area than that obtained through CONAFOR (Fig. 3). While CONAFOR reports 27,941 ha with fires in the last 10 years in the Huasteca Potosina, the burned areas acquired with the MODIS imagery total 99,959 ha between January 2010 and December 2019. The differences are significant, as previously mentioned. According to Bautista Vicente et al. (2014), CONAFOR only considers the total number of forest fires that are documented by this government agency body during the year; however, by using satellite images it is possible to calculate the number and magnitude of the fires,

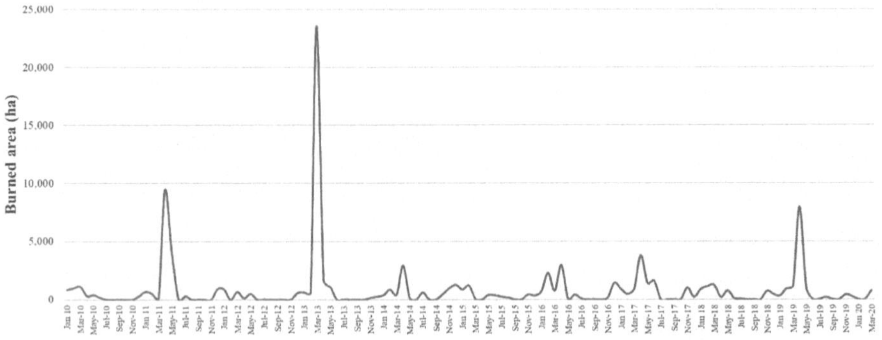

Fig. 7 Timeline of burned areas in the Huasteca Potosina

thereby providing a real-time option to make up for the lack of information on undocumented fires due to lack of budget, personnel or access to burned areas (Fig. 9).

The MODIS images used have a spatial resolution of 500 m per pixel, which allows a projection of the regional analysis of the fires in the Huasteca Potosina region. However, for local analysis, this scale is insufficient, due to the low capacity to separate burned land cover. Therefore, Landsat 8 OLI images were used, which present a high spectral resolution (0.43–12.5 μm) and high spatial resolution (30 m per pixel). Of these images, only bands 5 and 7 corresponding to the near infrared (NIR) band and SWIR 1 respectively were used. The Normalized Burn Ratio (NBR) was applied to these bands, which consists of highlighting burned areas in fire zones, through the low reflectance of the infrared band and the high reflectance in the SWIR 1 (Cardil et al. 2019). When executing the NBR index, values below 0 represent the areas with fire damage and the measurement of their intensity (Key and Benson 2005).

Figure 10 shows two different dates from the same site: the image on the right corresponds to the date of February 14, 2019 and the one on the left to an image of April 3 of the same year. On those dates, severe fires were recorded in areas surrounding sugarcane crops; MODIS images report damage to almost 7000 ha and CONAFOR reports indicated fires on 1200 ha. The image on the right shows the burned areas corresponding to the sugarcane plots and their burning process (harvest season); here, the index is very well defined. On the other hand, the image on the left, which indicates the period during or after a fire, shows significant damage to the rainforest areas. The proximity of the sugarcane to the natural vegetation means that it was the harvest fire that helped the wildfires to start. With MODIS images, the damaged area totals 6300 ha; with Landsat imagery, the damage is calculated to be 6137 ha. These figures indicate that the low spatial resolution image (MODIS) is not far from the calculation of the fire-affected area, while the high spatial resolution image (Landsat) specifically shows the location of the damage, where territorial elements can be associated with the affected area, in addition to specifying the type of vegetation that was affected by the forest fire.

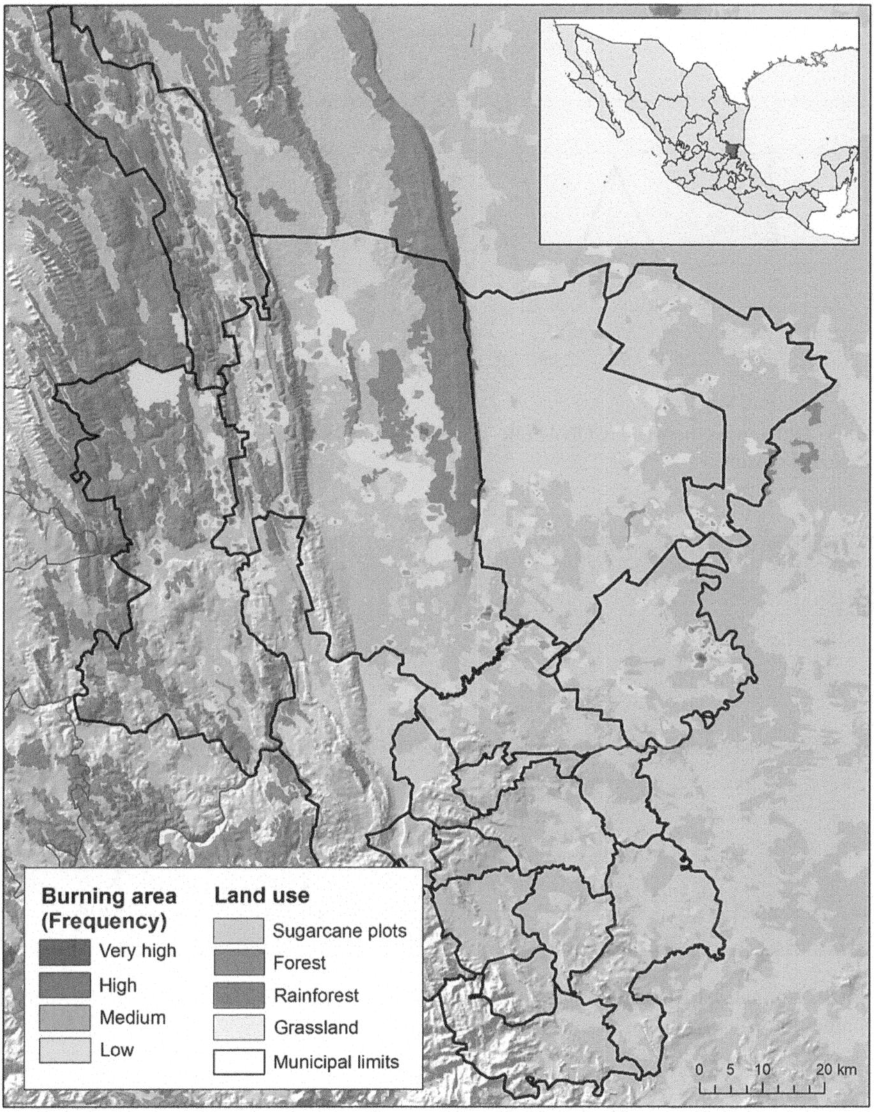

Fig. 8 Fires detected with MODIS imagery, fire frequency and relationship with land uses

Sugarcane could contribute significantly to increasing food, feed, bioproducts and energy needs under a changing climate, but this requires strategic planning of production and processing systems for efficient and sustainable use of natural resources, especially during the harvest season. Effective planning for the sustainable future requires reliable spatial information and a combination of innovative agronomic practices that includes land preparation, planting techniques, healthy

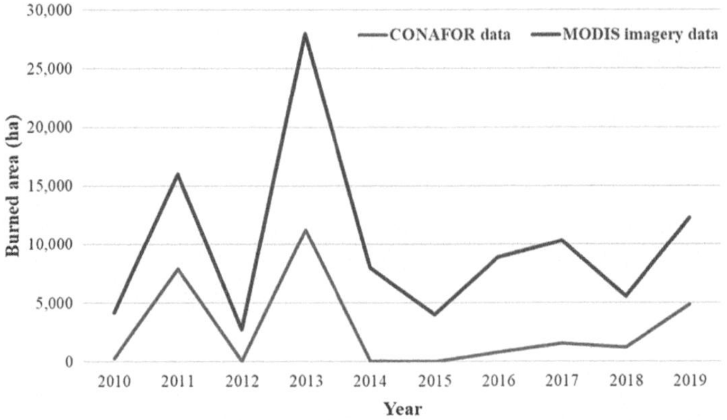

Fig. 9 Differences between CONAFOR data and MODIS imagery for burned areas in the Huasteca Potosina

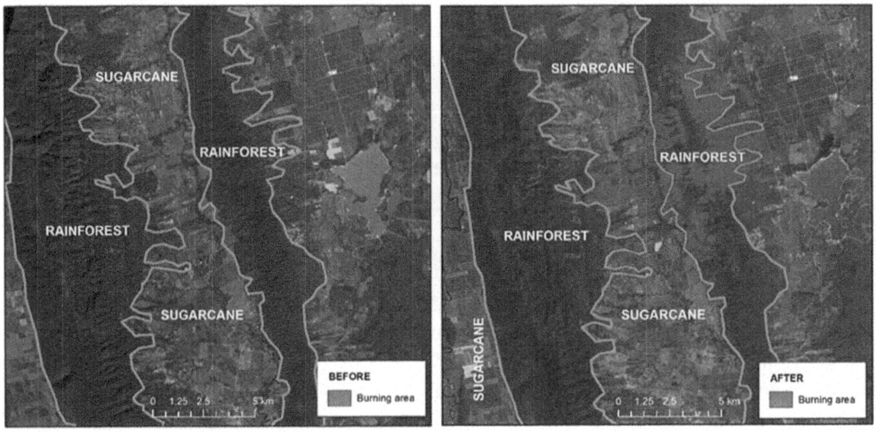

Fig. 10 Normalized Burn Ratio index in Landsat imagery. Left—before the fire, Right—during and after the wildfires

planting material of improved varieties, soil health and nutrition management, integrated pest management and other good crop husbandry practices with the integration of precision farming techniques mainly at harvest.

Therefore, it is necessary to promote research to convert sugarcane biomass into value-added products, thereby avoiding the risks associated with burning sugarcane plantations. This could be accomplished by establishing collaboration among research providers and using the input of sugar companies and other stakeholders to direct research priority for maximum benefit.

Factors such as the gradual adoption of mechanical harvesting, driven by public environmental policies that prohibit the burning of cane fields before and after harvesting, the scarcity of labor for manual harvesting, the perception of the energy and economic value of waste by entrepreneurs derived from projections based on the circular bioeconomy, the 2030 sustainable development agenda, international certifications such as BONSUCRO and the need to diversify the use of sugarcane straw in energy and bioproduct projects are being evaluated by the actors of the agro-industrial sugarcane chain in the producing countries (Negrete 2019).

6 Conclusions

Sugarcane farming and its processes should be changed to have a lower environmental impact, a transformation that will require support from the right partners in the engineering, financial, and institutional sectors. It is time to modernize the crop's entire value chain, since burning is used from before harvest to almost the end of the industrialization process. Due to its character as a raw material, this crop is of social and economic importance that in its value chain must be transformed in various stages into a final basic product such as sugar. Currently, the harvesting system through the burning and re-burning of residues emits atmospheric pollutants and GHGs, which only increase as the sugarcane area expands, affecting the survival of plant and animal ecosystems in addition to generating public health problems, mainly respiratory problems in nearby communities, and causing forest and grassland fires. The transformation to sustainable agriculture implies complex and dynamic challenges, where all existing technological tools must be used in the search for more environmentally friendly options.

The results confirm that the use of high and low resolution satellite images, especially MODIS and Landsat 8 OLI ones, as well as spectral indices, as geotechnical tools, provides a robust evaluation of the measurement and monitoring of wildfires, both in forests and grasslands, as well as of burns in sugarcane and other crop fields; therefore, these geotechnologies are useful for estimating biomass losses and fire risk, even when monitoring is carried out over large and heterogeneous areas.

The use of technology is essential, both to prevent the burning of sugarcane (and other crops) and to prevent the spread of fires to forest areas. The use of personnel in the field is important, but human capacity is exceeded when wildfires are of great magnitude. Moreover, such fires are also dangerous for the people in charge of surveillance. For this reason, the use of geotechnologies, particularly remote sensing ones, are very useful, both for identifying the area of the fire and the impact it generates and for determining the possible reasons for the fire. This could help prevent fires in protected nature areas and reduce pollutant emissions into the atmosphere.

Huasteca Potosina is an important area for sugarcane production, having 4 sugar mills and thousands of people dependent on cultivation and processing. However, producers and industry must seek sustainable alternatives to avoid burning sugarcane, since many wildfires in the area are caused by this practice. The area is rich in forests and rainforests, which are home to ecologically important species that could disappear if wildfires go unchecked.

References

Adolf C, Wunderle S, Colombaroli D, Weber H, Gobet E, Heiri O, …, La Mantia T (2018) The sedimentary and remote-sensing reflection of biomass burning in Europe. Glob Ecol Biogeogr 27(2):199–212

Aguilar-Rivera N, Michel-Cuello C, Cárdenas-González JF (2019) Green revolution and sustainable development. Encycl Sustain High Educ, 833–850

Aguilar-Rivera N, Rodríguez LDA, Enríquez RV (2012) The Mexican sugarcane industry: overview, constraints, current status and long-term trends. Sugar Tech 14:207– 222

Bautista Vicente F, Carbajal N, Pineda Martínez LF (2014) Estimation of total yearly CO_2 emissions by wildfires in Mexico during the period 1999–2010. Adv Meteorol

Cámara de Diputados del H. Congreso de la Unión, México (2015) Ley General para la Prevención y Gestión Integral de los Residuos

Cardil A, Mola-Yudego B, Blázquez-Casado Á, González-Olabarria JR (2019) Fire and burn severity assessment: Calibration of Relative Differenced Normalized Burn Ratio (RdNBR) with field data. J Environ Manage 235:342–349

Cardoso TF, Watanabe MDB, Souza A, Chagas MF, Cavalett O, Morais ER, ..., Bonomi A (2019) A regional approach to determine economic, environmental and social impacts of different sugarcane production systems in Brazil. Biomass Bioenerg120:9–20

Carvalho DJ, Veiga JPS, Bizzo WA (2017) Analysis of energy consumption in three systems for collecting sugarcane straw for use in power generation. Energy 119:178–187

Chagas MF, Bordonal RO, Cavalett O, Carvalho JLN, Bonomi A, La Scala Jr N (2016) Environmental and economic impacts of different sugarcane production systems in the ethanol biorefinery. Biofuels Bioprod Biorefin 10(1):89–106

CONAFOR (2019) Reporte de resultados de incendios forestales 2019. Programa Nacional de Manejo del Fuego. Comisión Nacional Forestal, México. https://snigf.cnf.gob.mx/wp-content/uploads/Incendios/Informes%20anuales/2018-CIERRE.pdf

Cherubin MR, Oliveira DMDS, Feigl BJ, Pimentel LG, Lisboa IP, Gmach MR, ..., Paiva SRD (2018) Crop residue harvest for bioenergy production and its implications on soil functioning and plant growth: a review. Scientia Agricola75(3):255–272

Crowe J, Wesseling C, Solano BR, Umaña MP, Ramírez AR, Kjellstrom T, ..., Nilsson M (2013) Heat exposure in sugarcane harvesters in Costa Rica. Am J Ind Med 56(10):1157–1164

Curtis PG, Slay CM, Harris NL, Tyukavina A, Hansen MC (2018) Classifying drivers of global forest loss. Science 361(6407):1108–1111

Davies J (1998) The causes and consequences of cane burning in Fiji's sugar belt. J Pac Stud 22(1 & 2):5–12

Du C, Kulay L, Cavalett O, Dias L, Freire F (2018) Life cycle assessment addressing health effects of particulate matter of mechanical versus manual sugarcane harvesting in Brazil. Int J Life Cycle Assess 23(4):787–799

FAO (2017) FAOSTAT online Database. Emissions agriculture—agriculture total. FAO, Rome. http://www.fao.org/faostat/en/#data/GT

Fernández-Guisuraga JM, Suárez-Seoane S, Calvo L (2019) Modeling *Pinus pinaster* forest structure after a large wildfire using remote sensing data at high spatial resolution. For Ecol Manage 446:257–271

Ferreira LEN, Muniz BV, Bittar TO, Berto LA, Figueroba SR, Groppo FC, Pereira AC (2014) Effect of particles of ashes produced from sugarcane burning on the respiratory system of rats. Environ Res 135:304–310

Figueroa Rodríguez KA, García García AMT, Mayett Moreno Y, Hernández Rosas F, Figueroa Sandoval B (2015) Factores que explican el rendimiento de caña de azúcar a nivel municipal en México. Revista Mexicana De Ciencias Agrícolas 6(6):1345–1358

Flores-Jiménez DE, Carbajal N, Algara-Siller M, Aguilar-Rivera N, Álvarez-Fuentes G, Ávila-Galarza A, García AR (2019) Atmospheric dispersion of methane emissions from sugarcane burning in Mexico. Environ Pollut 250:922–933

França D, Longo K, Rudorff B, Aguiar D, Freitas S, Stockler R, Pereira G (2014) Pre-harvest sugarcane burning emission inventories based on remote sensing data in the state of São Paulo, Brazil. Atmos Environ 99:446–456

França DDA, Longo KM, Neto TGS, Santos JC, Freitas SR, Rudorff BF, ..., Carvalho JA (2012) Pre-harvest sugarcane burning determination of emission factors through laboratory measurements. Atmosphere 3(1):164–180

Giglio L, Justice C, Boschetti L, Roy D (2015) MCD64A1 MODIS/Terra+aqua burned area monthly L3 Global 500 m SIN Grid V006. NASA EOSDIS land processes DAAC. Accessed on 19 May 2020 from https://doi.org/10.5067/MODIS/MCD64A1.006

Holder AL, Gullett BK, Urbanski SP, Elleman R, O'Neill S, Tabor D, ..., Baker KR (2017) Emissions from prescribed burning of agricultural fields in the Pacific Northwest. Atmos Environ 166:22–33

Huo H, Jiang X, Song X, Li ZL, Ni Z, Gao C (2014) Detection of coal fire dynamics and propagation direction from multi-temporal nighttime Landsat SWIR and TIR data: a case study on the Rujigou coalfield, Northwest (NW) China. Remote Sens 6(2):1234–1259

Kalluri ROR, Zhang X, Bi L, Zhao J, Yu L, Kotalo RG (2020) Carbonaceous aerosol emission reduction over Shandong province and the impact of air pollution control as observed from synthetic satellite data. Atmos Environ 222:117150

Kato S, Kouyama T, Nakamura R, Matsunaga T, Fukuhara T (2018) Simultaneous retrieval of temperature and area according to sub-pixel hotspots from nighttime Landsat 8 OLI data. Remote Sens Environ 204:276–286

Key CH, Benson NC (2005) Landscape assessment: remote sensing of severity, the normalized burn ratio and ground measure of severity, the composite burn index. In: FIREMON: Fire effects monitoring and inventory system Ogden. USDA Forest Service, Rocky Mountain Research Station, Utah

Leal MRL, Galdos MV, Scarpare FV, Seabra JE, Walter A, Oliveira CO (2013) Sugarcane straw availability, quality, recovery and energy use: a literature review. Biomass Bioenerg 53:11–19

Leite MR, Zanetta DMT, Trevisan IB, Burdmann EA, Santos UP (2018) Sugarcane cutting work, risks, and health effects: a literature review. Rev Saude Publica 52:80

Linnenluecke MK, Zhou C, Smith T, Thompson N, Nucifora N (2020) The impact of climate change on the Australian sugarcane industry. J Clean Prod 246:118974

Markham B, Barsi J, Kvaran G, Ong L, Kaita E, Biggar S, ..., Helder D (2014) Landsat-8 operational land imager radiometric calibration and stability. Remote Sens 6(12):12275–12308

Mugica-Álvarez V, Hernández-Rosas F, Magaña-Reyes M, Herrera-Murillo J, Santiago-De La Rosa N, Gutiérrez-Arzaluz M, ..., González-Cardoso G (2018) Sugarcane burning emissions: characterization and emission factors. Atmos Environ 193:262–272

Mugica-Alvarez V, Hernández-Moreno A, Valle-Hernández BL, Espejo-Montes FA, Millán-Vázquez F, Torres-Rodríguez M (2017) Characterization and modeling of atmospheric particles from sugarcane burning in Morelos, Mexico. Hum Ecol Risk Assess Int J 23(5):1056–1071

Mugica-Alvarez V, Santiago-de la Rosa N, Figueroa-Lara J, Flores-Rodríguez J, Torres-Rodríguez M, Magaña-Reyes M (2015) Emissions of PAHs derived from sugarcane burning and processing in Chiapas and Morelos México. Sci Total Environ 527:474–482

Negrete JC (2019) Reduction of pollution and diseases through a law to prohibit the burning of sugarcane. Biomed J Sci Tech Res 16(5):12286–12287

Neto TS, Carvalho JA Jr, Cortez EV, Azevedo RG, Oliveira RA, Fidalgo WRR, Santos JC (2011) Laboratory evaluation of Amazon forest biomass burning emissions. Atmos Environ 45 (39):7455–7461

Nguyen TH, Jones SD, Soto-Berelov M, Haywood A, Hislop S (2018) A spatial and temporal analysis of forest dynamics using Landsat time-series. Remote Sens Environ 217:461–475

Phairuang W, Hata M, Furuuchi M (2017) Influence of agricultural activities, forest fires and agro-industries on air quality in Thailand. J Environ Sci 52:85–97

Prato DF, Huertas JI (2019) Determination of the area affected by agricultural burning. Atmosphere 10(6):312

Ramos D, Pestana PRS, Trevisan IB, Christofaro DGD, Tacao GY, Coripio IC, ..., Ramos EMC (2019). The impact of sugarcane burning on hospitalization due to respiratory diseases. Ciencia Saude Coletiva 24(11):4133–4140

Rasch R, McCaffrey S (2019) Exploring wildfire-prone community trust in wildfire management agencies. Forest Sci 65(5):652–663

Reid CE, Brauer M, Johnston FH, Jerrett M, Balmes JR, Elliott CT (2016) Critical review of health impacts of wildfire smoke exposure. Environ Health Perspect 124(9):1334–1343

Remer LA, Kaufman YJ, Tanré D, Mattoo S, Chu DA, Martins JV, ..., Eck TF (2005) The MODIS aerosol algorithm, products, and validation. J Atmos Sci 62(4):947–973

Ribeiro H (2008) Sugar cane burning in Brazil: respiratory health effects. Rev Saude Publica 42:370–376

Sahu O (2018) Assessment of sugarcane industry: suitability for production, consumption, and utilization. Ann Agrarian Sci 16(4):389–395

Santiago-De La Rosa N, González-Cardoso G, Figueroa-Lara JDJ, Gutiérrez-Arzaluz M, Octaviano-Villasana C, Ramírez-Hernández IF, Mugica-Álvarez V (2018) Emission factors of atmospheric and climatic pollutants from crop residues burning. J Air Waste Manag Assoc 68(8):849–865

Santiago-De la Rosa N, Mugica-Álvarez V, Cereceda-Balic F et al (2017) Emission factors from different burning stages of agriculture wastes in Mexico. Environ Sci Pollut Res 24:24297–24310

SEGAM (2013) Inventario de emisiones a la atmosfera Estado de San Luis Potosí, México. Reporte Final. Secretaria de Ecología y Gestión Ambiental del Estado de San Luis Potosí. http://beta.slp.gob.mx/SEGAM/Documentos%20compartidos/ESTUDIOS%20PROGRAMAS%20Y%20PROYECTOS/InventarioEstataldeEmisiones_SLP-2011.pdf

SIAP (2019) Anuario estadístico de la Producción Agrícola, Caña de azúcar. Servicio de Información Agroalimentaria y Pesquera. https://nube.siap.gob.mx/cierreagricola/

Silalertruksa T, Gheewala SH (2019) Competitive use of sugarcane for food, fuel, and biochemical through the environmental and economic factors. Int J Life Cycle Assess 1–13

Silveira HCS, Schmidt-Carrijo M, Seidel EH, Scapulatempo-Neto C, Longatto-Filho A, Carvalho AL, ..., Saldiva PHN (2013) Emissions generated by sugarcane burning promote genotoxicity in rural workers: a case study in Barretos, Brazil. Environ Health 12(1):87

Singh N, Banerjee T, Raju MP, Deboudt K, Sorek-Hamer M, Singh RS, Mall RK (2018) Aerosol chemistry, transport, and climatic implications during extreme biomass burning emissions over the Indo-Gangetic Plain. Atmosp Chem Phys 18(19)

Stephens SL, Collins BM, Fettig CJ, Finney MA, Hoffman CM, Knapp EE, ..., Wayman RB (2018) Drought, tree mortality, and wildfire in forests adapted to frequent fire. BioScience 68 (2):77–88

Souza RA, Telles TS, Machado W, Hungria M, Tavares Filho J, de Fátima Guimarães M (2012) Effects of sugarcane harvesting with burning on the chemical and microbiological properties of the soil. Agr Ecosyst Environ 155:1–6

Tubiello FN, Salvatore M, Rossi S, Ferrara A, Fitton N, Smith P (2013) The FAOSTAT database of greenhouse gas emissions from agriculture. Environ Res Lett 8(1):015009

Yin S, Wang X, Zhang X, Zhang Z, Xiao Y, Tani H, Sun Z (2019) Exploring the effects of crop residue burning on local haze pollution in Northeast China using ground and satellite data. Atmos Environ 199:189–201

Youssef AM, Abu Abdullah MM, Pradhan B, Gaber AFD (2019) Agriculture sprawl assessment using multi-temporal remote sensing images and its environmental impact, Al-Jouf, KSA. Sustainability 11:4177

Wildfire Risk Mitigation in a Changing Climate

John Mervyn Rolfe, Michael Brian Wassing, and Lochlan Morrissey

1 Introduction

Climate change is increasing the severity and frequency of natural disasters in Australia, as well as instigating shifts in their geographic distribution (Coronese et al. 2019; CSIRO 2018; Reisinger et al. 2014). While society generally has made progress toward climate adaptation, the pace and scale of adaptation needs to significantly increase if we are to mitigate the exponentially rising levels of physical climate risk, particularly given the severity of recent predictions (Schwalm et al. 2020), and particularly in the case of bushfire risk in Australia (Emergency Leaders for Climate Action and The Climate Council 2020). Adaptation may well entail rising costs and difficult decisions that will include whether or not to invest in expensive mitigation options or to relocate people and assets (McNamara et al. 2018). Certainly, the long-term sustainability of some communities is being called into question (Woetzel et al. 2020).

Rainfall has experienced a clear downward trend since the 1990s in some parts of eastern Australia (Commonwealth of Australia 2018a). Extremely hot, dry conditions, underpinned by years of reduced rainfall and a severe drought, set the scene for the 2019–2020 summer's unprecedented fires (Filkov et al. 2020). The wildfire (bushfire) season was the worst on record in terms of sheer scale, the number of properties lost and the amount of area burned. Around 21% of Australian temperate broadleaf and mixed forests was burnt (Davey and Sarre 2020). The bushfires are estimated to have emitted between 650 million and 1.2 billion tonnes of carbon dioxide into the atmosphere (Steffen et al. 2019a). That is equivalent to the annual emissions from commercial aircraft worldwide and is far higher than Australia's annual emissions of around 531 million tonnes (Steffen et al. 2019a). Secondary effects of the bushfires were also severe, including most prominently

J. M. Rolfe (✉) · M. B. Wassing · L. Morrissey
Queensland Fire and Emergency Services, Brisbane, Australia
e-mail: john.rolfe@qfes.qld.gov.au

W. Leal Filho et al. (eds.), *Sustainability in Natural Resources Management and Land Planning*, World Sustainability Series,
https://doi.org/10.1007/978-3-030-76624-5_30

negative effects for public health (Borchers Arriagada et al. 2020; Yu et al. 2020). In Queensland, fires that threatened and destroyed housing and other buildings around Sarabah, Stanthorpe, and Peregian Springs were particularly fierce (Inspector-General Emergency Management 2020).

Australia has experienced overall upwards trends in fire danger indices spanning several decades with a marked upward shift more recently—we discuss these indices in Sect. 3. The risk of bushfires has and is expected to continue to change, including the geographical shift in bushfire prone areas. Due to the thermal inertia of the earth system, warming will continue for some time, even if net-zero emissions are reached. Managing that risk will require significantly more than just moving to a "new normal", but rather preparing for ongoing adaptation (Tiernan et al. 2019).

Even though substantial challenges remain in fully assessing and mitigating disaster risk, significant progress across multiple sectors has occurred. There is a deeper understanding on the part of fire and land management agencies as well as community and government more generally that bushfire risk management requires many partners working cooperatively (CSIRO 2020). A range of initiatives to harness climate science and understand disaster and bushfire risk assessment and mitigation have been undertaken by Queensland Fire and Emergency Services (QFES) in collaboration and partnership with a range of stakeholders. This paper reports on these measures being, with the aim of ensuring that observed and projected climate change is factored appropriately into disaster and bushfire risk mitigation. We aim to illustrate that successful fire risk management, especially within the context of a changing climate, requires a coordinated, multifaceted approach that involves proactively working at a Local, State and National level. We point to some strategies that have been successful in this space.

2 Australia's Rapidly Changing Fire Weather

Australia comprises of a wide diversity of climates, ranging from moist tropical monsoonal, moist temperate, alpine and arid conditions. Natural climatic variability is very high in the region, particularly with regard to rainfall. The El Niño-Southern Oscillation (ENSO) is the most fundamental driver however the Southern Annular Mode (SAM), Indian Ocean Dipole (IOD), and the Inter-decadal Pacific Oscillation are also important regional drivers. The very nature of this climate variability poses challenges for projecting anthropogenic climate change and its impacts in the region. Any changes, have the potential to significantly influence temperature, rainfall, bushfires, droughts, marine conditions and tropical cyclones (Reisinger et al. 2014).

The Intergovernmental Panel on Climate Change (IPCC) in 2014 reaffirmed that climate change will increase the severity and rate of natural disasters in Australia (Reisinger et al. 2014). The IPCC report noted that the frequency and duration of droughts will increase in addition to extreme fire weather. In addition, the

Commonwealth Science and Industrial Research Organisation (CSIRO) report that the frequency and severity of natural disasters climate change will increase in addition to indicating that shifts in the geographic distribution of natural disasters are also likely (Commonwealth of Australia 2016). In 2016, and the Australian Bureau of Meteorology (BOM) reported that large parts of Australia have experienced an increase in the duration, frequency and intensity of extreme heat events (Deloitte Access Economics and Australian Business Roundtable 2017).

In 2017, the Australian Business Roundtable estimated the total economic cost of natural disasters in Australia will reach $39 billion per year by 2050 (Deloitte Access Economics and Australian Business Roundtable 2017). An average of two events per year have incurred insured costs of more than $500 million, in 2017 prices. Some of the most impactful disasters have occurred in recent years with the tourism sector alone projected to lose at least $4.5 billion due to the recent bushfires. It is estimated that there was a 10–20% drop in international visitors booking holidays to Australia. More than 23,000 bushfire related insurance claims were lodged between the States of New South Wales, Queensland, South Australia and Victoria between November and February 2020, totalling an estimated value of $1.9 billion (Hughes et al. 2020).

In January 2020, the McKinsey Global Institute published an extensive analysis and report on global climate risk with a particular emphasis on physical hazards and socio-economic impacts. The socio-economic impacts of climate change will also shift as system thresholds are exceeded and have secondary effects for communities. Like physical systems, many economic and financial systems have been designed in a manner that renders them vulnerable to shocks and stressors from dynamic change. As an example, production systems globally, such as supply chains, have optimized cost efficiency over redundancy, which makes them vulnerable to failure if critical production and or distribution hubs are impacted by intensifying hazards (Woetzel et al. 2020). Theoretically, adaptation can be carried out at a rapid rate for some systems however the current rate of warming, which is at least an order of magnitude greater than any in the past 65 million years of paleoclimate records, means that some systems will be unable to evolve fast enough to keep pace (Intergovernmental Panel on Climate Change 2019a). Specific observations regarding the physical rate of change and their ensuing impacts will briefly be described in the remainder of this section and in Sect. 3.

Rainfall has experienced a clear downward trend since the 1990s in some parts of eastern Australia (Commonwealth of Australia 2018a). The extent and duration of the recent dry conditions in eastern Australia are virtually unprecedented. Whilst there have been individual years in the last 100 years with rainfall similar to or less than that in 2018, only twice since 1900 have such dry conditions been sustained for a period of nearly two years across the Murray-Darling Basin (Commonwealth of Australia 2018a). Rainfall on 2- to 3-year timescales over some parts of this region have been near or below previous record lows, many of which were set during the Federation Drought in 1900−1902. A notable feature of the rainfall deficits of the last three years is that they have been predominantly in the cooler seasons. Rainfall for the April to September period was well below average in 2017,

2018 and 2019 in most of New South Wales, and Queensland south of the Tropic of Capricorn (Commonwealth of Australia 2020a).

More generally, recent years in Australia can be characterised by prolonged heatwaves, record hot days and unprecedented bushfires yet the country still experiences an interleaving of violent storms, less frequent but more intense cyclones with in some cases record deluges of rainfall and flooding. In 2018, Australia experienced extreme heat across many parts of the country, and severe bushfires affected parts of New South Wales, Queensland and Victoria. Yet, early in 2019 an intense Monsoon event triggered severe flooding in north Queensland (Climate Council of Australia Ltd. 2019a). The climate is changing quickly evidenced by record temperatures broken successively in relatively short timeframes. Summer temperatures have significantly increased over recent decades across Australia, with above average temperatures since the late 1970s. For example, it is noteworthy that during the 2018–2019 summer, more than 206 temperate records were broken around Australia within a period of 90 days (Climate Council of Australia Ltd 2019a). This increase in the frequency of hot days is likely to continue going forward (Trancoso et al. 2020).

In order to better understand the changing climate, QFES collaborated with the BoM in 2018 to produce a detailed report as to the changing fire weather conditions in Queensland which was published in 2019 (Commonwealth of Australia 2019a). In addition, to further understand changes to the Queensland climate, QFES also led the development of the State Heatwave Risk Assessment in 2019 (Queensland Fire and Emergency Services 2019a) which was a collaborative effort between multiple stakeholders including Queensland Health. Long-term climate change projections were provided by the Climate Science Division of the Department of Environment and Science. The aforementioned collaboration was unique and represented the first such assessment within an emergency management risk assessment in Australia. The scientific basis of this collaboration underpins State agencies and disaster management groups planning for future heatwave risk (Queensland Fire and Emergency Services 2019a).

2019 was the warmest and driest year on record for Australia as a whole, and spring was also the driest on record nationally with record low rainfall occurring over large areas of inland Australia (Commonwealth of Australia 2020a). This resulted in very low soil moisture levels over most of the continent leading into December 2019 (Commonwealth of Australia 2020b). Extremely hot, dry conditions, underpinned by years of reduced rainfall and a severe drought, set the scene for the 2019–2020 summer's unprecedented fires. The fire weather of 2019 was made much more severe by record excursions of the IOD. The average effect of the IOD is small, but the anomaly was so large that this explains about one third of the anomalous drought in July–December 2019. The other factor was the SAM, which was also anomalous during this time, explaining another one third of the July–December drought. Both were predicted well and gave good warning of the high fire risk in late 2019.

Efforts to mitigate against the bushfire risk using controlled burning were hampered by the very high fire risk due to weather factors shrinking the window in which controlled burning can be safely executed. Overall, notwithstanding the unprecedented scale of the recent bushfires, Australia has been regarded as one of the most prepared countries in the world to manage such events. The impacts from the recent bushfires could have been dramatically worse if not for the risk informed preparedness measures and the early warning systems in place.

3 Forced Escalation of the Fire Danger Index

The degree of risk of bush and grass fires in Australia is measured by the Forest Fire Danger Index (FFDI) and Grass Fire Danger Index (GFDI). These indices are also used by the BoM and QFES to issue advice and warnings where required. After the 2009 Black Saturday bushfires in Victoria the categories or ratings these indices inform were revised. The FFDI was originally designed on a scale from 0 to 100, with 50 to 100 being categorised as Extreme. Due to the intensity of the 2009 fires the categories were revised to Severe (50–75), Extreme (75–100), and a new category for conditions exceeding the existing scale: Catastrophic (100 +) (Commonwealth of Australia 2018a). The State of Queensland has experienced successive years of elevated fire danger due to high temperatures, low humidity and strong westerly winds, coupled with dry conditions. The FFDI was Extreme across large parts of the State, and reached Catastrophic levels at some locations in November 2018. Rockhampton aerodrome reached a record peak FFDI over 130. Several areas recorded their highest November FFDI in the gridded dataset, which extends back to 1950 (Commonwealth of Australia 2020a).

Indices of this level are not anomalous, for Australia has experienced overall upwards trends in fire danger indices spanning several decades. In particular, the annual accumulated FFDI has been increasing across most of eastern Australia, including Queensland. The trend in accumulated FFDI demonstrates both an increase in the frequency and severity of dangerous fire weather conditions (Commonwealth of Australia 2020a). One of the more notable effects within the recent extreme bushfire seasons has been an increase in the number of fire-caused storms or pyrocumulonimbus event. These events, previously considered rare, occur when bushfires couple with the atmosphere and generate their own weather such as thunderstorms. These storms can include strong downdrafts and lightning making bushfire behaviour very unpredictable and dangerous. Australia experienced only two confirmed and two possible fire-caused storms between 1978 and 2001 however since 2001, 78 fire-caused storms have been recorded, including a 33% increase in 2019 (Steffen et al. 2019b). Fire researchers estimate that an additional 30 fire-caused storms have occurred since September 2019, with a further 15 fire-caused storms being investigated. This represents a very significant increase in the frequency of these events (Hughes et al. 2020).

In 2019, the bushfire season commenced early in Queensland with high temperatures throughout the year rendering wide spread areas primed for serious bushfire risk. The hot conditions combined with the dry landscape and strong winds to produce dangerous fire weather conditions during December 2019 and January 2020. There were also individual days with FFDI values very much above average or highest on record for December across large areas of the country. During December, daily FFDI values of 100 or above (Catastrophic) were observed in all mainland States and the Northern Territory. More than 95% of Australia by area had spring accumulated FFDI values that were very much above average (i.e., within the highest decile of recorded years), including almost 60% of the country that set new records for spring FFDI values (Commonwealth of Australia 2019b). The warmest period on record for maximum daytime temperature (1.71 °C above average) was observed across Australia as was the second-warmest for mean temperature (1.30 °C above average) behind 2016 (Steffen et al. 2019b). Dangerous fire weather conditions continued throughout the 2019–2020 summer, with December accumulated FFDI values highest on record across large areas of the country. Accumulated FFDI values for December were more than twice the average over large areas of Australia and the accumulated FFDI value for December was highest on record. That included the highest accumulated FFDI for any month in Queensland, New South Wales, the ACT, and South Australia. For Queensland, the Northern Territory, and Western Australia, December 2019 continued a run of three consecutive months of highest FFDI on record (Commonwealth of Australia 2020b).

Annual accumulated FFDI has risen across most of Queensland. The largest subregional changes in percentages are in the southeast—in particular, 51% for the South East Coast, 42% for Wide Bay and Burnett, and 27% for the Central South subregion. Annual highest daily FFDI has risen strongly in the south of the State, with increases of 15 FFDI points and higher seen along the southern border with New South Wales. The annual incidence of FFDI $\geq$ 25 days has risen across most of the State. Queensland is now experiencing an additional FFDI $\geq$ 25 on average 29 days per year. The largest subregional changes in percentages terms are in the southeast—South East Coast 254%, Wide Bay and Burnett 187%. For the Cape York Peninsula subregion, it is 109% (Commonwealth of Australia 2019a).

Flammable ecosystems are generally adapted to their specific fire regimes (Keeley et al. 2011). A fire regime shift alters vegetation and soil properties in complex ways with consequences for carbon stock changes and the ultimate biological capacity of the burnt land. A fire-driven shift in vegetation from a forested state to an alternative stable state such as a grassland with much less carbon stock is a distinct possibility for some areas. The risk of bushfires has and is expected to continue to change, thereby signalling geographical shifts in fire prone areas (Intergovernmental Panel on Climate Change 2019). Fire weather seasons have already lengthened by 18.7% globally between 1979 and 2013, with statistically significant increases across 25.3% of Earth's land surface covered with vegetation and decreases only across 10.7%. Correspondingly, the global area experiencing long fire weather seasons has increased by 3.1% per annum or 108.1% between 1979 and 2013. Fire frequencies under 2050 conditions are projected to increase by approximately 27% globally, relative to the year 2000 levels, with changes in future fire meteorology playing the most important role in enhancing global land scape fires (Intergovernmental Panel on Climate Change 2019a).

Although pivotal, climate is only one driver of a complex set of environmental, ecological and human factors in influencing fire regimes. Human exposure to bushfires will increase due to population expansion into areas with varying degrees of exposure. There is also need for integrating various fire management strategies, such as fuel-reduction treatments in natural and planted forests, with other environmental and societal considerations to achieve the goals of carbon emissions reductions, maintain water quality, biodiversity conservation and ultimately human safety (Intergovernmental Panel on Climate Change 2019a).

4 Impacts of Multiple Stressors on Land and Human Systems

As the Earth continues to warm, physical climate risk is manifesting exponentially. Climate models predict that further warming is ensured over the next decade due to inertia in the geophysical system, and that the temperature will continue to increase for decades to come. Climate science indicates that further warming and risk increase can only be arrested by achieving zero net greenhouse gas emissions. Therefore, warming will continue for some time even if net-zero emissions are reached. Mitigating that risk will require significantly more than just moving to a "new normal", which is not a helpful term, but rather preparing for a world of constant change therefore ongoing adaptation (Intergovernmental Panel on Climate Change 2019b). Increasing impacts on land are projected under all future Green House Gas emission scenarios. Some regions will face higher risks, while some regions will face risks previously not anticipated. Cascading risks with impacts on multiple systems and sectors also vary across regions While land-related response options make important contributions to adaptation and mitigation, in depth understanding of complex adaptive systems and the impact of climate risk on these systems limits their contribution to mitigation (Intergovernmental Panel on Climate Change 2019b).

A relatively new practice within risk assessment is moving to multi-hazard/multi-risk assessments recognising compounding events and risks (Steffen et al. 2009). Compound events refer to events that are characterised by multiple hazard sources that can amplify overall risk and/or cause cascading impacts. These impacts may be triggered by multiple hazards that occur coincidently or sequentially and can lead to substantial disruption of natural or human systems. A poignant example of this is the occurrence of heatwave and bushfires which have so overtly manifested recently across Australia. Further to this, climate change contributes to the potential for compound events exceeding capability and coping capacity. Concepts and methods for addressing compound events and cascading impacts have a solid foundation in disaster management response arrangements designed for all hazards however existing arrangements are limited in practice for responding to and recovering from concurrent and compounding events. Trends in geophysical and meteorological extreme events and their interaction with more complex social, economic and environmental vulnerabilities could easily overwhelm existing governance and institutional capacities because of the cumulative and cascading impacts (Malhi et al. 2020).

In human systems, incremental adjustments of current risk management approaches can increase resilience to climate variability and change, especially in the short term. However, a purely incremental approach which generally aims to preserve current objectives, governance, and institutional arrangements, can make transformational changes increasingly difficult when there is an entrenched culture of practice driven more by compliance to standards as opposed to optimal performance. Local communities are best placed to advise on fit-for-purpose disaster risk reduction and resilience opportunities. While physical vulnerability to disasters can be determined by engineers and planners, social vulnerability to natural disasters is more complex and depends on a range of socioeconomic factors. Better collaboration and consultation could accelerate the adoption of mitigation, adaptation and risk reduction initiatives. Physical preparedness by government can only mitigate part of the disaster risk, broader community measures that engender sustainable resilience remain (Deloitte Access Economics and Australian Business Roundtable 2017).

5 Land Use Planning as Part of an Integrated Disaster Management Strategy

The United Nations Office for Disaster Risk Reduction note that legislative and regulatory planning frameworks, when used effectively, can deter settlements from hazard-prone lands, enable the provision of safe land and security of tenure and establish risk-reducing design and construction standards (United Nations Office for Disaster Risk Reduction 2017). Queensland's State Planning Policy (SPP) is a key component of Queensland's land use planning system (The State of Queensland 2017). An effective planning system plays a critical role in keeping communities safe. Informed planning can ensure that the potential impacts of hazards caused by natural and socio-natural events are avoided or minimised. The consideration of climate change projections is integral when planning for natural hazards by using a scientific evidence-based approach to underpin risk assessments and planning.

Within Queensland, all of the following state interest policies must be appropriately integrated in planning and development outcomes, where relevant:

(1) Natural hazard areas are identified, including bushfire prone areas;
(2) A fit-for-purpose risk assessment is undertaken to identify and achieve an acceptable or tolerable level of risk for personal safety and property in natural hazard areas;

(3) Development in bushfire, flood, landslide, storm tide inundation or erosion prone natural hazard areas: (a) avoids the natural hazard area; or (b) where it is not possible to avoid the natural hazard area, development mitigates the risks to people and property to an acceptable or tolerable level;
(4) Development in natural hazard areas: (a) supports, and does not hinder disaster management capacity and capabilities (b) directly, indirectly and cumulatively avoids an increase in the exposure or severity of the natural hazard and the potential for damage on the site or to other properties (c) avoids risks to public safety and the environment from the location of the storage of hazardous materials and the release of these materials as a result of a natural hazard (d) maintains or enhances the protective function of landforms and vegetation that can mitigate risks associated with the natural hazard; and
(5) Community infrastructure is located and designed to maintain the required level of functionality during and immediately after a natural hazard event.

QFES plays a key role in preparing the community for bushfires. In support of Queensland's SPP, a range of technical guidance documents have been developed and published or are currently in development and will be released before the commencement of the 2020–2021 bushfire season. For new developments, QFES partnered with the Department of State Development Infrastructure and Planning to produce 'The SPP Bushfire Guidance—Natural Hazards, Risk and Resilience—Bushfire' document which provides guidance for new and existing developments in communities (The State of Queensland 2019). Further, QFES as the lead bushfire hazard agency, produced the 'Bushfire Resilient Communities' (BRC) technical document which includes more granular practical information about preparing properties for bushfire season in the context of structure, landscape design, access and vegetation advice for new development (Queensland Fire and Emergency Services 2019b).

The BRC provides technical guidance and the policy positions of QFES to state agencies, local governments, and practitioners engaged in land use planning for bushfire hazard and development activities that may be affected by bushfire hazard. It outlines factors affecting bushfire hazard and potential bushfire risks and impacts as well as the methodology used to prepare the state wide mapping of bushfire prone areas included in the State Planning Policy Interactive Mapping System. The BRC also provides technical guidance on procedures for reviewing bushfire prone area mapping, undertaking a Bushfire Hazard Assessment and Vegetation Hazard Class Assessment, calculating asset protection zone provisions and information on how to prepare a Bushfire Management Plan and Landscape Maintenance Plan (Queensland Fire and Emergency Services 2019b).

For existing housing and infrastructure developments, the Queensland Reconstruction Authority, in collaboration with QFES, CSIRO and numerous other relevant agencies, have led the development of Bushfire Resilient Building Guidance for Queensland Homes (The State of Queensland and CSIRO 2020). The guideline provides a suite of best practice building and landscaping measures, using tailored, site-specific solutions to adapt buildings for bushfire resilience. The guideline is based on extensive research into the attack mechanisms of bushfires and the different ways that buildings and gardens may be vulnerable to these attacks. The advice provided in the guideline can be used to design and build new homes or retrofit existing buildings. It can also be used to design or upgrade the landscape immediately surrounding the home, for better overall bushfire outcomes. It provides guidance on bushfire resilient design principles, constructions details, types of material, landscaping, and highlights the importance of maintenance and preparation in adapting homes and gardens to be bushfire resistant. This guidance is in advanced stages of development and will be published in 2020.

The procedural similarities of disaster risk management, building design and land use planning allow them to work together and complement each other if appropriate sequencing occurs and forums for decision-making are created (Australian Institute for Disaster Resilience 2020). Informed land use planning and building measures are a key element of an integrated disaster management strategy toward mitigating disaster risk and climate change impacts to acceptable or tolerable levels and enhancing resilience.

A key element toward mitigating bushfire risk has been the production of bushfire prone area mapping which assists in sustainable development of land and can contribute to reducing the negative impacts of multiple stressors, including climate change, on ecosystems and societies (Intergovernmental Panel on Climate Change 2019). Changes in land conditions, either from land-use or climate change, affect global and regional climate. At the regional scale, changing land conditions can reduce or accentuate warming and affect the intensity, frequency and duration of extreme events such as bushfires (Intergovernmental Panel on Climate Change 2019). Mutually supportive climate and land policies have the potential to save resources, enhance social resilience, support ecological restoration, and foster engagement and collaboration between multiple stakeholders. Sustainable land management can be improved by increasing the availability and accessibility of data and information increasing the efficiency of land use (Intergovernmental Panel on Climate Change 2019). Subsequently, a methodology to develop bushfire prone area mapping for Queensland was developed between the CSIRO and QFES in 2014. The Bushfire Prone Area Mapping[1] scales bushfire hazard based on the potential fire-line intensity of severe bushfires, and can be used to calculate the radiation profile of areas adjacent to potentially hazardous vegetation and indicate potential impact buffers.

Potential fire-line intensity is also a useful indicator of the level of safety needed for resident egress and firefighter access. A fire line intensity of 4000 kW/m is estimated to be the approximate threshold for the effective and safe direct attack on a head fire with aircraft and machinery. The threshold for continuous crown fire is 10,000 kW/m (Queensland Fire and Emergency Services 2019b). Control of fires of this intensity is extremely difficult, direct attack is rarely possible and likely to fail.

[1]See https://www.data.qld.gov.au/dataset/bushfire-hazard-area-bushfire-prone-area-mapping-methodology-for-queensland.

Suppression action must be restricted to the flanks and back of the fire. Fire line intensity greater than 30,000 kW/m is commonly understood as 'blow-up' conditions. Intensities exceeding 30,000 kW/m were a defining feature of the 2009 Black Saturday Fires in Victoria (Queensland Fire and Emergency Services 2019b).

In addition, Fire Weather Severity (FWS) for land use planning in Queensland is determined using three inputs under the methodology.[2] These are weather variables, FFDI and climate change projections out to 2050. The FFDI is used for bushfire hazard assessments and in regulations such as the Australian Standard 3959–2018: Construction of buildings in bushfire-prone areas. FFDI values for Queensland that arise during 5% Annual Exceedance Probability (AEP) fire weather events have been estimated from a gridded (83 kms, three-hourly resolution) prediction of FFDI from long-term spatial weather products produced by the BoM. Projected average FWS values for Queensland vary from 50 (Extreme) in Southeast Queensland and Cape York bioregions to 130 (Catastrophic) in the south-western parts of the state (Queensland Fire and Emergency Services 2019b). This represents a large increase in these values from the present and presents a serious and critical concern for fire risk managers going into the future.

6 Reducing Risk and Sustaining Resilience Through Effective Risk Governance

The United Nations 2030 Agenda for Sustainable Development outlines 17 Sustainable Development Goals (SDGs) (United Nations 2015). The agenda sets the international platform for sustainably managing the planets natural resources and promotes taking urgent action on climate change so that we can support the needs of present and future generations. Managing disaster risk and climate adaptation are key elements in planning for achieving SDGs as shocks and stresses can reverse years of mitigation actions and financial investment in development achievements. Working toward the achievement of the SDGs would contribute to risk reduction however poor management and development of land can be a key driver of risk such as development approvals that place communities in hazard impact zones or construction of exposed and or vulnerable critical infrastructure. Using holistic approaches toward understanding risk in depth and applying this understanding in development plans and investments is a key to long term resilience (United Nations Office for Disaster Risk Reduction 2017).

Establishing effective governance mechanisms is another key to reducing systemic risk. Indeed, strengthening disaster risk governance is Priority 2 of the Sendai Framework for Disaster Risk Reduction (United Nations Office for Disaster Risk

[2]Enhancement to this methodology is currently being investigated. Mapping data can be obtained in raster format from the Queensland Government data portal under the title 'Bushfire hazard area—Bushfire-prone area—inputs—Queensland'.

Reduction 2015). Governance implies that governments do not make decisions in isolation and negotiate policies and practices with those who are part of or affected by their decisions. Disaster risk governance affects the distribution of exposure, vulnerability and residual risk, among different groups of people (United Nations Office for Disaster Risk Reduction 2017). Good governance also entails improving accountability, transparency and meaningful participation throughout the procedures and practices. Negotiating, building consensus and reaching agreements comprise both formal and explicit mechanisms such as legislation, policies, standards and procedures that mediate social, economic and political relations. In places where there is a proactive, responsive and accountable system of governance that works with local actors, the possibilities of resilience are much higher (United Nations Office for Disaster Risk Reduction 2017).

In 2016, QFES conducted a literature review of international natural hazard risk assessment methodology. This review led to the development of the Queensland Emergency Risk Management Framework (QERMF). The QERMF uses scientific hazard data and geospatial intelligence to analyse the spatial and temporal manifestation of exposures, vulnerabilities and subsequent risk (Queensland Fire and Emergency Services 2017). QFES is responsible in accordance with the Queensland State Disaster Management Plan to prepare the State Natural Hazard Risk Assessment. The QERMF actions the Sendai Framework's first and second priorities. Similarly, the United Nations Words in Action guidelines are enacted by the QERMF by providing consistent guidance to assess and understand disaster risk acting as a means by which risk information can be made publicly available and to also inform the identification of mitigation priorities. By taking this approach a secondary benefit has been the establishment of a framework for sharing information on climate and disaster risk management. Collaboration of this nature can link State level policy direction with local level action which in turn can produce tangible enhancements to the safety and resilience of the community (Queensland Fire and Emergency Services 2017).

In 2016 QFES contributed to the development of guidelines on National Disaster Risk Assessment under the United Nations Office for Disaster Risk Reduction (UNISDR) "Words into Action" initiative. The guidelines were developed with the input from more than 100 leading experts from across the world including government, nongovernment organisations, academia and the private sector (United Nations Office for Disaster Risk Reduction 2017). QFES also contributed to the development of Australia's National Disaster Risk Reduction Framework (NDRRF) in 2018. The NDRRF is designed to guide Australia's efforts to reduce disaster risk associated with natural hazards. It translates the first three Sendai Framework priorities into action for the Australian context; though the strategies outlined in this framework are applicable to disaster preparedness and recovery efforts, the fourth priority of the Sendai Framework is largely progressed through other national strategies, primarily the Australian Disaster Preparedness Framework (Commonwealth of Australia 2018b).

A key aspect of multi layered risk governance is the communication of risk between levels of government will need to address the passage and accountability for mitigating residual risk. Appropriate management of residual risk requires escalation, therefore requiring transparency and appropriate accountability and responsibility in decision making to authorise the respective levels of funding, resources and action to effect large scale mitigation across the spectrum of prevention, preparedness, response and recovery (United Nations Office for Disaster Risk Reduction 2017). Implementation of disaster risk assessments is ideally carried out through established coordination bodies that include a variety of relevant stakeholders with an appropriate level of authority to make decisions and agree to plans for mitigation and disaster risk management planning measures. In a post pandemic world where fiscal impacts have been global it is best to manage climate and disaster related risk through existing intergovernmental coordination structures where possible.

Queensland's Disaster Management Arrangements, in accordance with the State legislation of the Disaster Management Act 2003, are built very deliberately with a dual bottom up—top down perspective. The 77 local governments across Queensland have primary responsibility for the assessment of disaster risk, the development of disaster plans and in the management of disaster events via Local Disaster Management Groups (LDMG). Sitting over this from a coordination perspective are 23 Disaster Districts that provide support and coordination if and when required via Disaster District Management Groups. A State operational overview is provided through the State Disaster Coordination Group (SDCG) comprising whole-of-government agency level representatives and other relevant stakeholders which in turn reports to and enacts strategic direction from the Queensland Disaster Management Committee (QDMC) which comprises the Premier of the State, Ministers, Police and Fire and Emergency Services Commissioners. In 2017 the QDMC endorsed the QERMF as Queensland's approach to managing disaster risk. Further to this, in 2019 the SDCG endorsed expanding its role to include the consideration of matters identified as residual risk communicated from LDMGs through DDMGs if the capability and capacity at those respective levels cannot mitigate the risk to an acceptable level.

Similarly, Queensland's bushfire management arrangements are characterised by partnerships and shared responsibility between land managers, the community, service providers, fire management groups, disaster management groups, committees at a regional and state level and government at the local, state and Commonwealth level. The intention is to enable a coordinated, risk-based approach to the management of bushfire throughout the state and facilitate linkages to Queensland's disaster management arrangements. Similar to the disaster management arrangements, Queensland's bushfire management arrangements are also coordinated by groups at the local, regional and state level. These groups provide support to Disaster Management Groups and land managers to manage bushfire risk. This support is reflected in the functions of these groups as outlined below. The Queensland State Bushfire Plan (Queensland Fire and Emergency Services 2020), approved by Queensland Cabinet in July 2020, formalises the establishment of these groups.

The state is comprised of Locally Specific Fire Management Groups, Area Fire Management Groups and seven Regional Inter-Departmental Committees for Bushfire. These groups report to the State Inter-Departmental Committee for Bushfire. All fire management groups have the following functions:

- Identification and consensus on areas of particular bushfire risk;
- Provision of advice on fuel reduction and other mitigation activities;
- Fostering effective and harmonious working relationships between partners, stakeholders and the community;
- Enhancing partner and stakeholder cooperation and resource sharing at bushfire incidents and hazard reduction activities;
- Providing a forum for clarifying and disseminating information on regulatory requirements and best practice principles for bushfire management and bushfire risk mitigation;
- Supporting cooperation and coordination in delivering community education activities to build community resilience; and
- Development of a Bushfire Risk Mitigation Plan.

The Sendai Framework for Disaster Risk Reduction 2015–2030 marked a crucial shift from managing disasters to managing disaster risk which is an inherently more proactive approach. The use of the classic risk management concepts, which describe risk in terms of likelihood and consequence, on their own are manifestly inadequate. Mainstreaming disaster risk reduction and climate change adaptation within and across sectors is essential to ensure administrative coordination and coherence across sectoral plans and (United Nations Office for Disaster Risk Reduction 2017). Given that observed climate change varies due to the difference in the climate across the State and the uneven distribution of observational data, obtaining a reliable climate outlook for long term decision making depends on obtaining high-resolution downscaled climate models. As such, the Climate Science Division of the Queensland Department of Environment and Sciences (DES) have developed a comprehensive set of data and resources (https://www.longpaddock.qld.gov.au/) for Queensland to underpin the Queensland Government's Climate Change Response. Sector Adaptation Plans are a key element Queensland's Climate Adaptation Strategy (Q-CAS) 2017–2030.

There are eight sector plans in Queensland to help to prioritise climate change adaptation activities across the key sectors of the community. They have been developed in consultation with sector and industry stakeholders to reflect the needs and priorities of each sector. They identify emerging opportunities, share knowledge and encourage collaboration. QFES has published a position on climate change to guide an integrated approach to actions on Climate Change. QFES is a committed supporter of the Queensland Government's Climate Change Response and the impact of a changing climate has been recognised as a strategic risk for QFES. The position paper also provides an overt commitment of the Department's intentions (Queensland Fire and Emergency Services 2018a). QFES in partnership with the National Climate Change Adaptation Research Facility and DES developed the Emergency Management Sector Adaptation Plan (EM-SAP) for Climate Change.

The EM-SAP notes sector-specific climate risks, existing activities and knowledge gaps as well as barriers to adaptation for the emergency management sector. This plan outlines a vision, principles and a series of priorities that intend to guide emergency management sector climate change adaptation activities to ensure that it is fully engaged in addressing the risks and opportunities associated with the changing climate. The eight priorities contained within the EM-SAP endeavour to embed climate change into sector planning at all levels and to remain a trusted broker of climate-related risk data and information for communities (Queensland Fire and Emergency Services 2018b).

In sum, limiting the risk from the impact of extreme events and abrupt changes leads to successful adaptation to climate change if climate-affected sectors and disaster management relevant agencies coordinate well. Transformative governance, including successful integration of disaster risk management and climate change adaptation, empowerment of vulnerable groups, accountability of governmental decisions, and longer-term land use planning promotes climate-resilient development pathways (Intergovernmental Panel on Climate Change 2019b). A wide range of adaptation and mitigation responses have the potential to make positive contributions to sustainable development, enhancement of ecosystem functions and services and other societal goals (Commonwealth of Australia 2009).

7 Transformative Focus of Climate and Disaster Risk Management as an Enabler to Bushfire Resilience

To improve our understanding of disaster risk there must be a more significant focus on transformative processes. Therefore, innovative tools and practices such as those found within cross-sectoral partnerships and networks are required to improve the uptake of research and knowledge for disaster risk management. Queensland's approach supports the uptake of complex scientific data by translating it into usable information but also works in earnest to allow affordance for locally led place-based solutions. This approach has also been adapted in the ongoing search for innovative mitigation options that contribute to the overall management of bushfire risk (Montserrat et al. 2019). An example poster used to generate discussion and distil very complex climate change and the ensuring effects on risk mitigation in the bushfire setting is reproduced in Fig. 1.

In addition to the scientific inputs, Queensland's Bushfire Management Arrangements are underpinned by consultation and engagement across a wide range of stakeholders, many being land managers. As every individual has a different understanding of risk, stakeholders communicate risk differently, have differing organisational and legal requirements, and vastly differing financial and physical support to engage within bushfire risk mitigation (United Nations Office for

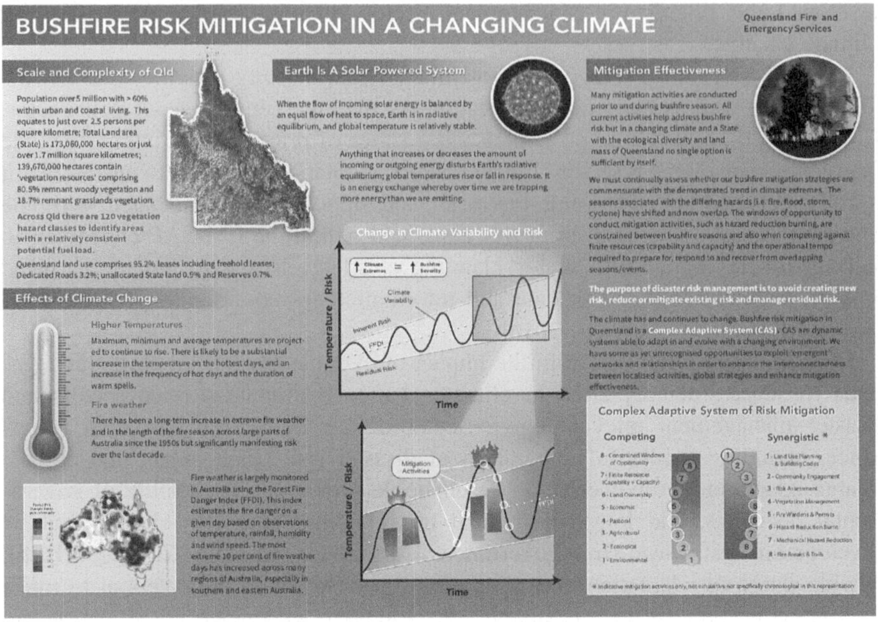

Fig. 1 A poster commissioned by QFES that distils some complexities of climate change and bushfire risk mitigation used for internal education and awareness. Produced by Kara Perilli Scientific Illustration and Design for the Queensland Fire and Emergency Services

Disaster Risk Reduction 2017). The involvement of multiple stakeholders requires that plans simultaneously appeal to multiple co-benefits and stakeholder motivations. This increases the likelihood of implementation and the overall effectiveness of any plans and arrangements (Roelich and Giesekam 2019).

A pivotal first step is understanding the inherent risk associated with the manifestation of any hazard through science and modelling if possible, then assessing the effectiveness of the capability and capacity of local resources to manage the risk is critical for identifying residual risk. The vulnerability of our infrastructure, systems and networks is a critical aspect of disaster risk management that may or may not be able to be managed at a local level. Further exaggerating the complexities of infrastructure systems is that they are highly interconnected and at times share dependencies (Johansson 2010). Arguably, the interconnectedness of infrastructure systems has rendered them more vulnerable as a result of the rapidly changing climate. Once infrastructure systems experience external shocks and or failures, they can then disrupt other infrastructure networks with the consequence of broad system failure so this must be addressed in a proactive manner (Wang et al. 2012).

When assessing broad geographical areas such as a local government area in Queensland the infrastructure and system/network vulnerability analysis becomes vitally important. Vulnerability and Business continuity management and to a lesser extent, vulnerability analysis, are common tools for infrastructure owners and operators. When systems operators assess their own systems the application of the consequences to the broader community can vary considerably between sectors. Broad geographical area risk assessment when considering infrastructure systems becomes more interdisciplinary and cross institutional (Wang et al. 2012). More recent approaches that do encompass broader societal factors have been developed within Australia that are indeed promising such as the Resilience, Adaptation Pathways and Transformation Approach (RAPTA). Version 2 of RAPTA has been published in 2019 by CSIRO to design, implement and evaluate interventions for achieving sustainability goals within highly uncertain and rapidly changing decision contexts (CSIRO 2019). In addition, CSIRO have also published a set of guidance documents in 2019 to assist in including climate and disaster risk into investment and strategic planning decisions. There are four sections to the guidance: Governance, Vulnerability, Scenarios and Prioritisation https://knowledge.aidr.org.au/resources/strategic-disaster-risk-assessment-guidance/.

Further compelling corporate incorporation of climate risk is guidance from the Australian Prudential Regulation Authority which denotes climate risk as distinctly financial with many risks foreseeable and actionable now (Martijn Wilder et al. 2017). The threat of climate litigation demands attention from government and business alike. There is also an increasing trend in litigation concerning climate risk disclosure. Company directors who fail to demonstrate they have adequately considered climate change risks, could be liable for breaching their duty of care and failing to exercise due diligence in the future (Martijn Wilder et al. 2017).

Conducting risk assessments and planning to address complex endeavours such as bushfire over broad landscapes must be seen through the lens of participants at a local level as well as from the perspective of the State, owners and operators of infrastructure, non-government organisations and relevant community representatives. When considering large scale mitigation strategies there is a tendency is to impose solutions in a top down manner (Alberts 2007). This can be seen via historical risk assessment efforts being driven from a top-down perspective but failing to result in completion of adequate risk assessments at a local level other than compliance-based outputs. Ideally, useful information is produced with a clear line of sight through hazard—exposure—vulnerability—risk—engendering the production of risk-based plans that are actionable.

Holistic bushfire risk mitigation requires an apolitical focus—synergy between individuals and organisations is required by leveraging the available information and expertise that would otherwise not be attainable if individuals and organisations act unilaterally (Alberts 2007). The notion of focusing a collective is certainly not new and should be concerned with engendering convergence of effort (Alberts 2007). Convergence if combined with focus is about moving in an agreed direction as a collective. An important note of distinction is that convergence does not necessarily imply control of one entity by another (Alberts 2007). QFES have evolved to recognise that bushfire risk mitigation efforts need to be recognised for what the endeavour is, collaborative problem solving. Within collaborative problem solving, individuals come together via the creation of common intent and then turning this intent into coordinated action toward a stated purpose or goals (Hesse et al. 2015; Organisation for Economic Cooperation and Development 2017).

It is a continuous improvement and adaptive learning culture that enables QFES to span the complexities of climate and bushfire science, policy and government/ governance and yet remain focussed on community and locally led or place-based solutions. Lessons management refers to collecting, analysing, disseminating and applying learning experiences. These experiences may include examples of good practice and those that need to improve. Where the preservation of life and the protection of property are the primary goals of any organisation, continuous improvement is highly pertinent. Appropriate lessons management can result in more efficient and effective practices, improved safety, and improved capture and mobilisation of knowledge such as evidenced in the range of previously mentioned initiatives (Australian Institute for Disaster Resilience 2019). QFES' approach enables continuous improvement through collaborative problem solving and learning. The process allows for the scalability of local solutions implemented at the local level through to broad scale solutions implemented as formal strategic projects. A new debrief and lessons management process has been developed to ensure QFES adapts to meet the needs of staff, volunteers, stakeholders and community as a result of operational activity. The debrief and lessons management process provides QFES personnel with the opportunity to contribute to the continuous improvement of QFES, to ensure the department develops, adapts and transforms to meet the needs of the workforce, stakeholders and the Queensland community (Queensland Fire and Emergency Services 2019c).

8 Conclusion

This paper has reported on some of the key means by which QFES is managing the risk posed by bushfires, especially in the context of a warming climate. Queensland is entering an unprecedented time of bushfire risk, in which the intensity and geographic extent of fires is rapidly increasing. This has been illustrated by increases in measurements of bushfire risk and proneness. This calls for a reappraisal of how bushfire risk is managed going forward. In this paper, we have identified some novel and innovative practices that QFES have utilised around bushfire risk, including:

- Developing novel methods of calculating bushfire risk and intensity, to provide increased mitigation and operational capacity;
- Incorporating bushfire risk mitigation into land use and planning policies at the State level;
- Proposing novel methods of expressing risk, especially through the Queensland Emergency Risk Management Framework, that can be communicated across levels of government, to make risk governance more consistent and efficient.

The role that QFES is undertaking in mitigating bushfire risk is that of transformational leadership. Transformational leadership in the sense that QFES is securing the commitment of a wide variety of stakeholders to work toward the attainment of common goals in spite of competing priorities. Bushfire risk mitigation in a changing climate is a complex, challenging yet ultimately rewarding endeavour enabling the emergence of resilient communities.

References

Australian Institute for Disaster Resilience (2019) Lessons management handbook

Alberts DS (2007) Agility, focus, and convergence: the future of command and control. Int C2 J 1:1–30

Australian Institute for Disaster Resilience (2020) Land use planning for disaster resilient communities

Borchers Arriagada N, Palmer AJ, Bowman DM, Morgan GG, Jalaludin BB, Johnston FH (2020) Unprecedented smoke-related health burden associated with the 2019–20 bushfires in eastern Australia. Medical J Australia. https://doi.org/10.5694/mja2.50545

Commonwealth of Australia (2016) Climate change in Australia [WWW Document]. Accessed 14 May 2020. https://www.climatechangeinaustralia.gov.au/en/

Commonwealth of Australia (2018a) Special climate statement 66—an abnormally dry period in eastern Australia. Bureau of Meteorology

Commonwealth of Australia (2018b) National disaster risk reduction framework. Department of Prime Minister and Cabinet

Commonwealth of Australia (2019a) Changes to fire weather in Queensland. Bureau of Meteorology

Commonwealth of Australia (2019b) Special climate statement 72—dangerous bushfire weather in spring 2019. Bureau of Meteorology

Commonwealth of Australia (2020a). Special climate statement 70 update—drought conditions in Australia and impact on water resources in the Murray–Darling basin. Bureau of Meteorology

Commonwealth of Australia (2020b) Special climate statement 73—extreme heat and fire weather in December 2019 and January 2020. Bureau of Meteorology

Coronese M, Lamperti F, Keller K, Chiaromonte F, Roventini A (2019) Evidence for sharp increase in the economic damages of extreme natural disasters. Proc Natl Acad Sci USA 116:21450–21455. https://doi.org/10.1073/pnas.1907826116

CSIRO (2019) Resilience adaptation pathways and transformation approach

CSIRO (2020) CSIRO report on climate and disaster resilience

CSIRO (2018) Earth systems and climate change hub (ESCC Hub) [WWW Document]. Accessed 13 May 2020. URL https://www.csiro.au/en/Research/OandA/Areas/Assessing-our-climate/ESCC-Hub

Davey SM, Sarre A (2020) Editorial: the 2019/20 black summer bushfires. Aust for 83:47–51. https://doi.org/10.1080/00049158.2020.1769899

Deloitte Access Economics, Australian Business Roundtable (2017) Building resilience to natural disasters in our states and territories. Deloitte Access Economics

Emergency Leaders for Climate Action, The Climate Council (2020) Australian bushfire and climate plan

Filkov AI, Ngo T, Matthews S, Telfer S, Penman TD (2020) Impact of Australia's catastrophic 2019/20 bushfire season on communities and environment. Retrospective analysis and current trends. J Saf Sci Resilience 1:44–56. https://doi.org/10.1016/j.jnlssr.2020.06.009

Hesse F, Care E, Buder J, Sassenberg K, Griffin P (2015) A framework for teachable collaborative problem solving skills. In: Griffin P, Care E (eds) Assessment and teaching of 21st century skills: methods and approach, educational assessment in an information age. Springer Netherlands, Dordrecht, pp 37–56. https://doi.org/10.1007/978-94-017-9395-7_2
Hughes L, Steffen W, Mullins G, Dean A, Weisbrot E, Rice M (2020) Summer of crisis. The Climate Council.
Inspector-General Emergency Management (2020) Queensland Bushfires review report 2: 2019–20
Intergovernmental Panel on Climate Change (2019a) Climate change and land. Intergovernmental Panel on Climate Change
Intergovernmental Panel on Climate Change (2019b) IPCC special report on the ocean and Cryosphere in a changing climate. Intergovernmental Panel on Climate Change
Johansson J (2010) Risk and vulnerability analysis of interdependent technical infrastructures: addressing socio-technical systems, Thesis/doccomp. Lund University
Keeley JE, Pausas JG, Rundel PW, Bond WJ, Bradstock RA (2011) Fire as an evolutionary pressure shaping plant traits. Trends Plant Sci 16:406–411. https://doi.org/10.1016/j.tplants.2011.04.002
Malhi Y, Franklin J, Seddon N, Solan M, Turner MG, Field CB, Knowlton N (2020) Climate change and ecosystems: threats, opportunities and solutions. Philosophical Trans Royal Soc b: Biological Sci 375:20190104. https://doi.org/10.1098/rstb.2019.0104
McNamara KE, Bronen R, Fernando N, Klepp S (2018) The complex decision-making of climate-induced relocation: adaptation and loss and damage. Clim Policy 18:111–117. https://doi.org/10.1080/14693062.2016.1248886
Montserrat MF, Poljanšek K, Clark I, De Groeve T (2019) Science for disaster risk management 2017: knowing better and losing less. The Disaster Risk Management Knowledge Centre
Organisation for Economic Cooperation and Development (2017) PISA collaborative problem-solving framework. Organisation for Economic Cooperation and Development
Queensland Fire and Emergency Services (2020) The Queensland state bushfire plan
Queensland Fire and Emergency Services (2019a) Queensland state heat wave risk assessment 2019. Queensland Fire and Emergency Services
Queensland Fire and Emergency Services (2019b) Bushfire resilient communities. Queensland Fire and Emergency Services
Queensland Fire and Emergency Services (2019c) 2018–2019 annual report
Queensland Fire and Emergency Services (2018a) Climate change position statement
Queensland Fire and Emergency Services (2018b) Emergency management sector adaptation plan for climate change
Queensland Fire and Emergency Services (2017) Queensland state natural hazard risk assessment
Reisinger A, Kitching R, Chiew F, Hughes L, Newton P, Schuster S, Tait A, Whetton P, Barnett J, Becken S, Blackett P, Boulter S, Campbell A, Collins D, Davies J, Dear K, Dovers S, Finlay K, Glavovic B (2014) Australasia. In: Barros VR, Field CB, Dokken DJ, Mastrandrea MD, Mach KJ (eds) Climate change 2014: impacts, adaptation and vulnerability. Cambridge University Press, Cambridge, pp 1371–1438. https://doi.org/10.1017/CBO9781107415386.005
Roelich K, Giesekam J (2019) Decision making under uncertainty in climate change mitigation: introducing multiple actor motivations, agency and influence. Clim Policy 19:175–188. https://doi.org/10.1080/14693062.2018.1479238
Schwalm CR, Glendon S, Duffy PB (2020) RCP8.5 tracks cumulative CO_2 emissions. Proc Natl Acad Sci USA. https://doi.org/10.1073/pnas.2007117117
Steffen W, Burbidge AA, Hughes L, Kitching R, Lindenmayer D, Musgrave W, Smith MS, Werner PA (2009) Australia's biodiversity and climate change: a strategic assessment of the vulnerability of Australia's biodiversity to climate change. CSIRO Publishing, Canberra
Steffen W, Dean A, Rice M, Mullins G (2019a) The angriest summer. The Climate Council
Steffen W, Hughes L, Mullins G, Bambrick H, Dean A, Rice M (2019b) Dangerous summer: Escalating bushfire, head and drought risk. The Climate Council.

The State of Queensland (2019) Natural hazards, risk and resilience—bushfire: state planning policy

The State of Queensland (2017) State planning policy

The State of Queensland, CSIRO (2020) Bushfire resilient building guidance for Queensland homes

Tiernan A, Drennan L, Nalau J, Onyango E, Morrissey L, Mackey B (2019) A review of themes in disaster resilience literature and international practice since 2012. Policy Design Pract 2:53–74. https://doi.org/10.1080/25741292.2018.1507240

Trancoso R, Syktus J, Toombs N, Ahrens D, Wong KK-H, Pozza RD (2020) Heatwaves intensification in Australia: A consistent trajectory across past, present and future. Sci Total Environ 742:140521. https://doi.org/10.1016/j.scitotenv.2020.140521

United Nations (2015) Transforming our world: The 2030 agenda for sustainable development

United Nations Office for Disaster Risk Reduction (2017) Words into action guidelines—implementation guide for land use and urban planning

United Nations Office for Disaster Risk Reduction (2015) Sendai framework for disaster risk reduction 2015–2030

Wang S, Hong L, Chen X (2012) Vulnerability analysis of interdependent infrastructure systems: a methodological framework. Physica A: Stat Mech Appl 391:3323–3335. https://doi.org/10.1016/j.physa.2011.12.043

Wilder M, Venuti S, Chatterjee S (2017) Recommendations of the task force on climate-related financial disclosures—review of local relevance. Baker McKenzie

Woetzel J, Pinner D, Samandari H, Engel H, Krishnan M, Boland B, Powis C (2020) Climate risk and response—physical hazards and socioeconomic impacts

Yu P, Xu R, Abramson MJ, Li S, Guo Y (2020) Bushfires in Australia: a serious health emergency under climate change. Lancet Planetary Health 4:e7–e8. https://doi.org/10.1016/S2542-5196(19)30267-0

Challenges in Sustainable Land Use Management

Walter Leal Filho, Ulisses Miranda Azeiteiro, and Andréia Faraoni Freitas Setti

1 Introduction

Biodiversity continues to decline all around the world, significantly reducing nature's ability to contribute to our well-being due to a series of common pressures: the uncontrolled, destructive use of natural resources; air, land and water pollution; and climate change, among others. This alarming trend puts economies, livelihoods, food safety and people's quality of life at risk (IPBES 2018).

The adequate use of natural resources requires a compatibility between their use —serving human needs—and environmental conservation, assuring that they will be there for future generations as well and maintaining their ecological functions. Using land in consonance with its capacity is crucial to maintaining the integrity and vitality of natural resources. Based on soil capacity knowledge, soil use and occupation can be planned adequately, alongside a set of conservational recommendations and practices aimed at the protection and improvement of natural resources (Lehmann and Stahr 2010; Pereira and Gomes 2017; Visser et al. 2019).

W. Leal Filho (✉)
European School of Sustainability Science and Research, Hamburg University of Applied Sciences, Hamburg, Germany
e-mail: walter.leal2@haw-hamburg.de

W. Leal Filho
Department of Natural Sciences, Manchester Metropolitan University, Chester Street, Manchester M1 5GD, UK

U. M. Azeiteiro · A. F. F. Setti
Department of Biology and CESAM Centre for Environmental and Marine Studies, University of Aveiro, 3810-193 Aveiro, Portugal
e-mail: ulisses@ua.pt

A. F. F. Setti
e-mail: andreiasetti@ua.pt

W. Leal Filho et al. (eds.), *Sustainability in Natural Resources Management and Land Planning*, World Sustainability Series,
https://doi.org/10.1007/978-3-030-76624-5_31

Mathematical climate models predict tipping points—a rupture of the balance of the natural system caused by the cumulative effect of disturbances, amplified by the intensity, duration, extension and frequency of interventions, exceeding the resilience of the planet (Lovejoy and Nobre 2018; Veraart et al. 2012). When nearing that point, the probability of a faster or rougher transition to a new point of balance (likely harsh or unfeasible to most species adapted to the previous balance) increases exponentially (Lenton et al. 2019).

Implementing strategies to incentivize sustainable land use, responsible forestry, the restoration of degraded lands, the expansion of research and development in innovation, and the conservation and restoration of ecosystems is fundamental for humanity in order to tackle the challenges imposed by the climate crisis, thus contributing to the achievement of the sustainable development goals (SDG 15 above all).

2 Future Trends

Sustainable development is a growing field that has increased in urgency over recent years. Sustainable land use management is imperative to ensuring the achievement of sustainable development. In most instances, agricultural land use is targeted to ensure maximum production with preservation of the land. However, this is not always possible, and lower income countries lack the knowledge, infrastructure, and resources to practice sustainability (Miheretu and Yimer 2017).

Ensuring that sustainable land use is initiated is largely dependent on the financial capacity of the landowners. Furthermore, government support in terms of finance and resources is required to ensure sustainability. This is difficult in low-income countries such as in Africa due to other problems being prioritised. The inability to acquire financial investment prevents the sustainable use of land (Dallimer et al. 2018).

Additionally, generational gaps pose a challenge to the adoption of new sustainable land use methods. Older generations tend to be unresponsive to new innovative ideas and to use their old methods, which may not be as effective. Younger land owners are more likely to use sustainable practices (Miheretu and Yimer 2017). This generational gap prevents the progression toward sustainability. In other instances, sustainability education is not widely promoted, causing people to be unaware of the importance of sustainability practices. The lack of education causes inadequate practices that lead to land degradation (Mango et al. 2017).

Furthermore, climate change has induced variability in weather patterns and has increased the frequency of extreme weather events and natural disasters. Such occurrences have had adverse effects on land use by causing soil erosion and land degradation (Blake et al. 2018; Tarnavsky and Bonifacio 2020). Already existing sustainable practices need to be adapted to changing climates in order to prevent this issue. However, the unpredictability of these events makes it difficult to develop methods to combat the problem. Therefore, new prediction systems need to be designed to assist in the creation of sustainable land use management (Zambrano et al. 2018; Issahaku and Abdul-Rahaman 2019).

In urban areas, the sustainable use of land is dependent on government planning. Various stakeholders are recruited in the process, and local and national plans are needed to ensure sustainability. Poor planning results in unsustainability. Usually, development guidelines are provided, but non-compliance results in aberrant development that threatens sustainability. It is therefore noted that planning in conjunction with pre- and post-evaluation is needed. Furthermore, education and resource allocations are imperative for successful land use management (Dambeebo and Jallo 2018).

3 Conclusions

Sustainable land use is a critical research and actuation (researchers and practitioners) area that addresses multiple dimensions (e.g., agroforestry, agro-ecology, habitat fragmentation and landscape ecology, soil erosion and land degradation and the restoration and urban areas). In the beginning of *The UN Decade on Ecosystem Restoration,* and in order to succeed in the *2030 Agenda* efforts towards sustainability, we are challenged to do more. In a context of global change, challenges are identified and must be addressed.

References

Blake WH, Rabinovich A, Wynants M, Kelly C, Nasseri M, Ngondya I, …, Boeckx P (2018) Soil erosion in East Africa: an interdisciplinary approach to realising pastoral land management change. Environ Res Lett 13(12):124014. https://doi.org/10.1088/1748-9326/aaea8b/meta

Dallimer M, Stringer LC, Orchard SE, Osano P, Njoroge G, Wen C, Gicheru P (2018) Who uses sustainable land management practices and what are the costs and benefits? Insights from Kenya. Land Degrad Dev 29(9):2822–2835. https://doi.org/10.1002/ldr.3001

Dambeebo D, Jallo C (2018) Sustainable urban development and land use management: Wa Municipality in perspective, Ghana. J Sustain Dev 11(5):235–248. https://doi.org/10.5539/jsd.v11n5p235

IPBES (2018) The IPBES assessment report on land degradation and restoration. In: Montanarella L, Scholes R, Brainich A (eds) Secretariat of the intergovernmental science-policy platform on biodiversity and ecosystem services, Bonn, Germany, 744 p

Issahaku G, Abdul-Rahaman A (2019) Sustainable land management practices, off-farm work participation and vulnerability among farmers in Ghana: is there a nexus? Int Soil Water Conserv Res 7(1):18–26. https://doi.org/10.1016/j.iswcr.2018.10.002

Lehmann A, Stahr K (2010) The potential of soil functions and planner-oriented soil evaluation to achieve sustainable land use. J Soils Sediments 10:1092–1102. https://doi.org/10.1007/s11368-010-0207-5

Lenton TM, Rockström J, Gaffney O, Rahmstorf S, Richardson K, Steffen W, Schellnhuber HJ (2019) Climate tipping points—too risky to bet against. The growing threat of abrupt and

irreversible climate changes must compel political and economic action on emissions. Nature 575:592–595. https://doi.org/10.1038/d41586-019-03595-0
Lovejoy T, Nobre C (2018) Amazon tipping point. (Editorial). Sci Adv 4(2) 21/II/2018. https://doi.org/10.1126/sciadv.aat2340
Mango N, Makate C, Tamene L, Mponela P, Ndengu G (2017) Awareness and adoption of land, soil and water conservation practices in the Chinyanja Triangle, Southern Africa. Int Soil Water Conserv Res 5(2):122–129. https://doi.org/10.1016/j.iswcr.2017.04.003
Miheretu BA, Yimer AA (2017) Determinants of farmers' adoption of land management practices in Gelana sub-watershed of Northern highlands of Ethiopia. Ecol Process 6(1):19. https://doi.org/10.1186/s13717-017-0085-5
Pereira LC, Gomes MAF (2017) Capacidade de uso das terras e desenvolvimento sustentável. Jornal Dia de Campo. Available in: www.diadecampo.com.br
Tarnavsky E, Bonifacio R (2020) Drought risk management using satellite-based rainfall estimates. In: Satellite precipitation measurement, vol 69. Springer, pp 1029–1053
Veraart A, Faassen E, Dakos V et al (2012) Recovery rates reflect distance to a tipping point in a living system. Nature 481:357–359. https://doi.org/10.1038/nature10723
Visser S, Keesstra S, Maas G, Cleen M, Molenaar C (2019) Soil as a basis to create enabling conditions for transitions towards sustainable land management as a key to achieve the SDGs by 2030. Sustainability 11(23):6792. https://doi.org/10.3390/su11236792
Zambrano F, Vrieling A, Nelson A, Meroni M, Tadesse T (2018) Prediction of drought-induced reduction of agricultural productivity in Chile from MODIS, rainfall estimates, and climate oscillation indices. Remote Sens Environ 219:15–30. https://doi.org/10.1016/j.rse.2018.10.006

The manufacturer's authorised representative in the EU is Springer Nature Customer Service Centre GmbH, Europaplatz 3, 69115 Heidelberg, Germany. If you have any concerns regarding our products, please contact ProductSafety@springernature.com

Printed and bound by CPI Group (UK) Ltd, Croydon, CR0 4YY
30/09/2026
02225690-0018